HANDBOOK OF APPLIED CHEMISTRY

HANDBOOK OF APPLIED CHEMISTRY

Facts for Engineers, Scientists, Technicians, and Technical Managers

VOLLRATH HOPP
Hoechst AG

INGO HENNIG
Bayer AG

⬤ HEMISPHERE PUBLISHING CORPORATION
Washington   New York   London

McGRAW–HILL BOOK COMPANY
New York   St. Louis   San Francisco   Auckland   Bogotá
Hamburg   Johannesburg   London   Madrid   Mexico
Montreal   New Delhi   Panama   Paris   São Paulo
Singapore   Sydney   Tokyo   Toronto

HANDBOOK OF APPLIED CHEMISTRY
Facts for Engineers, Scientists,
Technicians, and Technical Managers

English translation of the German edition of
Chemie Kompendium für das Selbstudium
© by Kaiserlei Verlag, Offenbach, FRG.

1 2 3 4 5 6 7 8 9 0   B C B C   8 9 8 7 6 5 4 3

Library of Congress Cataloging in Publication Data

Hopp, Vollrath.
   Handbook of applied chemistry.

   Translation of: Chemie Kompendium für das Selbstudium.
   Includes bibliographical references and index.
   1. Chemistry, Technical—Handbooks, manuals, etc.
I. Hennig, Ingo. II. Title
TP151.H5813 1983   660.2   83-310
ISBN 0-07-030320-7

CONTENTS

Preface

Increasing demands for quality, safe handling, and application of chemical intermediates and end products have created a need for greater knowhow and accountability on the part of chemical industry employees. It no longer suffices for the technological relationships to be understood by chemists alone. The experts in raw material procurement, sales, and engineering must also possess adequate basic knowledge of the conversion of materials and energy. They must understand industrial processes and the use of substances in a wide range of products. And they should be familiar with the economic and business environment in which capital, raw materials, energy, and products circulate.

In order to meet these requirements, this book was published, first in German in 1972, in collaboration with several chemical concerns: BASF Aktiengesellschaft, Ludwigshafen; Bayer Aktiengesellschaft, Leverkusen; Chemische Werke Hüls, Aktiengesellschaft, Marl; Degussa, Frankfurt (Main); Hoechst Aktiengesellschaft, Frankfurt (Main); and Veba Chemie Aktiengesellschaft Gelsenkirchen-Horst. The authors are Prof. Dr. Vollrath Hopp and Oberstudienrath Ingo Hennig; the co-authors are Dr. Ludwig Hüter, Dr. Werner Leutsch, and Dr. Fritz Merten.

The demand for the book was so great that the first edition was soon sold out. Following a thorough revision, the second German edition was published in 1976. The chemical industry enjoys close international links: raw materials have to be imported and finished products exported; different transportation systems have to be coordinated; and quality requirements and safety regulations have to be drawn up on an international basis. The third edition has thus been published in English, with the individual chapters having been revised accordingly and adapted to the information requirements of the world market.

Chapters I and III cover, in concise but easily understandable form, the fundamentals of inorganic and organic chemistry. Chapters II and IV are concerned with industrial production processes for base and primary chemicals. The subject of technical service and development is also discussed in detail in a number of separate sections, taking typical examples such as plastics, textile fibres, detergents, and nuclear chemistry. The importance of crude oil and coal as chemical feedstocks is dealt with in particular detail, and a section on biopolymers has been added to Chapter III.

Data on the enthalpies of reaction is provided in order to give the energy side of typical reaction processes. And the economic importance of the production processes and chemical products is indicated by listings of production volumes whenever possible.

In Germany, this book has been a successful means of home study. To provide readers with means of checking their progress, each section closes with a number of self-evaluation questions and the corresponding answers. The book is therefore directed not only to employees of the chemical sector and its related industries, but also to all interested persons who wish to familiarize themselves with the basic fundamentals of chemistry.

Thanks are expressly due to those persons in the German chemical industry and its organization, the Association of the Chemical Industry, Frankfurt (Main) who, through their advice, suggestions, and procurement of informational material, have assisted in preparation of this book. Special thanks are due to Director Rudolf von der Heyde, Hoechst Aktiengesellschaft. Only through his encouragement was it possible to overcome all the organizational difficulties connected with publishing this English edition. Thanks are also due to Mr. B. Carroll for careful proofreading and to Mr. Ing. (grad.) Hans W. Naundorf for design and drawing of the flow charts.

Vollrath Hopp
Ingo Hennig

Introduction

Chemistry is the science of the conversion of materials and energy and the characterization of substances. The flourishing of chemistry as a natural science in the eighteenth century was followed in the mid-nineteenth century by the establishment of a chemical industry in Europe, particularly in Switzerland, Germany, and Britain. In the United States, the chemical industry did not begin to develop until around the turn of the century, but then it expanded by leaps and bounds.

The chemical industry's worldwide sales in 1980 were $725 billion. With $163 billion, the U.S. chemical industry occupied the leading position, followed by Japan, with $76 billion and the Soviet Union with $68 billion. The Federal Republic of Germany occupied fourth place with sales of just under $60.4 billion. In the same period, the British chemical industry turned over $41.5 billion.

The performance of a chemical industry is reflected first in the value of its production and second in the range of products it manufactures. The latter may consist of large-tonnage products such as plastics or fertilizers or highly sophisticated preparations such as pharmaceuticals or biocatalysts.

The task of industrial chemistry is to supply products that help to make life easier for people throughout the world. Every day, 4.5 billion people need to eat, drink, clothe themselves, protect themselves against disease, and build dwellings. In the year 2000 there will be over 6 billion people on our planet for whom the necessities of life will need to be provided.

Provisions must be made for this, and the chemical industry can make a major contribution. For example, it can produce the fertilizers, crop-protection agents, and preservatives necessary to safeguard the food supply; 30 percent of the world's grain harvest is still being destroyed by vermin.

Chemical industry research, technical service and development, production, procurement, and sales staff work closely together in order to meet the challenge of the future. This challenge lies in the ability to produce sophisticated end products from simple raw materials such as crude oil, natural gas, coal, wood, or inorganic minerals such as rock salt, rock phosphates, sulphur and atmospheric nitrogen, oxygen, and water. Products manufactured from them, in addition to those already mentioned, include colorants, fibres, plastic film, paint, cleaning agents, and many others. The chemical industry is a research-intensive sector, and research activities continually call for development and modification of working methods for preparation and analysis of new substances. Substances have to be tested for their chemical behaviour, their physical properties, and toxic action.

In suitable technical laboratories, chemical production processes or individual process steps progress from a laboratory scale via a number of stages of varying length to the semi-industrial scale and from there are brought to the production stage.

Technical service and development is a link between research and production, on the one hand, and the manufacturer and the user, on the other. For the user and consumer, directions on how to handle the end product and how to proceed when using it have to be drawn up. This includes service and necessitates close cooperation among research, production, and sales. Active substances and materials that have been thoroughly tested and found suitable are tailored to specific uses and fields of application. For example, pigments are tested for suitability for colouring plastics and textile fibres or printing inks for high-speed rotary presses. Plastics are tested for their application as materials in the building sector, automobile industry, etc.

Production is, above all, concerned with the manufacture of active substances and materials from raw materials and intermediates. The actual manufacturing process is frequently followed by separation and purification processes such as distillation, filtration, extraction, and drying.

The optimization of chemical processes is nowadays determined by four principal factors:

- economic use of raw materials;
- reaction control adapted to environmental requirements;
- high energy utilization; and
- a safe process cycle.

When allowance is made for these factors, the products still have to be cost competitive in the world market. Sales are responsible for bringing the end products and intermediates to the consumer rapidly and in a form suitable for immediate use. This requires the proper means of transportation and storage. Here, too, the appropriate safety measures have to be observed.

The flow of capital, material, energy, and production determines the rhythm of a technologically highly developed chemical industry. The direction and optimum coordination of the former for the benefit of humanity impose major requirements with respect to the knowledge and accountability of the employees.

I
Fundamentals of General and Inorganic Chemistry

1 Chemistry and Physics.
Substances. States of Aggregation

1. Branches of science

Chemistry, physics and biology represent three sciences which are concerned with processes in the material world. The contents and method of working of these sciences are frequently closely intertwined.

2. Substances

The term **substance** signifies the material of which objects are composed, irrespective of their size or form. Thus knives and scissors consist of the same substance, steel, and have the characteristic properties of this substance. Similarly, tables, cupboards and chests are all composed of the same substance, wood, and possess its particular properties.

3. Chemistry and chemical processes

Chemistry deals with the nature of substances: it is the science of substances and their transformations. When wood, coal or oil burns to produce gases and ash or soot, when iron in moist air turns into flaky rust, or when plastics are made from natural gas, the substances originally present are completely transformed. Entirely new substances are formed, having properties which are quite different from those of the original substances. Transformations of this kind are known as **chemical processes**. Examples are the following:

Original substances		Final products
Coal and air	⟶	Gases, ash and soot
Rock salt and water	⟶	Caustic soda, chlorine and hydrogen
Carbohydrates (sugar, starch)	⟶	Alcohol and carbon dioxide

4. Physics

By contrast the sister science of **physics** deals predominantly with the properties of individual substances and their changes of state. Thus it describes the laws of motion, of free fall, of rotation and of optical behaviour such as the emission and refraction of light and the phenomenon of colour: it also deals with the principles governing sound as well as thermal, electrical and magnetic properties. In all these physical phenomena the substance itself remains unchanged.

5. States of aggregation

Physics thus deals with the three **states of aggregation** of matter, and with changes in the state of aggregation of a given substance, for example ice — water — water vapour. In such changes we are dealing with the same substance, independent of whether it is present in the **solid, liquid or gaseous state**.

6. Solids, particle shape

All substances are made up of very small particles. In **solids** the particles of which they are composed are held together by very strong forces of attraction. In solids these very small particles are arranged in a highly regular manner. If this regularity is apparent in the external appearance of the solid it is said to be crystalline. A solid can be converted into smaller fragments by breaking, grinding, sawing, filing, pulverizing, and it can be imagined that eventually a limit will be reached for any further mechanical subdivision. The ultimate small particles vary in shape — for example, they may be spherical or thread-shaped. The smallest particles of water, the water molecules, are approximately spherical.

7. Melting and freezing

If energy is supplied, for example the heat of combustion of fuels, the particles are set into more and more vigorous motion, until they become loosened and can move relative to one another. The binding of the solid is then broken and the substance **melts**, ie it becomes liquid. The regular arrangement of the particles is broken down. The supply of thermal energy has brought the substance to its melting point.
When a molten substance **freezes** this occurs at the same temperature, ie melting point = freezing point. The heat supplied in order to bring about melting is given out again on freezing.

8. Liquids

The particles of a liquid are in a state of continuous random motion. There is, however, a loose attachment between them. For this reason when a liquid is poured out it does not break up into innumerable tiny drops, but forms a coherent liquid layer. However, liquids do not have any fixed shape, but adapt themselves to the shape of the container.

9. Viscosity

If the particles of a flowing liquid interact strongly so as to hinder their relative motion the liquid is said to be viscous, or to have a high **viscosity**.

10. Evaporation, boiling and condensation

On the supply of more energy the movement of the particles increases and some of them shoot through the surface of the liquid: the substance **evaporates**. If still more energy is supplied the remaining forces of attraction are overcome and, at a certain temperature, the substance reaches its boiling point: It **boils**. If now the vapour is cooled the substance once more becomes liquid: it **condenses**. This happens at the same temperature as boiling, ie boiling point = condensation point.
The heat energy supplied in order to bring about boiling must be evolved again on condensation.

11. Melting point and boiling point as characteristic properties of substances

The **melting point** (mp) and the **boiling point** (bp) represent constants which are used to characterize substances. The freezing point (melting point) of water is taken as $0°C$ ($32°F$), and its boiling point (condensation point) as $100°C$ ($212°F$) under normal atmospheric pressure.

12. Gases, gas pressure, diffusion

The particles of a **gas** fly about at random with high speeds. A gas therefore has no fixed shape. Its particles collide frequently with one another and with the walls of the container.
The **pressure of a gas** is the resultant of all the collisions with the wall of the vessel.
In the absence of a containing wall the particles of a gas will occupy any available space. Gases therefore mix together spontaneously: they **diffuse** into one another.
The spreading of a perfume in air is an example of diffusion.

13. Energy content

Apart from the nature of a substance, the question of whether it exists as a solid, liquid or gas is determined by the temperature of its surroundings. A substance has a higher **energy content** in the gaseous state than in the liquid or solid state, since heat energy must be used in order to melt or vaporize it.

14. Vapours and gases

The term **vapour** is used when a substance which is liquid or solid at room temperature ($20°C$ or $68°F$) has been converted into a gas by heating: for example, water vapour.
Substances which exist in the gaseous state at room temperature are described as **gases**: for example, air or natural gas.

15. Sublimation

Some substances on heating pass directly from the solid to the gaseous state, thus passing over the liquid state; they are said to **sublime**. The same process can take place in the reverse direction. Thus snow is formed directly from water vapour and, conversely, washing will dry even when it is frozen. Iodine and naphthalene are further examples of substances which sublime and which can be purified by sublimation.

Freeze drying represents a technical application of the sublimation of ice.

16. Survey of the three states of aggregation

The following scheme gives a summary of the behaviour of the constituent particles in the three states of aggregation:

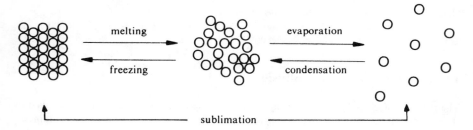

State of aggregation	Solid	Liquid	Gas
Distance between particles	Very small	Small	Very great
Order	Highly ordered	Slightly ordered	Completely disordered
Compressibility	Extremely small	Very small	Very great
Energy content	Low	High	Very high
Type of motion	Oscillations about an equilibrium position	Random, very short intervals between collisions	Random, short intervals between collisions
Cohesive forces	Completely effective	Partly effective	Usually very small

17. Behaviour of substances consisting of very large particles

A different type of behaviour on heating and cooling is exhibited by many substances,
particularly those composed of comparatively very large intertwined threadlike particles, for
example Nylon, polyethylene and polyvinyl chloride (PVC). These materials have no sharp
melting point, but soften over a range of temperatures and are then plastically deformable.
Glass behaves in the same way.
Certain other solids, for example wool, polyacrylonitrile and wood, do not soften and melt
on heating. The cohesive forces between their particles are relatively high, being greater
than the intrinsic stability of the particles. Substances of this kind decompose on heating
and are chemically degraded.

18. Summary

18.1.   **Chemistry** is the science of substances and their transformations.

18.2.   The common term **substance** is applied to objects (solids, liquids or gases) which differ only in their size and shape, but are otherwise identical in their chemical and physical properties.

18.3.   The term **chemical process** is used to describe a transformation of substances. The initial substances are converted into new ones with different properties.

18.4   **physics** is the science of the state of substances and of changes of state.

18.5.   A physical process involves only a change in (for example) the state of aggregation of a substance, the substance itself retaining its identity.

18.6.   The term **state of aggregation** refers to the solid, liquid and gaseous states.

18.7.   The **melting point**, mp (identical with the **freezing point**, fp) is the temperature at which a substance changes from the solid to the liquid state, or vice versa.

18.8.   **Viscosity** is a measure of the resistance to flow of a liquid, arising from the internal friction between its particles.

18.9.   Evaporation is the emergence of particles from the surface of a liquid below the boiling point.

18.10. The **boiling point**, bp (identical with the **condensation temperature**) is the temperature at which a substance passes from the liquid to the gaseous state, or vice versa.

18.11   The gas **pressure** is the force per unit area on the wall of a container exerted by the elastic collisions of the gas particles.

18.12. **Diffusion** is the spontaneous mixing of gases (or liquids) caused by the thermal motion of their particles.

18.13   The three states of aggregation differ in their **energy content**. For a given substance the energy content decreases in the order gas > liquid > solid.

18.14. A **vapour** is the gaseous state of a substance which is liquid or solid at room temperature.

18.15. **Sublimation** is the direct vaporization of a solid, or the reverse process. The liquid state is passed over.

19. Self-evaluation questions

1 Does a chemical change take place in

A   the magnetization of iron?   Yes — No
B   the rusting of iron?   Yes — No

2 Does the nature of the substance change when

   A   water freezes at 0°C?   Yes — No
   B   water boils at 100°C?   Yes — No

3 Which has the greater energy content: water vapour or ice?

Give reasons for your answer.

4 What do we call the processes which occur when

   A   mist is formed from water vapour?
   B   hoar frost is formed from water vapour?

5 Mark all physical processes with a cross.

   A   the shaping of thermoplastics
   B   the combustion of petroleum
   C   the melting of ice
   D   the occurrence of a rainbow

6 Which substances are described as "vapours" in their gaseous form?

   A   carbon dioxide
   B   water
   C   petrol (gasoline)
   D   air

7 Which processes in which the physical state of a substance changes are correctly named?

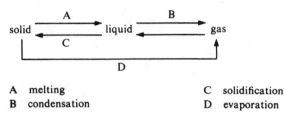

   A   melting   C   solidification
   B   condensation   D   evaporation

8 The formation of water vapour directly from snow is known as

A   condensation
B   crystallization
C   sublimation
D   drying

9 What increases when a solid substance first becomes liquid and then gaseous?

A   the distance between the particles of the substance
B   the arrangement of the particles
C   the energy content of the particles

10 What decreases when a solid first becomes liquid and then gaseous?

A   the energy content of the particles
B   the speed of the moving particles
C   the cohesion between the particles

11 The term "viscosity" connotes

A   the friction between solids, eg motor vehicles tires and roadway
B   the friction between liquid particles as they pass each other
C   the resistance of liquids to agitation
D   the elasticity of solids

12 Which of the following statements on the change in the physical state of substances are correct?

A   A substance evaporates when it passes from the liquid phase into the gas phase at its boiling point.
B   The quantity of heat released when a substance melts has, conversely, to be reintroduced into the substance when it solidifies.
C   The melting point of a substance is a criterion of purity.
D   The melting and solidification of a substance occur at the same temperature. The melting point is thus identical to the solidification point.

2 Classification of Materials. Mixtures, Elements and Compounds

1. Division into pure substances and mixtures

If we investigate to what extent a material can be separated into other substances we find that all materials can be divided between the two following groups:

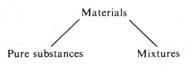

2. Pure substances

The criterion for a **pure substance** is that when it is successively divided into smaller and smaller amounts all these amounts continue to exhibit the same properties. A piece of iron can be mechanically divided into extremely small portions. These portions all possess the same properties, characteristic of iron, and iron is thus a pure substance. The same is true of, for example, copper, sulphur or common salt.

3. Mixtures

On the other hand, if finely powdered iron is stirred together with finely powdered sulphur we obtain a **mixture** of the two substances. Rocks, which are composed of diverse substances, are a natural example of mixtures. If a mixture is divided mechanically, particles having different properties will be obtained. The number of types of particles having different properties is equal to the number of pure substances contained in the mixture.

Thus in a mixture of iron and sulphur the dark grey iron particles can be easily separated by drawing them out with a magnet, or the yellow sulphur dissolved out with a suitable solvent.

3.1. Composition of mixtures

The **composition of a mixture**, ie the proportions in which the pure constituents are mixed, can be varied quite arbitrarily. A mixture of iron and sulphur can consist of a large amount or iron with a small amount of sulphur, or a large amount of sulphur with a small amount of iron.

3.2. Homogeneous and heterogeneous systems

On superficial examination a mixture may appear to be **homogeneous** (uniform), as with white coffee or brass, or alternatively examination with the naked eye or with a lens may reveal that it has a **heterogeneous** (non-uniform) structure, as with a mixture of iron and sulphur powder, with granite, or with speckled marble.

3.3. Types of mixture

The separate constituents of a mixture may be present in any of the three states of aggregation. We can therefore distinguish between the following types of mixture:

Heterogeneous mixtures:

State of aggregation	Examples
Solid/solid	Rocks, concrete
Solid/liquid	Pastes and doughs
	Sludges and suspensions
	Dispersed plastics
Solid/gaseous	Smoke
Liquid/liquid	Emulsions, milk
	Floor polish
Liquid/gaseous	Mist and clouds
	Soap suds and soap bubbles
	Sprays

Homogeneous mixtures:

State of aggregation	Examples
Solid	Alloys, e.g. brass, bronze, steel
Liquid	Petroleum solutions, e.g. sea water, sugar syrup
Gaseous	Air, natural gas

3.4. Methods of separation

Most mixtures can be separated into their constituents by simple physical methods. Examples of such **methods** are the following:

Filtration	—	Coffee filter, oil filter
Centrifuging	—	Spin dryer
Melting	—	Sulphur from rock
Dissolving out (extraction)	—	Oil from oil-seeds
Evaporation	—	Salt from sea water
Distillation	—	Gasoline from crude oil

3.5. Distillation, fractional distillation

Distillation consists of the vaporization of a liquid followed by condensation of the vapour.
If a liquid is heated to boiling it vaporizes and, if the resulting vapour is cooled, it
condenses, ie it once more becomes liquid.
Mixtures of liquids having different boiling points can be separated by distillation. When a
liquid mixture boils the component having the lowest boiling point vaporizes the most
vigorously, and can be separated from the other components by condensing the first portion
or **fraction** of the vapour. In the same way the mixture can be separated into further
fractions: this is known as **fractional distillation**. For example, crude petroleum can be
separated by fractional distillation into gasoline, diesel fuel, heating oil and lubricating oils.

4. Air, a mixture of gases

A typical mixture of gases is **air**, which represents one of the most important raw materials
for chemical industry. Dry air consists of:

ca. 78	vol.%	nitrogen
ca. 21	vol.%	oxygen
ca. 1	vol.%	rare gases
ca. 0.03	vol.%	carbon dioxide

Vol.% (volume per cent) means the percentage by volume: thus 100 litres of air contain 78
litres of nitrogen, or 100 quarts of air contain 78 quarts of nitrogen.
The atmosphere also contains varying amounts of water vapour. The higher the
temperature, the larger the quantity of water vapour which the air can take up. 1 litre of
air weighs 1.29 g at normal temperature and pressure (NTP), ie at $0°C$ (273 K or $32°F$) and
1013 mbar $\approx$ 1.0 bar (14.5 lb/in^2).

5. Classification of pure substances

Pure substances can be divided into **elements**, of which all substances are ultimately
composed, and **compounds**, which are formed by the chemical combination of elements:

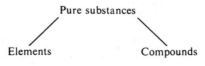

I Fundamentals of general and inorganic chemistry

6. Elements

An **element** cannot be broken down into any simpler substances.

6.1. Metals, nonmetals and metalloids

The elements can be divided into **metals**, such as gold, silver, copper and aluminium, and **nonmetals**, such as carbon, hydrogen, oxygen, nitrogen, sulphur and chlorine. Of the 105 known elements 81 are metals and 17 nonmetals, while 7 are not clearly either metals or nonmetals, and are therefore termed **metalloids**.

Out of the total number of elements little more than 20 are of much technological importance.

6.2. States of aggregation

At room temperature (20°C or 68°F) the 105 elements include:

11 gases	—	hydrogen, oxygen, nitrogen, fluorine and the so-called rare or inert gases helium, neon, argon, krypton, xenon and radon.
2 liquids	—	mercury and bromine.
92 solids	—	all the remaining elements.

6.3. The symbols for the elements

For the sake of simplicity the names of the elements are often replaced by **symbols**. These symbols are mostly derived from the Latin names for the elements and are the same in all languages, so that they can be used internationally.

The following list of the most important elements with their symbols indicates whether they are metals, nonmetals or metalloids. An alphabetical list of all the elements is given in Section I.9.

Names, symbols and nature of some important elements

Name of element	Symbol	Metal (Me), nonmetal (NMe) or metalloid (Md)
Aluminium	Al	Me
Argon	Ar	NMe
Arsenic	As	Md
Barium	Ba	Me
Boron	B	Md
Bromine	Br	NMe

Name of element	Symbol	Metal (Me), nonmetal (NMe) or metalloid (Md)
Calcium	Ca	Me
Carbon	C	NMe
Chlorine	Cl	NMe
Chromium	Cr	Me
Cobalt	Co	Me
Copper (Cuprum)	Cu	Me
Fluorine	F	NMe
Germanium	Ge	Md
Gold (Aurum)	Au	Me
Helium	He	NMe
Hydrogen	H	NMe
Iron (Ferrum)	Fe	Me
Iodine	I	NMe
Krypton	Kr	NMe
Magnesium	Mg	Me
Mercury (Hydrargyrum)	Hg	Me
Neon	Ne	NMe
Nickel	Ni	Me
Nitrogen	N	NMe
Oxygen	O	NMe
Phosphorus	P	NMe
Platinum	Pt	Me
Potassium (Kalium)	K	Me
Selenium	Se	Md
Silicon	Si	Md
Silver (Argentum)	Ag	Me
Sodium (Natrium)	Na	Me
Sulphur	S	NMe
Tin (Stannum)	Sn	Me
Titanium	Ti	Me
Uranium	U	Me
Zinc	Zn	Me

6.4. Abundance

The terrestrial **abundance of the elements** varies greatly. The Table below shows that oxygen, with about 50 wt. %, is by far the most abundant. Oxygen occurs as constituent of a mixture in air, and combined with other elements in water, minerals and other compounds. The second most abundant element is silicon, with about 25 wt. %. Silicon is contained in most minerals.

Only eight elements make up almost 98% of the accessible material world.

It is remarkable that carbon contributes such a small percentage, although the whole animal and vegetable world is composed of carbon compounds.

The following summary also gives, as an interesting comparison, the abundance of the elements in the human body.

The abundance of the elements

Elements	Wt. % in the air, the oceans, and accessible parts of the earth's crust	Wt. % in the human body
Oxygen	49.4	65.0
Silicon	25.8	0.002
Running total	75.2	
Aluminium	7.5	0.001
Iron	4.7	0.010
Calcium	3.4	2.01
Sodium	2.6	0.109
Potassium	2.4	0.265
Magnesium	1.9	0.036
Running total	97.7	
Hydrogen	0.9	10.0
Titanium	0.58	—
Chlorine	0.19	0.16
Phosphorus	0.12	1.16
Carbon	0.08	18.0
Nitrogen	0.03	3.0
Running total	99.6	99.753
All other naturally occurring elements	0.4	0.247
Total	100	100

7. Compounds

The properties of **compounds** differ completely from those of the elements of which they are composed. For example, common salt is a compound of sodium and chlorine, but the following comparison shows how great a difference there is between the properties of the compound and those of the constituent elements:

Sodium	Chlorine	Common salt
Sodium is a soft white metal which reacts vigorously with water	Chlorine is a green poisonous gas	Common salt is a solid colourless non-poisonous crystalline substance

It is not possible to resolve compounds into their component elements by simple physical methods of separation, such as filtration, centrifuging, etc.

7.1. Organic compounds

In classifying compounds the large number of carbon compounds, about one million, are treated separately under the heading of **Organic chemistry**.

7.2. Inorganic compounds

Inorganic chemistry includes the compounds of all the other elements, numbering about 100 000.

8. Scheme of classification

All substances can therefore be included in the following quite simple scheme:

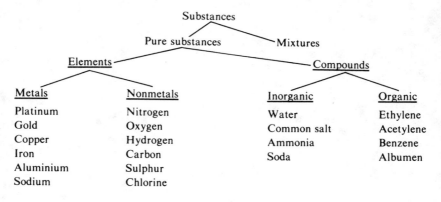

9. Summary

9.1. **Pure substances** can be divided into very small amounts which have the same characteristic properties.

9.2. **Mixtures** contain pure substances mixed in arbitrary proportions, and possess the properties of their constituents. Mixtures can often be separated by simple physical means.

9.3. **Distillation** is a physical method of separation, consisting of vaporization followed by condensation.

9.4. **Fractional distillation** is a method of separating liquid mixtures, in which the constituents are collected separately as fractions.

9.5. **Air** is a mixture of gases.

9.6. **Pure substances** consist of either an element or a compound. The elements are divided into metals, nonmetals and metalloids, and the compounds into organic and inorganic.

9.7. **Elements** are the ultimate constituents of which all other substances are composed. They cannot be broken down into simpler substances.

9.8. The **symbols** for the elements are abbreviated forms of their names.

9.9. **Compounds** are substances made up of elements. Their properties differ from those of their constituent elements.

9.10. **Inorganic compounds** include the compounds of all elements with one another, except those of carbon.

9.11. **Organic compounds** are compounds of carbon with other elements.

10. Self-evaluation questions

13 Which of the following can be classified as pure substances?

A   elements
B   mixtures
C   alloys
D   compounds

14 A mixture is

A   a collection of various substances
B   a compound such as water, since this is composed of the elements hydrogen and oxygen
C   usually capable of being separated into its individual constituents by simple physical methods
D   a substance that possesses properties completely different from those of its individual constituents

15 Which of the following materials are pure substances, and which are mixtures?

Common salt, wine, butter, air, sugar, mercury, synthetic resin varnish.

Pure substances: _____

Mixtures: _____

16 Place a cross against all the homogeneous mixtures.

A   air
B   alloys
C   petroleum
D   seawater

17 Classify as homogeneous or heterogeneous mixtures:

Coal, glass, petrol, jam, jelly, brick.

Homogeneous mixtures: _____

Heterogeneous mixtures: _____

18 How can the following mixtures be separated?

A   Oil and water
B   Brine
C   Muddy river water

I Fundamentals of general and inorganic chemistry

19 Which of the following statements are correct in connection with fractional distillation?

A   Even pure substances can be separated from one another by fractional distillation.
B   The first fraction of a liquid mixture to be distilled off is that with the lowest boiling point.
C   During distillation all the constituents of a liquid mixture are present at some time in each of the three physical states.
D   Provided they are first liquefied, even gas mixtures can be separated into their different constituents by fractional distillation.

20 The following can be separated by the means stated:

A   oil from oil fruits by distillation
B   gasoline from petroleum by evaporation
C   sulphur from sulphur-bearing rock by melting
D   salt from seawater by centrifuging

21 Which is the approximate composition of pure air in % by volume?

	Oxygen	Noble gases	Nitrogen	Carbon dioxide
A	78	0.03	21	1
B	78	1	21	0.03
C	21	0.03	78	1
D	21	1	78	0.03

22 Which of the following statements in connection with air are correct?

A   Air is a weightless gas.
B   Air is one of the most important raw materials in the chemical industry.
C   In winter air can absorb more water vapour than in summer.
D   Air can be liquefied by cooling and compression.

23 Chemical basic materials are

A   elements
B   pure substances
C   compounds
D   homogeneous mixtures

24 Classify as elements or compounds:
Sugar, phosphorus, nickel, sand, lead, water.

Elements: _____

Compounds containing
more than one element: _____

25 Which row consists entirely of non-metals?

A	C	—	Ne	—	F	—	I
B	Al	—	Kr	—	Cl	—	O
C	H	—	N	—	P	—	Ca
D	He	—	Br	—	Hg	—	S

26 Which element is not a gas at room temperature?

A   carbon
B   nitrogen
C   oxygen
D   fluorine

27 Which elements are liquid at room temperature?

A   krypton
B   bromine
C   sulphur
D   mercury

28 Which element in the form of its compounds is the most abundant in the world when calculated on a weight basis?

A   carbon
B   oxygen
C   nitrogen
D   hydrogen

3 Chemical Reactions and the Laws Governing Them. Energy

1. Chemical reactions

Chemical reactions are transformations of substances involving:
  A   compounds and compounds;
  B   compounds and elements; or
  C   elements and elements.

In a chemical reaction a compound can be broken down into simpler compounds or into elements. In every case the final products differ in properties from the initial substances.

2. Synthesis

The building up of a compound from other substances, elements or other compounds is termed **synthesis**. A synthesis is often carried out in steps by way of intermediate compounds.

3. Analysis

Analysis is the investigation of the constituents of a chemical compound or a mixture. **Qualitative analysis** determines the nature of the constituents and **quantitative analysis** their amounts. Analysis is also concerned with the measurement of melting points, boiling points, and other characteristics of substances, such as their density*, colour, crystal form, solubility, electrical conductivity and heat conductivity.

4. The laws of chemical combination

While the constituents of mixtures can be present in arbitrary proportions, elements combine to form compounds according to quantitative laws. The **law of chemical combination** states: Elements always combine in definite proportions by weight. Thus water always contains the elements hydrogen (H) and oxygen (O) in the proportion 1:8 by weight.

5. The law of the conservation of mass

This law states that in a chemical reaction the total mass (or weight) of the substances taking part remains unchanged.

---

* Density = mass/volume. It is designated by the Greek letter ρ (rho). The unit of density is g/cm^3 or lb/ft^3. The density of water is 1 g/cm^3 (ie, 1 cm^3 of water weighs 1 g) or 62.41 lb/ft^3.

I Fundamentals of general and inorganic chemistry

The amount of a material can be expressed in terms of either the weight (or the mass) or number of particles. The unit of weight (or mass) used by scientists is the kilogram (kg). The unit of the amount of substance is the mole representing a certain number of particles (6.10^{23}).

6. Forms of energy

Every chemical process is accompanied by transformations of energy. Energy, ie the capacity for doing work, occurs in various forms. Thus we distinguish between chemical, electrical, thermal and kinetic energy: foodstuffs contain chemical energy, while a rolling cheese possesses both chemical and kinetic energy. The forms of energy can be interconverted.

7. The law of the conservation of energy

Energy is never lost. This is the practical content of the law of the conservation of energy.

8. Interconversion of different forms of energy

Important types of energy transformation are the following:

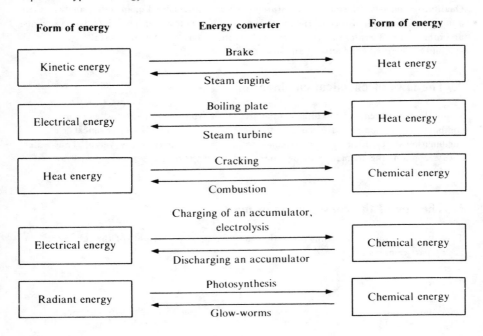

Form of energy	Energy converter	Form of energy
Kinetic energy	Brake → / ← Steam engine	Heat energy
Electrical energy	Boiling plate → / ← Steam turbine	Heat energy
Heat energy	Cracking → / ← Combustion	Chemical energy
Electrical energy	Charging of an accumulator, electrolysis → / ← Discharging an accumulator	Chemical energy
Radiant energy	Photosynthesis → / ← Glow-worms	Chemical energy

9. Heat of reaction

The energy transformation associated with a chemical reaction often appears in the form of heat, the **heat of reaction.** Quantities of heat are measured in Joules (J) or British thermal units (Btu). 4 187 J are required to raise the temperature of 1 kg of water from 15°C to 16°C, or 1 000 J (1 kJ) will raise the temperature of 0.239 kg of water from 15°C to 16°C.

9.1. Exothermic reactions

A process which continuously gives out heat is called an **exothermic reaction,** for example the combustion of fuels such as coal or oil, and also human metabolism. The combustion of 1 kg of heating oil produces about 41 870 kJ. Similarly, 1 lb of heating oil produces about 18 000 Btu.

9.2. Endothermic reactions

If a continuous supply of heat is needed to keep a process going it is termed an **endothermic reaction.** Examples are the calcination of limestone to obtain quicklime, and the smelting of iron ores.

10. The energy content of substances

Since energy is transformed in all chemical reactions, the **energy content** of the products differs from that of the reactants. In an exothermic reaction the products contain less energy than the reactants, for example:

| Wood + atmospheric oxygen | Combustion → | Products of combustion (+ heat energy)* |

In endothermic reactions the products contain more energy than the reactants. Thus high-energy and hence very reactive chlorine is made industrially from rock salt (NaCl):

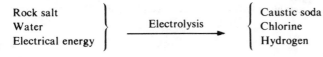

In endothermic reactions the products contain more energy than the reactants. Thus high-energy and hence very reactive chlorine is made industrially from rock salt (NaCl):

Rock salt
Water   Electrolysis →   Caustic soda
Electrical energy   Chlorine
                                     Hydrogen

---

* Since this heat energy is liberated during the reaction and is given up to the surroundings it is usual to refer to it as a negative enthalpy of reaction = $-\Delta H$. In an endothermic reaction the heat absorbed is $+ \Delta H$.

11. Energy requirements of the chemical industry

In the high industrialized countries the chemical industry is easily the largest industrial consumer of energy. This consumption is expected to increase in the future. The generation and distribution of energy at favourable prices is thus of prime economic importance. It is therefore not surprising that the chemical industry shows great interest in the supply of energy from coal, oil, natural gas and nuclear energy.

12. Reversible reactions

Many chemical reactions can proceed in opposite directions, ie the reaction products can react to re-form the original reactants. In such instances it is necessary to distinguish between the forward and the back reactions. The **forward reaction** is the reaction between the initial reactants to give the final products, while the **back reaction** is the reverse process:

Reactants	Forward reaction	Products
A + B	$\rightleftharpoons$	C + D
(left-hand side)	back reaction	(right-hand side)

Many chemical reactions are coupled to varying extents with back reactions, and it is frequently impossible to suppress completely the return of the products to give the initial reactants.

12.1. Chemical equilibrium, yield

If a reaction reaches a state in which the number of molecules of the reactants which react to give products is equal to the number of molecules of the products which react to give reactants, then the system is said to be in a state of **chemical equilibrium**. The amount of product formed then ceases to increase. (However, this does not necessarily imply that only 50% of the initial materials has reacted.) If a reaction is in a state of equilibrium this is indicated in the chemical equation by double arrows pointing in opposite directions.
Even when the equilibrium position lies far to the left it is often possible in chemical industry to obtain a satisfactory yield by carrying out the reaction in a circulatory system and removing the end product continuously from the system (cf. Section II.3, Synthesis of ammonia).

12.2. Factors affecting the position of equilibrium

The equilibrium position or extent of conversion can be affected by varying: 1. the temperature; 2. the pressure (when the reaction involves gases); and 3. the concentrations of the reactants.

13. Catalysts

In addition, most syntheses employ **catalysts**. A catalyst is a substance which is added to the reaction mixture. It affects the course of the reaction but is not itself consumed. Catalysts can act in the following ways:

a.   to initiate the reaction;
b.   to accelerate the progress of the reaction;
c.   to lower the temperature required in the absence of the catalyst;
d.   to steer the reaction towards the desired end-products.

The selection of reaction conditions which are the most favourable from the economic point of view represents a difficult problem in optimization.

13.1. Technical catalysts

Catalysts used in industry are frequently finely divided metals, such as platinum or nickel, or special compounds of metals. Such catalysts are usually deposited on carriers or supporting substances. Catalysts are used, for example, in the manufacture of ammonia, sulphuric acid and methanol.

13.2. Biocatalysts

Biocatalysts are active in living organisms and their cells. Well-known examples are enzymes, vitamins and hormones, and chlorophyll in photosynthesis in green plants.

13.3. Inhibitors (stabilizers)

Substances which retard or prevent a reaction are known as **inhibitors** or stabilizers. Preservatives are stabilizers. They act as negative catalysts.

13.4. Catalyst poisons

Catalyst poisons are substances which (often in very small amounts) affect or destroy the action of catalysts. Examples are hydrogen cyanide and hydrogen sulphide.

I Fundamentals of general
   and inorganic chemistry

14. Summary

14.1. **Chemical reactions** involve the interconversion of compounds and/or elements with the production of new substances. Some reactions involve building up and others breaking down.

14.2. **Synthesis** means the building up of a compound from other substances, usually simpler ones, often in stages.

14.3. **Analysis** means the determination of the nature and amounts of the constituents of a compound or a mixture.

14.4. Elements always combine in fixed proportions by weight.

14.5. According to the **law of the conservation of mass** the total mass (or weight) of the initial substances in a reaction is equal to the total mass of the products formed.

14.6. All chemical reactions are accompanied by **transformations of energy**.

14.7. Energy exists in different forms, which are interconvertible. The **law of the conservation of energy** states that energy is never lost.

14.8 Most chemical reactions are accompanied by heat changes. An **exothermic reaction** is a chemical process in which heat is liberated, and an **endothermic reaction** is a chemical process which proceeds only when energy is continuously supplied.

14.9. Many reactions lead to **chemical equilibrium**, which prevents the reaction from going to completion.

14.10 By an optimum of choice of reaction conditions (temperature, pressure, time of contact with the catalyst) the optimum **yield** may be obtained.

14.11. A **catalyst** is a substance which initiates or accelerates a chemical reaction, but which itself remains unchanged at the end of the reaction.

14.12. **Stabilizers** or **inhibitors** are substances which retard or prevent a reaction, but which themselves remain unchanged.

15. Self-evaluation questions

29 The chemical reaction in which the elements nitrogen and hydrogen combine to form
the compound ammonia is termed _____

30 The investigation of the purity and chemical composition of a product is described
as _____

31 What transformations of energy occur when

A   an electric light bulb is switched on?
B   an oil heating plant is working?
C   a hydro-electric plant is running?
D   an explosive blows up?

32 Classify as exothermic or endothermic reactions:

A   the explosions of the gasoline-air mixture in a car engine
B   the growth of crops
C   the metabolism of glucose in the human body

Exothermic reactions _____

Endothermic reactions _____

33 For which chemical reactions is the conversion of one form of energy into another
correctly represented?

A   Thermal energy   cracking   chemical energy
B   Light energy   plant growth   chemical energy
C   Chemical energy   charging of a battery   electrical energy
D   Chemical energy   combustion   thermal energy

34 A chemical reaction is described as exothermic if

A   heat has to be supplied uninterruptedly for it to continue
B   the end products created have less energy than the starting products employed
C   heat of reaction is released, ie the ΔH of the reaction is negative

35 Catalysts may act in chemical reactions in such a way that they

A   initiate reactions
B   shift the chemical equilibrium
C   are employed as participants in a reaction
D   steer the reaction to the desired end products

36 A chemical reaction is described by the following reaction equation

$$A + B \xrightleftharpoons{Pt} C, \text{ giving out thermal energy}$$

This reaction proceeds

A   catalytically
B   reversibly
C   endothermically
D   completely

37 Which substance has the opposite action to that of a catalyst?

A   activator
B   stabilizer
C   catalyst poison
D   vitamin

38 Deep-frozen foods can be kept without spoilage for relatively long periods. What is the main reason for this?

A   The cold acts as a catalyst
B   There is no constant contact with fresh oxygen in the air
C   Chemical reactions proceed more slowly at low temperatures
D   Substances prone to spoilage are encapsulated in ice

4 Atoms and Molecules. Chemical Formulae. Valency. Chemical Equations

1. Elements, atoms

The very small particles of which all substances are made have already been mentioned (I, 1-6). The fundamental constituents of an element are called **atoms**. An element is a substance which consists of only one kind of atom.

2. Compounds, molecules

The smallest self-contained particles of a compound are called molecules. A molecular consists of at least two atoms joined together. A compound is a substance which consists of only one kind of molecule. Water is a substance consisting of only one kind of molecule. Most molecules contains atoms of more than one kind.

2.1. Low-molecular-weight compounds

In organic chemistry one speaks of **low-molecular-weight compounds** when their molecules contain up to about 1000 atoms, predominantly C- and H-atoms.

2.2 High-molecular-weight compounds. Macromolecules

High-molecular-weight compounds with molecules containing more than 1000 atoms occur in plastics and synthetic fibres, which may contain molecules with tens of thousands of atoms, mainly C- and H-atoms.

Such large molecules are called **macromolecules**.

3. Compounds, ions

Some compounds consist of not molecules but of **ions**. Most inorganic compounds fall into this category. Ions are electrically charged atoms or groups of atoms.

Common salt is a substance which consists of ions.

Ions will be dealt with in Section I, 5.5

4. Orders of magnitude

In order to illustrate the extremely small size of atoms, molecules and ions, the following survey compares some orders of magnitude:

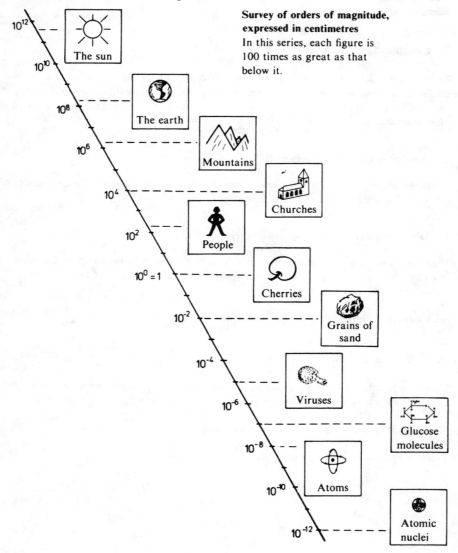

Survey of orders of magnitude, expressed in centimetres
In this series, each figure is 100 times as great as that below it.

4 Atoms and molecules. Chemical formulae. Valency. Chemical equations

The inconceivably small size of, for example, water molecules is shown by the following calculation. A glass of water contains about 100,000,000,000,000,000,000,000,000 water molecules. If it were possible to count them like peas, the task would occupy a million people for 100 million years!

5. Chemical formulae

Chemical formulae represent a shorthand way of designating chemicals. They contain the symbols for the elements of which the compound is composed.

CH_4, methane, consists of carbon C and hydrogen H.

The chemical formula also shows that 1 C-atom is combined with 4 H-atoms in CH_4. If a molecule contains several atoms of the same element, the symbol for that element bears a subscript Arabic numeral, eg:

H_2O: two H-atoms combined with one O-atom form one water molecule
C_3H_8:   one propane molecule consists of three C-atoms and eight H-atoms.

6. Valency

In any molecule there is a fixed whole-number ratio between the numbers of atoms combined together. These ratios arise from an important property of atoms, their **valency**. Hydrogen is a monovalent element, ie one H-atom cannot combine with more than one atom of another element.

The valency of an element indicates how many atoms of hydrogen one atom of that element can combine with.

In this way the valency of other elements can be deduced. In hydrogen chloride, HCl, one monovalent H-atom combines with one Cl-atom. Cl is therefore monovalent in this compound.

In H_2O oxygen must be divalent, since it is combined with two monovalent H-atoms. Similarly, the nitrogen in NH_3 is trivalent and the carbon in CH_4 is tetravalent.

In their compounds: H is always monovalent
                       O is always divalent
                       C is always tetravalent (except in CO).

Further, one H-atom can be replaced by another monovalent atom. Thus in common salt, NaCl, the hydrogen atom in HCl has been replaced by one Na-atom, and in silver chloride, AgCl, by one Ag-atom. It therefore follows that the atoms of all elements which can replace one H-atom are monovalent (Na, K, Ag) and can also combine with other monovalent atoms, for example Na and Ag with Cl in NaCl and AgCl respectively.

Divalent elements comprise all those whose atoms can replace two H-atoms (Mg, Ca) or can combine with other divalent atoms, as Ca with O in CaO.

The valency therefore also indicates how many H-atoms one atom of the element in question can replace.

It is easy to construct the formula of a compound if the valencies of the atoms which it contains are known. Conversely, the valencies of the atoms of any element can be deduced from the formulae of its compounds.

Many elements have different valencies in different compounds:

S is divalent	in H_2S (hydrogen sulphide)
tetravalent	in SO_2 (sulphur dioxide) and
hexavalent	in SO_3 (sulphur trioxide).

The valencies of such elements are indicated in the names of their compounds by Roman numerals in parentheses after the name of the element, for example.

FeO	=	iron (II) oxide
Fe_2O_3	=	iron (III) oxide
SO_2	=	sulphur (IV) oxide or sulphur dioxide
P_2O_5	=	phosphorus (V) oxide or diphosphorus pentoxide.

The older names, such as sulphur dioxide, are still used, and are formed by indicating the numbers of atoms by Greek numerical prefixes (see I.9 p.4).

The most important valencies exhibited in their compounds by elements of interest in technology are given in the following Table:

Name of element	Symbol	Valencies in compounds
Aluminium	Al	always III
Bromine	Br	I
Calcium	Ca	always II
Carbon	C	IV
Chlorine	Cl	I (V, VII)
Chromium	Cr	III and VI
Copper	Cu	I and II
Flourine	F	I
Hydrogen	H	always I
Iron	Fe	II and III
Lead	Pb	II and IV
Magnesium	Mg	always II
Nitrogen	N	III and V
Oxygen	O	always II
Phosphorus	P	III and V
Potassium	K	always I
Silicon	Si	always IV
Sodium	Na	always I
Sulphur	S	II, IV and VI
Titanium	Ti	always IV
Zinc	Zn	always II

7. Molecular formulae, structural formulae

The formulae used so far for compounds are **molecular formulae**. They indicate only the number and nature of the atoms in the molecule.

Structural formulae give a simplified picture of the spatial arrangement of the atoms in the molecule. The bonds, corresponding to the valencies, are indicated by lines joining the atoms. Structural formulae are particularly important for compounds consisting of molecules. They are of prime importance in organic chemistry.

Molecular formula	Structural formula
HCl	H – Cl
H_2O	$\begin{array}{c}H \\ H\end{array}\!\!>\!O$
NH_3	$\begin{array}{c}H \\ | \\ N \\ H \quad H\end{array}$
CH_4	$\begin{array}{c}H \\ | \\ H-C-H \\ | \\ H\end{array}$
CO_2	O=C=O
H_2	H – H
Cl_2	Cl – Cl
O_2	O=O
N_2	N≡N

The technically important gaseous elements hydrogen, oxygen, nitrogen and chlorine occur in general only as diatomic molecules.

8. Atomic mass (atomic weight)

The molecular formula H_2O indicates only that one H_2O molecule is made up of two H-atoms and one O-atom, and gives no information about its composition by weight.

The H-atom is the lightest of all atoms and is arbitrarily assigned a mass of 1. The masses of all other elements are expressed as multiples of the mass of an H-atom. (This is practically equivalent to the more precise definition, in which the comparison is made with one-twelfth of the mass of one C-atom of the carbon isotope ^{12}C.)

For example, one O-atom is 16 times as heavy as one H-atom, and one O-atom is therefore assigned a mass of 16. This is the meaning of the statement that H and O have **atomic masses** (atomic weights) of 1 and 16 respectively. Similarly, the atomic mass of C is 12, of Fe 56, and of Hg 201. An alphabetical list of the elements with their atomic masses (atomic weights) is given in Section I.9, p.5.

9. Molecular mass (molecular weight)

The **molecular mass** (molecular weight) of a substance is obtained by adding together the atomic masses of all the atoms in the molecule. Water has a molecular mass (molecular weight) of 18, calculated as follows:

$$
\begin{array}{ll}
2 \times H = & 2 \\
1 \times O = & 16 \\
\hline
H_2O \quad = & 18
\end{array}
$$

Similarly, the molecular mass of CO_2 is given by:

$$
\begin{array}{ll}
1 \times C = & 12 \\
2 \times O = & 32 \\
\hline
CO_2 \quad = & 44
\end{array}
$$

A CO_2 molecule is thus 44 times as heavy as one H-atom (or one-twelfth the atomic mass of a carbon atom).

10. The mass of single atoms and molecules

The actual mass (or weight) of a single atom or molecule is much too small for practical calculations. For example, the mass of an H-atom is 1.67×10^{-24} g (2.30×10^{-28} oz) and that of an H_2O-molecule is 30.6×10^{-24} g (4.21×10^{-28} oz).

11. The mole, the molar mass

The amount of substance that contains a particular number (6×10^{23}) of entities (e.g. atoms, molecules, ions) is called one mole. The mass of this particular number of atoms, molecules or ions is called the molar mass. For example 6×10^{23} H-atoms weighs 1 g (or 0.0022 lb). The molar mass of H-atoms is 1 g/mol. The same number of O_2-molecules weighs 32 g (or 0.071 lb), and 1 mol of H_2O molecules weighs 18 g (or 0.040 lb).

12. Molar volumes

In practice, quantities of gases or vapours are rarely specified as masses (weights), but are more often expressed by their volumes. The **molar volume** of a gas is the volume occupied by one mole (i.e. 6×10^{23} particles of the substance as a gas). At normal temperature and pressure (NTP: $0°C$ and 1 bar or $32°F$ and 14.5 lb/in^2) the molar volume of all gases is 22.4 litres.

Thus 44 g CO_2 (carbon dioxide), 16 g CH_4 (methane) and 17 g NH_3 (ammonia) each occupy a volume of 22.4 litres at NTP.

13. Chemical reactions and equations

Chemical reactions occur when atoms or molecules and their reaction partners meet one another. This can lead to addition, regrouping or cleavage, with the formation of molecules of different composition or the splitting off of atoms.

Chemicals equations are an abbreviated way of writing down chemical reactions.

A chemical equation conveys the following information:

1. Which substances (reactants) react together, and which new substances (products) are formed.

2. The numerical proportions in which the particles of the reactants react with one another, and the relation between the numbers of particles of the products.

3. The proportions of the substances by weights for 100% conversion (ie 100% yield).

Thus the equation $S + O_2 \longrightarrow SO_2$ tells us:

1. Reactants and products: sulphur burns in oxygen to give sulphur dioxide.

2. Each atom of sulphur reacts with one molecule of oxygen to give one molecule of sulphur (IV) oxide (sulphur dioxide).

3. Statements about weights (expressed in grams):

S	+	O_2	$\longrightarrow$	SO_2
32g sulphur		32g oxygen		64g sulphur dioxide
		$(2 \times 16g = 32g)$		

The equation satisfies the law of the conservation of mass, ie total mass of reactants = total mass of products.

The reaction or production of several atoms or molecules of the same substance is shown in the equation by numerical prefixes (factors), eg:

$$2\,H_2 \quad + \quad O_2 \longrightarrow 2\,H_2O$$

Water is formed from hydrogen and oxygen. The factors in this equation show that two hydrogen molecles always react with one oxygen molecule to give two water molecules.

Statements about weights (expressed in grams):

$2\,H_2$	+	O_2	$\longrightarrow$	$2\,H_2O$
$2 \times 2g = 4g$		32g		$2 \times 18g = 36g$

The arrow (———▶ or ◀———) is used, thus showing the direction of the reaction. For reversible (equilibrium) reactions (cf. I, 3.12) double arrows are used (⇌ or ⇌).

14. Stoichiometric calculations

The formulae and the chemical equations make it possible to carry out chemical (**stoichiometric**) **calculations**.

1. Example:

What is the percentage of iron in iron (III) oxide, Fe_2O_3?
Calculation of molar masses, in grams:

$Fe = 56g/mol$	$2\ Fe = 2 \times 56g = 112g$
$O = 16g/mol$	$3\ O = 3 \times 16g = \ \ 48g$
	$Fe_2O_3 \qquad\quad = 160g$

160g Fe_2O_3 contain   112g Fe
100g Fe_2O_3 contain $112 \times 100/160 = $ 70g Fe
Fe_2O_3 thus contains <u>70% of iron</u>

2. Example:

How much quicklime (CaO) can be made from 150 t of pure limestone (calcium carbonate)?

The equation for the reaction is:

$$CaCO_3 \quad\longrightarrow\quad CaO \quad + \quad CO_2$$

calculation of molecular masses:

$$40 + 12 + (3 \times 16) = \qquad 40 + 16 \quad + \quad 12 + (2 \times 16)$$
$$100 \quad = \qquad\qquad 56 \qquad + \qquad\quad 44$$

100 g $CaCO_3$ will give 56 g CaO
150 t $CaCO_3$ will give $56 \times 150/100 = $ <u>84 t CaO</u>

15. Summary

15.1. An **atom** is the smallest particle of an element.

15.2. Compounds may be composed of **molecules** or **ions**.

15.3. Molecules are uncharged particles consisting of at least two atoms. Most molecules contain more than one kind of atom.

15.4. The term **macromolecule** is used to describe very large molecules, consisting of tens of thousands of atoms.

15.5. The chemical **formula** of a compound specifies which elements it contains, and also the numerical relation between the numbers of different atoms in its molecule. The number of atoms is indicated by a subscript following the symbol for the element (except when the number is one).

15.6. The **valency** of an element is the number which specifies how many H-atoms an atom of the element can combine with or replace.

15.7. **Molecular formulae** give information about the nature and number of atoms in a molecule.

Structural formulae are simplified pictures of the spatial arrangement of the atoms in a molecule.

15.8. The **atomic mass (atomic weight)** of an element is the number which specifies how many times heavier one atom of the element is than one atom of hydrogen.

15.9. The **molecular mass (molecular weight)** of a compound is the number which specifies how many times heavier one molecule of the compound is than one atom of hydrogen. The molecular weight is equal to the sum of the atomic weights of the elements which make up the compound.

15.10. The amount of substance which contains 6×10^{23} particles (atoms, molecules) is known as one **mole**. The mass of one mole of a substance is called molar mass. The molar mass is numerically equal to its molecular mass.

15.11. The **molar volume** of a gas is the volume occupied by one mole. At normal temperature and pressure (NTP) every gas has a molar volume of 22.4 litres.

15.12 A **chemical equation** is an abbreviated description of a chemical reaction, from which the porportions by weight of the reacting substances can be derived. The numbers of atoms or molecules taking part in the reaction are indicated by numerical factors preceding them in the equation.

15.13 **Stoichiometric calculations** employ chemical equations together with the corresponding molar masses.

16. Self-evaluation questions

In answering the following questions, use should be made of the Tables of atomic masses (Section I.9, pp. 5) and valencies (Section I.4, p. 7).

39 What is the mass of one mole of mercury (II) oxide, HgO?

40 What percentage of nitrogen is contained in urea, $CO(NH_2)_2$?

41 How many litres and how many moles of O_2 are contained in 1 m³ of air at NTP?

42 What is the equation for the formation of iron sulphide, FeS, from its elements?

43 How many grams of iron are required to convert 64 g sulphur completely into FeS, and how many grams of iron sulphide will be obtained?

44 In what proportions by volume and by weight must hydrogen (H_2) and chlorine (Cl_2) be mixed so that both gases are consumed completely when hydrogen chloride (HCl) is formed?

45 The smallest unitary particles of a compound are termed

    A molecules
    B isotopes
    C atoms
    D elements

46 Which substance consists of ions?

    A petrol (gasoline)
    B sulphur
    C common salt
    D titanium dioxide

47 Molar mass corresponds to

    A the total number of atoms that together form one molecule
    B the mass of a molecule
    C the mass of 6×10^{23} molecules of a compound
    D the mass of 1 litre of a gas in its normal state

4 Atoms and molecules. Chemical formulae. Valency. Chemical equations

48 The molecular mass of sulphur dioxide (SO_2) is 64 and the atomic mass of oxygen (O) is 16.
What must be the atomic mass of the sulphur?

A 32
B 16
C 48
D 80

49 How many atoms of the elements involved are contained in one molecule of ethanol, CH_3-CH_2OH?

	carbon atoms	hydrogen atoms	oxygen atoms
A	1	3	1
B	2	6	1
C	2	5	1

50 If several atoms join together to form a new, electrically uncharged particle, these create

A a molecule
B an ion
C a mol
D a new atom

51 From a chemical reaction equation it is possible to ascertain

A in which direction a reaction proceeds
B the speed at which a reaction proceeds
C the mass ratios in which the starting materials react with each other
D the percentage conversion (= yield) in a reaction

52 From which of the following formulae are the valencies of the elements linked with oxygen correctly stated?

A FeO;   iron is monovalent
B N_2O_3;   nitrogen is trivalent
C PbO_2;   lead is quadrivalent
D H_2O;   hydrogen is bivalent

53 Which of the empirical formulae are correct?

A Iron (III) chloride = Fe_3Cl
B Lead (II) oxide = PbO_2
C Potassium chloride = KCl
D Carbon dioxide = CO_2

54 What is the mass, expressed in grams, of 1 mol of dimethylamine, $(CH_3)_2 NH$?

    A   45 g
    B   42 g
    C   28 g
    D   27 g

5 Periodic System, Atomic Structure and Isotopes. Chemical Bonds. Ions

1. The periodic system of the elements

If the elements are arranged in order of increasing atomic mass and, or according to their chemical properties, we obtain a diagram which is known as the **periodic system of the elements** (PSE) or the **periodic table.**

1.1. Groups, periods and atomic numbers

In this arrangement each element has a serial number, the **atomic** number (AN). In the PSE elements with similar properties are collected together in **groups**, and appear one above the other.

This classification leads to seven horizontal rows, called **periods**.

1.2. Main groups and sub-groups

The vertical groups of the PSE are subdivided into main groups and sub-groups. The names of the eight **main groups** are as follows:

I.	The alkali metal group
II.	The alkaline earth metal group
III.	The boron-aluminium group
IV.	The carbon group
V.	The nitrogen group
VI.	The oxygen-sulphur group
VII.	The halogen group
VIII.	The rare gas group

These groups contain all the non-metals and metalloids, together with some of the metals. The **sub-groups** contain the majority of the metals.

1.3. The complete periodic system

The complete periodic system contains all the elements, with their atomic numbers and atomic masses (see p. 4 in this chapter).

1.4. The shortened periodic system

The shortened periodic system contains only the main groups, and is sufficient for considering fundamental principles (see p. 5). The atomic number appears as a subscript before the symbol, with the atomic mass (to the nearest integer) above it.

Periodic system of the elements
(Rounded values of atomic masses)

Main and sub-groups

Period	I	II	III	IV	V	VI	VII	VIII
1st period	$^1_1$H							$^4_2$He
2nd period	$^7_3$Li	$^9_4$Be	$^{11}_5$B	$^{12}_6$C	$^{14}_7$N	$^{16}_8$O	$^{19}_9$F	$^{20}_{10}$Ne
3rd period	$^{23}_{11}$Na	$^{24}_{12}$Mg	$^{27}_{13}$Al	$^{28}_{14}$Si	$^{31}_{15}$P	$^{32}_{16}$S	$^{35.5}_{17}$Cl	$^{40}_{18}$Ar
4th period	$^{39}_{19}$K $^{63.5}_{29}$Cu	$^{40}_{20}$Ca $^{65}_{30}$Zn	$^{45}_{21}$Sc $^{70}_{31}$Ga	$^{48}_{22}$Ti $^{73}_{32}$Ge	$^{51}_{23}$V $^{75}_{33}$As	$^{52}_{24}$Cr $^{79}_{34}$Se	$^{55}_{25}$Mn $^{80}_{35}$Br	$^{56}_{26}$Fe $^{59}_{27}$Co $^{59}_{28}$Ni $^{84}_{36}$Kr
5th period	$^{85.5}_{37}$Rb $^{108}_{47}$Ag	$^{88}_{38}$Sr $^{112}_{48}$Cd	$^{89}_{39}$Y $^{115}_{49}$In	$^{91}_{40}$Zr $^{119}_{50}$Sn	$^{93}_{41}$Nb $^{122}_{51}$Sb	$^{96}_{42}$Mo $^{128}_{52}$Te	$^{97}_{43}$Tc $^{127}_{53}$I	$^{101}_{44}$Ru $^{103}_{45}$Rh $^{106}_{46}$Pd $^{131}_{54}$Xe
6th period	$^{133}_{55}$Cs $^{197}_{79}$Au	$^{137}_{56}$Ba $^{201}_{80}$Hg	$^{*}_{57-71}$ $^{204}_{81}$Tl	$^{178}_{72}$Hf $^{207}_{82}$Pb	$^{181}_{73}$Ta $^{209}_{83}$Bi	$^{184}_{74}$W $^{210}_{84}$Po	$^{186}_{75}$Re $^{210}_{85}$At	$^{190}_{76}$Os $^{192}_{77}$Ir $^{195}_{78}$Pt $^{222}_{86}$Rn
7th period	$^{223}_{87}$Fr (111)	$^{226}_{88}$Ra (112)	$^{**}_{89-103}$ (113)	$^{261}_{104}$Ku (114)	(105) (115)	(106) (116)	(107) (117)	(108) (109) (110) (118)

*) Lanthanides: $^{139}_{57}$La $^{140}_{58}$Ce $^{141}_{59}$Pr $^{144}_{60}$Nd $^{145}_{61}$Pm $^{150}_{62}$Sm $^{152}_{63}$Eu $^{157}_{64}$Gd $^{159}_{65}$Tb $^{162.5}_{66}$Dy $^{165}_{67}$Ho $^{167}_{68}$Er $^{169}_{69}$Tm $^{173}_{70}$Yb $^{175}_{71}$Lu

**) Actinides: $^{227}_{89}$Ac $^{232}_{90}$Th $^{231}_{91}$Pa $^{238}_{92}$U $^{237}_{93}$Np $^{242}_{94}$Pu $^{243}_{95}$Am $^{247}_{96}$Cm $^{247}_{97}$Bk $^{249}_{98}$Cf $^{254}_{99}$Es $^{253}_{100}$Fm $^{256}_{101}$Md $^{254}_{102}$No $^{257}_{103}$Lw

Shortened periodic system of the elements
(Rounded values of atomic masses)

Main groups	I	II	III	IV	V	VI	VII	VIII
1st period	$^{1}_{1}H$							$^{4}_{2}He$
2nd period	$^{7}_{3}Li$	$^{9}_{4}Be$	$^{11}_{5}B$	$^{12}_{6}C$	$^{14}_{7}N$	$^{16}_{8}O$	$^{19}_{9}F$	$^{20}_{10}Ne$
3rd period	$^{23}_{11}Na$	$^{24}_{12}Mg$	$^{27}_{13}Al$	$^{28}_{14}Si$	$^{31}_{15}P$	$^{32}_{16}S$	$^{35.5}_{17}Cl$	$^{40}_{18}Ar$
4th period	$^{39}_{19}K$	$^{40}_{20}Ca$	$^{70}_{31}Ga$	$^{73}_{32}Ge$	$^{75}_{33}As$	$^{79}_{34}Se$	$^{80}_{35}Br$	$^{84}_{36}Kr$
5th period	$^{85.5}_{37}Rb$	$^{88}_{38}Sr$	$^{115}_{49}In$	$^{119}_{50}Sn$	$^{122}_{51}Sb$	$^{128}_{52}Te$	$^{127}_{53}I$	$^{131}_{54}Xe$
6th period	$^{133}_{55}Cs$	$^{137}_{56}Ba$	$^{204}_{81}Tl$	$^{207}_{82}Pb$	$^{209}_{83}Bi$	$^{210}_{84}Po$	$^{210}_{85}At$	$^{222}_{86}Rn$
7th period	$^{223}_{87}Fr$	$^{226}_{88}Ra$						

2. Atomic structure. The Bohr atom

The chemical properties of the elements are determined by the structure of its atoms. This can be elucidated by the use of model concepts. The simplest atomic model is that of Bohr:*

Electrons, having a negative electric charge, move around the positively charged atomic nucleus.

Both the nucleus and the electrons consist of fundamental particles

2.1. Atomic nuclei, protons and neutrons

The **atomic nucleus** consists of nucleons, ie protons and neutrons.

Protons carry a single positive electric charge. **Neutrons** are electrically neutral. The number of charges on the nucleus is equal to the number of protons which it contains. It is also equal to the atomic number in the PSE.

The atoms of different elements differ in the numbers of protons in their nuclei.

The modern definition of an element is therefore as follows:

All the atoms of a chemical element have the same nuclear charge, and hence the same number of protons in their nuclei.

Every proton and every neutron has unit mass. The total number of protons and neutrons in the nucleus therefore gives the atomic mass (atomic weight).

2.2. Extranuclear electrons

Each of the electrons outside the nucleus has unit negative charge. Their mass is negligible, so they do not contribute to the mass of the atom. The number of extranuclear electrons is always equal to the number of protons in the nucleus.

Every atom is therefore electrically neutral as a whole.

2.3. Electron shells, atomic models

According to the Bohr model atoms are roughly spherical. The electrons move around the nucleus in definite orbits, known as **shells**.

---

*Niels Bohr, 1885–1962. Danish physicist, worked with Rutherford, Nobel Prize for physics 1922

The following illustrations show some atomic models:

Hydrogen

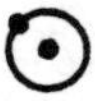

Neon

$^{1}_{1}$ H

$^{20}_{10}$ Ne

Electrons		
in the 1st shell	1	2
in the 2nd shell	–	8
Nucleus:		
Protons	1	10
Neutrons	–	10

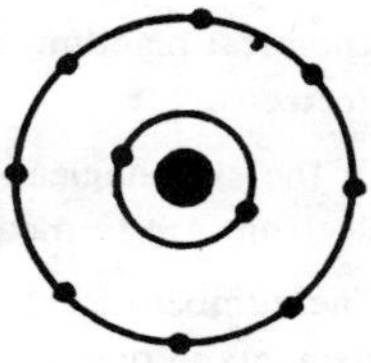

Sodium

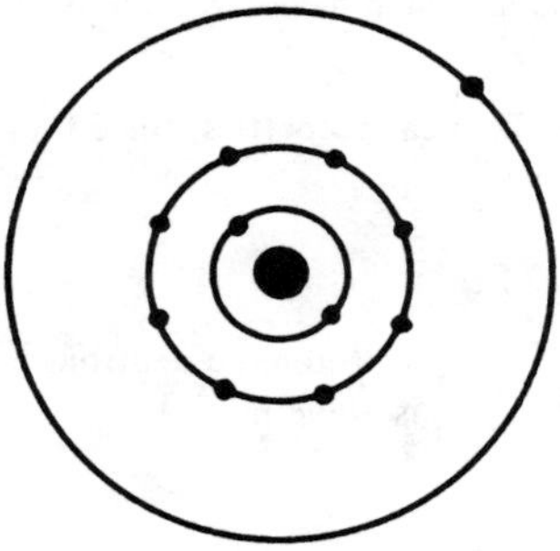

Sulphur

$^{23}_{11}$ Na

$^{32}_{16}$ S

Electrons		
in the 1st shell	2	2
in the 2nd shell	8	8
in the 3rd shell	1	6
Nucleus:		
Protons	11	16
Neutrons	12	16

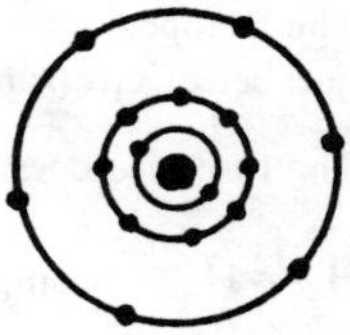

I Fundamentals of general and inorganic chemistry

2.4. Chemical reactions, valence electrons

Chemical reactions **do not involve the atomic nuclei**, but exclusively the **extranuclear electrons**.

Of the extranuclear electrons only the outermost ones are involved. These outermost electrons are termed **valence electrons.**

The number of valence electrons determines the maximum valency of an element. All the elements in one main group (see I.5, p. 3) have the same number of valence electrons in their atoms.

The number of the group is the same as the number of valence electrons. The group number therefore in general corresponds to the maximum valency.

For example, sodium (Na) is in the first main group, has one valence electron, and is therefore monovalent. Sulphur (S) is in the sixth main group and therefore has a maximum valency of six.

3. Isotopes

All the atoms of a given element have the same number of protons, but the number of neutrons may vary, thus changing the atomic mass. This leads to the existence of **isotopes**. The isotopes of a given element contain atoms which have the same nuclear charge, but different masses.

The isotopes of an element have the same chemical properties, since they possess the same number of electrons arranged in the same way.

The three isotopes of hydrogen are:

$H = {}_1^1H$ Composition of nucleus: 1 proton, no neutrons
Hydrogen   isotopic mass 1

$D = {}_1^2H$ Composition of nucleus: 1 proton +1 neutron
Deuterium   isotopic mass 2
(heavy hydrogen)

$T = {}_1^3H$ Composition of nucleus: 1 proton + 2 neutrons
Tritium   isotopic mass 3

The compound of heavy hydrogen (D) which corresponds to H_2O is D_2O, deuterium oxide or 'heavy water'.

Most elements consist of a mixture of isotopes. For this reason most atomic masses are not whole numbers, but represent a mean of the isotopic masses, weighted according to their natural abundance.

Chlorine and its compounds contain two isotopes of masses 35 and 37. These are always present in the proportions 3:1, giving an average atomic mass (atomic weight) of 35.5.

4. Nuclear disintegration

Certain elements, especially those with high nuclear charges, are unstable and undergo continuous spontaneous **nuclear disintegration**. They are **radioactive**, and are converted into other elements with the emission of radiation.

The atomic nucleus of uranium (U) disintegrates with the production of so much energy that it is used industrially in nuclear reactors. Some of the products of the fission of the uranium nucleus, for example iodine (I) and caesium (Cs) are used as radioactive isotopes in radiation medicine (cf. Nuclear chemistry, II.9).

5. Chemical bonds

Separate uncombined atoms are not stable, and they join together to form united atoms, since forces of attraction act between them. In this way **chemicals bonds** are formed.

The tendency of atoms to form compounds varies greatly. The elements whose atoms exhibit the least tendency to react are called the inert gases (or rare gases).

5.1. Covalent bonds

The electric force field of the positively charged nucleus extends outside the atoms. Atoms can therefore attach extra electrons when two atoms of the same or different kinds approach one another. When two atoms meet the electrons of one atom can enter the attractive region of the other.

In this way two atoms can be held together.

The electrons go preferentially into the region between the two atomic nuclei, since it is here that the forces of attraction by the two nuclei are greatest. Bonding between the atoms is thus brought about by the electrons between the nuclei. In this way the two atoms combine to form a molecule, a new independent unit.

The simplest molecule is H_2, formed from two H-atoms:

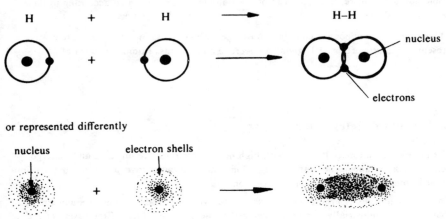

or represented differently

In the structural formula H–H the line joining the two H-atoms represents the shared bonding electron pair.

This type of bond is called a **covalent** or **homopolar bond**.

5.2. Single, double and triple bonds

The atoms of nonmetals combine predominantly by means of covalent bonds. Thus two Cl-atoms combine to form a Cl_2-molecule by sharing one pair of electrons, Cl-Cl. When there is only one shared pair of electrons we speak of a **single bond**.

The O_2-molecule, $O=O$, contains a **double bond** with **two** shared electron pairs.

The N_2-molecule, $N\equiv N$, contains a **triple bond**.

The number of electrons which an atom contributes to a bond is equal to its valency.

Carbon atoms in organic compounds may be bound by single, double or triple bonds, for example:

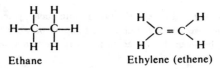

Ethane   Ethylene (ethene)   Acetylene (ethyne)

5.3 Polar covalent bonds. The water dipole

If a molecule contains two atoms of the same nonmetal the bonding electron pair lies exactly at the centre of the molecule. On the other hand, if a molecule is formed from two atoms of different nonmetals, the two atoms will exert different attractive forces on the bonding electron pair, which will lie nearer to the atom exerting the stronger attraction.

In the H_2O-molecule the bonding electron pair lies nearer to oxygen than to hydrogen. This gives the O-atom a slight excess of negative charge, while the H-atoms have small positive charges. When the binding electron pair is asymmetrically situated we speak of a **polar covalent bond**.

Furthermore, the H_2O molecule is bent:

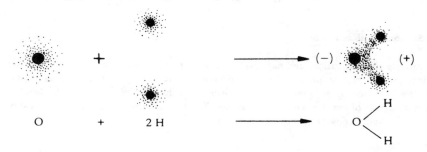

The water molecule has a positive end (at the hydrogens) and a negative end (at the oxygen): it is a **dipolar molecule**.

Bonds between atoms of different nonmetals are always more or less polar. Bonds of specially high polarity occur in molecules containing atoms of F, O, Cl, N and S.

Molecules containing polar bonds are more reactive than those with nonpolar bonds. This fact is made use of in the synthesis of organic compounds. The reactivity of organic molecules is considerably increased by introducing halogens, oxygen, nitrogen or sulphur.

5.4. Electron transfer. Ions, cations, anions

In contract to nonmetals, the atoms of metals attract further electrons only very weakly; on the other hand, their own valence electrons are only weakly bound. If the atoms of nonmetals such as O or Cl encounter metal atoms such as Na or Fe, the former pull off the loosely bound outer electrons of the latter. This transfer of electrons causes the atoms to become electrically charged.

These new types of particle are called **ions**: in them the total number of extranuclear electrons is no longer equal to the number of protons in the nucleus.

I Fundamentals of general and inorganic chemistry

Metals are converted by loss of electrons into positively charged ions, or **cations**, for example:

$$Na \quad - \quad e^- \quad \longrightarrow \quad Na^+ \qquad (e^- = 1 \text{ electron})$$

Nonmetals are converted by accepting electrons into negatively charged ions, or **anions**, for example:

$$Cl \quad + \quad e^- \quad \longrightarrow \quad Cl^-$$

Ions may consist not only of single atoms, but also of groups of atoms.

5.5. Ionic lattices and bonds

Positively charged cations and negatively charged anions attract each other very strongly. Since these attractive forces act in all directions in space, each anion tends to surround itself with cations in all directions, and conversely each cation with anions. This leads to regular arrays of ions known as **ionic lattices**. The crystal lattice of common salt (sodium chloride, NaCl) consists of ions and not of molecules:

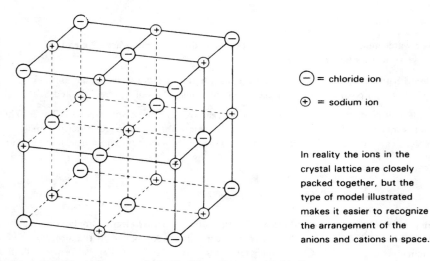

$\ominus$ = chloride ion

$\oplus$ = sodium ion

In reality the ions in the crystal lattice are closely packed together, but the type of model illustrated makes it easier to recognize the arrangement of the anions and cations in space.

This type of binding is known as **ionic binding** or heteropolar binding.

The transfer of electrons causes the elements to lose their characteristic properties. Thus while chlorine is a poisonous yellowish-green gas, the colourless anion Cl^- (the chloride ion) is a constituent of solid common salt.

5.6.   Metallic binding

Metals consist of a regular lattice of cations. These cations are atomic cores, ie atoms
without their valence electrons. The valence electrons move freely between the cations as a
so-called electron gas. The electrons 'glue' the cations of the metal lattice together, and one
speaks of **metallic binding**.

5.7.   Atomic and molecular species

There are thus three kinds of particles of which all matter is composed:

1. **Atoms**
2. **Molecules**
3. **Ions**

The nature of these species determines the chemical properties of a substance.

6.   Binding and the physical properties of substances

The physical properties of a substance depend to a considerable extent on whether it
consists of uncombined atoms, of nonpolar or polar molecules, or of anions and cations.

6.1.   Ionic compounds

Compounds which consist only of **ions** are always solids at room temperature. The forces of
attraction between anions and cations are so strong that high temperatures are required (usually)
between 500 and 1000°C or 932 and 1832°F) to loosen the ionic lattice and to convert it into
a liquid. Solids of this kind have a crystalline structure, for example common salt.

6.2.   Intermolecular forces

Forces of attraction also exist between molecules. They are considerably weaker than those
involved in covalent or ionic bonds.

Intermolecular forces are particularly weak between small nonpolar molecules. The melting
and boiling points of such compounds are hence usually fairly low. Substances of this kind
are gases at room temperature; thus nitrogen (N_2) boils as low as $-196°C$ ($-321°F$).

The more polar the molecules of a substance, the stronger are the intermolecular binding
forces. Substances consisting of small but highly polar molecules thus tend to have higher
melting and boiling points. For example, water is a liquid at room temperature because of
the strong forces of attraction between the H_2O-molecules.

6.3. Hydrogen bonds

The aggregation (association) of polar H_2O-molecules involves the juxtaposition of the positive end of one H_2O-molecule and the negative end of another.

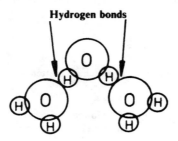

As shown in the sketch, the H_2O-molecules are joined through hydrogen atoms: this is termed **hydrogen bonding**. Other polar molecules containing H-atoms (eg hydrogen fluoride, HF, and polyamide synthetic materials) can also associate through hydrogen bonds.

6.4. Macromolecular substances

Intermolecular forces increase with increasing molecular size. For this reason **macromolecular substances**, for example plastics, are usually solids. Both the nature of the macromolecules and the way in which they are arranged with respect to one another are of importance in this context.

The meshing together and intertwining of macromolecules has the effect of increasing the intermolecular forces. This serves to increase the strength of the material. Macromolecular substances are often strengthened by hydrogen bonding, for example in synthetic polyamide fibres.

If the macromolecules in a plastic material are highly ordered the structure becomes more or less crystalline. Plastics of this kind soften at higher temperatures (see plastics, IV.7).

6.5. Plasticizers

Plastics which are hard solids in virtue of their structure can be softened by incorporating **plasticizers**. Plasticizers are liquids of high boiling point. The molecules of plasticizer are inserted between the macromolecules, which are held further apart. This has the effect of decreasing the intermolecular forces. Plasticized macromolecular substances therefore cover a wide range of conditions between the solid and the liquid state, according to the nature and quantity of the plasticizer.

6.6.   Melting and vaporization

In melting and vaporization the attractive forces between the particles (atoms, molecules or ions) have been overcome, so that the particles are separated but not destroyed. This process alters the physical properties, while the chemical properties remain unchanged.

7. **Summary**

7.1. In the **periodic system** (**PSE**) the elements are arranged in order of increasing nuclear charge.

7.2. The periodic system is divided into eight **groups**, each containing chemically similar elements.

7.3. Every atom consists of a positive charge **nucleus** surrounded by negative charge.

The charges on the nucleus and its surroundings are equal, so that the atom as a whole is electrically neutral.

7.4. The atomic nucleus consists of nucleons, ie **protons** and **neutrons,** each of unit mass. Their sum gives the atomic mass (atomic weight) of the element. The proton bears a positive charge, while neutrons are uncharged.

7.5. The charge on the nucleus of an element, equal to the **atomic number**, gives the serial number of the element in the periodic system. It represents both the number of positive charges on the nucleus and the number of electrons surrounding it.

7.6. An **element** is a substance in which all the atoms have the same nuclear charge. Different elements have different nuclear charges.

7.7. The negative charge surrounding the nucleus consists of **electrons**, which have negligible mass and each of which bears a single negative charge.

7.8. According to Bohr's model of the atom the electrons move in orbits round the nucleus. Electrons in orbits approximately the same distance from the nucleus are said to constitute an electron shell.

7.9. Chemical reactions involve only the extranuclear electrons, and not the atomic nucleus.

7.10. Electrons in the outermost shell of an atom are termed **valence** electrons: they determine the maximum valency of an element.

7.11. The number of neutrons in the nucleus can vary even when the number of protons remains the same. This behaviour leads to the existence of **isotopes**.

The isotopes of an element contain atoms with the same nuclear charge and the same chemical properties, but with different numbers of neutrons and hence different masses.

7.12. The atoms of some elements of high atomic weight undergo spontaneous **nuclear disintegration**. They emit radiation and are said to be radioactive.

7.13. **Covalent binding** means the holding together of atoms by means of shared electron pairs. The molecules thus formed may have bonds containing either one, two or three shared electron pairs.

A covalent bond is polar if the shared electron pair lies unsymmetrically between the two atoms.

7.14. A **dipolar molecule** is one which possesses a positive and negative end.

7.15. **Ions**, ie electrically charged atoms, are formed by the transfer of electrons from one atom to another.

Positively charged ions are called **cations**, and negatively charged ions **anions**.

7.16. The regular arrangement of ions to form a crystal is termed an **ionic lattice**.

7.17. **Ionic binding** means the binding together of oppositely charged ions to form an ionic lattice.

7.18. The term **metallic bonding** refers to a lattice of metallic cations held together by their valency electrons (an 'electron gas').

7.19. Matter is composed of only three kinds of particles, **atoms**, **molecules** and **ions**.

7.20. The physical properties of substances depend on the forces of attraction between their particles.

7.21. The forces of attraction between molecules are termed **intermolecular forces** (or van der Waals forces).

7.22 The association of polar molecules through their H-atoms is known as **hydrogen bonding**. It represents a particularly strong type of intermolecular bonding.

8. Self-evaluation questions

55 The elements are arranged in the periodic system according to

A   increasing atomic mass
B   increasing atomic radius
C   increasing number of neutrons
D   increasing atomic number (nuclear charge)

56 The electrical charges in an atom cancel one another out, since the number
of _____ in the nucleus is equal to the number of _____ outside the nucleus.

57 Complete the following Table:

Symbol of element	Number of protons	Atomic weight	Number of neutrons
$_{1}^{1}$ H	1	1	0
$_{11}^{23}$ Na			12
$_{6}^{12}$ C			
$_{8}^{16}$ O			
$_{92}^{238}$ U			

58 What kinds of particles are contained in

A   compounds between metals and nonmetals, eg NaCl, CaO
B   compounds of nonmetals with one another, eg CH_4, H_2O?

59 The following are identical to the atomic number of an element:

A   the total number of electrons in the atomic shell
B   the nuclear charge of the atom
C   the number of neutrons in the atom
D   the mass of the atom

60 The symbols representing elements are often given indices. What is the significance of this in the case of $^{39}_{19}K$?

   A   The atom of the element potassium has 19 portions in the nucleus.
   B   The atom of the element potassium is 39 times as heavy as a carbon atom.
   C   The atom of the element has 39 neutrons in the nucleus.
   D   The atom of the element potassium has $(39-19)$, ie 20 electrons in the atomic shell.

61 Which of the following statements in conjunction with atoms are correct?

   A   All atoms of an element have the same nuclear charge.
   B   The electrons have a mass of 1.
   C   Only the nucleus of the atom contains elemental particles.
   D   Only the noble gases consist of individual, unlinked atoms.

62 Isotopes of an element differ from one another

   A   in their nuclear charge
   B   in their atomic mass
   C   in their chemical properties
   D   in their number of neutrons

63 The forces of attraction that act between particles are

   A   greater in the case of ions with opposite charges than ions with small polar molecules
   B   greater with non-polar molecules than with polar molecules
   C   greater with small molecules than with large molecules

64 The usual difference between compounds consisting of ions and those consisting of molecules is that the ionic compounds

   A   have higher melting and boiling points
   B   cannot be crystallized
   C   are harder and more brittle
   D   are more readily soluble in water

65 Which two properties have to be present simultaneously for a compound to be highly volatile (ie, have a low boiling point)?

   A   low molar mass
   B   high molar mass
   C   non-polar atomic linkage
   D   polar atomic linkage

66   Which of the following substances has a non-polar atomic linkage?

A   common salt, NaCl
B   chlorine, Cl_2
C   copper, Cu
D   water, H_2O

67   Which combinations of substance and type of bond are correct?

A   methane (CH_4) – ionic bond
B   rock salt (NaCl) – polar interatomic bond
C   copper (Cu) – metallic bond
D   water (H_2O) – polar interatomic bond

68   If atoms release one or more electrons from the atomic shell, which of the following will result?

A   negatively charged ions, ie anions
B   positively charged ions, ie cations
C   isotopes of the type of atom involved
D   negatively charged groups of atoms, ie complex ions

69   Valency electrons are

A   the electrons least securely linked in an atom
B   electrons on the outermost shell of an atom
C   the electrons that determine the chemical properties of an element
D   the electrons whose number determines, in a metal atom, what charge the atom can carry as a cation

70   Ionic compounds are based on

A   the neutral electrostatic attraction of anions and cations
B   a common pair of electrons
C   the mutual mass attraction between ions
D   the attraction between ions and atoms

71   Which linkages have the character of a polar interatomic bond?

A   O=O
B   O-H
C   F-F
D   C-Cl

72   Compounds may have a dipolar character if they

A   are built up from ions with opposite charges
B   have relatively low volatility or are difficult to melt
C   possess only non-polar interatomic bonds in the molecule
D   possess polar interatomic bonds in the molecule.

73   In which cases are the type of compound and form of bond correctly matched?

	Type of compound	Example	Form of bond
A	metallic/non-metallic	TiO_2, titanium dioxide	ionic bond
B	metallic/non-metallic	KCl, potassium chloride	atomic bond
C	non-metallic/non-metallic	CH_4, methane	atomic bond
D	non-metallic/non-metallic	H_2O, water	ionic bond

6 Important Metals and Non-metals. Oxidation and Reduction

I Fundamentals of general and inorganic chemistry

1. Metals, non-metals and metalloids

Of the 105 known elements 81 are metals, 17 non-metals and 7 metalloids.

1.1. Metals and alloys

Metals are very important for use as materials. Their useful properties can be improved by fusing pure metals together to make homogeneous mixtures known as **alloys**. All metals and alloys have certain characteristic properties which can be explained in terms of the metallic binding present in the lattice of a metal.

They are good conductors of heat and electricity. This is due to the mobile electron gas. When an electric current flows through a metallic conductor the electrons move in a definite direction.

All metals are solids, with the exception of mercury (Hg). Some of them are very hard on account of the strong forces of attraction between the metal cations and the electron gas.

1.2. Mechanical properties

The mechanical properties of metals are particularly important for their use as materials. They can be deformed by bending, stretching, rolling, hammering and forging. The metal lattice is not broken down in these processes, its components being merely displaced relative to one another.

It requires very large forces to break down the cohesion of the lattice. This can be done by cutting, sawing, drilling, turning, milling or planing.

2. Light and heavy metals

Metals are classified as being light or heavy.

The **light metals** have densities up to about 5 g/cm³ (312 lb/ft³). They occur mainly in the first three main groups of the periodic system. The light metals of the greatest industrial importance are magnesium (Mg), aluminum (Al) and titanium (Ti).

I Fundamentals of general and inorganic chemistry

The most important **heavy metals** are iron (Fe) together with the non-ferrous metals copper (Cu), silver (Ag), lead (Pb), cadmium (Cd), zinc (Zn), cobalt (Co), nickel (Ni), platinum (Pt), mercury (Hg), gold (Au), chromium (Cr), manganese (Mn), molybdenum (Mo), tungsten (W) and tin (Sn). In the periodic system the metals are to be found in the fourth and fifth main groups and the sub-groups.

3. Important alloys

These include the following:

Steel	:	Fe + C
Special steels	:	steel + (mainly) Cr, Ni, W, Ti, V
Electron alloy	:	Mg + Al
Brass	:	Cu + Zn
Bronze	:	Cu + Sn
Amalgams	:	Hg-alloys
German silver	:	Cu + Zn + Ni

4. Chemical properties of metals

The most important chemical properties of metals are their stability towards air and water. A metal is said to be noble when it is highly resistant to chemical action.

4.1. Noble and semi-noble metals

Noble metals include gold (Au), platinum (Pt), palladium (Pd) and silver (Ag).

The **semi-noble metals** include copper (Cu), nickel (Ni), cobalt (Co) and chromium (Cr).

4.2. Base metals, corrosion

Base metals are by contrast susceptible to chemical attack. The destructive effect on the metal lattice, usually on the surface, is known as **corrosion**. The commonest example of corrosion is the rusting of iron in moist air. This causes a considerable wastage of iron: in the USA this is estimated to be about 40% of the annual production of pig iron. However, scrap iron is also an important source of raw material for the production of iron.

The choice of materials and the prevention of corrosion is of particular importance in chemical plants. In order to reduce corrosion the surface of materials can be protected by painting, coating with rubber or with noble or semi-noble metals (chromium and silver plating), or by chemical treatment in which unreactive surface layers shield the underlying metal.

Examples of surface protection by chemical means are:

A anodic oxidation, in which aluminium surfaces are coated electrolytically with a layer of aluminium oxide:
B phosphate treatment, which deposits a stable and impermeable layer of metal phosphate on the surface.

4.3. Catalysts

Some metals are used as catalysts in chemical reactions. In order to achieve a high activity the metal, usually in a finely divided state, is deposited on the surface of a carrier. The most important metal catalysts are iron (Fe), nickel (Ni), cobalt (Co), platinum (Pt) and copper (Cu).

5. Metals of industrial importance

5.1. Iron

Iron (Fe) is by far the most important metal for practical use. It owes its pre-eminent position partly to the existence of extensive deposits of pure ore from which it can be extracted relatively easily, and partly to its superlative strength and the wide range of properties which can be produced by alloying (especially with carbon to give steel) or by heat treatment (for example tempering).

Iron is extracted from iron oxide ores in the blast furnace. It then contains on average 4% carbon.

The term **pig iron** (or crude iron) is generally used to describe iron containing more than 1.7 carbon. It is brittle, cannot be forged, and on heating softens suddenly rather than gradually. Cast iron is a variety of crude iron.

It becomes malleable and can be welded when it is converted by a process of decarbonization into **steel**, which contains less than 1.7% C.

Pure iron, free from carbon, is of no industrial importance.

Iron and structural steels can be magnetized. By alloying with other metals the toughness, hardness, resistance to chemical attack and other properties can be modified to suit particular uses.

5.2. Copper

On account of its high electrical conductivity **copper (Cu)**, in the form of very pure electrolytic copper, is used in large quantities in electrical engineering for conducting wires.

On account of its high thermal conductivity Cu is used in heat exchanges, refrigerating plant and distilling plant. As a semi-noble metal Cu is resistant to chemical attack and is therefore used in constructing equipment, for example for the brewing industry. Cu is used extensively as a basis for alloys (see I.6, p. 4).

5.3. Aluminium

Aluminium (Al) is the most important light metal. Al and Al-alloys are used extensively for making light components in aeroplanes, containers, railway trucks, cars (eg crank-cases and pistons), in ship-building and house-building (eg window-frames). The hardness of aluminium can be increased by alloying with copper (in duraluminium). Other considerable outlets for aluminium are conductors in electrical engineering and as a packaging material in the form of foil, tubes, removable closures for cans, etc.

Al is used as a reducing agent in the extraction of base metals by the so-called aluminothermic process.

5.4. Magnesium

Magnesium (Mg), also a light metal, is used almost exclusively as a component of alloys, especially with aluminium. Mg is also used as a reducing agent in the production of titanium. The higher the production of aluminium and titanium, the more Mg is manufactured.

Magnesium is predominantly used in the form of its alloys.

5.5. Cadmium and zinc

Cadmium (Cd) and **zinc (Zn)** are stable in air and are therefore used for coating other metals, especially for protecting steel (cadmium and zinc plating).

In terms of consumption Zn occupies the third place among the non-ferrous metals, after Cu and Al. The corrosion resistance of zinc depends primarily on its purity. Coating steel surfaces with pure zinc makes it considerably more economical to use steel for building, scaffolding, power-line masts and bridge railings. Many new possibilities for using zinc have been opened up by die-casting, in which the liquid metal is forced into a mould. For example, pure zinc has been used in the automobile industry for door handles, pump barrels, brake cylinders, manufacturers names and emblems, as well as model cars for children.

5.6. Lead

On account of its resistance to corrosion **lead (Pb)** is used for pipes and sheet metal.

A future market for lead is assured by its use for the plates of storage batteries and for lead-plated steel, which is used as a cladding and construction material in chemical industry.

On the other hand there is no future for the use of lead in making lead tetraethyl, an anti-knock gasoline additive, since environmental regulations are becoming progressively stricter.

There is a small but steadily increasing market for lead as a protection against radiation in laboratories and atomic energy installations. Alloys of lead are used for making lead shot.

5.7. Tin

Tin (Sn) is primarily of importance as an ideal material for coating steel sheet. About one half of the world production of tin is used in the tin-plate industry for making cans for food products. Sn is very resistant to corrosion.

On account of their low melting point Sn-alloys are used for soldering, especially in the electrical industry, and also as bronzes.

5.8. Cobalt and nickel

Cobalt (Co) and **nickel (Ni)** are used mainly in the manufacture of stainless steel. Nickel in particular is expensive and in short supply, but there are many modern techniques in which there are no substititues for Co and Ni.

For example, in aircraft construction the turbine blades of jet engines are made of Ni-Cr-Fe alloys: one jumbo jet requires about 5000 kg of Ni.

Ni is also used in coinage alloys.

Ni-Fe and Ni-Co-Fe alloys can be magnetized and are used to make permanent magnets. On account of the high corrosion resistance of nickel, nickel-plated steel is often employed. Ni also servies as a catalyst in hydrogenation, and is used in nickel-cadmium batteries.

5.9. Titanium

Titanium (Ti) is gaining increasing importance as a corrosion-resistant material in the construction of aircraft, ships, chemical apparatus and rockets. It is also used in the steel industry to prepare alloys which are particularly resistant to blows and to impact.

5.10. Silver

Silver (Ag) is used for silver-plating vessels and for making mirrors. On account of its softness it is alloyed with copper when used for coinage metal or articles for daily use. Silver is also used to make light sensitive compounds for coating photographic plates, film and paper. In the electrotechnical industry it is used for making electrical contacts which will not oxidize.

5.11. Platinum

Platinum (Pt) in the form of a finely meshed grid is used as a catalyst. On account of its chemical inertness and high melting point ($1770°C$ or $3218°F$) platinum is used both in the laboratory and in industry as a material for crucibles and electrodes. Pt is also a constituent of dental gold alloys.

5.12. Mercury

Mercury (Hg) is used on account of its high density (13.6 g/cm^3 or 849 lb/ft^3) for filling manometers and on account of its high thermal expansion for thermometers. Hg-vapour and organic Hg-compounds are very poisonous.

It is used as an electrode material in alkali-chlorine electrolysis (see II.2).

In dentistry a plastic and slowly setting silver amalgam is used as a filling for teeth.

The annual production of Hg is small compared with that of Cu, Al, Zn, Sn or Pb.

6. Occurrence and extraction of metals

Only the noble metals occur native, ie in the elementary state, uncombined with other elements. All other metals occur only as compounds with non-metals. Industrial **extraction** involves almost exclusively the compounds of metals with oxygen (oxides) or with sulphur (sulphides). Metals are mainly obtained from their oxides by reduction (see I, 6.9.2. and 6.12.5) with coke, hydrogen or a direct electric current.

7. Important non-metals

The most important **non-metals** are oxygen, nitrogen, the rare gases, hydrogen, the halogens (fluorine, chlorine, bromine and iodine), sulphur, phosphorus and carbon.

8.   Air

Air is a homogeneous mixture of gases (see I, 2.4.).

Nitrogen N_2	ca. 78	vol. %
Oxygen O_2	ca. 21	vol. %
Rare gases	ca. 1	vol. %
Carbon dioxide CO_2	ca. 0.03	vol. %

Air becomes a liquid if cooled below −196°C (−321°F). Liquid air can be separated into its components by fractional distillation. In this way nitrogen, oxygen and the rare gases (especially argon) are obtained industrially on a large scale (see, II.1).

9. Oxygen

Oxygen is the most abundant of all the elements (see I, 2.6.4.). Apart from its presence as the diatomic molecule O_2 in air, it occurs combined with hydrogen in water, H_2O, with metals in oxide ores, and with other elements in minerals. In addition many organic compounds contain combined oxygen.

Oxygen (O_2) is a colourless, odourless gas. It will not burn but is necessary for combustion.

9.1.   Oxidation and oxides

All chemical reactions in which oxygen takes part are termed **oxidations**. Oxygen can combine with both metals and nonmetals. The resulting compounds are called **oxides**.

$$\text{C} \quad + \quad O_2 \quad \longrightarrow \quad CO_2 \quad ; \Delta H = -393.7 \text{ kJ/mol*}$$
coke   atmospheric oxygen   carbon dioxide

$$\text{S} \quad + \quad O_2 \quad \longrightarrow \quad SO_2 \quad ; \Delta H = -296.9 \text{ kJ/mol}$$
sulphur   atmospheric oxygen   sulphur dioxide

Oxidation are exothermic reactions. The combustion of coke or sulphur is a rapid oxidation; on the other hand the rusting of iron is a slow oxidation.

In human or animal metabolism the energy for vital processes comes from a series of oxidation reactions. The final product of the combustion of foodstuffs is carbon dioxide (CO_2), which is exhaled.

Oxygen, and compounds which readily lose oxygen, are known as oxidizing agents.

---

* The values of ΔH given refer to the numbers of moles in the chemical equation. In this instance it refers to the formation of 1 mole of carbon dioxide.

9.2. Reduction

Chemical reactions in which oxygen is removed from a compound are known as de-
oxidations, or **reductions**.

Reduction is thus the reverse of oxidation, for example:

$$CuO \quad + \qquad\qquad H_2 \longrightarrow Cu \quad + \quad H_2O$$
Copper (II) oxide   hydrogen   copper   water

Many O-compounds, such as ores, can be reduced by reducing agents such as H_2 or coke.
Reducing agents are substances which can remove oxygen from other compounds by
binding it tightly themselves.

9.3. Redox reactions

When iron oxide ores lose their oxygen to coke, a reducing agent, in the blast furnace, the
ore is reduced to iron. Simultaneously the coke is oxidized to carbon monoxide (CO).

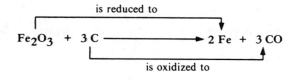

Thus an oxidation cannot occur without a simultaneous reduction. The overall process is
termed a **redox reaction** (ie a reduction-oxidation reaction).

9.4. Applications

In the chemical industry oxidation is effected by means of air or air enriched with oxygen.
Oxygen is also used in the cutting and welding of metals with a torch. These applications
employ the high heat of combustion of the reactions C_2H_2 (acetylene) $+ O_2$ or $H_2 + O_2$ to
melt the metal.

Pure oxygen is transported either as a liquid in low-temperature containers or as a gas at
high pressures up to 150 bar in steel cylinders painted blue. In chemical works the pipelines carrying oxygen gas are also painted blue.

9.5. Ozone

Ozone (O_3) consists of triatomic oxygen molecules. It is formed from atmospheric oxygen
by supplying large amounts of energy, as in lighting (spark discharges), and by the action
of UV radiation. The ozone molecule decomposes easily into $O_2 + O$. O signifies atomic

oxygen, which when freshly formed (in a nascent state) has a very high energy content, and is therefore more reactive than molecular oxygen. Ozone is thus a very powerful oxidizing agent. It is used for the sterilization of water.

10. Nitrogen

The diatomic molecule N_2 of **nitrogen** is the major constituent of air. Like oxygen it is a colourless and odourless gas, but it will neither burn nor support combustion.

Nitrogen occurs in combination with other elements in the different kinds of saltpetre and in plant and animal protein (16-17% N). The nitrogen in coal and lignite comes from such proteins.

10.1. Reactivity

In contrast to the oxygen molecule O_2 the nitrogen molecule is very unreactive (inert), since the bond in the N_2-molecule is extremely strong. Nitrogen will therefore react with other elements only when energy is supplied at high temperatures and with the participation of catalysts.

Although nitrogen is an indispensable constituent of all forms of life, neither plants, animals nor human beings can obtain their nitrogen requirements directly from the air. The nitrogen must therefore be supplied in the form of more reactive nitrogen compounds. Plants obtain these compounds from artificial or natural fertilizers and convert them into proteins. Human beings and animals transform these proteins further for building up their own bodies. Only some leguminous plants (eg lupins) can produce N-compounds directly from atmospheric nitrogen by the action of nodular bacteria attached to the their roots.

Some kinds of proteins can be completely synthesized, using petroleum and atmospheric nitrogen as raw materials.

10.2. Applications

For chemical industry atmospheric nitrogen represents an important raw material which is available in unlimited quantities. It is combined with H_2 to give ammonia (NH_3), the starting material for all nitrogen compounds.

Nitrogen gas is also used in industries dealing with chemicals, metals, minerals and plant construction as a protective inert gas, and also for purging. Thus boilers and tanks which have contained combustible materials are flushed out with nitrogen rather than with compressed air, so as to avoid the production of explosive gas mixtures. On account of its inertness nitrogen is used as a protective gas in chemical reactions where the presence of atmospheric oxygen could lead to undesirable oxidation processes.

Pure nitrogen is available commercially, compressed up to 150 bar (2200 lb/in^2), in steel cylinders painted green. In chemical works pipe-lines carrying nitrogen are also painted green.

11. The rare gases

The **rare gases** (also known as inert or noble gases) occur in the atmosphere and in natural gas. The six rare gases are:

Helium	He	up to 1 vol. % in natural gas
Neon	Ne	
Argon	Ar	0.9 vol. % in air
Krypton	Kr	
Xenon	Xe	
Radon	Rn	

The atoms of the rare gases are particularly stable. The rare gases therefore occur naturally only as monatomic species and form hardly any compounds. They are used for filling electric light bulbs and fluorescent tubes, and also as protective and purging gases in the metallurgical industry.

12. Hydrogen

Another non-metallic element used in large quantities by chemical industry is **hydrogen**, which occurs naturally only in combination with other elements.

12.1. Manufacture

Hydrogen can be obtained from cheap raw materials, especially from natural gas and the lightest petroleum fractions, which consist of hydrocarbons, ie C-H compounds. Hydrogen is also obtained from the reaction between water vapour (H_2O) and incandescent coke (C).

12.2. Oxy-hydrogen, explosions

Hydrogen (H_2) is a colourless, odourless, combustible gas. Mixtures of hydrogen with air or oxygen are explosive over a wide range of compositions.

An **explosion** is a very rapid exothermic reaction. The hot gases thus produced create a shock wave.

12.3. Applications

H_2 -gas is compressed and marketed under pressure (up to 150 bar or 2200 lb/in²) in steel cylinders, painted red and with a left-handed thread to the output nozzle. It is used for welding and cutting metals. Large quantities of hydrogen are used for hydrogenation by chemical industry.

12.4. Hydrogenation and dehydrogenation

Hydrogenations are reactions in which hydrogen combines with elements or compounds, for example:

$$N_2 \quad + \quad 3H_2 \longrightarrow 2NH_3 \qquad \Delta H = -92.1 \text{ kJ/mol}$$
nitrogen $\qquad$ hydrogen $\quad$ ammonia

All hydrogenations are reductions: this means that the products are poorer in oxygen or richer in hydrogen than the reactants.

The reverse of hydrogenation, ie the removal of hydrogen, is termed **dehydrogenation**, for example:

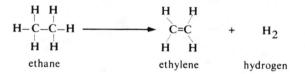

ethane $\qquad\qquad$ ethylene $\qquad$ hydrogen

All dehydrogenations are oxidations. The result of an oxidation is to produce substances which are poorer in hydrogen or richer in oxygen.

12.5. General aspects of oxidation and reduction

A closer consideration of the processes of oxidation and reduction leads to the generalization that these reactions involve an exchange of electrons. Thus in the combustion of Mg to give MgO the magnesium loses electrons which are taken up by the oxygen:

Separate processes: $\qquad 2\,Mg \longrightarrow 2\,Mg^{2+} \quad + \quad 4\,e^-$
$\qquad\qquad\qquad\qquad$ magnesium $\qquad$ magnesium $\quad$ electrons
$\qquad\qquad\qquad\qquad\qquad\qquad$ cations

$$O_2 \quad + \quad 4\,e^- \longrightarrow 2\,O^{2-}$$
oxygen $\quad$ electrons $\qquad\qquad$ oxide anions

Overall process: $\quad 2\,Mg + O_2 \longrightarrow 2\,MgO$

In this redox reaction the magnesium is oxidized to magnesium cations. **Oxidation thus implies loss of electrons.** The oxygen is reduced to oxide anions. **Reduction therefore implies gain of electrons.**

These rules apply quite generally, even when neither oxygen nor hydrogen take part in the reaction. For example, the formation of iron sulphide from the elements iron and sulphur is a redox reaction:

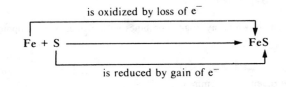

In this instance the separate processes are:

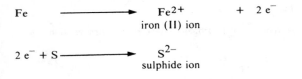

12.6. Hydrogen peroxide

In addition to water (H_2O, an oxide of hydrogen) there exists another compound of hydrogen and oxygen, **hydrogen peroxide**, H_2O_2 (previously known as hydrogen superoxide).

It has the structural formula H-O-O-H.

H_2O_2 readily loses atomic oxygen, O, and therefore acts as a powerful oxidizing agent, being used for bleaching and disinfecting.

13. The halogens

The seventh main group of the periodic system contains the **halogens**, fluorine (F), chlorine (Cl), bromine (Br) and iodine (I). They all have the same number of valency electrons, which gives them similar chemical properties. At room temperature fluorine and chlorine are gases, bromine a liquid and iodine a solid. Since they are extremely reactive elements they occur naturally only as compounds. The most important compounds, from which they are obtained, are fluorspar (CaF_2), rock salt (NaCl), magnesium bromide ($MgBr_2$) and sodium iodate ($NaIO_3$).

13.1. Fluorine

Fluorine is an extremely reactive and poisonous gas.

Compounds of fluorine with metals – fluorides – are used as fluxes. The addition of fluxes to ionic compounds, such as aluminium oxide, lowers their melting point.

Reactions in which fluorine is introduced into a compound are called fluorinations. They play a certain part in organic chemistry (see halogenation, IV, 5).

13.2. Chlorine, hydrogen chloride, hydrochloric acid

By far the most important halogen is **chlorine**, a yellowish-green poisonous gas. It is obtained by electrolysis, by passing a direct electric current through an aqueous solution of NaCl (see alkali-chlorine electrolysis, II.2.).

The introduction of chlorine into inorganic or organic compounds is termed chlorination. Chlorine is also used for bleaching paper and cellulose and as a disinfectant. Liquid chlorine is solid in steel cylinders painted grey.

Gaseous **hydrogen chloride**, HCl, is formed from the elements in an exothermic reaction:

$$H_2 + Cl_2 \longrightarrow 2\,HCl; \qquad \Delta H = -184.5 \text{ kJ/mol}$$

If hydrogen chloride is passed into water, **hydrochloric acid** (aqueous HCl) is formed.

13.3. Bromine

Bromine is used for brominations in organic synthesis. At room temperature it is a reddish-brown evil-smelling liquid.

13.4. Iodine

Iodine exists at room temperature as greyish violet crystalline leaflets. The thyroid hormone contains an iodine compound. In organic synthesis the use of iodination is of only minor importance.

14. Sulphur

Sulphur is found together with oxygen in the sixth main group of the periodic system. Sulphur and oxygen both possess six valency electrons and are therefore chemically related.

14.1. Occurrence and extraction

Sulphur occurs as the element, for example in Texas, Louisiana, Japan and Poland. It is extracted from rocks by melting and after purification is obtained as a brittle yellow solid.

Important S-compounds which occur in nature are the sulphide ores, such as the industrially important iron pyrites (FeS_2), galena (PbS), and zinc blends (ZnS). Other sulphur-containing minerals are gypsum and heavy spar.

Hydrogen sulphide (H_2S) is a foul-smelling gas which is formed by the decay and putrefaction of plant and animal protein. This is the origin of the sulphur in coal and the H_2S-content of natural gas.

14.2. Applications

Sulphur is used mainly for the manufacture of **sulphuric acid**, H_2SO_4, and other sulphur compounds. It is also used for vulcanizing rubber, for pest control, and for cosmetic purposes (see II,5).

15. Phosphorus

Phosphorus occurs together with nitrogen in the fifth main group of the periodic system. It never occurs naturally as the element but always combined in minerals.

Two forms of the element can be prepared from these compounds. The first is red phosphorus, which is non-poisonous and stable in air, and the second is white (or yellow) phosphorus, which is soft, waxy, poisonous, and easily catches fire. In a finely divided form it is spontaneously inflammable even at room temperature, and must therefore be stored under water.

The commenest and most important mineral is rock phosphate, or phosphorite, $Ca_3(PO_4)_2$. It is insoluble in water, but is the raw material for conversion to soluble phosphate fertilizers and for the manufacture of phosphoric acid (H_3PO_4) and other phosphorus compounds (see II.6).

16. Carbon

Carbon is the most important element in the fourth main group of the periodic system, and occurs native in two modifications having different crystal structures.

16.1. Diamond, graphite, carbon black

In **diamond** the C-atoms are packed very densely and regularly, which is responsible for its extreme hardness and high refractive index.

In **graphite** the C-atoms are packed less densely in layers. Because the layers can be displaced relative to one another graphite is soft. In contrast to diamond, graphite can conduct electricity on account of its layered structure. It is therefore used for making electrodes.

Carbon black consists of extremely small leaflets of graphite.

16.2. Coal, coke, carbonate minerals

The different varieties of **coal** (lignite, hard coal, anthracite) do not consist of pure carbon, but are complicated mixtures, consisting mainly of carbon compounds containing a high proportion of carbon. **Coke**, made from coal, is fairly pure carbon, and the same is true of charcoal, prepared from wood.

Minerals such as limestone, chalk and marble ($CaCO_3$) and magnesite ($MgCO_3$) contain carbon in a combined form. Dolomite has the composition $CaCO_3.MgCO_3$.

16.3. Carbon dioxide, dry ice

Carbon dioxide, CO_2, is the end product in the complete combustion of all kinds of C-compounds. In human and animal metabolism CO_2 is also the final product of the step-wise oxidation of foodstuffs. CO_2 is a colourless and odourless gas.

Dry ice (solid CO_2) is used for cooling. It sublimes at $-78.5°C$ ($-109.3°F$).

16.4. Carbon monoxide

When fuels are incompletely burned, in particular in stoves with a poor draught and in car engines which are ticking over, the gases formed contain not only CO_2 but also the odourless and extremely poisonous **carbon monoxide**, CO.

When breathed, even at concentrations as low as 0.2 vol.%, it causes poisoning which may be fatal.

17. Metalloids, semiconductors

Like carbon (graphite) the **metalloids** boron (B), silicon (Si), selenium (Se) and germanium (Ge) are termed **semiconductors**.

This term is used to describe substances which conduct electricity only slightly or not at all at low temperatures, but which become good conductors at high temperatures or after the addition of very small amounts of other elements (doping). This contrasts with conduction by metals, which is decreased by increasing the temperature or by the presence of small quantities of impurities.

Special importance attaches to semiconductors in which the incorporation of various additives produces a system in which current can pass in only one direction (rectifiers). Transistors constructed from semiconductors are used to amplify electrical signals, and by virtue of their minute size have revolutionized communication techniques.

18. Summary

18.1.   The main importance of **metals** lies in their use as materials. In this context their physical properties are particularly important.

18.2.   **Alloys** are made by fusing metals together. In this way their useful properties can be improved.

18.3.   Metals are good conductors of heat and electricity.

18.4.   Metals can be divided into **light metals** ($p < 5$ g/cm^3 or 312 lb/in^3) and **heavy metals** ($p > 5$ g/cm^3 or 312 lb/in^3).

18.5.   Metals are classified as **noble metals** or **base metals** according to their resistance to chemical attack.

18.6.   The term **corrosion** means the destruction of base metals by chemical attack.

18.7.   **Oxygen** is an incombustible gas, but is necessary for all kinds of combustion processes, including life.

18.8.   Reactions involving oxygen are called **oxidations**, and the products are termed **oxides**.

18.9.   The reverse of oxidation, the removal of oxygen, is called deoxidation or **reduction**.

18.10.   An oxidation is always accompanied by a reduction. The combined process is known as a **redox reaction**.

18.11.   **Nitrogen** is a very inert gas and does not take part in combustion.

18.12.   The nitrogen requirements of plants are largely supplied by synthetic fertilizers.

18.13.   The starting material for the synthesis of all nitrogen compounds is **ammonia**, NH_3.

18.14.   The **rare gases** form hardly any compounds.

18.15.   **Hydrogen**, the lightest of all substances, is a combustible gas. Hydrogen forms a highly explosive mixture with air or oxygen.

18.16.   The reaction of hydrogen with other substances is termed **hydrogenation**. A hydrogenation is a reduction.

18.17. The reverse of hydrogenation, the removal of hydrogen, is termed **dehydrogenation**. A dehydrogenation is an oxidation.

18.18. In a more general sense **oxidation** implies a loss of electrons, and **reduction** a gain of electrons.

18.19. The **halogens**, fluorine, chlorine, bromine and iodine, are extremely reactive elements. The most important halogen is chlorine.

18.20. The introduction of halogens into other compounds is called **halogenation**.

18.21. **Sulphur** is chemically related to oxygen. Sulphur is used chiefly for the manufacture of sulphuric acid.

18.22. **Phosphorus** occurs in minerals, some of which are used for making phosphate fertilizers and P-compounds.

18.23. **Carbon dioxide** is the end product of the complete combustion of all C-compounds

18.24. **Carbon monoxide** is produced in the incomplete combustion of fuels. It is a very dangerous gas which can easily lead to fatal poisoning.

18.25. **Semiconductors** are elements on the borderline between metals and non-metals. Doped semiconductors are very important as rectifiers for alternating current and in transistors.

19. Self-evaluation questions

74   Which properties are typical of metals?

A   ductility
B   high melting points
C   good thermal conductivity
D   good electrical conductivity

75   Which groups are correctly represented?

	Light metal	Heavy metal	Base metal	Semi-precious metal	Precious metal
A	sodium	lead	zinc	copper	stainless steel
B	aluminium	copper	potassium	iron	platinum
C	magnesium	mercury	iron	nickel	silver
D	silver	zinc	mercury	chromium	gold

76   Which combinations of alloy and composition are correct?

A   steel = iron and carbon
B   brass = copper and zinc
C   bronze = copper and tin

77   Which group of metals corrodes most in moist air?

A   base metals
B   semi-precious metals
C   precious metals

78   Which of the following are base metals which will corrode in moist air?

A   gold
B   tin
C   iron
D   titanium

79   In the following processes which substances are oxidized and what are the products of oxidation?

A $4\,Fe \quad + \quad 3\,O_2 \longrightarrow 2\,Fe_2O_3$

B $CH_4 \quad + \quad 2\,O_2 \longrightarrow CO_2 + 2\,H_2O$

80 In the following reactions which substances are reducing agents?

A $Fe_2O_3 + 3H_2 \longrightarrow 2Fe + 3H_2O$

B $SiO_2 + C \longrightarrow Si + CO_2$
quartz-
sand

C $TiO_2 + 2Mg \longrightarrow Ti + 2MgO$

81 Which statements about metals are correct?

A Mercury vapour is toxic.
B Steel contains about 4% carbon.
C Lead is used to give protection against radiation.
D Aluminium is used by the electrical industry as a conductor material.

82 Which statements about oxygen are correct?

A Compounds of oxygen with other elements are called oxides.
B Oxygen is insoluble in water.
C Oxygen is used in the welding of metals.
D Oxygen is transported and handled in red-painted cylinders.

83 Which statements about nitrogen are correct?

A Nitrogen is only very slightly reactive.
B Nitrogen is used in the chemical industry as a protective gas.
C Nitrogen cannot be taken direct from the air by humans and animals for metabolism.
D Nitrogen is the second most plentiful constituent of air (after oxygen).

84 Which statements about hydrogen are correct?

A Hydrogen is the lightest of all elements.
B Hydrogen is supplied under pressure in red-painted steel cylinders with left-hand threads.
C Hydrogen forms highly explosive mixtures with air and oxygen
D Hydrogen gives natural gas its characteristic smell.

85 Which substance is oxidized in the following reaction?

$$\underset{A}{CuO} + \underset{B}{H_2} \dashrightarrow \underset{C}{Cu} + \underset{D}{H_2O}$$

86   Which statements about halogens are correct?

   A   The elements fluorine, chlorine, bromine and iodine are classed as halogens
   B   The compounds of chlorine are recognizable by their green colour
   C   Chlorine is used as a disinfectant.
   D   Gaseous chlorine is toxic.

87   Which statements about sulphur and sulphur compounds are correct?

   A   Sulphur is highly reactive and therefore occurs only in compounds
   B   Colloidal sulphur is used for vulcanizing rubber.
   C   Sulphur is non-combustible.
   D   Petroleum and natural gas contain compounds with a sulphur content.

88   Which substances consist of pure carbon?

   A   diamond
   B   graphite
   C   wood
   D   coal

89   With which substances are oxides reduced in industry?

   A   carbon
   B   hydrogen
   C   chlorine

7 Acids, Bases and Salts

1. Acids

The name **acid** derives from the sour (acid) taste of dilute acids, for example citric acid in lemon juice, acetic acid in vinegar, hydrochloric acid in the abdominal fluid. Concentrated acids are highly corrosive and must be handled with great care.

The following acids are of industrial importance:

$\underline{H}Cl$ hydrochloric acid
$\underline{H}_2SO_4$ sulphuric acid of sulphuric (VI) acid
$\underline{H}NO_3$ nitric acid or nitric (V) acid
$\underline{H}_3PO_4$ phosphoric acid or phosphoric (V) acid
$\underline{H}CN$ prussic acid or hydrocyanic acid

1.1. Acidic hydrogen, hydrogen ions, acid reaction

The characteristic ingredient of all acids is **acidic hydrogen**, underlined in the above formulae. The acqueous solutions of all acids contain the whole or a part of this acidic hydrogen as **hydrogen ions**, or protons, H^+. If H^+ can be detected in a solution it is said to have an acid reaction. The higher the concentration of hydrogen ions the more concentrated and strongly acidic is the solution.

1.2. Indicators

H^+ can be detected by means of **indicators**, ie substances which undergo a characteristic colour change in the presence of acids.

1.3. Anions of acids

The other constituent of any acid is the negativey charged acid anion. Thus hydrochloric acid consists of H^+ and Cl^- (chlorine ions), and sulphuric acid of 2 H^+ and SO_4^{2-} (sulphate ions).

1.4. The definition of acids

Many characteristic properties of acids depend on the fact that they can lose protons. The modern definitions of an acid is therefore:

An acid is a substance which can lose protons; it is a proton donor.

1.5. Mono- and polybasic acids

A distinction is made between **mono- and polybasic acids** according to the number of acidic hydrogen atoms in their molecules:

HCl	is a monobasic acid,
H_2SO_4	is a dibasic acid,
H_3PO_4	is a tribasic acid.

1.6. Acid anhydrides

There is a distinction between oxygen-free acids, such as HCl and HCN, and those which contain oxygen, such as H_2SO_4 and HNO_3. If water is removed from an acid containing oxygen, an **acid anhydride** is obtained:

$$H_2SO_4 \xrightarrow{\;-H_2O\;} SO_3$$

Thus sulphur (VI) oxide (sulphur trioxide) is the anhydride of sulphuric (VI) acid (sulphuric acid).

Conversely, the addition of water to oxides of nonmetals gives oxyacids:

$$SO_2 \xrightarrow{\;+H_2O\;} H_2SO_3$$

Sulphur (IV) oxide (sulphur dioxide) and water give sulphuric (IV) acid (sulphurous acid).

1.7. Important inorganic acids and their anhydrides

The following Table lists some important inorganic acids and their anhydrides:

Acid		Anhydride	
Name	Formula	Formula	Name
Sulphurous acid	H_2SO_3	SO_2	Sulphur dioxide
Sulphuric acid	H_2SO_4	SO_3	Sulphur trioxide
Carbonic acid	H_2CO_3	CO_2	Carbon dioxide
Nitrous acid	HNO_2	N_2O_3	Dinitrogen trioxide, or nitrogen (III) oxide
Nitric acid	HNO_3	N_2O_5	Dinitrogen pentoxide, or nitrogen (V) oxide
(Ortho)phosphoric acid	H_3PO_4	P_2O_5	(Di)phosphorus pentoxide, or phosphorus (V) oxide

2. Bases

Bases are the counterparts of acids.

2.1. Hydroxides, alkalis

Bases include all **hydroxides**, of which the following are industrially important:

NaOH	Sodium hydroxide or caustic soda; aqueous solution sometimes called soda lye.
KOH	Potassium hydroxide or caustic potash; aqueous solution sometimes called potash lye.
Ca(OH)$_2$	Calcium hydroxide or slaked lime; aqueous solution sometimes called lime water.

Dilute alkali solutions have a soapy feeling. The solid hydroxides and their concentrated solutions are highly corrosive and must be handled with great care.

2.2. Hydroxide ions, basic or alkaline reaction

The characteristic constituent of all hydroxides and alkalis is the **hydroxide ion** OH^-. The other constituent is the positively charged metal cation. Thus sodium hydroxide is composed of equal numbers of the ions Na^+ and OH^-.

In aqueous solutions the presence of hydroxide ions is detected by the characteristic colours which they give to indicators: these differ from the colours of the indicators in acid solution. If an indicator demonstrates the presence of OH^- the solution is said to show an alkaline or basic reaction. The higher the concentration of hydroxide ions in a solution, the more strongly alkaline it is.

2.3. Mono- and polyacidic bases

A distinction is drawn between **mono- and polyacidic bases** according to the number of OH^- ions in the formula:

NaOH	is a mono-acid base
Ca(OH)$_2$	is a di-acid base

2.4. The definition of bases

While acids are proton donors, bases are characterised by their ability to take up protons. The modern definition of a base is therefore:

A base is a substance which can take up protons; it is a proton acceptor.

I Fundamentals of general and inorganic chemistry

2.5. Neutralization, salts

Hydroxide ions (HO^-) will capture protons (H^+) with the formation of water:

$$H^+ \quad + \quad OH^- \longrightarrow H_2O; \Delta H = -55.7 \text{ kJ/mol}$$

This reaction is termed **neutralization**, since the acidic properties of H^+ and the basic properties of OH^- cancel one another out.

However, when caustic soda is completely neutralized with hydrochloric acid water is not the only product, since the Na^+ from the sodium hydroxide and the Cl^- from the hydrochloric acid form an aqueous solution of a salt. When this salt solution is evaporated down solid common salt NaCl remains:

NaOH	+	HCl	⟶	H_2O	+	NaCl
caustic soda		hydrochloric acid		water		sodium chloride

or more accurately in terms of ions:

$Na^+ + OH^-$	+	$H^+ + Cl^-$	⟶	H_2O	+	$Na^+ + Cl^-$
caustic soda		hydrochloric acid		water		sodium cloride in solution

Hydrochloric acid is monobasic, and one HCl neutralizes just one OH^-. Caustic soda is thus a mono-acid base.

2.6. The ammonium ion

Ammonia (NH_3) is also a base, since it can accept H^+:

NH_3	+	H^+	⟶	NH_4^+
ammonia		proton		ammonium ion

An aqueous solution of the base has an alkaline reaction (spirits of ammonia).

The ammonium ion behaves like a monovalent metal cation. Thus the reaction between ammonia and sulphuric acid produces the salt ammonium sulphate:

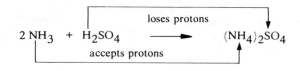

$$2\,NH_3 \quad + \quad H_2SO_4 \quad \longrightarrow \quad (NH_4)_2SO_4$$

loses protons

accepts protons

3. Salts

Salts are solids consisting of metal cations and the anions of acids

3.1. Salt formation

The following are the most important methods for obtaining salts:

1. Neutralization of acids by alkalis

KOH	+	HNO_3	$\longrightarrow$	KNO_3	+	H_2O
caustic potash		nitric acid		potassium nitrate		water

or in general:

alkali   +   acid $\longrightarrow$ salt   +   water

2. Reaction of acid anhydrides (oxides of nonmetals) with alkalis

CO_2	+	2 NaOH	$\longrightarrow$	Na_2CO_3	+	H_2O
carbon dioxide		caustic soda		sodium carbonate (washing soda)		water

3. The reaction of base metals with dilute acids

Fe	+	2 HCl	$\longrightarrow$	$FeCl_2$	+	$H_2\uparrow$
iron		hydrochloric acid		iron (II) chloride		hydrogen

Noble and seminoble metals do no react with dilute acids.

4. Reactions between metals and non-metals

Fe	+	S	$\longrightarrow$	FeS
iron		sulphur		iron sulphide

5. Double decomposition between salts

In these reactions the anions and the cations are exchanged. One of the newly formed salts is less soluble and is precipitated from solution as a solid.

$AgNO_3$	+	KBr	$\longrightarrow$	$AgBr\downarrow$	+	KNO_3
silver nitrate		potassium bromide		silver bromide (precipitates)		potassium nitrate

3.2. Acid salts

If only part of the acidic hydrogens of a polybasic acid are replaced by metal cations an **acid salt** is obtained:

$NaOH$	+	H_2SO_3	$\longrightarrow$	$NaHSO_3$	+	H_2O
				sodium hydrogen sulphite		

Sodium hydrogen sulphite consists of Na^+ and HSO_3^-. The anion thus still contains acidic hydrogen.

$NaHSO_3$ is termed a primary salt, since only the first atom of acidic hydrogen has been replaced by a metal.

Na_2SO_3 is a secondary salt, since both the acidic hydrogen atoms have been replaced by a metal.

3.3. Basic salts

Conversely, if the OH^--groups of a polyacidic base are only partially replaced by anions, the product is a **basic salt**:

$Mg(OH)_2$	+	HCl	$\longrightarrow$	$Mg(OH)Cl$	+	H_2O
				magnesium hydroxide chloride		

Basic salts thus still contain hydroxide ions.

3.4. Important inorganic acids, their anions and their salts

Acid		Anion	Salts	
Formula	Name	Formula and valency	Formula	Name (M = monovalent metal)
Hydracids				
HF	Hydrofluoric acid	F^-	MF	fluoride
HCl	Hydrochoric acid	Cl^-	MCl	chloride
BHr	Hydrobromic acid	Br^-	MBr	bromide
HI	Hydriodic acid	I^-	MI	iodide
HCN	Prussic acid Hydrocyanic acid	CN^-	MCN	cyanide
H_2S	Hydrogen sulphide	S^{2-}	M_2S	sulphide
Oxo-acids				
H_2CO_3	Carbonic acid	HCO_3^-	$MHCO_3$	hydrogen carbonate
	and	CO_3^{2-}	M_2CO_3	carbonate
HNO_2	Nitrous acid	NO_2^-	MNO_2	nitrite
HNO_3	Nitric acid	NO_3^-	MNO_3	nitrate
H_3PO_4	(Ortho)phosphoric acid	$H_2PO_4^-$	MH_2PO_4	primary phosphate
	and	HPO_4^{2-}	M_2HPO_4	secondary phosphate
	and	PO_4^{3-}	M_3PO_4	tertiary phosphate
H_2SO_3	Sulphurous acid	HSO_3^-	$MHSO_3$	hydrogen sulphite
	and	SO_3^{2-}	M_2SO_3	sulphite
H_2SO_4	Sulphuric acid	HSO_4^-	$MHSO_4$	hydrogen sulphate
	and	SO_4^{2-}	M_2SO_4	sulphate
$H_2S_2O_3$	Thiosulphuric acid	$S_2O_3^{2-}$	$M_2S_2O_3$	thiosulphate
$HClO$	Hypochlorous acid	ClO^-	$MClO$	hypochlorite
$HClO_2$	Chlorous acid	ClO_2^-	$MClO_2$	chlorite
$HClO_3$	Chloric acid	ClO_3^-	$MClO_3$	chlorate
$HClO_4$	Perchloric acid	ClO_4^-	$MClO_4$	perchlorate
$HBrO_3$	Bromic acid	BrO_3^-	$MBrO_3$	bromate
HIO_3	Iodic acid	IO_3^-	MIO_3	iodate

3.5. Water of crystallization

Many salts contain **water of crystallization**, for example washing soda, $Na_2CO_3 \cdot 10\ H_2O$ and blue vitriol (copper sulphate) $CuSO_4 \cdot 5\ H_2O$. The dipolar water molecules are incorporated in the ionic lattice, especially round the cation. Like all other salts, these salts are solid crystalline substances. Strong heating is necessary to remove the water from the crystal lattice as water vapour.

3.6. Double salts

Double salts represents a special class of salts, for example the alums. Potassium chromium alum has the formula $KCr(SO_4)_2 \cdot 12\ H_2O$. In alums, as in double salts in general, two different metal cations are surrounded by the same kind of anions in a uniform crystal lattice. Most double salts also contain water of crystallization.

3.7. Complex ions and salts

Ions which do not consist of charged atoms (eg Na^+ and Cl^-) but of charged groups of atoms (eg NH_4^+ and SO_4^{2-}) are known as **complex ions**. This term is frequently used in a narrower sense for complicated ions such as $[Fe(CN)_6]_4^-$ and $[Cu(NH_3)_4]^{2+}$, and the salts formed from such ions are called **complex salts**, for example

$K_4[Fe(CN)_6]$	potassium hexacyanoferrate (II)
	(common name potassium ferrocyanide)
$[Cu(NH_3)_4]SO_4$	copper (II) – tetrammine sulphate

4. Summary

4.1. **Acids** are substances which can lose a proton; they are proton donors. Their characteristic constitutent is **acidic hydrogen atoms** H.

4.2. Solutions with an acid reaction are those in which H^+ can be detected. The other component of an acid solution is the anion.

4.3. Acids are termed monobasic or polybasic according to the number of acidic hydrogen atoms in the molecule.

4.4. The removal of H_2O from an acid containing oxygen produces an **acid anhydride**.

4.5. **Bases** are substances which can take up a proton; they are proton acceptors.

4.6. **Hydroxides** represent an important class of base. Solutions of hydroxides are called **alkalis**. The characteristic constituent of bases and alkalis is the **hydroxide** ion, OH^-, which is combined with a metallic cation.

4.7. If OH^- ions can be detected in a solution it is said to have a basic or alkaline reaction.

4.8. Bases are described as monoacidic or polyacidic according to the number of hydroxide ions in their formula.

4.9. H^+ and OH^- can be detected by means of **indicators**.

4.10. The reaction of H^+ with OH^- to give H_2O is called **neutralization**.

4.11. **Salts** are solid substances composed of metal cations and the anions of acids.

4.12. The ammonium ion NH_4^+ behaves like a metallic cation and can form salts with the anions of acids.

4.13. In acid salts the anion still contains acidic hydrogen. Basic salts still contain hydroxide ions.

4.14. H_2O molecules which are incorporated in the ionic lattice of a salt are termed water of crystallization.

4.15. In double salts two different metal cations are surrounded by the same anions in a uniform crystal lattice.

4.16. Complex salts are composed of complicated groups of atoms bearing a charge (complex ions).

5. Self-evaluation questions

90 Mark the following Table with crosses in the appropriate columns:

Formula	Acid	Anion of acid	Acid anhydride	Alkali	Salt
HCl					
SO_3					
SO_3^{2-}					
SO_4^{2-}					
K_2SO_4					
KOH					
$Ca(OH)_2$					
NH_4Cl					
H_3PO_4					
CO_2					
Na_2CO_3					
Cl^-					
HNO_3					
NaOH					

91 Which statements about acids are correct?

A Acids are compounds capable of accepting H^+ ions from an aqueous solution.
B The characteristic constituent of acids is oxygen.
C The corrosive action of acids increases with a rise in temperature.
D The valency of acids depends on the number of hydrogen atoms in the molecule which can be split off.

92 Which acid is bivalent?

A H_2SO_3
B HNO_2
C H_3PO_4
D HCl

93 The anhydride of sulphuric acid is

A SO_2
B S_2O_3
C SO_3
D SO_4

94 Which of the following compounds are hydroxides?

A CH_3OH
B C_2H_5OH
C $NaOH$
D NH_4OH

95 Which statements about hydroxides are correct?

A Hydroxides are compounds produced from metallic oxides and water.
B The aqueous solutions of the hydroxides are called lyes.
C With indicators, bases display characteristic colour reactions.
D Hydroxides have a highly caustic action.

96 Which combinations of name and formula are correct?

A HCl – chloric acid
B H_3PO_4 – phosphoric acid
C H_2CO_3 – carbonic acid
D H_2NO_3 – nitric acid

I Fundamentals of general and inorganic chemistry

97 In neutralization reactions

    A   acids and alkalis destroy each other
    B   salts are formed
    C   water is split off from an acid molecule
    D   the pH value of the solution changes.

98 Which acid/salt combinations are correct?

    A HCl – chloride
    B HNO_3 – nitrate
    C H_2SO_4 – sulphonate
    D H_2CO_3 – carbonate

99 Which two ions really have a different charge?

    A Cl^{2-}
    B SO_4^{2-}
    C Na^+
    D Fe^+

8 Water as a Solvent. Electrolytic Dissociation, pH Values, Electrolysis. Disperse Systems

8 Water as a solvent. Electrolytic dissociation, pH values, electrolysis. Disperse systems

1. Water as a solvent

Four-fifths of the surface of the earth is covered with water. The air also contains water as invisible water vapour. Plants and animals consist largely of water; about 70% of the human body consists of water. Water is indispensable as a solvent and as a transport medium for all forms of life.

1.1. Naturally occurring water

Naturally occurring water always contains dissolved gases and salts, as well as undissolved matter. In particular the surface water of rivers and lakes contains impurities of vegetable, animal, mineral and human origin. Pure **spring water** with a fairly high content of salts and/or gases is called mineral water. **Sea water** contains on the average 3% common salt and 0.5% other salts.

Drinking water ought to contain small quantities of salts, since these are necessary for building up the body, and also improve the taste: it must however be clear and free from bacteria.

1.2. Hard and soft water

The more Ca- and Mg- salts water contains, the "harder" it is. **Hard water** is unsuitable for washing, which is better carried out with **soft water** such as rain water or water which has been artificially softened (partly demineralized).

1.3. Distilled water and demineralized water

Distilled, sterile water has been rendered chemically pure and free from microorganism and destroyed microorganism by distillation. It is used in the production of pharmaceutical products. Demineralized water is free from salts. It is used in the laboratory and in the production of pure chemicals. Distilled and demineralized water are both unsuitable for drinking (see II, 1)

1.4. Solubility and absorption of gases

Water is by far the most important solvent not only in nature but also in chemical industry. The solubility of different substances in water covers a very wide range.

The solubility of **gases** in water and other solvents increases with increasing gas pressure and with decreasing solvent temperature. Everyday examples of this are provided by beer and other effervescent drinks. Air is also soluble in water, the oxygen — necessary for the survival of fishes — being more soluble than the nitrogen. The dissolution of gases in liquids is called **absorption**.

1.5. Solubility of liquids

Liquids consisting of nonpolar molecules, such as gasoline and benzene, are practically insoluble in the polar solvent water. The solubility of liquids in water increases with increasing polarity of their molecules and with decreasing molecular size. Thus the alcohols methanol and ethanol are miscible with water in all proportions.

1.6. Solubility of solids, dissolution, hydration, saturated solutions

Among **solids** macromolecular organic compounds are as a rule insoluble in water. Smaller organic molecules are water-soluble if they contain polar groups, as is the case for sugar and washing agents.

Many salts consisting of ionic lattices are soluble in water, especially inorganic salts.

In the process of **dissolution** the small water molecules force their way between the molecules or ions of the solid. This weakens the attractive forces between the particles of the solid, their cohesion is loosened, and the particles are separated and dissolve. In solution the ions or molecules are surrounded by H_2O-molecules; they are **hydrated**.

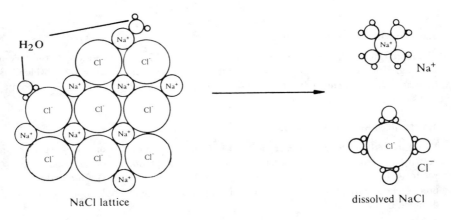

NaCl lattice   dissolved NaCl

The solubility of most solids increases with rise of temperature. When no more solid will dissolve the solution is said to be **saturated**. If a hot saturated solution is cooled, part of the dissolved substance separates out in a crystalline form. This process of recrystallization is used to purify crystalline substance.

2. Electrolytic dissociation

When an ionic lattice dissolves in water the solution contains mobile hydrated anions and cations. This separation of the ions is known as **electrolytic dissociation**.

The following processes take place on solution:

Salts produce metal ions and the anions of acids:

$$NaCl \xrightarrow{\text{dissociates into}} Na^+ \text{ and } Cl^- \text{ (both hydrated)}$$

Hydroxides produce metal ions and hydroxide ions:

$$NaOH \xrightarrow{\text{dissociates into}} Na^+ \text{ and } OH^- \text{ (both hydrated)}$$

Fused salts and hydroxides also contain mobile ions.

In contrast to salts and hydroxides **acids** produce ions only after dissolution:

$$H_2SO_4 \xrightarrow{+H_2O} 2H^+ + SO_4^{2-} \text{ (all hydrated)}$$

2.1. Degree of dissociation

The extent to which acids dissociate in solution varies greatly. The **degrees of dissociation** is a measure of the fraction of the acid molecules which are split up into ions by H_2O.

2.2. Strong and weak acids

Strong acids such as hydrochloric, sulphuric and nitric acids are almost 100% dissociated in diluted solution. In these acid solutions the hydrogen is present predominantly as H^+.

On the other hand **weak acids** such as hydrogen sulphide and acetic acid are only slightly dissociated. In their aqueous solutions only a small proportion of the acidic hydrogen is present as H^+, the greater part remaining bound to the anion of the acid.

3. pH-values, pH-scale, pH-paper

The concentration of H^+ ions in a solution is expressed by its so-called **pH-value.** The usual **pH-scale** extends from 0 to 14: the value pH 7 represents a neutral reaction, while all pH-values < 7 signify an acidic reaction and all pH-values > 7 an alkaline reaction.

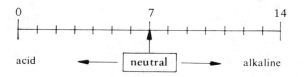

The pH-value of a solution can be determined very simply by means of indicator papers designed for different pH-ranges, sometimes known as pH-papers. The pH-value can be measured accurately by using a special instrument known as a pH-meter.

4. Electrolytes, ionic migration, electrolysis

Since aqueous solutions of salts, hydroxides and acids contain freely mobile ions, these solutions can conduct a direct electric current. They are therefore known as **electrolytes.**

When a direct electric potential is applied the cations move towards the negatively charged electrode, the cathode, and the anions towards the positively charged electrode, the anode. Chemical reactions take place at the anode and the cathode (see illustration in 4.2).

The decomposition of a compound by the action of an electric current is known as **electrolysis.** The material of which the electrodes are composed is of prime importance. It may remain unchanged (eg Pt), or may take part in the reaction.

4.1. Electronic and ionic conductors

The conduction of electricity in metals involves the migration of electrons, and they are therefore known as **electronic conductors.**

In electrolytes, on the other hand, the current is conducted by the movement and discharge of ions, and they are therefore called **ionic conductors.**

4.2. Electrolysis of melts

The electrochemical reactions which take place at the electrodes may be illustrated by an example of the **electrolysis of a melt.** The products of the electrolysis of molten common salt. NaCl, are metallic sodium and chlorine gas:

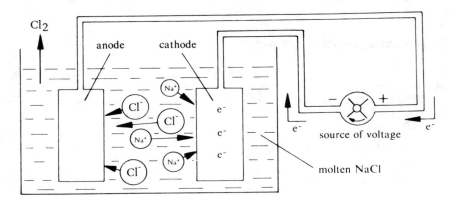

The cathode (— v pole) is negatively charged by an excess of electrons, and the anode (+ v pole) is positively charged because of a deficit of electrons.

The Na^+ cations migrate to the cathode, where they receive electrons from the electric current and become neutral atoms of metallic Na:

$$Na^+ + e^- \longrightarrow Na$$

The Cl^- anions migrate to the anode, where their excess electrons are removed. This produces neutral Cl-atoms, which then unite and form bubbles of Cl_2 gas:

$$2\,Cl^- - 2\,e^- \longrightarrow Cl_2$$

These electrons pass into the anode as an electric current. The voltage source thus acts as a pump for electrons.

4.3. Non-electrolytes

A solution of sugar is **non-electrolyte,** since it contains no ions, but only uncharged molecules.

Chemically pure water conducts electricity extremely poorly, since it is only dissociated to a minute extent

5. Disperse systems, disperse phases, dispersing medium

When substances such as sodium chloride or sugar are disolved in water, either ions (Na^+ Cl^-) or molecules (sugar molecules) are distributed throughout the water. The substance thus distributed is called the **dispersed substance** or the disperse phase. The liquid in which the dispersed substance is distributed is known as a **dispersion medium.** The dispersed substance and the dispersion medium constitute together a **disperse system.** As an example one may quote a dispersion of an artificial polymer, in which insoluble solids or liquids are dispersed in water.

Disperse systems are classified according to the size of the particles of the substance dispersed in the liquid dispersion medium, as follows:

	Diameter of particles	
Coarsely disperse systems	greater than	10^{-4} cm
Colloidal disperse systems	10^{-5} cm to	10^{-6} cm
Molecular or ionic disperse systems	less than	10^{-7} cm

I Fundamentals of general and inorganic chemistry

5.1. Coarsely disperse systems, suspensions, emulsions, emulsifiers

The distribution of an insoluble solid throughout a liquid is called a **suspension,** and the distribution of a liquid throughout another liquid in which it is insoluble is called an **emulsion.** An example of a coarsely disperse suspension is a slurry of chalk in water, and an example of a coarsely disperse emulsion is an emulsion of oil in water, which can only be maintained by shaking or stirring.

In a **coarsely disperse suspension** the dispersed insoluble particles quickly fall to the bottom. Coarsely disperse emulsions separate out. Substances which hinder this separation, and hence stabilize the emulsion, are known as **emulsifiers.**
There exists both water-in-oil and oil-in-water emulsions, according to which of the two substances in the disperse phase, and which the dispersion medium.

5.2. Colloidal disperse systems, colloidal solutions, coagulation, gels, peptization, adsorption, reversible and irreversible colloids, protective colloids, aerosols

If a solid or an insoluble liquid forms part of a **colloidal disperse system,** the solid does not settle out, or sediments only very slowly, and the liquids do not separate into two layers. The particles are so small that they cannot be distinguished with a magnifying glass, although a **colloidal solution** (or sol) is usually cloudy and opalesces, especially when light is incident at an angle (the Tyndall effect). Colloids cannot be separated by simple filtration, for example:

glue, gelatine, starch solution, adhesives,
liquid cosmetics,
white of egg.

The colloidal particles can be made to separate out by boiling or by adding precipitating agents: they then stick together or coagulate.

If coagulation produces a gelatinous mass containing a large amount of water of hydration, then a **gel** has been formed. These gels can be reconverted into sols by soaking in water. This process is called **peptization.**

$$\text{sol} \; \underset{\text{peptization}}{\overset{\text{coagulation}}{\rightleftarrows}} \; \text{gel}$$

Colloids which can be peptized are known as **reversible.**

Gels possess large surface areas, and are therefore able to **adsorb** gases, liquids or solids, ie to hold them fast on the surface.

When gels are used as drying agents their large surface serves to retain moisture. Gels can also be used as carriers for catalysts.

By contrast, **irreversible colloids** are those which cannot be reconverted to sols after coagulation, which produces a permanent increase in particle size (for example, coagulated white of an egg).

Colloids can be stabilizes by the addition of protective **colloids,** which prevent them from coagulating.

There are also colloidal disperse systems in which the dispersion medium is a gas rather than a liquid. For example, suspended solid particles may be present in a gas, as in smoke.

The term mist is used to describe a liquid dispersed in a gas, for example small water droplets in air. The general term for a colloidal dispersion of solids or liquids in a gas is **aerosol,** as in hair sprays.

5.3. True solutions

The dispersion of a substance in a liquid as molecules or ions is termed a **true solution.** The dispersion medium is a solvent fior the substance, as in solutions of sugar or salts.

6. Summary

6.1.   Naturally occurring water always contains dissolved and insoluble matter.

The content of dissolved salts is high in **hard water** and low in **soft water.**

6.2   **Distilled water** and demineralised water are chemically pure water. Both are free from salts

6.3.   **Water is by far the most important solvent.** The solubility of different substances in water varies widely.

6.4.   **Absorption** is the dissolution of gases in liquids. The dissolution of gases is favoured by increase of gas pressure and decrease of the temperature of the solvent.

6.5.   When solids dissolve in water the H_2O-molecules force their way between the molecules or ions, thus separating them and **hydrating** them (ie surrounding them by H_2O-molecules).

6.6.   The separation of the ions of an ionic lattice on dissolving in H_2O is called **electrolytic dissociation.**

6.7.   **Salts** dissociate in aqueous solution into metal ions and the anions of acids.

6.8.   **Hydroxides** dissociate in aqueous solutions into metal **ions** and hdroxide ions.

6.9.   Fused salts and hydroxides also contain freely mobile ions.

6.10.   On dissolving in water **acids** dissociate into H^+-ions and the anions of the acid.

6.11.   The **degree of dissociation** is the ratio of the number of particles which have dissociated to the total number of particles.

6.12.   **Strong acids** are almost 100% dissociated in diluted solution, **weak acids** only to a very small extent.

6.13.   The **pH-value** is a measure of the concentration of H^+-ions in a solution. The practical **pH-scale** extends from 0 to 14. Its meaning is:

pH   below   7:   acidic
pH   =   7:   neutral
pH   above   7:   alkaline

6.14. Aqueous solutions of acids, hydroxides and salts, and also fused solid salts and hydroxides, can conduct electricity. They are termed **electrolytes** or **ionic conductors.**

6.15. On applying an electric potential the ions migrate to the electrodes, where they are discharged. The decomposition of a substance in this way is called **electrolysis.**

6.16. A **non-electrolyte** is a solution of a substance which does not contain mobile ions and therefore cannot conduct electricity.

6.17. The term **disperse system** is used to describe a mixture in which the dispersed substance (or phase) is distributed throughout another substance, the dispersion medium.

6.18. According to the particle size of the substance dispersed in liquid dispersion media the system is classified as coarsely disperse, colloidal disperse, or molecular or ionic disperse.

6.19. A **suspension** consists of an insoluble solid distributed in a liquid.

6.20. An **emulsion** is a dispersion of one insoluble liquid in another.

Substances which hinder the separation of a coarsely disperse emulsion into two layers are termed emulsifiers.

6.21. A sol or **colloidal solution** is a colloidal disperse system.

6.22. **Coagulation** means the precipitation of colloidal dissolved substances from a sol. Coagulated masses which are strongly hydrated are called gels. On account of their large surface area gels can adsorb other substances.

6.23. **Gels** can be reconverted to sols by soaking in water: they are reversible colloids. This process is called peptization.

6.24. Irreversible colloids are those which cannot be reconverted to sols after coagulation.

6.25. The coagulation of a colloid can be hindered by adding a **protective colloid.**

6.26. An **aerosol** is a system in which a solid or a liquid is colloidally dispersed in a gas.

6.27. In a **true solution** a substance is dispersed in a solvent as molecules or ions.

I Fundamentals of general
and inorganic chemistry

7. Self-evaluation questions

100 Why does beer or soda-water taste flat after standing for a long time in an open vessel?

101 Which statements about water are correct?

A Water is a non-polar solvent.
B The hardness of water is caused by magnesium and calcium salts.
C The dissolving power of the water for solids usually decreases with a rise in temperature.
D Distilled water is unsuitable as drinking water.

102 The solubility of gases in water increases

A with increasing gas pressure
B with decreasing gas pressure
C with rising tempertaure
D with falling temperature

103 When an acid affluent is brought to pH 7 by adding an alkali the process is called ____

104 A criterion for the strength of an acid is

A the degree of dissociation
B its corrosive action
C its odour
D its concentration

105 Which series contains exclusively accurate information about the pH value of the liquids?

	pH = 1	pH = 7	pH = 13
A	water	acid	lye
B	lye	water	acid
C	water	lye	acid
D	acid	water	lye

8 Water as a solvent. Electrolytic dissociation, pH values, electrolysis. Disperse systems

106 Which substances are electrically conductive

A all metals
B all non-metal
C aqueous salt solutions
D aqueous acid solutions

107 The process consisting of the migration and discharge of ions termed ____________

108 The decomposition by an electric current of

A molten sodium chloride gives ____________ and ____________

B aluminium oxide dissolved in fused cryolite gives ____________ and ____________

109 Electrolyses are energy-consuming (ie endothermic) reactions. Considerable amounts of energy are consumed in the electrolysis of fused NaCl. This energy must turn up again in the reaction products. How can it be shown that sodium and chlorine contain large amounts of chemical energy?

110 Put crosses in the appropriate columns of the following Table:

Substance	True solution	Colloidal solution	Suspension
A beaten up slurry			
Wine			
Latex			
Sea water			
Stirred up water colours			

9 Appendix

1. Summary of contents

In industry and in science it is often necessary to work with very small or very large numbers. A table of the commonest of these is therefore given. Some of the most frequently used terms and their abbreviations have been appended, together with the Greek alphabet and the first few Greek numerical prefixes. This is followed by a list of the elements with their atomic weights and a shortened version of the periodic system.

Finally, the answers to the questions on the preceding eight chapters are given.

2. Powers of ten

10^0 = 1

10^1 = 10

10^2 = 100 = 10 · 10

10^3 = 1 000 = 10 · 10 · 10

10^6 = 1 000 000 = (1 million)

10^9 = 1 000 000 000 = (1 billion – thousand million)

$10^{-1} = \dfrac{1}{10^1} = \dfrac{1}{10}$

$10^{-2} = \dfrac{1}{10^2} = \dfrac{1}{100}$

$10^{-3} = \dfrac{1}{10^3} = \dfrac{1}{1000}$

etc.

3. Measures of length

1 m = 10^2 cm = 10^3 mm = 10^6 μm = 10^9 nm = 10^{10} Å

10^{-2} m = 1 cm = 10 mm = 10^4 μm = 10^7 nm = 10^8 Å

10^{-3} m = 10^{-1} cm = 1 mm = 10^3 μm = 10^6 nm = 10^7 Å

10^{-6} m = 10^{-4} cm = 10^{-3} mm = 1 μm = 10^3 nm = 10^4 Å

10^{-9} m = 10^{-7} cm = 10^{-6} mm = 10^{-3} μm = 1 nm = 10 Å

10^{-10} m = 10^{-8} cm = 10^{-7} mm = 10^{-4} μm = 10^{-1} nm = 1 Å

μm = micrometre

nm = nanometre

Å = Angström unit

I Fundamentals of general and inorganic chemistry

4. Common abbreviations and symbols

Kilo-	=	thousand (km, kg, kJ)
Mega-	=	million (MW = megawatts)
Giga-	=	billion
%	=	per cent = parts per 100 parts
ppm	=	parts per million parts
ppb	=	parts per billion parts
1 Mass.ppm	=	1 mg per kg, or 1 oz per 1 million oz, etc.
1 Vol.ppm	=	$1\ cm^3$ per m^3, or 1 pint per 1 million pints, etc.
<	=	less than
>	=	greater than
≙	=	equivalent to

5. The Greek alphabet

Alpha	A	α	a	Nu	N	ν	n
Beta	B	β	b	Xi	Ξ	ξ	x
Gamma	Γ	γ	g	Omikron	O	o	o
Delta	Δ	δ	d	Pi	Π	π	p
Epsilon	E	ε	e	Rho	P	ϱ	r
Zeta	Z	ζ	z	Sigma	Σ	σ, ς	s
Eta	H	η	ē	Tau	T	τ	t
Theta	Θ	ϑ	th	Upsilon	Y	υ	y
Iota	I	ι	i	Phi	Φ	φ	ph
Kappa	K	$\varkappa$	k	Chi	X	χ	ch
Lambda	Λ	λ	l	Psi	Ψ	ψ	ps
Mu	M	μ	m	Omega	Ω	ω	o

6. The Greek numerals

mono	1	hexa	6
di	2	hepta	7
tri	3	octa	8
tetra	4	nona	9
penta	5	deka	10

7. List of the elements with their atomic masses

The atomic masses are given relative to $^{12}C = 12$ (1970 values)

Element	Symbol	AN*	Atomic mass	Element	Symbol	AN*	Atomic mass
Actinium	Ac	89	227	Mendelevium	Md	101	256
Aluminium	Al	13	26.9815	Mercury	Hg	80	200.59
Americium	Am	95	243	Molybdenum	Mo	42	95.94
Antimony	Sb	51	121.75	Neodymium	Nd	60	144.24
Argon	Ar	18	39.948	Neon	Ne	10	20.179
Arsenic	As	33	74.9216	Neptunium	Np	93	237.0482
Astatine	At	85	210	Nickel	Ni	28	58.71
Barium	Ba	56	137.34	Niobium	Nb	41	92.9064
Berkelium	Bk	97	247	Nitrogen	N	7	14.0067
Beryllium	Be	4	9.01218	Nobelium	Mo	102	255
Bismuth	Bi	83	208.9806	Osmium	Os	76	190.2
Boron	B	5	10.811	Oxygen	O	8	15.9994
Bromine	Br	35	79.904	Palladium	Pd	46	106.4
Cadmium	Cd	48	112.40	Phosphorus	P	15	30.9738
Caesium	Cs	55	132.9055	Platinum	Pt	78	195.09
Calcium	Ca	20	40.08	Plutonium	Pu	94	242
Californium	Cf	98	251	Polonium	Po	84	209
Carbon	C	6	12.01115	Potassium	K	19	39.102
Cerium	Ce	58	140.12	Praseodymium	Pr	59	140.9077
Chlorine	Cl	17	35.453	Promethium	Pm	61	145
Chromium	Cr	24	51.996	Protactinium	Pa	91	231.0359
Cobalt	Co	27	58.9332	Radium	Ra	88	226.0254
Copper	Cu	29	63.546	Radon	Rn	86	222
Curium	Cm	96	247	Rhenium	Re	75	186.2
Dysprosium	Dy	66	162.50	Rhodium	Rh	45	102.9055
Einsteinium	Es	99	254	Rubidium	Rb	37	85.4678
Erbium	Er	68	167.26	Ruthenium	Ru	44	101.07
Europium	Eu	63	151.96	Samarium	Sm	62	150.35
Fermium	Fm	100	253	Scandium	Sc	21	44.9559
Fluorine	F	9	18.9984	Selenium	Se	34	78.96
Francium	Fr	87	223	Silicon	Si	14	28.086
Gadolinium	Gd	64	157.25	Silver	Ag	47	107.868
Gallium	Ga	31	69.72	Sodium	Na	11	22.9898
Germanium	Ge	32	72.59	Strontium	Sr	38	87.62
Gold	Au	79	196.9665	Sulphur	S	16	32.064
Hafnium	Ht	72	178.49	Tantalum	Ta	73	180.9479
Helium	He	2	4.00260	Technetium	Tc	43	98.9062
Holmium	Ho	67	164.9303	Tellurium	Te	52	127.60
Hydrogen	H	1	1.00797	Terbium	Tb	65	158.9254
Indium	In	49	114.82	Thallium	Tl	81	204.37
Iodine	I	53	126.9045	Thorium	Th	90	232.0381
Iren	Fe	26	55.847	Thulium	Tm	69	168.9342
Iridium	Ir	77	192.22	Tin	Sn	50	118.69
Krypton	Kr	36	83.80	Titanium	Ti	22	47.90
Kurtschatovium	Ku	104	261	Tungsten	W	74	183.85
Lanthanum	La	57	138.9055	Uranium	U	92	238.024
Lawrencium	Lw	103	257	Vanadium	V	23	50.9414
Lead	Pb	82	207.19	Xenon	Xe	54	131.30
Lithium	Li	3	6.941	Ytterbium	Yb	70	173.04
Lutetium	Lu	71	174.97	Yttrium	Y	39	88.9059
Magnesium	Mg	12	24.305	Zinc	Zn	30	65.37
Manganese	Mn	25	54.9380	Zirconium	Zt	40	91.22

*AN atomic numbers

8. Shortened periodic table
(Atomic masses rounded off)

Main groups	I	II	III	IV	V	VI	VII	VIII
1st period	$^{1}_{1}$H							$^{4}_{2}$He
2nd period	$^{7}_{3}$Li	$^{9}_{4}$Be	$^{11}_{5}$B	$^{12}_{6}$C	$^{14}_{7}$N	$^{16}_{8}$O	$^{19}_{9}$F	$^{20}_{10}$Ne
3rd period	$^{23}_{11}$Na	$^{24}_{12}$Mg	$^{27}_{13}$Al	$^{28}_{14}$Si	$^{31}_{15}$P	$^{32}_{16}$S	$^{35.5}_{17}$Cl	$^{40}_{18}$Ar
4th period	$^{39}_{19}$K	$^{40}_{20}$Ca	$^{70}_{31}$Ga	$^{73}_{32}$Ge	$^{75}_{33}$As	$^{79}_{34}$Se	$^{80}_{35}$Br	$^{84}_{36}$Kr
5th period	$^{85.5}_{37}$Rb	$^{88}_{38}$Sr	$^{115}_{49}$In	$^{119}_{50}$Sn	$^{122}_{51}$Sb	$^{128}_{52}$Te	$^{127}_{53}$I	$^{131}_{54}$Xe
6th period	$^{133}_{55}$Cs	$^{137}_{56}$Ba	$^{204}_{81}$Tl	$^{207}_{82}$Pb	$^{209}_{83}$Bi	$^{210}_{84}$Po	$^{210}_{85}$At	$^{222}_{86}$Rn
7th period	$^{223}_{87}$Fr	$^{226}_{88}$Ra						

10 Answers to the Self-evaluation Questions on Sections 1–8

1   A No
    B Yes

2   A No
    B No

3   Since heat energy must be used to melt ice and to heat and vaporize water, water vapour must have a much greater energy content than ice.

4   A Condensation
    B Sublimation

5   A, C, D

6   B, C

7   A, C

8   C

9   A, C

10 C

11 B, C

12 C, D

---

13 A, D

14 A, C

15 Pure substances: common salt, sugar, mercury.
    Mixtures: wine, butter, air, synthetic resin varnish.

16 A, B, C, D

17 Homogeneous: coal, glass, gasoline, jelly.
    Heterogeneous: jam, brick.

18 A By decanting or ladling off.
    B The water is separated from the salt by evaporation.
    C By filtration, for example through thick layers of gravel.

19 B, D

20 C

21 D

22 B, D

23 A

24 Elements: phosphorus, nickel, lead.
    Compounds: sugar, sand, water.

25 A

26 A

27 B, D

28 B

29 Synthesis

30 Analysis

31   A   Electrical energy   ⟶   radiant energy and heat energy

     B   Chemical energy   ⟶   heat energy and radiant energy (flames)

     C   Kinetic energy of water   ⟶   electrical energy

     D   Chemical energy   ⟶   kinetic energy (pressure and shock waves)

                                         radiant energy (explosive flash)

                                         and heat energy

32   A Exothermic
    B Endothermic (the energy of the sun is collected and stored)
    C Exothermic (the solar energy stored in the glucose is converted into thermal and kinetic energy).

33   A, B, D   34   B, C, D

35   A, D   36   A, B

37   B   38   C

---

39   $Hg = 201$
    $+ O = 16$
    $HgO = 217$
    One mole of HgO weights 217 g

40   $C = 12$
    $O = 16$
    $2 N = 28$
    $4 H = 4$
    ———
        60

    60 g urea thus contains 28 g nitrogen
    100 g urea contains $28 \times 100/60 = 46.7$ g nitrogen
    Urea contains 46.7% N.

41   Air contains 21 vol. % of oxygen, ie 1 m^3 (1000 litres) contains 210 l oxygen. At normal temperature and pressure 22.4 l oxygen = 1 mole O_2
    210 l O_2 thus contain $210/22.4 = 9.4$ moles O_2

42   $Fe + S \longrightarrow FeS$

43 Fe + S ⟶ FeS

56 g + 32 g ⟶ 88 g

64 g S = 2 × 32 g S require 2 × 56 g Fe = 112 g Fe, and 176 g FeS are produced

44 H_2 + Cl_2 ⟶ 2 HCl

1 mol + 1 mol ⟶ 2 mol

22.4 l   22.4 l   so that the ratio of volumes is 1:1

2 × 1 g   2 × 35.5 g   so that the ratio of weights is 1:35.5

45 A   46. C, D

47 C   48. A

49 B   50. A

51 A, C   52. B, C

53 C, D   54. A

55 D

56 Protons, electrons

57

Symbol	Number of protons	Atomic weight	Number of neutrons
$_1^1 H$	1	1	0
$_{11}^{23} Na$	11	23	12
$_6^{12} C$	6	12	6
$_8^{16} O$	8	16	8
$_{92}^{238} U$	92	238	146

58 A from ions
    B from molecules

59 A, B

60 A

61 A, D

62 B, D

63 A

64 A, C, D

65 A, C

66 B

67 C, D

68 B

69 A, B, C, D

70 A

71 B, D

72 B, D

73 A, C

---

74 A, B, C, D

75 C

76 A, B, C

77 A

78 C

79 A Iron is oxidized to iron (III) oxide
    B Methane (CH_4) is oxidized. The products of oxidation are carbon dioxide and water vapour.

80 A Hydrogen is the reducing agent
    B Carbon is the reducing agent
    C Magnesium is the reducing agent since these elements remove oxygen.

81 A, C, D

82 A, C

83 A, B, C

84 A, B, C

85 B

86 A, C, D

87 B, D

88 A, B

89 A, B

90

Formula	Acid	Anion of acid	Acid anhydride	Alkali	Salt
HCl	x				
SO_3			x		
$SO_3{}^{2-}$		x			
$SO_4{}^{2-}$		x			
K_2SO_4					x
KOH				x	
$Ca(OH)_2$				x	
NH_4Cl					x
H_3PO_4	x				
CO_2			x		
Na_2CO_3					x
Cl^-		x			
HNO_3	x				
NaOH				x	

91   C, D   92   A

93   C   94   C, D

95   A, B, C, D   96   B, C

97   A, B, D   98   A, B, D

99   A, D

100   Soda water and beer contain dissolved carbon dioxide, CO_2, which gives the beverage a slightly acid taste. In an open vessel the beverage is no longer under pressure, and CO_2 is evolved, especially if the beverage is warm. This affects the flavour.

101   B, D   102   A, D

103   Neutralization   104   A

105   D   106   A, C, D

107   Electrolysis.

108   A Sodium and chlorine
       B Aluminium and oxygen.

109   Because of their high reactivity

110

Substance	True solution	Colloidal solution	Suspension
A beaten up slurry		(x)	x
Wine	x		
Latex		x	
Sea water	x		
Stirred up water colours			x

II
Inorganic Raw Materials and Large-Scale Products: Large-Scale Industrial Processes

1 Water and Air.
Fractionation of Air

1. Water

1.1.   Occurrence

Water is one of the main constituents of our animate and inanimate surroundings.

It covers 4/5 of the earth's surface and about 70% of the human body consists of water. Fruits and vegetables often contain more than 90% of water.

The atmosphere can contain up to 4 vol.% of water in the form of water vapour, and changes of pressure or temperature can precipitate this water as fog, clouds, rain, hoar frost, snow or hail.

Minerals often contain chemically bound water of crystallization.

1.1.1.   Chemical and physical properties

Water, H_2O, contains 88.9% oxygen and 11.1% hydrogen by weight. It is formed by the combustion of hydrogen or compounds containing hydrogen (eg natural gas, gasoline) in an atmosphere containing oxygen.

	hydrogen	oxygen		water	
	$2\,H_2$	$+\quad O_2$	$\longrightarrow$	$2\,H_2O$	$;\Delta H = -571.5\ kJ/mol*$
	4 g	32 g		36 g	

Water is also formed in neutralization and other chemical reactions.

Neutralization and the combustion of hydrogen are exothermic reactions, ie energy is liberated. This is related to the fact that the product, water, is a low-energy compound.

Low-energy compounds are chemically stable. Conversely, energy must be supplied in order to decompose water into its elements, hydrogen and oxygen.

Because of their high energy content hydrogen and oxygen are very reactive.

Pure water is an odourless, tasteless transparent liquid, which exhibits a bluish tinge in thick layers.

---

*The value of ΔH refers to the numbers of moles represented in the chemical equation.

II Inorganic raw materials and large-scale products: Large-scale industrial processes

Physical properties:

Freezing point:	$0°C$ ($32°F$)
Boiling point:	$100°C$ ($212°F$)
Heat of fusion:	334.96 kJ/kg (700.0 Btu/lb)
Heat of vaporization:	2256.8 kJ/kg (4716.7 Btu/lb)
Maximum density (at $4°C$ or $39.2°F$):	1 kg/cm³ (57.33 lb/ft³)
Increase in volume when water freezes to ice at $0°C$ ($32°F$) ie ice has a smaller density than liquid water	9%

1.1.2. Physical constants

The mass of 1 cm³ of water at $4°C$ ($39.2°F$) is by definition termed 1 gram (g). In the Celsius (or Centigrade) scale of temperature the boiling point and freezing point of water are $0°C$ and $100°C$, respectively. In the Fahrenheit temperature scale, water boils at $212°F$ and freezes at $32°F$. These temperatures are at normal pressure, ie 1013 mbar or 14.5 lb/in².

Quantities of heat can be defined in terms of the properties of water: 1 kJ is the amount of heat required to heat 0.239 kg (0.527 lb) H_2O from 15 to $16°C$ (59 to $60.8°F$), and is equal to 0.948 Btu.

1.2. Water as a raw material

The manufacture of water is unnecessary, since water occurs in sufficient quantities in nature and can be treated so as to make it suitable for use. The quantity of water available is conveniently expressed in the units used in meteorology, ie as mm of precipitation.

> 1 mm of precipitation corresponds to 1 litre per m²
> 1 inch of precipitation corresponds to 27,152 gals/acre

The average annual precipitation in Western Europe is about 800 mm (31.5 in) distributed as follows:

280 mm (11.0 in)	rivers
400 mm (15.7 in)	evaporation
120 mm (4.8 in)	groundwater

10 mm (0.4 in) (mostly groundwater) is used for domestic purposes. 40 mm (1.6 in) (more than half of which is from rivers) is used by industry and agriculture. The general balance sheet for water thus looks quite favourable, although difficulties may arise in densely populated areas. The daily consumption of water per head was only about 25 l (6.6 gals) in 1900, but today amounts to 200 to 300 l (53 to 79 gals) in industrialized countries.

1.2.1. The water cycle

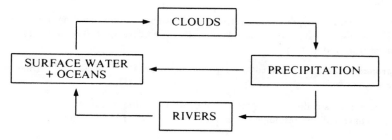

1.2.2. Types of water

Rain water: this is the purest naturally occurring water, since it has passed through a natural process of distillation. It contains dissolved gases and particles of dust.
Spring and river water: the content of solids (0.01 to 0.2%) consists mainly of calcium and magnesium which determine the hardness of the water. In industrialized countries surface water (rivers and lakes) is polluted by the products of civilization. For example, at present about 10% of the water in the Rhine consists of effluents.
Mineral waters: these consist of spring water containing considerable quantities of gases or salts. They are sometimes hotter than ordinary water.
Sea water: this contains about 3.5% of salts, of which 3% is common salt. The remainder consists of compounds of about 50 different elements.
Drinking water: spring water is the best for this purpose. Failing this purified surface or river water is used. Dissolved minerals improve the taste of drinking water.

1.3. Treatment of water for industrial use

The three largest industrial consumers of water are:
1. chemical industry;
2. the iron and smelting industries;
3. the paper industry.

Naturally occurring water which is to be used in steam raising boilers must undergo the following treatment:
1. precipitation of suspended and colloidal matter;
2. softening and desalination;
3. removal of oxygen, carbon dioxide and other chemically reactive gases.

1.3.1. Distillation

The purification of water by distillation is now used only for pharmaceutical and medicinal purposes, if sterile water free from destroyed germs and bacteria is requested. Before being distilled the water will be demineralized.

1.3.2. Softening

Hardness caused by the presence of calcium, magnesium or iron salts often interferes with the industrial use of water. Methods have therefore been developed to remove these substances from water.

The hardness of water is classified as temporary or carbonate hardness, and permanent hardness.

Carbonate hardness is due to the presence of dissolved hydrogen carbonates (bicarbonates), which are converted into insoluble carbonates when the water is boiled and are deposited as boiler scale.

$$Ca(H\,CO_3)_2 \xrightarrow{\text{boiling}} CaCO_3 \downarrow + H_2O + CO_2 \uparrow$$

calcium hydrogen   calcium
carbonate   carbonate

Permanent hardness which can not be destroyed by boiling is due mainly to the content of calcium sulphate, $CaSO_4$. The sum of the carbonate hardness and the permanent hardness gives the total hardness, expressed in degrees of hardness.

9.6 mg of calcium oxide (CaO) per litre (or 0.567 grain/gal) of water (or the equivalent amount of other salts causing hardness) corresponds to 1 degree of hardness on the US scale; $1°$ Ah = 9.6 mg/l CaO (0.567 grain/gal).
8.0 mg of calcium oxide per litre (or 0.472 grain/gal) of water corresponds to 1 degree of hardness on the English scale; $1°$ Eh = 8.0 mg/1 CaO.
10 mg of calcium oxide per litre (or 0.5904 grain/gal) of water corresponds to 1 degree of hardness on the German scale; $1°$ Gh = 10 mg/1 CaO.

The industrial softening of water is carried out to an increasing extent by means of ion exchangers, which involve the lowest expenditure and are simple to service. However, in order to increase their capacity a preliminary treatment to remove carbonate is usually carried out. The water, containing bicarbonate, is mixed with milk of lime in a rapid reactor, and is then passed over a layer of granules of calcium carbonate with diameters between 1 and 3 mm (0.039 and 0.12 in). The precipitated calcium carbonate adheres to the granules (see Fig. 1, p. 7).

Water flows continuously through the reaction bed, thus keeping the granules in a state of suspension. The heavier calcium carbonate granules sink to the bottom and can be continuously removed. The time required for the removal of dissolved carbonate is five to ten minutes.

Figure 1. Removal of carbonate by the contact method

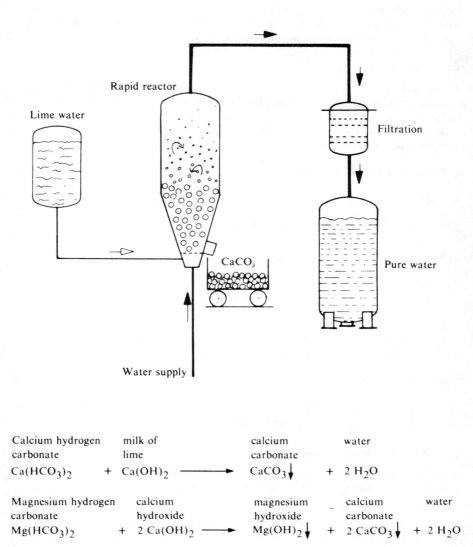

Calcium hydrogen carbonate	milk of lime		calcium carbonate	water

$$Ca(HCO_3)_2 \quad + \quad Ca(OH)_2 \quad \longrightarrow \quad CaCO_3\downarrow \quad + \quad 2\,H_2O$$

Magnesium hydrogen carbonate	calcium hydroxide		magnesium hydroxide	calcium carbonate	water

$$Mg(HCO_3)_2 \quad + \quad 2\,Ca(OH)_2 \quad \longrightarrow \quad Mg(OH)_2\downarrow \quad + \quad 2\,CaCO_3\downarrow \quad + \quad 2\,H_2O$$

Water softened by this method is suitable for cooling, since the substances which form scale have been removed.

Softening by means of polyphosphates

In the manufacture of washing agents polyphosphates are added in order to soften the water. Polyphosphates combine with Ca^{2+} and Mg^{2+} ions to form complex compounds which remain in solution and are not precipitated by heating or by detergents.

1.3.3.   Desalination

The complete desalination (demineralization) of water is effected by means of special synthetic resins which act as ion exchangers. The water is passed successively through a cation exchanger and an anion exchanger. The cations (eg Ca^{2+} and Mg^{2+}) are replaced by H^+-ions, and the anions (eg HCO_3^- and SO_4^{2-}) by OH^--ions. The H^+ and OH^- ions then combine to form neutral water.

1. Reactions on a cation exchanger

Acid form of resin	calcium ions	calcium form of resin	hydrogen ions
$2\,H - resin$ $+$ Ca^{2+}		$\longrightarrow$ $Ca(resin)_2$	$+\ 2\,H^+$

2. Reactions on an anion exchanger

Basic form of resin	sulphate ions	sulphate form of resin	hydroxide ions
$2\,resin - OH$ $+$ SO_4^{2-}		$\longrightarrow$ $(resin)_2SO_4$	$+\ 2\,OH^-$

The hydrogen ions liberated on the cation exchanger combine with the hydroxide ions liberated on the anion exchanger to give undissociated water,

$$2\,H^+ \quad + \quad 2\,OH^- \quad \longrightarrow \quad 2\,H_2O$$

The acidic ion exchanger can be regenerated with hydrochloric acid, and the basic one with caustic soda, so that they can be re-used repeatedly.

1.4.   Water in a chemical works

Water is an indispensable material for chemical industry, more as a solvent and a heat-transfer agent in heating and cooling than as a reactant (see II/1 page 18).

The supply of water is therefore a problem, and also the disposal of effluents. The water used in a chemical process is often recycled so as to save the expense of treating natural water and to avoid special problems in the purification of effluents. The manufacture of 1 t of high-quality paper requires about 400 t of water. By recycling the water through a purification plant the water consumption can be reduced by more than 50%. Recycling of cooling water is obviously appropriate, since apart from an increase in temperature the cooling water undergoes no change. The water is therefore re-cooled by running it down cooling towers. Because of the large quantities of dirty water carried by rivers it is accepted as self-evident that industries should purify their effluents. The decomposition of organic impurities is effected by selected micro-organisms. This biological purification of effluents is often assisted by blowing in air and adding nutrients for the bacteria.

Various factors can retard the biological degradation of organic substances, or bring it to a stands'ill, thus leading to contamination of the water, as follows:

1. Too low an oxygen content, or the complete removal of oxygen (remedied by bubbling with air)

2. Poisoning by non-biodegradable substances.

For example, the following quantities are sufficient to kill fish:

0.1 ppm CN^-	(cyanide ions)
0.2 ppm Cl_2	(chlorine)
1 ppm SO_2	(sulphur dioxide)
0.5 ppm Cu^{2+}	(copper ions)

1 ppm = 1 part per million, ie 1 mg in 1 litre of water, or 1 lb in 1 000 000 lb water.

NH_3 acts as a poison in slightly alkaline solutions (pH 8). Detergents which are not biodegradable can destroy the gill membranes of fish. Too high a salt concentration is harmful. The contamination of water by oil represents a further danger.

1.5.   Manufacture of hydrogen from water

1.5.1.   The water gas process

In order to prepare hydrogen from water the oxygen must be removed by a reducing agent, for example carbon,

Water vapour	carbon		hydrogen	carbon monoxide
H_2O	+ C	$\xrightarrow{1000°C}$	H_2	+ CO

The mixture of gases thus produced is called water gas: it can be freed from carbon monoxide by further chemical reaction and subsequent absorption.

However, the importance of this process is steadily decreasing, since it is possible today to produce hydrogen with a lower energy consumption from petrochemical raw materials as a by-product in cracking and re-forming (platforming) — see IV,1.

1.5.2. Electrolysis of water

A simple method of decomposing water into its constituents is electrolysis, ie by supplying electrical energy.

$$2 H_2O \quad + \quad \text{electrical energy} \quad \longrightarrow \quad 2 H_2 \quad + \quad O_2$$

Decomposition by electrolysis is thus an endothermic (ie energy-consuming) process.

This method of obtaining hydrogen is more expensive than other methods, for example the cracking of hydrocarbons from the distillation of petroleum to obtain hydrogen for the ammonia synthesis.

It is also cheaper to manufacture oxygen by the fractionization of air than by the electrolysis of water. The latter process is therefore used only when specially cheap electrical energy is available.

The largest amounts of hydrogen are used in the ammonia synthesis and for hydrogenating unsaturated organic compounds (for example for the refining of fats).

Hydrogen is also an important reducing agent in metallurgy and the processing of metals: it is also used for cooling dynamos and engines.

Despite these facts that have been mentioned the decomposition of water to produce hydrogen as a source of primary energy has been predicted to have a bright future. That will be true if economically feasible processes can be developed to decompose water into hydrogen and oxygen.

At present three processes are being discussed:

1. The action of solar radiation in solar cells.
2. The biotechnological decomposition of water with the help of microorganisms.
3. In the long term the utilization of the energy of nuclear fusion.

2. Air and its fractionation

2.1. Composition

Air is a mixture of gaseous elements. The average composition of pure dry air is as follows:

Nitrogen	N_2	78 vol.%
Oxygen	O_2	21 vol.%
Rare gas (mainly argon)		1 vol.%
Carbon dioxide	CO_2	0.03 vol.%

In addition the atmosphere contains water vapour (up to 4 vol.%) and, in the lower layers, solid dust particles.

2.1.1. The oxygen cycle

In spite of the many natural processes which produce or consume oxygen, nitrogen and carbon dioxide, the composition of air remains practically constant.
However, analyses have shown that the carbon dioxide content of the air has increased over the last decades owing to the use of fossil fuels as a source of energy. This is due to the existence of cyclic processes involving oxygen (coupled with carbon dioxide) and nitrogen.

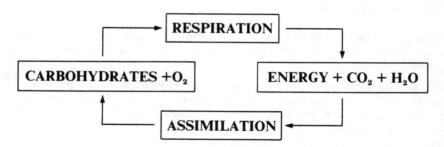

The processes which consume oxygen (animal respiration and decay) convert carbohydrates into CO_2, H_2O and energy. This is the reverse of the assimilation process of plants, in which the sunlight absorbed by the green leaf pigment (chlorophyll) converts carbon dioxide and water into carbohydrates (sugar and starch) which are stored in the plant. At the same time oxygen is produced.

Solar energy		carbon dioxide	water	carbohydrates	oxygen
Q	+	$6\,CO_2$	$+\ 6\,H_2O \longrightarrow$	$C_6H_{12}O_6$	$+\quad 6\,O_2$

$$\Delta H = +\ 2820\ kJ/mol$$

II Inorganic raw materials and large-scale products: Large-scale industrial processes

Plants serve as a source of energy for human beings and animals. In the body they react with the inhaled oxygen.

A parallel cycle involving nitrogen also takes place. Plants have the special ability to convert inorganic nitrogen compounds (eg nitrates) into organic nitrogen compounds, the plant proteins. Plants of the Leguminosae family (peas, beans, lupins, etc) have on their roots nodular bacteria which can convert even atmospheric nitrogen into useful proteins. When organic nitrogen compounds decompose or decay the nitrogen once more returns to the atmosphere as the element.

2.1.2. Physical properties

Weight of 1 litre of air		1.293 g/l (0.0807 lb/ft^3) at 1013 mbar (14.5 lb/in^2) and 0°C (32°F)
Oxygen:	Boiling point	−183°C (297°F)
	Freezing point	−218.9°C (362°F)
Nitrogen:	Boiling point	−195.8°C (320°F)
	Freezing point	−210.5°C (347°F)
Argon:	Boiling point	−186°C (303°F)
	Freezing point	−189°C (308°F)

All the values refer to a pressure of 1013 mbar, or 14.5 lb/in^2.

2.2. Air as a raw material

Air is the initial material for the production of pure oxygen and nitrogen. The rare gases of the atmosphere (helium, neon, argon, krypton and xenon) are also separated and purified by special processes.

Air in its natural state is also used industrially for a number of technological purposes. It is used for the transport of energy as a heating and cooling agent. It is also used for the transport of solids (for example blowing grain into a silo) and liquids (compressed air), for the separation of mixtures of solids (cyclones, impact separators), for drying moist materials, and for spraying liquids, dispersions and solids.

Atmospheric oxygen is necessary for all processes of combustion, and atmospheric nitrogen is used in chemical reactions such as the synthesis of ammonia.

2.2.1. Air pollution

The air may contain substances which are harmful to the environment, arising particularly from combustion processes. Thus the burning of coal, heating oil and gasoline produces sulphur dioxide (SO_2) and carbon monoxide (CO). The sulphur dioxide comes from sulphur compounds occurring as impurities in petroleum. The combustion of fuels in engines produces not only sulphur dioxide and carbon monoxide, but also nitrogen dioxide (NO_2) and lead compounds, the latter arising from the antiknock agent lead tetraethyl.

The most important atmospheric pollutants are:

a. sulphur dioxide;
b. carbon monoxide;
c. nitrogen oxides;
d. hydrocarbons;
e. solid dust particles.

In operations involving substances dangerous to health workers may be exposed locally to appreciable concentrations of these substances. Therefore checks are carried out in the chemical industry to ensure that the maximum permissible concentrations are not exceeded.

The maximum permissible concentration represents the concentration of a gas, vapour or dust which according to present knowledge will not affect the health of an employee who is exposed to it over long periods, assuming an 8-hour day and a working week of 45 hours.

In the chemical industry stringent controls ensure that these limits are not exceeded.

2.3.   Liquefaction and fractionation of air

2.3.1.   Physical principles

The principle of all methods for fractionating air involves essentially two consecutive processes:

1. the liquefaction of air,
2. the fractionation of liquid air by distillation.

1. The liquefaction of air depends on the Joule-Thomson effect, according to which compressed air is cooled when the pressure is released. This effect is used in all plants for the liquefaction of air.

2. The separation of air into its constituents, oxygen, nitrogen and the rare gases, depends on differences between their boiling points. The boiling point of nitrogen is lower than that of oxygen, while the rare gas argon has an intermediate boiling point.

2.3.2.   Industrial practice

Industrial processes for the fractionation of air can be divided into three steps:

1. compression;
2. liquefaction;
3. fractionation.

1. Compression

Air sucked in from the atmosphere is compressed to the required extent by a compressor. The heat of compression thus produced is completely removed by the circulation of cooling water.

2. Liquefaction by the Linde process

The air which has been compressed and cooled to the temperature of the cooling water is cooled further in a heat exchanger by a counter-current of cold air. It is then allowed to expand through a flow control valve. On account of the Joule-Thomson effect part of the air is cooled below the condensation point and is liquefied. The liquefied portion is collected in a vessel.

The unliquefied portion of the air is used for pre-cooling the compressed air (see Fig. 2, p. 15).

3. Fractionation

The liquefaction plant is connected directly to the fractionation plant, which consists essentially of distilling columns.

By repeated distillation of the liquid air the separate components, oxygen, nitrogen and the rare gases, can be obtained in the required state of purity.

When liquid air evaporates, the vapour rising in the fractionating column is always poorer in oxygen than the liquid remaining in the still, since oxygen has the highest boiling point of all the components and is therefore the least volatile.

In the fractionating column repeated condensation and evaporation leads to a progressive enrichment of nitrogen in the vapour phase and a concentration of oxygen in the liquid phase.

The purity of the residual oxygen can be varied from 60% to 99.8%, as required. Similarly, nitrogen of different degrees of purity can be obtained. Very pure nitrogen contains only a few parts per million of oxygen.

2.4. The manufacture and distribution of oxygen

In order to meet the increasing demand for pure liquid oxygen, additional installations for fractionating air are under construction or are planned. Large plants with a daily production of 1400 tons (t)* oxygen are not uncommon today. These plants are mainly set up in the neighbourhood of the industries consuming their products, for example in the Ruhr district of the Saar, where smelting and steel production are situated. In these large installations it is worth while collecting and separating the rare gases in adjacent special equipment. For example, the daily production of a helium liquefier can amount to 1200 litres (42.4 ft³). Temperatures down to $-253°C$ ($-423°F$) can be reached in these liquefiers.

*Throughout this book, ton indicates either metric or long.

Figure 2. Linde process for liquefying air

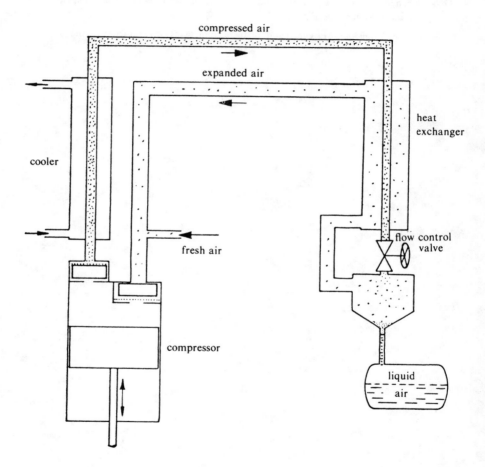

II Inorganic raw materials and large-scale products: Large-scale industrial processes

In the smelting industry the distribution of oxygen by pipelines is widely developed. The efficiently planned construction of pipelines has led to a network which ensures adequate supplies of oxygen to the chemical industry and to steel works. In particular the centres of production and consumption in the area Cologne — Krefeld — Duisburg — Oberhausen — Gelsenkirchen — Bochum — Dortmund are to a large extent connected by pipelines for both oxygen and nitrogen (see Fig. 3, p. 17).

A 35 km (22 mile) pipeline for oxygen and nitrogen connects Hürth and Dormagen.

Figure 3.   Oxygen network in the Ruhr area/West-Germany

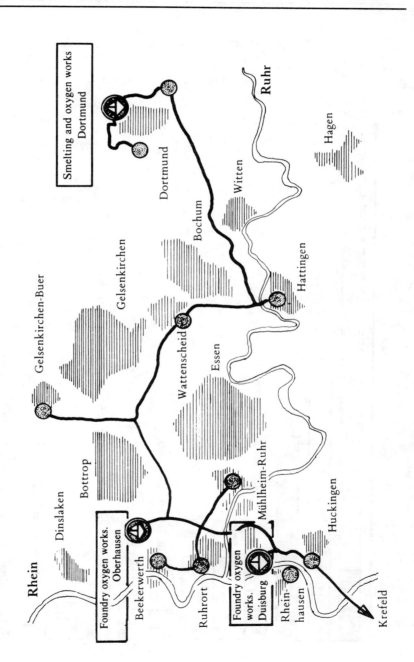

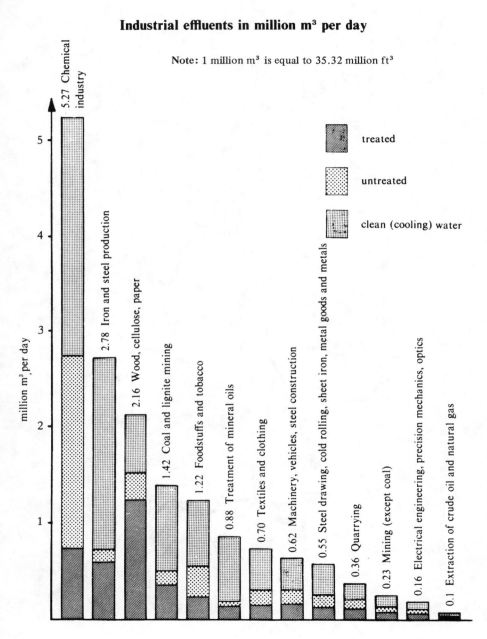

Industrial effluents in million m³ per day

Note: 1 million m³ is equal to 35.32 million ft³

treated

untreated

clean (cooling) water

million m³ per day

5.27 Chemical industry

2.78 Iron and steel production

2.16 Wood, cellulose, paper

1.42 Coal and lignite mining

1.22 Foodstuffs and tobacco

0.88 Treatment of mineral oils

0.70 Textiles and clothing

0.62 Machinery, vehicles, steel construction

0.55 Steel drawing, cold rolling, sheet iron, metal goods and metals

0.36 Quarrying

0.23 Mining (except coal)

0.16 Electrical engineering, precision mechanics, optics

0.1 Extraction of crude oil and natural gas

3. Self-evaluation questions

1 Which statements about water hardness are correct?

A   Industrial decarbonization serves to eliminate temporary hardness.
B   The ability to demineralize water depends on the type of ion exchangers.
C   Soft water is water from which all salts have been removed.
D   In the detergent industry, polysulphates are used to soften the water.

2 Which statements about the production of hydrogen from water are correct?

A   The removal of oxygen from H_2O molecules is an endothermic process.
B   The electrolysis of water is a relatively expensive way of producing hydrogen compared with other methods.
C   The vast majority of hydrogen produced in the chemical industry is obtained from water.
D   Obtaining hydrogen from water is called hydrogenation.

3 How must water be treated in order to meet medical requirements?

A   It must be demineralized with cation exchangers.
B   It must be demineralized with anion exchangers.
C   It must be demineralized with both cation and anion exchangers.
D   It must be distilled repeatedly.

4 Which two properties of water are unusual if compared with the properties of other liquid substances?

A   When water evaporates it forms invisible water vapour.
B   When water solidifies to form ice on cooling, the volume increases.
C   If water is heated from $0°C$ ($32°F$), the volume first decreases until the temperature reaches $4°C$ ($39.2°F$).
D   If water is subjected to increased air pressure, it does not boil until above $100°C$.

5 Which group contains only substances obtained from air as a raw material?

A   Hydrogen — oxygen — nitrogen
B   Rare gases — carbon dioxide
C   Nitrogen — rare gases — oxygen
D   Rare gases — oxygen

6 At which point in the drawing is the temperature at its lowest value?

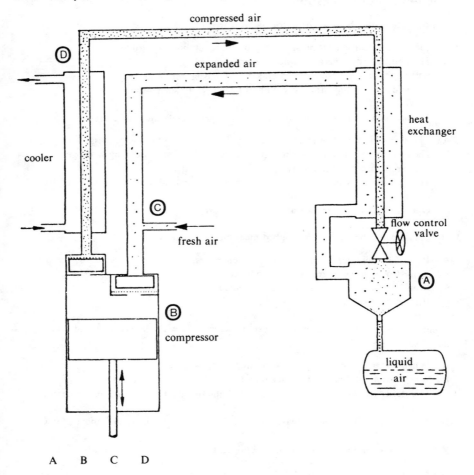

A   B   C   D

7 The MAC value provides information on

A   the maximum working time in a workplace environment inimical to the health
B   the mean concentration of air constituents inimical to the health at the workplace
C   the upper limit of the permissible concentration of substances inimical to the health
     at the workplace with eight working hours per day.

2   Alkali–Chlorine Electrolysis

II Inorganic raw materials and large-scale products: Large-scale industrial processes

1. Raw materials and products of alkali-chlorine electrolysis

1.1 Alkali metal chlorides

Chlorine is a high-energy element and is therefore very reactive. For this reason it does not occur naturally as the element, but only in combination with other elements. By far the most abundant chlorine compounds are the two salts sodium chloride (rock salt), NaCl, and potassium chloride (sylvine), KCl.

The electrolysis of the alkali metal chlorides requires three **raw materials**:
1. rock salt, NaCl (or sylvine, KCl);
2. water, H_2O;
3. direct electric current.

Three **products** are obtained:
1. chlorine, Cl_2;
2. a solution of caustic soda, NaOH (or caustic potash, KOH);
3. hydrogen, H_2.

1.2. Chlorine

Chlorine, Cl_2, is a yellowish-green gas with a pungent smell, which attacks the mucous membranes strongly. It can easily be liquefied by pressure and cooling, and liquid chlorine is marketed in steel cylinders and tank cars. In a large chemical works chlorine is conveyed from the production unit in pipelines. On account of the chemical reactivity, high volatility and poisonous nature of chlorine special protective and safety measures must be taken in its storage, filling and transport.

Chlorine boils at $-34°C$ ($-29.2°F$) and freezes at $-120°C$ ($-184°F$), both at 1013 mbar (14.7 lb/in²) pressure.

1.3. Caustic soda

The electrolysis of sodium chloride produces an aqueous solution of the solid substance sodium hydroxide (caustic soda), NaOH, usually containing 50% of NaOH by weight. A litre (0.264 gal) of this 50% caustic soda solution weighs 1.53 kg (3.37 lb); it is a strongly alkaline, corrosive liquid. When all the water is evaporated off, sodium hydroxide (caustic soda), NaOH is obtained as a white crystalline solid which is highly hygroscopic. Sodium hydroxide has a density of 2.1 g/cm³ (131.06 lb/ft³) and melts at 328°C (622.4°F). It is sold in the form of flakes, tablets, sticks, and also as grains of about 1 mm (0.039 in) diameter which are easily poured.

1.4. Hydrogen

Hydrogen, H_2, is a colourless combustible gas. It is used mainly in chemical industry for hydrogenation and reduction (see IV,3. hydrogenation, and II,3. ammonia synthesis). Hydrogen is also used together with oxygen for welding. Hydrogen is available commercially in steel cylinders as a gas under a high pressure (147 bars or 2130 lb/in²).

1.5. Economic importance

The production of sufficient quantities of chlorine and caustic soda (and also of hydrogen) constitutes the basis for the manufacture of many consumer goods. The range of products includes polyvinyl chloride, PVC, plant protection agents, and raw materials for detergents and soaps.

In 1978 the world production of chlorine was about 27.5 million tonnes, of which 9.94 million tonnes was produced in USA and 0.97 million tonnes in Great Britain (see Fig. 1).

In the electrolysis of alkali metal chlorides caustic soda solution is necessarily produced at the same time as chlorine and hydrogen. For each tonne of chlorine, caustic soda solution

Figure 1.   Production of chlorine in 1978

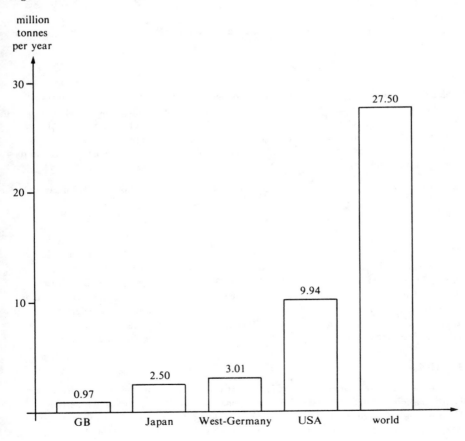

containing 1.13 tonnes of NaOH is obtained. The openings for disposing of chlorine and caustic soda depend on scientific and industrial developments, and at different times there have been alternating periods of surplus chlorine, surplus caustic soda, and periods when there has been a balance between demand for the two products.

Alkali-chlorine electrolysis meets with no difficulties as regards raw materials, though in Europe the high cost of electric power in comparison with the USA is a significant factor. Sodium chloride is one of the most widely distributed salts. The supplies of sodium chloride in rock salt deposits throughout the world are estimated at 1000 billion tonnes.
The oceans contain 36000 billion tonnes of dissolved NaCl. Not ónly in countries with a hot climate is sea water used for obtaining solid NaCl.

2. Industrial practice

There are two competing processes for obtaining chlorine, caustic soda and hydrogen by the electrolysis of aqueous sodium chloride solutions. At present (1979) more than 90% of the chlorine required in West Germany is made by the **mercury cell process**. The other method, the **diaphragm process**, is particularly important in the USA. A third method for making chlorine starts with hydrochloric acid, obtained as a by-product (see p. 11).

The decisive advantage which the mercury process possesses over the diaphragm process is that it produces very pure caustic soda, completely free from sodium chloride. However, the importance of this high degree of purity depends on the use to which the caustic soda is to be put.

By contrast, the diaphragm process produces a solution of caustic soda which still contains some sodium chloride. However, this process has the advantages of a lower capital investment, the avoidance of mercury, and the fact that it can employ sodium chloride solutions obtained by the subterranean dissolution of salt deposits.

2.1. Electrolysis in mercury cells

2.1.1. Chemistry of the process

Sodium chloride is dissociated in aqueous solution into sodium and chloride ions:

$$2\ NaCl \longrightarrow 2\ Na^+ \quad + \quad 2\ Cl^-$$

The chloride ions are discharged at the positive anode, where they lose electrons to form chlorine, which is evolved as a gas.

Process at the anode (positive pole):

$$2\ Cl^- \longrightarrow Cl_2 \quad + \quad 2\ e^-$$

The sodium ions are discharged at the negative cathode, where they take up electrons to form metallic sodium, which dissolves in the mercury as an amalgam.

II   Inorganic raw materials and large-scale products: Large-scale industrial processes

Process at the cathode (negative pole):

$$2\ Na^+ \quad + \quad 2\ e^- \quad \longrightarrow \quad 2\ Na$$

The sodium amalgam thus formed, an alloy of sodium and mercury, flows together with the liquid mercury into a separate decomposition chamber, where it reacts catalytically with water to give caustic soda solution and hydrogen:

$$2\ Na \quad + \quad 2\ HOH \quad \longrightarrow \quad 2\ NaOH \quad + \quad H_2\uparrow$$

These separate processes when added together give the following overall equation for the reaction:

electrical energy	rock salt	water		chlorine	caustic soda	hydrogen
$0.232\ kWh$	$+\ 2\ NaCl$	$+\ 2\ HOH \longrightarrow$		Cl_2	$+\ 2\ NaOH$	$+\ H_2$

The electrolysis of alkali metal chlorides is an endothermic reaction, ie it consumes energy. The voltage across the electrolytic cell varies between 4 and 5 volts according to the rate of production. Large installations have a daily production of several hundred tonnes of chlorine.

2.1.2. Process stages

The industrial process can be divided into six stages (see Fig. 2, p. 9).

1. dissolution;
2. purification of solutions;
3. electrolysis;
4. decomposition of amalgam;
5. treatment of chlorine;
6. energy supply.

2.1.2.1. Dissolution

In West Germany rock salt is mined and delivered by water. Rock salt contains 98–99% NaCl. The main impurities are small quantities of sea sand, particles of shells, gypsum ($CaSO_4.H_2O$), potassium chloride (KCl), magnesium chloride ($MgCl_2$) and traces of iron. The rock salt is dissolved in bunkers in the depleted salt solution returned from the electrolysis. Dissolution is carried out at 60–80°C (140–176°F), and the concentration of the saturated solution is about 315 g NaCl per litre (2.63 lb/gal).

2.1.2.2. Brine purification

A commonly used purification process is to precipitate the ions of the alkaline earth metals (Ca^{2+} and Mg^{2+}) and the ferric ions (Fe^{3+}) by adding caustic soda and soda solution. At the same time a suspension of barium carbonate is added in order to precipitate the sulphate ions, SO_4^{2-}.

The precipitated impurities are filtered off, after which the purified solution is acidified to pH 4–6 and fed into the electrolysis cells.

2.1.2.3. Electrolysis

The electrolysis cells consist of slightly sloping, long and narrow iron troughs, whose sides are coated with rubber internally. A thin layer of mercury, serving as the cathode (negative pole) flows over the uncoated iron bottom of the trough, which has an area of 10–30 m² (107–323 ft²). The anodes (positive poles) are held about 3 mm (0.118 in) above the surface of the mercury in iron sheaths coated with rubber.

The anode material is either graphite or an activated metal (eg titanium) coated with ruthenium oxide.

Application of a voltage of 4–5 volts per cell liberates chlorine at the anodes. Metallic sodium is liberated at the cathode, where it dissolves in the mercury to give sodium amalgam. The electrolysis cells are connected in series.

The depleted salt solution which runs out is dechlorinated by blowing through air or by evacuating, after which it is made alkaline with caustic soda and re-saturated in the dissolution vessels. In large installations the pumps for circulating the solution have capacities up to 2000 m³/h (70.640 ft³/h).

2.1.2.4. Decomposition of amalgam

The sodium amalgam is decomposed with softened water on a carbon or graphite catalyst to give caustic soda and hydrogen.

The water supply is regulated so that a solution containing 50 wt.% of NaOH leaves the decomposition vessel. It needs no further treatment, but is ready for use.

The hydrogen formed is cooled with water and then cooled to a low temperature in order to purify it. It is then distributed by pipelines for hydrogenations or reductions. The mercury leaving the decomposition vessels is returned to the electrolysis cells.

2.1.2.5. Purification of chlorine

The chlorine produced by electrolysis is about 99% pure. In addition to entrained droplets of solution it is contaminated with some hydrogen, carbon dioxide, nitrogen and oxygen from air which has leaked in, and small quantities of organic chlorine compounds formed by the chlorination of graphite.

The chlorine gas is purified by washing with water and by an electrostatic gas purifier, and is finally dried by passing it through concentrated sulphuric acid. Up to this point the chlorine must be handled in equipment made of glass or coated with rubber. However, dry chlorine does not corrode metals and can be treated further in equipment made of unprotected iron or steel. For this reason the turbocompressors which extract the gas are situated after the gas has been dried with sulphuric acid. The final purification of the chlorine is effected on the other side of the compressors by means of filters containing chlorinated active carbon. The chlorine is then passed to chlorinating units or is liquefied by cooling. Very pure chlorine containing 99.99% Cl_2 is obtained by evaporating liquid chlorine.

2.1.2.6. Energy supply

Modern electrolysis plants have a load of the order of 100 megawatts (100 million watts). Large power stations supply three-phase alternating current at about 100 kilovolts (100,000 volts). This supply is usually transformed down to 25 kV in local sub-stations, and is then converted to the working voltage of 500 to 1000 volts in the electrolysis plants.

The rectification of the alternating current is nowadays effected almost exclusively with silicon diodes, which can be connected in parallel to give currents of any required magnitude. There are electrolysis plants in Germany which operate with currents up to 200,000 amperes, and in the USA this fugure rises to 300,000 amperes.

The electrolysis cells are connected in series. The potential across each cell is about 4 volts, so that with a working voltage of 500 volts up to 125 cells can be connected in series. The current efficiency of electrolysis plants is nearly 98%.

2.2. Electrolysis in diaphragm cells

A diaphragm cell consists of two separate compartments, the cathode compartment with an iron cathode and the anode compartment with a graphite anode. The two compartments are separated by a porous semipermeable diaphragm.

The diaphragm serves to keep apart the hydrogen produced at the cathode and the chlorine produced at the anode. It also prevents the mixing of the electrolyte solutions produced at the cathode and the anode, which would otherwise react with one another. The diaphragms are constituted so that the OH^- ions migrate in the opposite direction to the flow of liquid. The fresh solution of rock salt is introduced into the anode compartment.

At the anode the chlorine ions of sodium chloride are discharged to give chlorine.

Sodium chloride		sodium ions		chloride ions
2 NaCl	$\longrightarrow$	2 Na^+	$+$	2 Cl^-

Figure 2.   Flow diagram for alkali-chlorine electrolysis by the amalgam process

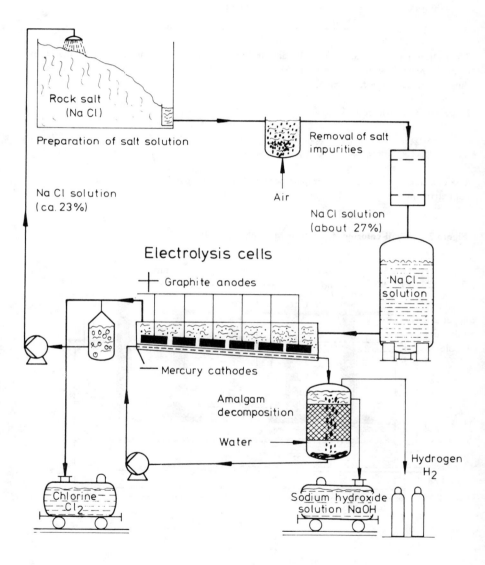

Process at the anode (+)

Chloride ions		chlorine		electrons
$2\ Cl^-$	$\xrightarrow{\text{discharge}}$	$.\quad Cl_2$	$+$	$2\ e^-$

The sodium ions migrate through the diaphragm into the cathode compartment, where they meet the hydroxide ions which have been formed by the electrolysis of water; the solution thus contains sodium hydroxide.

Process at the cathode (−)

Water		electrons			hydrogen		hydroxide ions
$2\ H_2O$	$+$	$2\ e^-$	$\longrightarrow$		H_2	$+$	$2OH^-$

The chlorine from the anode compartment and the hydrogen from the cathode compartment are drawn off separately and purified.

Figure 3. Alkali-chlorine electrolysis in diaphragm cells

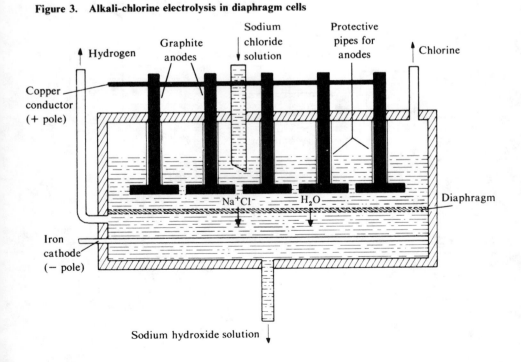

The construction of the diaphragm depends greatly on technological requirements.

The rate of flow of the solution through the diaphragm cell is adjusted so that the solution in the cathode compartment contains about 170 g NaCl per litre and 140 g NaOH per litre (1.17 lb/gal). The efficient use of the electrical energy precludes the production of a higher concentration of NaOH.

Subsequent repeated evaporation produces finally a 50% solution of caustic soda containing only 1–2% of NaCl. The reduction in the amount of NaCl depends on the fact that NaCl is less soluble in water than NaOH and therefore separates out on evaporation and can be removed.

3. Chlorine as a raw material

Chlorine is produced on a large scale and is the starting point for the manufacture of many important chemical products. The development of chlorine production therefore serves as a criterion for the productivity of a modern chemical industry based on a variety of end products. This is particularly true in highly industrialised countries where the chemical industry is tending towards processing in a number of stages.

In this type of processing industry chlorine is of much greater interest than in a heavy chemical industry, which is characterized by the production indices of sulphuric acid and caustic soda.

3.1. Chlorination reactions

The term chlorination is used to describe chemical reactions in which elementary chlorine reacts with organic or inorganic substances to give the corresponding chloro-compounds. In organic chlorinations hydrogen chloride gas is often produced as a by-product.

Hydrocarbon	chlorine		chlorohydro-carbon	hydrogen chloride gas
$R - H$	$+ Cl - Cl$	$\longrightarrow$	$R - Cl$	$+ H - Cl$

In these and similar chlorination reactions only one half of the chlorine employed is converted to the required chloro-compound, the other half appearing as hydrogen chloride, which is passed into water to produce hydrochloric acid.

In contrast to hydrogen chloride gas, aqueous hydrochloric acid consists of the ions H^+ and Cl^- in solution.

This hydrochloric acid can also be electrolysed to recover the chlorine:

Electrical energy		hydrochloric acid		chlorine		hydrogen
0.13 kWh	$+$	2 HCl	$\longrightarrow$	Cl_2	$+$	H_2

The electrode material is graphite. In order to keep the ohmic resistance of the electrolysis cells as low as possible 30% hydrochloric acid is used. The concentration of HCl is reduced to about 20% by the formation of H_2 and Cl_2, and the emerging acid is then reconcentrated by passing in gaseous hydrogen chloride. The heat of solution developed during the last process has the welcome effect of maintaining the temperature of the electrolyte at 60–70°C (140–158°F).

The electrolysis cell consists of a series of frames, in each of which a graphite electrode is cemented. The frames are assembled together as in a filter press, and are separated by PVC sheet.

Chlorine, 99% pure, is evolved at the anode, and hydrogen at the cathode. The potential across each cell is 2.3–2.5 volts. 1 t of chlorine requires 1800 to 2000 kWh of electricity.

3.2. Chlorinated organic products

Among the organic compounds whose chlorination is important are methane, acetylene, ethylene, benzene and acetic acid (see IV,5, halogenation).

3.2.1. Vinyl chloride, chloroprene, chlorobenzene, chloromethanes, etc

Most of the chlorine produced is used for chlorinating organic compounds. Some of the products obtained are mentioned in the title of this section.

Because of the great industrial importance of these products they are described separately in Section IV,5, under halogenation. The manufacture of vinyl chloride and polyvinyl chloride (PVC) is described in Section IV,7, under polymers.

Another important product of chlorination is phosgene. Its preparation and uses are summarized in the next section.

3.2.2. Phosgene

Chlorine reacts with gaseous carbon monoxide to give phosgene. The synthesis is carried out at 100–120°C (212–248°F) with active carbon as catalyst.

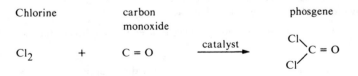

Phosgene, $COCl_2$, the dichloride of carbonic acid, H_2CO_3, is a gas which boils at +8°C (39–46°F) and is readily liquefied by compression. Since phosgene is a very dangerous respiratory poison special safety precautions are necessary in its manufacture, storage, transport and subsequent handling.

The importance of phosgene as an intermediate product depends on the high reactivity of the two chlorine atoms in the $COCl_2$-molecule. Phosgene reacts particularly readily with compounds which contain reactive H-atoms. It undergoes important synthetic reactions with amines, containing H_2N-groups, and with alcohols, containing HO-groups. These reactions give isocyanates and carbonate esters. Isocyanates are of great importance for making polyurethanes (cf IV,7, polymers, and IV,8, synthetic fibres). Carbonate esters are important as solvents and as the starting materials for making polycarbonates (cf IV,7, polymers).

3.3. Chlorinated inorganic products

3.3.1. Compounds of chlorine with sulphur

The following compounds containing sulphur and chlorine are prepared from elemental chlorine and are of industrial importance.

Disulphur dichloride (or sulphur chloride), S_2Cl_2, is a liquid used for the cold vulcanization of rubber blends.

Thionyl chloride $SOCl_2$, is a liquid used for chlorinating organic substances when this cannot be readily effected by elemental chlorine.

Sulphuryl chloride SO_2Cl_2, is a liquid used for the sulpho-chlorination of hydrocarbons and other substances (cf IV,6, detergents).

3.3.2. Compounds of chlorine with phosphorus

The following compounds containing phosphorus and chlorine are prepared from elemental chlorine and are of industrial importance.

Phosphorus trichloride PCl_3, is a liquid used in the preparation of $POCl_3$ and $PSCl_3$. It acts as a chlorinating agent by replacing OH-groups by chlorine. It is also used for preparing esters which act as plant protection agents.

Phosphorus pentachloride, PCl_5, is a solid which is used as a chlorinating agent in organic chemistry. It is used in particular for preparing acid chlorides by replacing OH-groups with chlorine.

Phosphorus oxychloride (or phosphoryl chloride), $POCl_3$, is a liquid which is used both as a chlorinating agent for hydroxy-compounds and for the preparation of phosphate esters which are used as plant protection agents.

Phosphorus thiochloride (or thiophosphoryl chloride), $PSCl_3$, is the starting material for phosphate esters used as insecticides, for example, methylparathione.

3.3.3. Compounds of chlorine with metals

The following metal chlorides are prepared by using elemental chlorine and are of industrial importance:

Titanium tetrachloride, $TiCl_4$, is a liquid which serves as an intermediate in the manufacture of titanium dioxide white pigments by the chloride process (cf II,8, inorganic pigments).

Aluminium chloride, $AlCl_3$, is a solid used as a catalyst for the Friedel-Crafts synthesis of alkylbenzenes and aromatic ketones, which in turn serve as starting materials for washing agents, dyestuffs, perfumes and pharmaceutical products.
The combustion of $AlCl_3$ in an oxyhydrogen flame produces very pure aluminium oxide, Al_2O_3, which has special applications as an absorbent, a pigment and a filler.

Iron (III) chloride (or ferric chloride), $FeCl_3$, is also a solid and a catalyst, for example in chlorinations and Friedel-Crafts syntheses in organic chemistry. It is also used in the treatment of water.

Tin tetrachloride (or stannic chloride), $SnCl_4$, is a liquid which is used to prepare organic tin compounds, which are important in the production of plant protection agents, PVC stabilizers, and protective materials for under-water paints.

3.3.4. Silicon tetrachloride

Silicon tetrachloride, $SiCl_4$, is used as as intermediate in the preparation of the very pure silicon required in semiconductor devices.

3.4. Bleaching solutions and chloride of lime

The most important industrial bleaching solution is made by passing chlorine into caustic soda solution at temperatures below $35°C$ ($95°F$):

Chlorine		caustic soda		soda bleach
Cl_2	$+$	$2\,NaOH$	$\longrightarrow$	$NaOCl + NaCl + H_2O$

The active bleaching and disinfecting agent is sodium hypochlorite, $NaOCl$.

This solution is used for bleaching in the paper, cellulose and textile industries, and also for sterilizing water and for decontaminating effluents containing cyanide. Sodium hypochlorite is also used for oxidizing ammonia, NH_3, to hydrazine, $H_2N\text{-}NH_2$.

For bleaching sodium hypochlorite is often used in combination with sodium chlorate, $NaClO_3$, or with chlorine dioxide, ClO_2, which is made from sodium chlorate.

Bleaching solutions are also obtained by passing chlorine into solutions of caustic potash or calcium hydroxide (lime).

If chlorine is passed into a suspension of the sparingly soluble calcium hydroxide in a little water (known as milk of lime), a solid called **chloride of lime** is precipitated. This is used as a bleaching and disinfecting agent, its active constituent being the hypochlorite ion, OCl^-,

Chlorine		milk of lime		chloride of lime		water
Cl_2	$+$	$Ca(OH)_2$	$\longrightarrow$	$Ca^{2+} \begin{matrix} Cl^- \\ OCl^- \end{matrix}$	$+$	H_2O

4. Caustic soda as a raw material

The importance of caustic soda as a raw material depends on its ability to form salts. Sodium salts are readily soluble in water.

4.1. Sodium polyphosphates

In conjunction with phosphoric acid caustic soda serves as a raw material for the manufacture of washing agents based on polyphosphates. The sodium polyphosphates act mainly as water softeners (see II,6, phosphoric acid, and IV,6, washing agents).

4.2. Treatment of borax, sodium perborate

Treatment of borax and other polyborate minerals with caustic soda produces sodium metaborate, $NaBO_2$, which is used together with hydrogen peroxide. H_2O_2, to produce the important bleaching and washing agent sodium perborate (or peroxoborate) — see II,7, inorganic oxo-compounds.

4.3. Treatment of bauxite, aluminium production

Naturally occurring bauxite contains in addition to $Al_2O_3.H_2O$ impurities of iron (III) oxide and silica. By treatment with hot caustic soda the aluminium can be separated in a soluble state, and after subsequent stages of treatment is finally obtained as the pure metal by the electrolysis of Al_2O_3, or, more recently, of $AlCl_4$.

4.4. Treatment of quartz sand, fillers

Treatment of quartz sand with hot caustic soda gives soluble sodium silicates, which are mainly treated further to produce fillers (see II,8, inorganic fillers).

4.5. Cellulose fibres

Treatment of cellulose with caustic soda solutions yields "viscose", a viscous solution which can be spun into cellulose fibres (see IV,8, artificial fibres).

4.6. Caustic soda as a neutralizing agent

Neutralization of inorganic acids with caustic soda gives the corresponding salts, for example sodium sulphate, Na_2SO_4 (with water of crystallization), sodium nitrate, $NaNO_3$, etc.

The solubility of organic compounds containing the sulphonic acid group can be increased by neutralization with caustic soda. This applies to many compounds used as colouring matters or for pharmaceutical purposes.

Compound with sulphonic acid group	sodium hydroxide	sodium sulphonate	water

$$R - CH_2 - SO_3^{\ominus}H^{\oplus} + Na^{\oplus}OH^{\ominus} \longrightarrow R - CH_2 - SO_3^{\ominus}Na^{\oplus} + H_2O$$

5. Review of the further treatment of chlorine and caustic soda

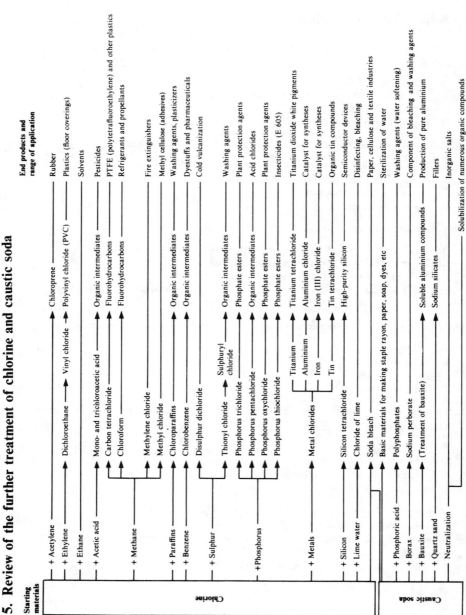

Starting materials	Intermediate	End products and range of application
Chlorine + Acetylene	Chloroprene	Rubber
+ Ethylene → Dichloroethane → Vinyl chloride	Polyvinyl chloride (PVC)	Plastics (floor coverings)
+ Ethane		Solvents
+ Acetic acid → Mono- and trichloroacetic acid	Organic intermediates	Pesticides
+ Methane → Carbon tetrachloride	Fluorohydrocarbons	PTFE (polytetrafluoroethylene) and other plastics
Chloroform	Fluorohydrocarbons	Refrigerants and propellants
Methylene chloride		Fire extinguishers
Methyl chloride		Methyl cellulose (adhesives)
+ Paraffins → Chloroparaffins	Organic intermediates	Washing agents, plasticizers
+ Benzene → Chlorobenzene	Organic intermediates	Dyestuffs and pharmaceuticals
+ Sulphur → Disulphur dichloride		Cold vulcanization
Thionyl chloride → Sulphuryl chloride	Organic intermediates	Washing agents
+ Phosphorus → Phosphorus trichloride	Phosphate esters	Plant protection agents
Phosphorus pentachloride	Organic intermediates	Acid chlorides
Phosphorus oxychloride	Phosphate esters	Plant protection agents
Phosphorus thiochloride	Phosphate esters	Insecticides (E 605)
Titanium	Titanium tetrachloride	Titanium dioxide white pigments
+ Metals — Aluminium	Aluminium chloride	Catalyst for syntheses
Metal chlorides — Iron	Iron (III) chloride	Catalyst for syntheses
Tin	Tin tetrachloride	Organic tin compounds
+ Silicon → Silicon tetrachloride	High-purity silicon	Semiconductor devices
+ Lime water → Chloride of lime		Disinfecting, bleaching
Soda bleach		Paper, cellulose and textile industries
Caustic soda	Basic materials for making staple rayon, paper, soap, dyes, etc	Sterilization of water
+ Phosphoric acid → Polyphosphates		Washing agents (water softening)
+ Borax → Sodium perborate		Component of bleaching and washing agents
+ Bauxite (Treatment of bauxite)	Soluble aluminium compounds	Production of pure aluminium
+ Quartz sand	Sodium silicates	Fillers
Neutralization		Inorganic salts
		Solubilization of numerous organic compounds

II Inorganic raw materials and large-scale products: Large-scale industrial processes

6. Self-evaluation questions

8 Which groups contain the substances required as raw materials in brine electrolysis or occurring as products?

 A soda — rock salt — chlorine
 B hydrogen — chlorine — caustic soda solution
 C water — rock salt
 D rock salt — oxygen — water

9 Which factors determine the price of electrolytically produced chlorine?

 A the cost of 1 kWh of electrical energy
 B the capacity utilisation factor for the plant
 C the purity of the rock salt
 D the density of the rock salt

10 Which processes take place in the electrolysis of brine by the mercury process?

 A $NaCl \longrightarrow Na^+ + Cl^-$, ions are formed from sodium chloride
 B $2Cl \longrightarrow Cl_2 + 2e^-$, chloride ions are discharged
 C $Na^+ + e^- \longrightarrow Na$, sodium ions are discharged
 D $NaCl + H_2O \longrightarrow NaOH + HCl$, sodium chloride reacts with water to give caustic soda solution

11 Which particles are attracted to the anode by electrical forces in brine electrolysis?

 A sodium ions
 B water molecules
 C mercury atoms
 D chloride ions

12 One advantage of the mercury method compared with other methods of brine electrolysis is that:

 A it provides purer caustic soda
 B it provides purer chlorine
 C it requires smaller capital expenditure
 D it causes no environmental protection problems

13 The total reaction occurring in brine electrolysis is:
electrical energy $+ 2NaCl + 2H_2O \longrightarrow Cl_2 + 2NaOH + H_2$
How many tonnes of chlorine can accordingly be produced from 117 t of pure sodium chloride?
Use the atomic mass table in section I/9 on page 5 to calculate the answer.

 A 71 t   C 117 t
 B 35.5 t   D 142 t

14 Which products do the individual vessels contain?

A ① contains chlorine   C ③ contains chlorine
B ② contains caustic soda D ① contains hydrogen

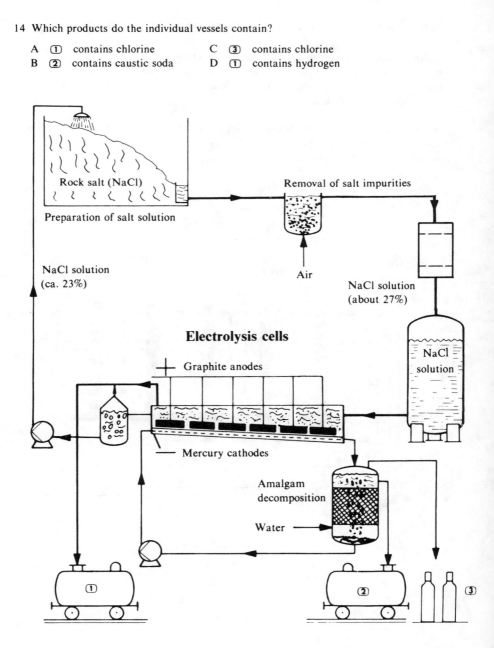

Rock salt (NaCl)

Preparation of salt solution

Removal of salt impurities

NaCl solution (ca. 23%)

Air

NaCl solution (about 27%)

Electrolysis cells

NaCl solution

Graphite anodes

Mercury cathodes

Amalgam decomposition

Water

① ② ③

II Inorganic raw materials and large-scale products: Large-scale industrial processes

15 In brine electrolysis using the diaphragm process the task of the diaphragm is

A   to prevent the formation of an explosive hydrogen/chlorine mixture
B   to prevent the migration of Na^+ ions into the cathode space
C   to separate the electrolysis cell into the cathode space and anode space
D   to prevent the caustic soda in the cathode space from mixing with the rock salt brine in the anode space

16 At the points indicated in the drawing, starting materials are added to, or end products removed from, the electrolysis cell.
Which of the product names is correct?

A   chlorine
B   common salt solution
C   hydrogen
D   caustic soda solution

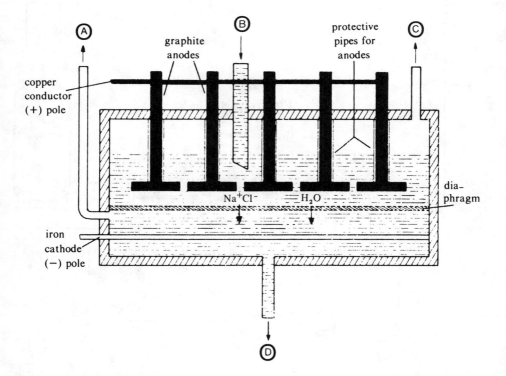

17 Which statements about chlorine as a raw material are correct?

A The level of chlorine production is the characteristic index for a chemical base materials industry.

B The main quantity of the chlorine produced is used for chlorination reactions with inorganic materials.

C Frequently, only 50% of the chlorine employed in chlorination reactions is reacted because hydrogen chloride forms as a by-product.

D Bleaching solutions are formed by passing chlorine into lyes.

18 Which chlorination products are inorganic compounds?

A $COCl_2$

B SO_2Cl_2

C PCl_5

D $POCl_3$

19 Which chlorination product is named correctly?

A $SOCl_2$ — sulphur chloride

B PCl_5 — phosphorus pentachloride

C $POCl_3$ — phosgene

D $TiCl_4$ — titanium dichloride

20 Which statements about the raw material caustic soda are correct?

A Caustic soda is a neutralizing agent.

B By reacting caustic soda with aluminium oxide one obtains elemental aluminium.

C A major feature of caustic soda as a raw material is its ability to form salts.

D Caustic soda is suitable for the dissolution of water-insoluble substances.

21 Which statements about sodium hypochlorite solution are correct?

A Sodium hypochlorite solution is a liquid.

B Sodium hypochlorite solution is important to the textile industry because of its sodium chloride content.

C Sodium hypochlorite solution can be used to detoxify effluent containing cyanide.

D Sodium hypochlorite solution occurs as a by-product in all brine electrolysis.

3 The Ammonia Synthesis

1. Ammonia production and requirements for fertilizers

The enormous importance of adequate supplies of ammonia arises from the necessity of producing sufficient supplies of food for a rapidly growing population. The greater part of the ammonia produced is used for making nitrogen fertilizers.

The foundations of today's fertilizer industry were laid in 1840 by the chemist Justus von Liebig. He analysed plant ashes, and concluded that the mineral constituents of plants were necessary for their growth. In soils deficient in minerals plants could fluorish only feebly or not at all.

Soon after this it was realized that plant nutrition dependended not only on the mineral constituents of plant ashes, but also on a number of other elements such as carbon, C, nitrogen, N, oxygen, O, hydrogen, H, phosphorus, P and sulphur, S.

Adequate applications of nitrogen compounds were found to be indispensable not only to prevent a decrease in the yields of agricultural plant products, but also to increase them several-fold. About half of current foodstuff production can be attributed to the correctly planned application of artificial fertilizers.

Phosphate, sulphate and potassium fertilizers can be procured by fairly simple methods on account of their natural occurrence. The provision of nitrogen fertilizers presents a more difficult problem. Elementary nitrogen, which constitutes 78% of the atmosphere by volume, is one of the most inert elements. It must therefore be converted into a more reactive compound suitable for use by plants. This is achieved industrially by the synthesis of ammonia from elementary nitrogen and hydrogen.

2. Properties of ammonia

Ammonia, NH_3, is composed of the elements hydrogen and nitrogen. It is one of the most important of all nitrogen compounds and is the starting material for making all other compounds containing nitrogen.

The catalytic oxidation of ammonia with atmospheric oxygen gives oxides of nitrogen, which can be easily converted to nitric acid. Ammonia and nitric acid are the basic chemical constituents of nitrogen fertilizers.

Ammonia is formed naturally by the decay of animal and vegetable material, which contains nitrogen in the form of proteins. It is also formed as a product of the metabolism of animal organisms. In this case it is combined as urea, which is excreted.

At room temperature ammonia is a colourless, poisonous gas with a characteristic odour. At normal pressure (1013 mbar) it melts at $-77°C$ and boils at $-33.5°C$. It is readily soluble in water: at 15°C 1 litre of water dissolves 727 litres of ammonia to give a solution with an alkaline reaction.

3. Manufacture of ammonia from elemental nitrogen and hydrogen

The scientific basis for the synthesis of ammonia from the elements nitrogen and hydrogen
was established by Fritz Haber by means of laboratory experiments. Subsequently Carl
Bosch of BASF built plant for large-scale industrial production in Oppau/Germany in 1913
and in Leuna, Germany in 1914–16. The Haber-Bosch process not only made it possible to
satisfy the growing demand for nitrogen fertilizers, but it also initiated a completely new
development in industrial chemistry, namely the use of high pressures and temperatures.
The ammonia synthesis served as a model for a number of other high-pressure catalytic
processes, such as the synthesis of methanol (see III.3, hydrogenation).

The raw materials for ammonia synthesis, atmospheric nitrogen and hydrogen from water,
would not have been recognized as raw materials a few decades ago. These are available
even to countries with very limited natural resources. Since the constructuon of the first
ammonia plant factories have been built in almost every country in the world. In the last 40
years the world production of synthetic nitrogen compounds has increased more than tenfold.
(See Fig. 1 p.5, World production and consumption of synthetic nitrogen.)

3.1. Chemical principles

Nitrogen   hydrogen   ammonia
$$N_2 \;+\; 3\,H_2 \;\rightleftharpoons\; 2\,HN_3; \quad \Delta H = -92.0 \text{ kJ/mol}^*$$

The opposed arrows in this chemical equation indicate that the synthesis of ammonia from
nitrogen and hydrogen is a reversible reaction.

This means that in the industrial process N_2 and H_2 cannot be converted completely into
ammonia. This fact governs the reaction conditions under which economic yields can be
obtained. The object is to obtain the greatest possible yield of ammonia per unit time at a
favourable cost. In spite of the small degree of conversion this is rendered possible by the
high reaction velocity. According to the catalyst employed, working temperatures of
400–500°C (752–932°F) and pressures of 300–400 bar (4350–5800 lb/in^2) are used.

The unreacted nitrogen and hydrogen are recirculated to the reactor and used again.

The conversion of materials is accompanied by a conversion of energy. According to the
above mentioned equation.

28 g (or lb) nitrogen + 6 g (or lb) hydrogen   34 g (or lb) ammonia

and the conversion of these quantities evolves heat, which must be removed.

Let us calculate the amount of heat liberated in the production of 1 t (2205 lb) of ammonia.

---

* The value of ΔH refers to the number of moles represented in the chemical equation.

Figure 1. World production and consumption of synthetic nitrogen

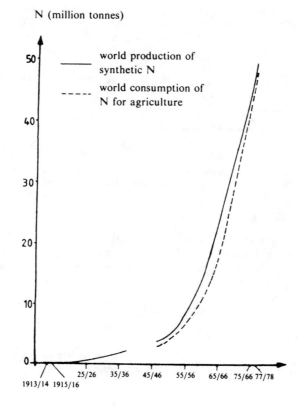

The formation of 34 g (0.075 lb) of ammonia liberates 92 kJ (87.2 Btu). The formation of 1 t (2205 lb) (10^6 g) of ammonia therefore liberates

$$\frac{92 \times 10^6}{34} = 2.7 \text{ million kJ}$$

This heat is removed from circulation in two ways:

1.   as useful heat by raising steam in a waste-heat boiler;
2.   as waste heat in cooling water or to the atmosphere.

II Inorganic raw materials and large-scale products: Large-scale industrial processes

The synthesis of ammonia has the following features:

i.   it is a reversible reaction, so that a process involving continuous circulation is required;
ii. it is an exothermic process, so that the heat liberated must be removed from circulation;
iii. it is a catalytic reaction, and therefore requires the correct choice of catalyst to ensure optimum economic production;
iv. it operates at high pressures and temperatures, which demands the development and selection of suitable equipment (eg compressors) and suitable materials for the reaction vessels.

3.2. Raw materials for the synthesis of ammonia

Production of hydrogen

The hydrogen required for the ammonia synthesis does not occur naturally in a free state. It must therefore first be prepared. Water is used as the starting material in almost all processes.

1.   Hydrogen from hydrocarbons and steam

The greater part of the hydrogen required is obtained nowadays from natural gas or from special petroleum fractions. The hydrocarbons are vaporized and allowed to react with water vapour on a catalyst at 800°C. This produces hydrogen, together with carbon monoxide, carbon dioxide and methane as by-products, which must be removed by special procedures.

$$
\underset{CH_4}{\text{Methane}} \quad + \quad \underset{H_2O}{\text{steam}} \quad \longrightarrow \quad \underset{CO}{\text{carbon monoxide}} \quad + \quad \underset{3\,H_2}{\text{hydrogen}}
$$

$$\Delta H = -96.0 \ kJ/mol$$

$$
\underset{Energy}{\text{Energy}} \ + \ C_nH_m \ + \ \underset{H_2O}{\text{steam}} \ \xrightarrow{\text{catalyst}}
$$

$$
\underset{H_2}{\text{hydrogen}} \quad + \quad \underset{CO}{\substack{\text{carbon} \\ \text{monoxide}}} \quad + \quad \underset{CO_2}{\substack{\text{carbon} \\ \text{dioxide}}} \quad + \quad \underset{CH_4}{\text{methane}}
$$

The method of producing hydrogen from naphtha and steam is known as **steam cracking** or **steam re-forming**.

2. Production of hydrogen by the electrolysis of water

$$
\underset{2\,H_2O}{\text{Water}} \quad \longrightarrow \quad \underset{2\,H_2}{\text{hydrogen}} \quad + \quad \underset{O_2}{\text{oxygen}} \quad ; \Delta H = +571.5 \ kJ/mol
$$

This method of producing hydrogen is used in countries possessing cheap electric power, such as Norway and, recently, Assuan in Egypt.

3.   Depending on the distribution of production in a chemical industry the electrolysis of alkali chlorides or of hydrochloric acid may serve as sources of hydrogen

Rock salt		water		caustic soda		chlorine		hydrogen
$2\,NaCl$	+	$2\,H_2O$ $\longrightarrow$		$2\,NaOH$	+	Cl_2	+	H_2

$\Delta H = +\,454.4\ kJ/mol$ as electrical energy

In a similar way hydrogen is formed in the electrolytic decomposition of hydrochloric acid.

Aqueous hydrochloric acid		chlorine		hydrogen
$2\,HCl + H_2O$ $\longrightarrow$		Cl_2	+	H_2

According to the pattern of production of a chemical undertaking the last two processes may be used as supplementary sources of hydrogen.

Production of nitrogen

The nitrogen is obtained from the atmosphere, which consists essentially of 78 vol.% of elementary nitrogen and 21 vol.% of oxygen. The remainder consists of the rare gases and carbon dioxide. The nitrogen is freed from oxygen by using the required amount of air in the combustion of methane, which is obtained (together with hydrogen) in the steam cracking of naphtha.

Methane		air		carbon dioxide		hydrogen		nitrogen
CH_4	+	$O_2 + 4\,N_2$ $\longrightarrow$		CO_2	+	$2\,H_2$		$+\,4\,N_2$

$$\Delta H = -71.5\ kJ/mol$$

Additional amounts of pure nitrogen are obtained by liquefying air by the Linde process and subsequent fractional distillation.

Purification of the mixed gases

Hydrogen and nitrogen are mixed in the proportions in which they react.

Before this crude mixture is passed into the reactor for the synthesis it must be freed from the following impurities:
   atmospheric oxygen,
   carbon monoxide, carbon dioxide, and
   sulphur compounds.

If the catalysts come into contact with these impurities they gradually become inactive and the synthesis becomes uneconomic.

The equipment for purifying the crude gases occupies the greater part of a modern ammonia plant.

In order to remove the carbon monoxide it is first converted to carbon dioxide by reaction with steam (the water gas shift reaction).

The water gas shift reaction:

Carbon monoxide		steam		carbon dioxide		hydrogen
CO	$+$	H_2O	$\longrightarrow$	CO_2	$+$	H_2

$$\Delta H = -41.0 \text{ kJ/mol}$$

The carbon dioxide can be mainly removed by absorption in water under pressure. Any remaining CO and CO_2 is reduced with hydrogen to give methane, which does not interfere with the synthesis.

The concentration of sulphur compounds or halogens in the synthesis gas must be reduced to less than 1 ppm (parts per million). The sulphur compounds are derived from the naphtha, and even from the oil used to lubricate the cylinders of the compressors.

The organically bound sulphur is catalytically reduced to hydrogen sulphide by the addition of hydrogen: the hydrogen sulphide is then absorbed by (for example) zinc oxide.

Hydrogenation:

Sulphur		hydrogen	catalyst	hydrogen sulphide
S	$+$	H_2	$\longrightarrow$	H_2S

Absorption of hydrogen sulphide:

Hydrogen sulphide		zinc oxide	zinc sulphide		water
H_2S	$+$	ZnO	$\longrightarrow \quad ZnS$	$+$	H_2O

Halogen impurities can be removed by adding hydrogen from the electrolysis of alkali metal chlorides. See Fig. 2, p. 9, schematic course of the ammonia synthesis.

Figure 2.   Schematic course of the ammonia synthesis

The course of the preparation of the synthesis gas, followed by the synthesis of ammonia, can be divided schematically into the following stages:

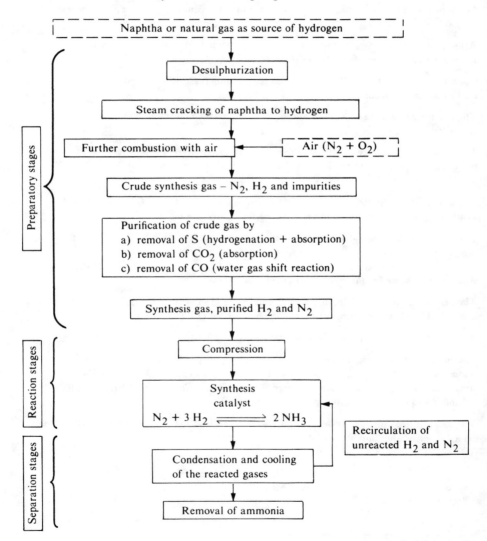

3.3. The synthesis*

After the crude gases have been freed from all accompanying impurities they are compressed to 200–400 bars (2900–5800 lb/in²) by huge high-capacity turbocompressors.

These turbocompressors rotate at 12,000 revs/min and require a power input of between 15,000 and 25,000 kW.

After compression the gas is fed into the synthesis loop, ie it is added to the unreacted hydrogen and nitrogen coming from the reactor.

The mixed gases are brought to the required temperature (ca. 500°C or 930°F) in heat exchangers and then converted to ammonia in the reactor.

The reactor contains catalyst beds and heat exchangers which serve to regulate the temperature.

The time of contact between the synthesis gas and the catalyst is short, being usually not more than 30 seconds. The gas leaves the reactor with an ammonia content of about 21 vol.%.

The ammonia gas formed on the catalyst in the reactor is separated from the unreated synthesis gas by liquefaction, and is removed from circulation.

For this purpose the gas, at a pressure of less than 300 bar (4350 lb/in²), is cooled to the temperature of the cooling water, about 30°C (86°F).

3.4. Flow scheme for the continuous circulation process

The production of 1 ton (2205 lb) of NH_3 requires 1975 m³ (69,757 ft³) of hydrogen and 658 m³ (23,241 ft³) of nitrogen at 0°C (32°F) and 1 bar (14.7 lb/in²).

A large plant can produce up to 2000 t of ammonia per day.

The energy and steam requirements of a synthetic plant depend on the throughput of materials. 1 t (2205 lb) of ammonia requires 34.3 million kJ (32.5 million Btu) of heat, used mainly to obtain elementary hydrogen. The exothermic formation of NH_3 liberates 2.7 million kJ (2.56 million Btu) per tonne of ammonia.

The catalyst used for the synthesis is microcrystalline iron together with alkali metal oxides on a carrier of aluminium oxide, Al_2O_3.

---

*Fig. 3, p. 11. Synthesis of ammonia by the continuous circulation process.

Figure 3.   The synthesis of ammonia

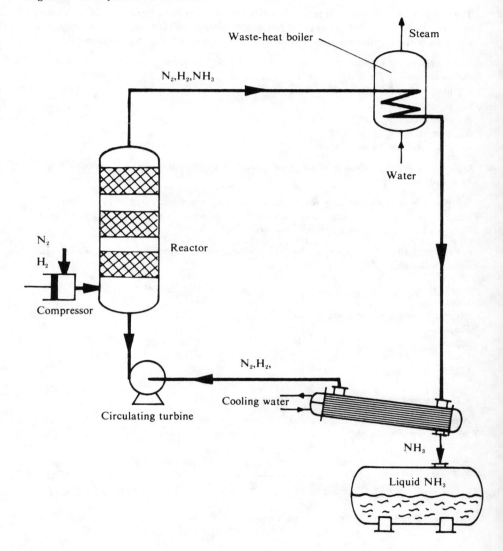

4. Uses of ammonia*

The demand for ammonia and products derived from it has risen steadily. This increase will continue, since world requirements for nitrogen fertilizers are far from being satisfied, and ammonia is the starting material for other industrially important nitrogen compounds.

75% was used for making fertilizers based on ammonia or nitrates, 20% was used in chemical industry for making:
a. fibre raw materials;
b. plastics;
c. adhesives;
d. dyes;
e. pharmaceutical products, etc.

Only 5% of world production of nitrogen through the synthesis of ammonia is used in the explosives industry.

For the year 1983 the capacity for ammonia production was estimated at 11.5 million tonnes; of this quantity, industrialized countries of the west accounted for 34.5%, Eastern Europe 33.8% and the development countries 31.7%.

For the year 1978 the capacity for ammonia production in different countries of the world was as follows:

Country	1,000 t N/year
Canada	2,343.3
Federal Republic of Germany (FRG)	1,955.4
France	2,035.0
Japan	2,879.3
USA	12,631.0
USSR	11,476.7

4.1. Starting material for the manufacture of nitric acid

Ammonia is the only starting material for making nitric acid. It is burned catalytically in the presence of atmospheric oxygen to form oxides of nitrogen.

$$\underset{4\,NH_3}{\text{Ammonia}} + \underset{5\,O_2}{\text{oxygen}} \xrightarrow{\text{catalyst}} \underset{4\,NO}{\text{nitrogen monoxide}} + \underset{6\,H_2O}{\text{water}}$$

$$\Delta H = -906.3 \text{ kJ/mol}$$

See Section II.4, for the details of the manufacture of nitric acid.

---

*See page 17 Review of products derived from ammonia.

4.2.   Ammonia as a base

On account of its basic character, which enables it to neutralize acids, ammonia (and also hydrazine) is added to water used for raising steam.

Water under pressure boils at temperature in excess of 100°C (212°F).

At these temperatures the dissociation of water into hydrogen and hydroxide ions is increased.

$$Energy \quad + \quad H_2O \rightleftharpoons H^+ + OH^-$$

Figure 4. Ammonia production

Production referred to 100 000 t nitrogen per year

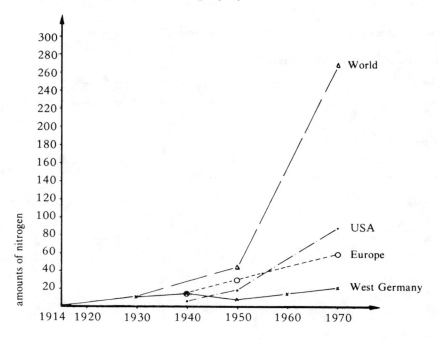

II Inorganic raw materials and large-scale products: Large-scale industrial processes

The hydrogen ions promote chemical corrosion, but are captured by the neutralizing effect of the ammonia.

$$NH_3 + H^+ \longrightarrow NH_4^+$$

Ammonia is an important neutralizing agent. The fertilizers nitrate of ammonia and sulphate of ammonia are obtained by neutralizing the corresponding acids.

Nitrate of ammonia (ammonium nitrate)

Ammonia		nitric acid		ammonium nitrate
NH_3	+	HNO_3	$\longrightarrow$	NH_4NO_3

Ammonium nitrate in aqueous solution is being used increasingly as a component of liquid fertilizers. In addition to facilitating the application of the fertilizer, this saves the costly processes of evaporating down and crystallising

Sulphate of ammonia (ammonium sulphate)

Ammonia		sulphuric acid		ammonium sulphate
$2\,NH_3$	+	H_2SO_4	$\longrightarrow$	$(NH_4)_2SO_4$

Ammonium sulphate is declining in importance, since sulphate-free fertilizers are now increasingly preferred. Ammonium fertilizers act quickly, but are not so short-lived as nitrate fertilizers. They are usually applied before sowing.

Both liquid ammonia and acqueous ammonia are often used as liquid fertilizers.

The quality of a fertilizer depends not only on its chemical composition, but also on how readily it can be applied.

Ammonium bicarbonate

A further example of neutralization is the reaction of ammonia with carbonic acid. The product is ammonium bicarbonate, used as a baking powder.

Ammonia		carbonic acid		ammonium bicarbonate		water
NH_3	+	$CO_2 + H_2O$	$\longrightarrow$	$(NH_4)HCO_3$	+	H_2O

4.3. Urea $H_2N-\underset{\underset{O}{\|}}{C}-NH_2$

The reaction between ammonia and carbon dioxide under pressure and at temperatures between 170 and 190°C (338 and 374°F) gives urea.

Ammonia   carbon   urea   water
                 dioxide

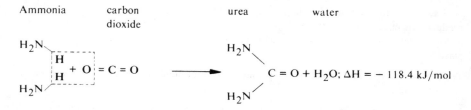

$$\Delta H = -118.4 \text{ kJ/mol}$$

The annual world production of urea is now (1978) more than 26 million tonnes. 87% of it is used for fertilizers.

As a fertilizer it is used either as a solid (pellets), or in aqueous solution. It has also proved possible to develop a fertilizer which releases its nitrogen to the plants gradually, over the whole period of growth. These products are based on aldehyde-ureas.

If urea is mixed in suitable proportions with cellulose or molasses it gives a fodder which produces protein when animals chew the cud.

Urea condenses with formaldehyde to give polycondensates, which are used industrially as hardening resins and as binders in the manufacture of varnishes and adhesives.

Urea is an intermediate stage in the manufacture of melamine resins and of hydrazine.

4.4. Hydrazine H_2N-NH_2 and hydroxylamine H_2N-OH

Hydrazine can be prepared either from ammonia, which is treated with chlorine and caustic soda solution, or from urea.

Ammonia	sodium hydroxide	chlorine	hydrazine	common salt	water

$$2NH_3 + 2NaOH + Cl_2 \longrightarrow H_2N-NH_2 + 2\,NaCl + 2H_2O$$

Hydrazine is a colourless liquid and a powerful reducing agent. For this reason it is used to remove residual oxygen from steam and from boiler feed water. The reaction between hydrazine and oxygen produces elementary nitrogen and water, which are not corrosive.

Hydrazine	oxygen	nitrogen	water

$$H_2N-NH_2 + O_2 \longrightarrow N_2 + 2H_2O$$

For the same reason hydrazine is added to anticorrosion agents for heating-oil tanks, boiler tubes and borehole pipes.

When combined with liquid oxygen, hydrazine is a valuable fuel for rocket propulsion. The reaction represented by the last equation is accompanied by a large heat of combustion, which is responsible for the high thrust.

Hydrazine is often used as a foaming agent for rubber, thermoplastics and other expanded products. Its effect is due to the nitrogen gas liberated. Derivatives of hydrazine are important intermediates in the synthesis of special dyes, pharmaceutical products and plastics.

The annual world production of hydroxylamine is several hundred thousand tonnes. It has a very wide range of uses.

In particular it plays an important part in the production of intermediates for the synthesis of polyamides such as Nylon.

It also has a stabilizing effect on products made from natural rubber or styrene.

Hydroxylamine is sometimes prepared industrially by the reduction of salts of nitrous acid, HO-N = O, but mainly by the catalytic reduction of NO or HNO_3 with hydrogen.

4.5. Hydrogen cyanide HCN

Hydrogen cyanide (also known as hydrocyanic acid or prussic acid) is a highly poisonous liquid smelling like bitter almonds. It is manufactured on a large scale from ammonia and methane by the process developed by Degussa.

Ammonia		methane		hydrogen cyanide		hydrogen
NH_3	+	CH_4	$\longrightarrow$	$H{-}CN$	+	$3\,H_2$

Its economic importance increases each decade. World production has almost doubled over the last ten years, and amounted to 450,000 t in 1970.

Only very few figures on the production of HCN are obtainable. Some 180,00 tonnes of HCN (100%) were produced in USA in 1977.

The addition to the original use of its salt (cyanides) for the extraction of gold, the main fields of applications of hydrogen cyanide are in the manufacture of plastics and synthetic fibres such as polymethacrylates, polyacrylonitrile and polyamides.

4.6.   Review of products derived from ammonia

Starting materials

End-products and field of application

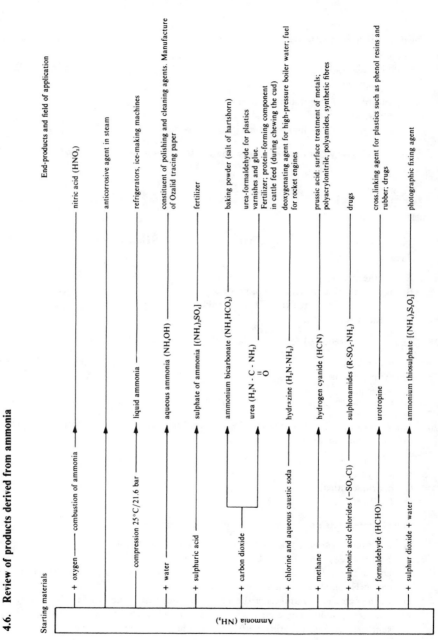

Ammonia (NH_3)

Starting materials		End-products and field of application
+ oxygen	combustion of ammonia → nitric acid (HNO_3)	nitric acid (HNO_3)
	compression 25°C/21.6 bar → liquid ammonia	anticorrosive agent in steam refrigerators, ice-making machines
+ water	aqueous ammonia (NH_4OH)	constituent of polishing and cleaning agents. Manufacture of Ozalid tracing paper
+ sulphuric acid	sulphate of ammonia [$(NH_4)_2SO_4$]	fertilizer
+ carbon dioxide	ammonium bicarbonate (NH_4HCO_3)	baking powder (salt of hartshorn)
	urea ($H_2N \cdot C \cdot NH_2$, =O)	urea-formaldehyde for plastics varnishes and glue. Fertilizer; protein-forming component in cattle feed (during chewing the cud)
+ chlorine and aqueous caustic soda	hydrazine ($H_2N\text{-}NH_2$)	deoxygenating agent for high-pressure boiler water; fuel for rocket engines
+ methane	hydrogen cyanide (HCN)	prussic acid: surface treatment of metals; polyacrylonitrile, polyamides, synthetic fibres
+ sulphonic acid chlorides ($-SO_2\text{-}Cl$)	sulphonamides ($R\text{-}SO_2\text{-}NH_2$)	drugs
+ formaldehyde (HCHO)	urotropine	cross-linking agent for plastics such as phenol resins and rubber; drugs
+ sulphur dioxide + water	ammonium thiosulphate [$(NH_4)_2S_2O_3$]	photographic fixing agent

II Inorganic raw materials and large-scale products: Large-scale industrial processes

5. Self-evaluation questions

22   Ammonia, NH_3, is

    A   toxic
    B   readily liquefiable under pressure
    C   odourless
    D   an important raw material for nitrogen production

23   Which substances can be used as raw materials for ammonia production?

    A   air
    B   water
    C   natural gas
    D   petroleum

24   The synthesis of ammonia, $N_2 + 3H_2 \rightleftharpoons 2\,NH_3$; $\Delta H = -92\ kJ/mol$, takes place

    A   exothermally
    B   as a discontinuous circulation process
    C   with a low yield involving a single reaction
    D   without reverse reaction

25   The hydrogen required for ammonia synthesis is obtained chiefly from

    A   water
    B   hydrocarbons
    C   water gas
    D   hydrogen chloride

26   The majority of nitrogen produced in the world by way of ammonia synthesis is used for the production of

    A   plastics
    B   explosives
    C   pharmaceuticals
    D   fertilizers

27   Which statements about ammonia synthesis are correct?

    A   Ammonia synthesis takes place with the aid of a catalyst.
    B   The price of ammonia is tied to the cost of liquefying air.
    C   Owing to the foreseeable saturation of the world market by nitrogen fertilizers, no new NH_3 plants are being constructed anywhere in the world.
    D   In the synthesis of ammonia the accompanying gases (such as oxygen from the air, carbon monoxide, etc.) have to be separated from the raw gas prior to the synthesis because separation after the synthesis is more expensive.

28   Urea can be obtained from ammonia. Which statements about urea are correct?

A   Urea is obtained by the reaction of ammonia with carbon monoxide.
B   Urea has the structural formula

C   Urea is used in fertilizers and feed agents.
D   Urea is used as a starting product for the manufacture of synthetic resins.

29   Which compound manufactured from ammonia is highly toxic?

A   hydrocyanic acid
B   urea
C   ammonium nitrate
D   ammonium hydrogen carbonate

30   In which two ways can ammonia be most economically transported?

A   Liquefied, under pressure
B   Liquefied, under high pressure and at elevated temperature
C   In the form of gas under low pressure
D   Dissolved in water under pressure

31   What is the percentage by weight of nitrogen in NH_3?

A   5.9%
B   20%
C   82%
D   100%

4 Manufacture of Nitric Acid. Fertilizers

1. Manufacture of nitric acid

1.1. Properties of nitric acid

Nitric acid is a strong monobasic acid with the formula HNO_3.

Pure anhydrous nitric acid is a colourless liquid which boils at 84°C (183°F) and freezes at −41.1°C (−42°F) to snow-white crystals, both at 1 bar (14.5 lb/in²) pressure.

However, if an aqueous solution of nitric acid is heated it forms an azeotropic mixture boiling at 121.8°C (251.2°F) and containing 69.2% nitric acid and 30.8% water. Nitric acid of this composition is termed "concentrated nitric acid".

On boiling, and also when exposed to light at room temperature the acid is partly decomposed into gaseous nitrogen dioxide, water and oxygen.

Nitric acid	nitrogen dioxide	water	oxygen
$4\,HNO_3 \longrightarrow$	$4\,NO_2$	$+\quad 2\,H_2O\quad +$	O_2

Part of the reddish-brown nitrogen dioxide remains dissolved in the liquid nitric acid and gives it a yellow or orange colour, according to its concentration.

Nitric acid which evolves brown fumes in air is called fuming nitric acid.

In addition to its acid properties nitric acid possesses strong corrosive and oxidizing powers. Because of its powerful oxidizing action, plant and vessels used for the transport or concentrated nitric acid must be made of special materials in order to minimise the economic consequences of corrosion.

A mixture of 1 volume of concentrated nitric acid with 3 volumes of concentrated hydrochloric acid is called aqua regia, since it will dissolve almost all nobel metals, even including gold, the king of metals.

The ability of aqua regia to dissolve noble metals such as gold depends on the production of active chlorine according to the following equation:

nitric acid	hydrochloric acid	nitrosyl chloride	active chlorine	water
$HNO_3\quad +$	$3\,HCl \longrightarrow$	$NOCl\quad +$	$2\,Cl\quad +$	$2\,H_2O$

On the other hand there are some base metals, such as aluminium, cobalt, nickel and chromium, which are not attacked by concentrated nitric acid. The oxidizing action of the acid produces on the surface of the metal a thin and coherent layer of oxide, which protects the underlying metal from further attack. This phenomenon is known as passivation.

II Inorganic raw materials and large-scale products: Large-scale industrial processes

1.2. Industrial importance

The world production of nitric acid in 1977 was 27.1 million tonnes; of this quantity, the following countries accounted for:

Federal Republic of Germany	2 849 000 t
Great Britain	2 623 000 t
Japan	646 000 t
Poland	2 113 000 t
USA	7 146 000 t

The largest fraction of this was converted to nitrogen fertilizers. Another large amount was used as 50–70% aqueous acid for the wet treatment of crude phosphate (see II,6).

When mixed with sulphuric acid 98–100% nitric acid is used as a nitrating agent for introducing nitro-groups into organic compounds. This process is important for making dyes, intermediates for pharmaceutical products, and explosives.

1.3. Manufacture of nitric acid from ammonia

In order to produce nitric acid it is necessary to make NO_2, the oxide of nitrogen which reacts readily with water to give nitric acid. Since elementary nitrogen is very inert, and reacts with oxygen to give oxides of nitrogen only under extreme conditions, a process of direct oxidation is not economic: this contrasts with the ease with which sulphur is oxidized to sulphur dioxide and trioxide, and phosphorus to phosphorus pentoxide. An indirect route is followed by the catalytic combustion of ammonia with atmospheric oxygen, which produces the required oxides of nitrogen.

1.3.1. Chemical principles

The reaction of ammonia with atmospheric oxygen, ie the combustion of ammonia, is a reaction of great industrial and economic importance.

Ammonia and molecular oxygen do not react together spontaneously to give oxides of nitrogen. However, by heating to a high temperature in the presence of catalysts they can be activated and caused to react.

Once the combustion of ammonia has started it continues with the evolution of heat. The reaction is therefore catalytic and exothermic, and can be represented by the following equations:

1) Ammonia   atmospheric oxygen   nitric oxide   water

$$4\,NH_3 \quad + \quad 5\,O_2 \quad \xrightarrow{\text{catalyst}} \quad 4\,NO \quad + \quad 6\,H_2O$$
$$\Delta H = -904.4 \text{ kJ/mol*}$$

---

*The value of ΔH refers to the number of moles represented in the chemical equation.

The nitric oxide (nitrogen monoxide) produced reacts further with atmospheric oxygen to give nitrogen dioxide. This second stage of the reaction is also exothermic:

2)　　Nitrogen　　　　atmospheric　　　　　nitrogen
　　　　monoxide　　　　　oxygen　　　　　　　dioxide

$$4\,NO \quad + \quad 2\,O_2 \quad \longrightarrow \quad 4\,NO_2$$
$$\Delta H = -228.5\ \text{kJ/mol}$$

The combustion of 68 g (0.15 lb) of ammonia in atmospheric oxygen produces 184 g (0.41 lb) of nitrogen dioxide and 1133 kJ (1074 Btu) of heat energy. This liberated heat is used for raising steam.

The nitrogen dioxide formed is absorbed in water, and reacts with it to give nitric acid. This reaction can be represented in a simplified form by the following equation.

3)　　Nitrogen　　　　water　　　　　　　nitric　　　　nitrogen
　　　　dioxide　　　　　　　　　　　　　　acid　　　　monoxide

$$3\,NO_2 \quad + \quad H_2O \quad \longrightarrow \quad 2\,HNO_3 \quad + \quad NO$$

The nitrogen monoxide thus formed is recirculated for oxidation to nitrogen dioxide.

The chemical equation for the total process is:

$$4\,NO_2 + 2\,H_2O + O_2 \quad \longrightarrow \quad 4\,HNO_3; \quad \Delta H = -347\ \text{kJ/mol}$$

A different procedure is followed for making 98–100% nitric acid, required for nitration in organic syntheses.

The nitrogen monoxide which has not reacted completely according to equation 2 is oxidized to nitrogen dioxide with liquid 98–100% nitric acid.

　　Residual nitrogen　　　98–100%　　　　　nitrogen　　　　water
　　　monoxide　　　　　　nitric acid　　　　　dioxide

$$NO \quad + \quad 2\,HNO_3 \quad \longrightarrow \quad 2\,NO_2 \quad + \quad H_2O$$

The NO_2 thus formed from the residual nitrogen monoxide is separated off as N_2O_4 and continuously oxidized to nitric acid in autoclaves. In contrast with other processes the effluent gases are practically free from oxides of nitrogen.

1.3.2. Industrial manufacturing process

1.3.2.1. Raw materials

Up to 1914, the beginning of the First world War, nitric acid was manufactured on a large scale by treating Chile saltpetre (sodium nitrate) with sulphuric acid.

Sodium nitrate		sulphuric acid		nitric acid		sodium sulphate
$2\,NaNO_3$	$+$	H_2SO_4	$\longrightarrow$	$2\,HNO_3$	$+$	Na_2SO_4

As soon as the synthesis of ammonia from the elements on a large scale had been achieved, ammonia was invariably used as the starting material for the maufacture of nitric acid. The necessary raw materials were then liquid ammonia, atmospheric oxygen and water. Nitric acid plants were often built in conjunction with ammonia synthesis installations.

1.3.2.2. Process details[*]

The manufacture of nitric acid by the combustion of ammonia can be divided into four stages:

1. Mixing the starting materials.
2. Reaction stages:
   a) combustion of ammonia to give nitric oxide;
   b) oxidation of nitric oxide to nitrogen dioxide.
3. Absorption of the nitrogen dioxide in water.
4. Purification of the effluent gases.

1. Mixing the starting materials
Liquid ammonia is evaporated in a vaporizer and then thoroughly mixed with filtered air for combustion. This air must be carefully purified by filtration, since even traces of impurities can act as catalyst poisons and hence considerably reduce the yield of nitrogen dioxide. the yield depends critically on the ratio ammonia/air. The proportion of ammonia is usually about 12.5 vol %. an upper limit for this ratio is set by the explosion limit of an ammonia-air mixture (about 14 vol.% NH_3 in air at 1 bar or 14.5 lb/in²).

2. Reaction stages
The ammonia-air mixture is burnt at temperatures between 800 and 930°C (1470 and 1706°F) on a platinum-rhodium gauze catalyst. Since all the reactions taking place are exothermic the heat content of the products of combustion is used for raising steam. This, together with subsequent cooling with water, is necessary in order to cool the gases sufficiently for absorption.

The mixture of nitric oxide, nitrogen and water vapour formed by the combustion of ammonia (see equation 1, p. 4) is cooled to 120 to 150°C (248 to 302°F) in a waste heat boiler (heat exchanger), and then to 20–35°C (58–95°F) in a second cooler. This causes the condensation of the water formed in reaction 1. The nitrogen present comes from the air used for combustion and plays no part in the reaction.

---

[*]See also Fig. 1, p. 9, Flow scheme for the manufacture of nitric acid. (This scheme does not apply to the manufacture of 98–100% acid.)

Finally, the nitric oxide is oxidized to nitrogen dioxide with secondary air according to equation 2, p. 5.

In different versions of the process the platinum content of the catalyst is between 90 and 95%. The catalyst gauze is very fine, with 1024 meshes per cm² (6605 per in²). According to the combustion system used for the life-time of a gauze varies between 3 and 18 months. At this point it has lost 40% of its original weight, mainly by loss of platinum. It is then replaced and sent for melting down.

Platinum is volatilized together with the nitrogen dioxide and is deposited in the cooler parts of the plant. Up to 80% of these platinum deposits can be recovered and worked up by special methods.

3. Absorption of the nitrogen dioxide in water
This occurs in absorption towers, in which the water flows in the opposite direction to the flow of gas.

The formation of nitric acid takes place according to equation 3, p. 5.

The absorption towers are fitted with Raschig rings or with gauze or bubble trays. Continuous cooling is necessary to remove the heat evolved on oxidation and absorption. Absorption produces 50 to 68% nitric acid, depending on the type of unit. It still contains some dissolved nitric oxide and nitrogen dioxide. These can be removed by blowing with air, leaving clear colourless nitric acid.
The water required for absorption is precooled by the cooling effect of the vaporization of liquid ammonia (first stage of the process).

4. Purification of the effluent gases

All installations are provided with units for purifying the effluent gases, so as to avoid unacceptable pollution of the atmosphere. They serve to remove the poisonous oxides of nitrogen from the exhaust gas, or at least to reduce their concentrations so much that no unpleasant or harmful effects are produced in the environment.
The oxides of nitrogen in the effluent gas are decomposed into their elements on a palladium catalyst. The temperature required for decomposition is produced by the combustion on the catalyst of fuel gases such as methane, propane, hydrogen or carbon monoxide with the oxygen remaining in the exhaust gas. The surplus heat of reaction thus produced is utilized in a waste heat boiler or a gas turbine. Other methods of purifying the effluent gas have been used as alternatives to the catalytic decomposition of the nitrogen oxides. They all have the common feature that they are very costly.

1.3.2.3. Materials

The powerful corrosive properties of nitric acid present a complex problem to the plant designer. In addition to being resistant against corrosion the materials must be easy to weld. For nitric acid of moderate concentrations austenitic chromium-nickel steels are satisfactory. (Austenitic steels are iron-carbon alloys having a particular type of crystal structure.)

For 98–100% nitric acid the most suitable materials are 99.8% pure aluminium, ferrosilicon, and austenites containing nitrogen. For boiling 98–100% nitric acid the very costly metal tantalum has to be used.

1.4. Different processes based on ammonia

The processes for manufacturing nitric acid from ammonia can basically be divided into four categories:

1. combustion of ammonia at atmospheric pressure and absorption of the nitrogen dioxide under pressure;
2. combustion of ammonia and absorption of nitrogen dioxide at the same pressure, which may be anything between 4 and 10 bar (72.5 and 145 lb/in²);
3. combustion of ammonia and absorption of nitrogen dioxide under pressure, but with different pressures for the two processes;
4. the special process for making 98–100% nitric acid, in which unreacted NO is oxidized to N_2O_4.

The choice of process for the manufacture of nitric acid depends on local conditions, on the costs of raw materials and energy, on investment costs, periods for writing off equipment, and other costs of servicing capital.

The balance between these costs must be optimized. The following are the requirements for making 1 tonne of 60–65% nitric acid, the exact figures depending on the details of the process:

Ammonia	between	0.2788	and	0.289 t
Cooling water (22°C)	between	150	and	170 t
Platinum catalyst	between	0.05	and	0.22 g
Electrical energy	between	10	and	14 kWh

About 75% of the production costs stem from the price of ammonia.

On account of the exothermic nature of the reaction between 0.21 and 0.53 t of steam at 400°C (752°F) and 40 bar (580 lb/in²) is also produced which appreciably offsets the cost of production.

Single-stream plants now being built are designed for a daily maximum production of between 1000 and 1200 t 100% nitric acid.

Figure 1. Flow scheme for the manufacture of nitric acid

1. Mixing of starting materials
2. Combustion of ammonia and production of steam
3. Oxidation and absorption

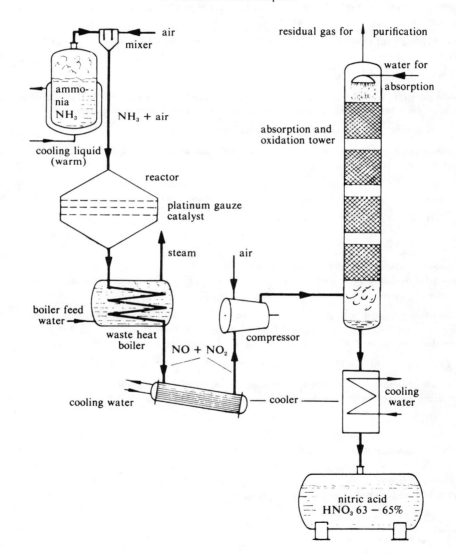

II Inorganic raw materials and large-scale products: Large-scale industrial processes

1.5. Uses of nitric acid*

1.5.1. Nitrogen fertilizers

The largest fraction of nitric acid is used (as 50–70% aqueous acid) for the production of nitrogen fertilizers. Fig. 2, p. 16 shows the increase in world consumption of nitrogen fertilizers over a period of 38 years.

In 1978 the total world production of nitrogen fertilizers was 49.4 million tonnes, calculated as N.

a) Sodium nitrate, $NaNO_3$;

sodium nitrate is used particularly as a spring top-dressing for sugar beet. It is also used as a basis for explosives and as an oxidizing agent in the glass and enamel industries.

b) Ammonium nitrate, NH_4NO_3, and calcium ammonium nitrate;

ammonium nitrate (or nitrate of ammonia) is obtained by neutralizing nitric acid with gaseous ammonia.

| Ammonia | nitric acid | | ammonia nitrate |
| gas | (liquid) | | (solid) |

$$NH_3 \quad + \quad HNO_3 \quad \longrightarrow \quad NH_4NO_3;$$
$$\Delta H = -145.7 \text{ kJ/mol}$$

Ammonium nitrate is readily soluble in water and is also strongly hygroscopic. It is a powerful oxidizing agent and is used for making mining explosives, characterized by a low explosion temperature.

Ammonium is used in large quantities as a fertilizer in West Germany it has to be mixed with at least 30% of an inert material. Ammonium nitrate mixed with at least 35% of powdered limestone (calcium carbonate) one of the most important nitrogen fertilizers.

c) Nitrate of lime, $Ca(NO_3)_2$, with not more than 8.5% ammonium nitrate.

Nitrate of lime is used mainly as a fertilizer. Its water of crystallization can be easily replaced by urea, giving the double-salt fertilizer $Ca(NO_3)_2.4 CO(NH_2)_2$.

1.5.2. Treatment of crude phosphate

The treatment of crude phosphate with nitric acid is of considerable importance. It is mainly used for producing phosphate fertilizers.

---

*See page 21, Review of the uses of nitric acid.

The reaction can be represented by the following overall equation:

Crude phosphate		nitric acid		calcium nitrate		phosphoric acid		hydrogen fluoride

$$Ca_5(PO_4)_3F \ + \ 10\,HNO_3 \longrightarrow 5\,Ca(NO_3)_2 \ + \ 3\,H_3PO_4 \ + \ HF$$

The hydrogen fluoride is absorbed as a silicate compound.

The calcium nitrate crystallizes out on cooling and is then centrifuged off.

This treatment with nitric acid is known as the Odda process. At present about 40 plants are operating this process in Europe, with a capacity of about 600 000 t P_2O_5, of which about half is in West Germany.

1.5.3. Nitric acid as a nitrating agent

When mixed with sulphuric acid, 98–99% nitric acid is used as a nitrating agent for introducing nitro-groups into organic compounds.

The nitration of benzene gives nitrobenzene (see III,4).

Benzene   nitrating acid   nitrobenzene   water

$$\text{Benzene} - H \ + \ HO-NO_2 \longrightarrow \text{Benzene} - NO_2 \ + \ H_2O$$

The reduction of nitrobenzene to aniline leads to the wide spectrum of azo-dyes.

The nitration of toluene gives trinitrotoluene, TNT, the basis of many explosives.

Toluene   nitrating acid   trinitrotoluene   water

$$\text{Toluene} - CH_3 \ + \ 3\,HO-NO_2 \longrightarrow O_2N-\text{(ring)}-CH_3 \ + \ 3\,H_2O$$

2. Fertilizers

2.1. Requirements of plants for nutrient elements

The building up of plants requires a number of chemical elements, known as nutrient elements. They are vital and indispensable for plant growth, and cannot be replaced by any other elements. In addition to the non-metals hydrogen H, boron B, carbon C, nitrogen N, ogygen O, sulphur S, phosphorous P and chlorine Cl, they include the following metals: magnesium Mg, potassium K, calcium Ca, manganese Mn, iron Fe, copper Cu, zinc Zn and molybdenum Mo. By means of experiments in aqueous cultures it has been shown unequivocally that these 16 nutrient elements are indispensable for higher plants. The nutrient elements are absorbed by plants in different and appropriate amounts, according to their specific functions in plant metabolism. The elements C, O and H, which constitute the basic components of carbohydrates, are obtained by plants from the compounds CO_2 and H_2O. Carbon dioxide, CO_2, is present in the atmosphere at a concentration of 0.033 vol. %.

The growth and yield of plants is limited by the nutrient element which is available in the smallest quantity relative to the amount required. This was first stated by J. Liebig (1803–1873) in his "law of the minimum".

2.2 Definitions and classification

Organic or inorganic materials which contain the nutrients required by plants in a form suitable for assimilation are known as fertilizers. A distinction is usually made between natural and artificial fertilizers. The natural fertilizers are all organic, and are obtained in agricultural practice as valuable by-products of stock-farming or crop cultivation. They include stable manure, liquid manure, straw, compost and crops for ploughing in, such as leguminosae. Their content of nutrients is relatively low compared with artificial fertilizers.

The average amounts of animal manure produced by various species (in tonnes per animal per year) are as follows: cows 11.5, horses 5.0, pigs 2.0 and sheep 0.7.

Artificial fertilizers include all the varieties which are commercially available. They are manufactured industrially and contain definite quantities of nutrients. A large variety of products is necessary in order to cover the specific requirements of individual plant crops and the nature of the soil in which they are cultivated.

Commercial fertilizers may be either inorganic or organic. The inorganic salts are also known as mineral fertilizers.

Commercial organic fertilizers include peat (with and without the addition of minerals), and meal made from horn, bones or fish. Their value lies mainly in their nitrogen content and their use is confined to horticulture.

Commercial organic fertilizers based on urea are gaining in importance.

The most important mineral fertilizers are the following:

	Nutrient element	specified as
Nitrogen fertilizers	N	% N
Phosphate fertilizers	P	% P_2O_5
Potash fertilizers	K	% K_2O
Lime fertilizers	Ca/Mg	% CaO/MgO
Double fertilizers	egN, P	% N, % P_2O_5
	N, K	% N, % K_2O
	P, K	% P_2O_5, % K_2O
Multiple fertilizers	N, P, K	% N, % P_2O_5, % K_2O

They may also be enriched with any trace elements required. Their production requires chemical or industrial plant.

2.3. Use of fertilizers

Fertilizers serve to replace the nutrients removed from the soils by the crops, especially nitrogen, phosphorus, potassium and magnesium. Abundant harvests require the application of increased amounts of fertilizers.

The optimum growth of a plant can be affected by fertilizers in three ways:

1. by replacing the necessary reserves of nutrients;
2. by regulating the pH of the soil;
3. by improving the soil structure.

In addition to carbon and oxygen, special importance attaches to the essential component nitrogen, which is necessary for synthesizing plant protein. Although the atmosphere contains 78 vol. % of nitrogen, only a few micro-organisms are able to assimilate elemental nitrogen directly. The leguminosae family of plants constitute an exception: they include lupins, peas, beans, vetches, lucerne and clover. They have a symbiotic relationship with nodular bacteria, which can extract nitrogen from the atmosphere for plant growth. A good field of lucerne or clover fixes 200 kg (440 lb) of nitrogen annually per hectare. For this reason leguminous plants are ploughed in and in this way serve as fertilizer.

All other plants have to take up nitrogen in a chemically combined form.

Recently (1972) it has proved possible to cross nodular bacteria with ordinary soil bacteria, so that the desired aim of making atmospheric nitrogen available to plants may be supported in future by the soil bacteria.

Chemically combined nitrogen is provided by mineral fertilizers in the form of nitrate ions, NO_3^-, ammonium ions, NH_4^+, or amino-groups, $-NH_2$.

II Inorganic raw materials and large-scale products: Large-scale industrial processes

The pH of soil determines which form is preferentially absorbed by the plants. In the neutral range pH 6.8–7.0 nitrogen and be assimilated either as NO_3^- or as NH_4^+, but in the acid range nitrate ions are utilized more readily than ammonium ions.

Compounds of phosphorus are vital constituents of plant cells and of their nuclei. They also play a decisive part in the metabolism of matter and energy. Plants absorb phosphorus from the soil as phosphate. The quality of a phosphate fertilizer depends not only on its content of P_2O_5, but also on its solubility in water and citrate solutions, ie its solubility under soil conditions.

The function of potassium in growth is not yet fully understood. It regulates the water balance in plants and promotes their exchange of carbohydrates. Potash fertilizers contain the potassium ion K^+ in a water-soluble form. They are obtained by processing crude potassium salts, which must be freed from accompanying substances.

Lime fertilizers fulfil a double function. They consist chemically of calcium carbonate, oxide, chloride and silicate, accompanied by magnesium carbonate, $MgCO_3$. Their action depends on raising the pH of the soil, improving the soil structure, and providing Ca^{2+} and Mg^{2+} nutrients for the plants.
A deficiency of calcium leads to poor root development.

Magnesium ions are indispensable for the formation of chlorophyll, the green leaf pigment.

The rationalization of agriculture has made it necessary to give the plants as many nutrients as possible in the correct proportions in a single distribution of fertilizer. This has led to the development of multiple fertilizers, which usually consists of a combination of nitrogen, phosphorus and potassium compounds, often with the addition of trace elements. They are known as N-P-K fertilizers. The proportions of nutrient elements are usually:

Nitrogen : phosphorus : potash = 1 : 1 : (0.8-1.6)

though a wide range of types is in use, for example with ratios of 3:1:1 and 2:1:2.

2.4 Fertilizers requirements

2.4.1. Uptake of nutrients

A good grain harvest averaging 3.5 tonnes per hectare (2.47 acres) extracts from the soil approximately the following quantities:

70 to 100 kg N/ha (154 to 221 lb/ha);
30 to   50 kg P_2O_5/ha (66 to 110 lb/ha);
90 to 140 kg K_2O/ha (198 to 309 lb/ha).

For a good potato harvest these figures are still higher. A harvest yield of 25 t/hectare implies a loss of:

100 to 150 kg N/ha (221 to 331 lb/ha);
 30 to   50 kg P_2O_5/ha (66 to 110 lb/ha);
160 to 260 kg K_2O/ha (353 to 573 lb/ha).

Adequate nourishment for the population can only be ensured by good harvest yields. In the long term these can only be obtained if the nutrient substances extracted from the soil are replaced by the application of sufficient amounts of fertilizers.

The application at the correct time of adequate amounts of multiple fertilizer, ie appropriate proportions of nitrogen, phosphate and potassium salts together with trace elements, leads to the following increases in yields per kg (2.205 lb) of nitrogen:

16 kg (35 lb) of grain (wheat, barley, rye, rice, etc);
80 kg (176 lb) of potatoes;
70 kg (154 lb) of sugar beet;
30 kg (66 lb) of hay.

2.4.2. Consumption of fertilizers

The annual world production of nitrogen fertilizers amounts at present to about 49.4 million t, calculated as 100% N. If these quantities were used exclusively to intensify the production of grain they would lead to an additional production of 790 million tonnes per year. This extra yield (compared with what would be obtained without the use of mineral fertilizers) could provide 4.0 thousand million people with additional daily bread ration of about 850 g (1.87 lb) corresponding to a calorific value of about 7704 kJ* (1840 calories), which represents almost 75% of the daily energy requirements of an adult. The consumption of fertilizers throughout the world is shown in Fig. 2, p. 16.

According to current estimates, in order to feed adequately the population expected by the year 2000, world production of nitrogen fertilizers alone must be increased to 90 to 95 million t/year.

The following are the 1978 production figures for the whole world:

nitrogen fertilizers	49.4	million t, calculated as	N
phosphorus fertilizers	30.0	million t, calculated as	P_2O_5
potassium fertilizers	26.0	million t, calculated as	K_2O

2.5. Fertilizer costs

In farming practice the costs of applying planned and sufficient amounts of fertilizers is made up of the variable costs of the fertilizers and the fixed costs of applying them. Fertilizers represent a means of increasing production, and their use must show a net financial profit. There are upper and lower limits within which the application of fertilizers is profitable. The maximum net return is obtained somewhere between these limits. If too little fertilizer is applied the fixed costs of spreading it in the field will have too great an effect in the balance sheet, and the maximum increase in yield will not be obtained. The lower limit for profitability may not thus be reached.

---

* In calculating these figures it was assumed that grain contains on average 15% moisture, and bread 45%.

Figure 2. World consumption of fertilizers

million tonnes

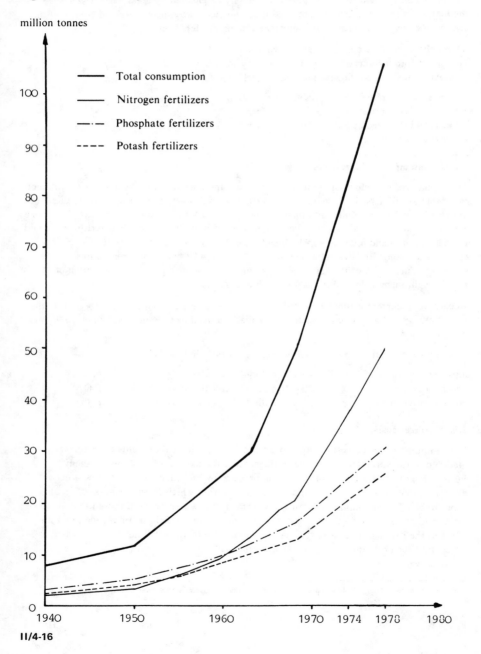

Financial losses may also be incurred if too much fertilizer is applied, since the amounts of fertilizer may bear no relation to the improvement in crop yields. In this case the upper limit for profitability will have been passed.

The relation between the costs of applying fertilizer and the net financial return is shown schematically in Fig. 3.

Figure 3.

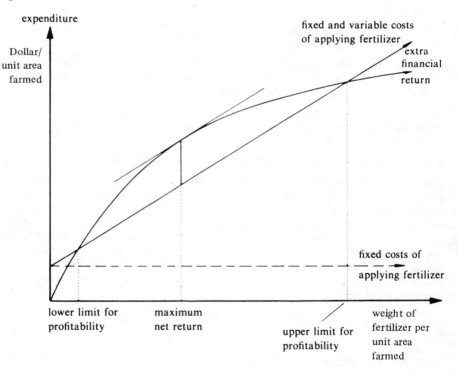

2.6 Sources of raw materials and production of mineral fertilizers

2.6.1. Nitrogen fertilizers (including urea)

The key product for nitrogen fertilizers is ammonia. The synthesis of ammonia has made it possible to obtain atmospheric nitrogen in a chemically combined form (see II, 3). Nitrogen fertilizers contain ammonium, nitrate or amino-compounds, with further additions, and are sold under various trade names according to their composition.

II Inorganic raw materials and large-scale products: Large-scale industrial processes

Lime/nitrate of ammonia (a mixture of nitrate of ammonia with powdered limestone)	$NH_4NO_3/CaCO_3$
Ammonium sulphate/nitrate (a mixture of ammonium nitrate with a large excess of ammonium sulphate)	$2\,NH_4NO_3/(NH_4)_2SO_4$
Nitrochalk	$Ca(NO_3)_2$ with not more than 8.5% ammonium nitrate
Nitrate of soda	$NaNO_3$
Urea	NH_2CONH_2
Calcium cyanamide	$CaNCN + C$

For the purpose of comparing different nitrogen fertilizers their percentage content of elemental nitrogen N is always given. Naturally occurring nitrates, such as Chile saltpetre ($NaNO_3$), are of only minor importance as fertilizers.

The same applies to guano, a naturally occuring nitrogen and phosphorus fertilizer. Guano consists of the dried droppings of sea-birds, accumulated over thousands of years on the coasts of Chile and Peru.

The use of urea as a nitrogen fertilizer is increasing year by year (see II,3, p. 15).

The development of urea fertilizers has led to products containing particularly high percentages of nitrogen, up to about 46%. Plants exhibit a high tolerance to urea, which is therefore applied through the leaves, often in conjunction with plant protection agents. Its initially delayed action is due to the necessity of converting the amide nitrogen into nitrate. Urea forms condensation compounds with aldehydes, such as formaldehyde, acetaldehyde and crotonaldehyde. These compounds act as reservoirs, and the release of the amide nitrogen for assimilation by the plant takes place only slowly. Loss by leaching is also considerably reduced. These high-grade and slowly acting fertilizers are used in commercial horticulture and for lawns.

2.6.2. Phosphate fertilizers

The starting materials for phosphate fertilizers are the deposits of crude phosphate which occur naturally in sufficient quantities. The crude phosphate, $Ca_5(PO_4)_3F$, is treated with nitric or sulphuric acid and converted into fertilizers, in which the phosphorus occurs as the phosphate ions PO_4^{3-}, HPO_4^{2-} and $H_2PO_4^{-}$.

The various types of fertilizers differ in the other compounds which they contain, and also in their solubility in moist soil, which determines whether they act immediately or after some delay. The composition of commercial products is usually expressed in terms of % P_2O_5 (phosphorus pentoxide).

The following are examples of current commercial products:

Superphosphate (calcium dihydrogen phosphate with calcium sulphate)	$Ca(H_2PO_4)_2 + CaSO_4$
Rhenania phosphate (calcium sodium silicophosphate)	$CaNaPO_4 . (Ca_2SiO_4)$
Ammonium phosphate (diammonium hydrogen phosphate)	$(NH_4)_2HPO_4$
Thomas phosphate	$\approx Ca_3P_2O_8; Ca_4P_2O_9; Ca_2SiO_4$

Thomas phosphate is a fertilizer obtained by treating the phosphate slag produced in the smelting of iron ores containing phosphate.

2.6.3. Potash fertilizers

Potash fertilizers are obtained from naturally occuring potash minerals such as carnallite, $KCl.MgCl_2.6H_2O$, kainite, $KCl.MgSO_4.3H_2O$, and hard salt, a mixture of rock salt, $NaCl$, kieserite, $MgSO_4.H_2SO_4.H_2O$ and sylvine, KCl. These potash minerals are mined.

The crude salts are freed from associated materials by special processes such as flotation and dissolution, which depend on the different velocities of dissolution or crystallization of the various components.

The potassium content of commercial products is usually expressed as % K_2O (potassium oxide).

2.6.4. Lime fertilizers

The addition of lime, having a basic action, to cultivated land is effected not only by lime fertilizers, but also by lime ammonium nitrate, calcium cyanamide and Thomas phosphate. Lime fertilizers usually consist of powdered limestone, which is mainly calcium carbonate ($CaCO_3$) contaminated with clay. Other lime fertilizers are metallurgical lime, quicklime, industrial waste lime, shell lime, etc. The calcium content of these fertilizers is expressed as CaO.

The following are commercial products:

Carbonate of lime, lime marl	$CaCO_3$
Slaked lime (calcium hydroxide)	$Ca(OH)_2$
Quicklime (calcium oxide)	CaO
Metallurgical lime	Calcium silicate

2.6.5. Liquid fertilizers

The use of liquid fertilizers has recently made considerable progress in the USA, where for instance in California huge areas are fertilized in this way.

Liquid fertilizers generally consist of nitrogen compounds such as ammonia solution and solutions of ammonium nitrate and urea.

2.7. Storage of fertilizers

The value of a fertilizer depends not only on its chemical composition, but also on how easily it can be stored and distributed in the field.

A fertilizer must not cake, harden or deliquesce on storage. It is therefore made if possible in the form of rounded granules of well-defined size. If necessary a powdered substance can be added to minimize the area of contact between the granules.

2.8 Review of the uses of nitric acid

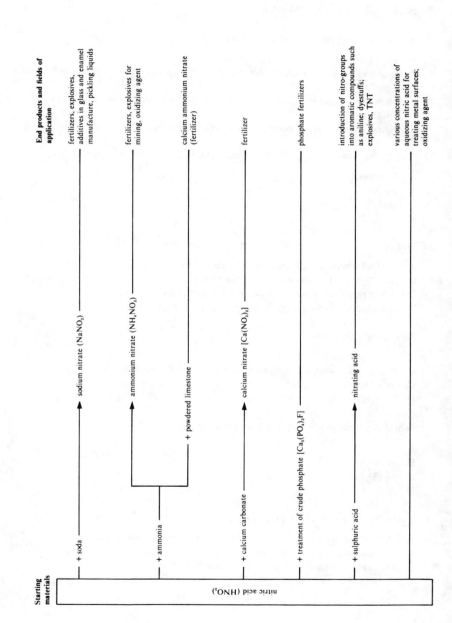

3. Self-evaluation questions

32 Which statement on nitric acid is correct?

A Pure nitric acid is an explosive liquid
B Nitric acid dissolves even noble metals such as gold and is therefore called aqua regia
C 100% nitric acid is impossible to manufacture industrially on account of its instability
D Nitric acid is classed among the oxidizing acids.

33 Which metal can be passivated by nitric acid, ie with a stable protective layer?

A aluminium
B copper
C gold
D steel

34 Production of HNO_3 by the combustion of ammonia

A is an exothermic reaction (like all combustion processes)
B takes place especially rapidly at low temperatures
C is a catalytic reaction (like nearly all industrial reactions)
D is a reaction in which the released heat is used to produce steam

35 Pure nitric acid is used

A for producing fertilizer salts
B in a mixture with concentrated sulphuric acid for nitration
C as a neutralizing agent
D for the removal of rust from steel

36 Which nitrogen compound is a brown toxic gas?

A N_2
B HNO_3
C HNO_2
D NO_2

37 The action of fertilizers in promoting growth is based on

A the breakdown of harmful organic and inorganic substances in the soil
B the catalytic action which enables the plants to assimilate the nitrogen in the air
C the supplementation of necessary nitrogen reserves
D the regulation of the soil pH

38 Which nitrogen compounds are used as components of fertilizers?

A $CO(NH_2)_2$, urea
B KCN, potassium cyanide
C $NaNO_3$, sodium nitrate
D $(NH_4)_2SO_4$, ammonium sulphate

39 Fertilizers contain nitrogen chemically linked as

A N_2
B NO_3^- ion
C NH_4^+ ion
D NH_2 group

40 In which order of importance do fertilizers rank in terms of volumetric consumption in the world?

	Nitrogen fertilizers	Phosphorus fertilizers	Potassium fertilizers
A	1	2	3
B	1	3	2
C	2	1	3
D	3	2	1

41 For which of the following fertilizer salts is the empirical formula given incorrectly? Note the valencies of the acid radicals!

A NH_4SO_4, ammonium sulphate
B NH_4NO_3, ammonium nitrate
C KNO_3, potassium nitrate
D KCl, potassium chloride

5 Sulphur and the Manufacture of Sulphuric Acid

1. Sulphur

1.1. Physical properties

At room temperature sulphur is a yellow brittle solid of density 2.0 g/cm³ (124.8 lb/ft³). Sulphur is tasteless and odourless. On heating it melts above 110°C (230°F) to a yellow mobile liquid. In this state it can be transported and loaded by pumping.

On further heating sulphur becomes viscous above 160°C (320°F) and acquires a reddish-brown colour. Above 250°C (482°F) the viscosity again decreases. Sulphur boils at 445°C (833°F) at 1 bar (14.5 lb/in²) pressure.

1.2. Occurrence

1.2.1. Elemental sulphur

Sulphur is one of the elements which occur in nature as such, ie uncombined with other elements.

However, elemental sulphur occurs only to a limited extent. Considerable deposits are found in limestone or ore formations. Known reserves amount to about 400 million tonnes, divided between Mexico, the USA (Texas and Louisiana), Poland, the USSR, Italy, Japan and Canada.

1.2.2. Hydrogen sulphide, H_2S

Naturally occurring compounds of sulphur can be divided into the following four groups:

a. hydrogen sulphide;
b. sulphide ores;
c. organic sulphur compounds, and
d. sulphates

Hydrogen sulphide is a gas at ordinary temperatures. It has an unpleasant foul smell. It is formed, for example, by the decay of organic compounds containing sulphur. Hydrogen sulphide is extremely poisonous. Natural gas can contain up to 20% of H_2S by volume and therefore cannot be supplied to the consumer without purification.

Hydrogen sulphide dissolves in water to give a weakly acid solution. Natural gas with a high H_2S content is therefore known as "acid natural gas".

1.2.3. Sulphide ores

The sulphide ores are compounds of sulphur with metals, ie inorganic sulphides, for example:

FeS_2	iron pyrites;
$CuFeS_2$	copper pyrites;
PbS	galena;
ZnS	zinc blende

Extensive deposits of sulphide ores are found ie in USA and Western Europe (Spain and Norway).

1.2.4. Organic sulphur compounds

Organic sulphur compounds of industrial importance occur in petroleum, in coal and in lignite.

They consist mainly of:

a. mercaptans (functional group -SH);
b. organic sulphides or thioethers (functional group -S-);
c. disulphides (functional group -S-S-); and
d. thiophens

In these compounds sulphur is attached to carbon atoms. Mercaptans and sulphides have an extremely unpleasant smell. They impart an odour to petroleum.

1.2.5. Sulphates

The most important sulphate minerals are:

$CaSO_4 \cdot 2\,H_2O$	gypsum
$CaSO_4$	anhydrite
$BaSO_4$	barytes (or heavy spar)

1.3. Extraction of sulphur

Sulphur is obtained both from the naturally-occurring element and also from H_2S and from the organic sulphur compounds contained in fossil raw materials. In spite of its considerable amount, sulphur removed from natural gas, petroleum or coal represents only a by-product.

1.3.1. Elemental sulphur from natural sources

Under favourable geological conditions elemental sulphur can be obtained by the Frasch process. By pumping in hot water and steam the sulphur is melted free from the rocks. The molten sulphur is then forced up to the surface by hot compressed air. The Frasch process is mainly used in the USA and in Mexico.

Figure 1.   Extraction of sulphur by the Frasch process

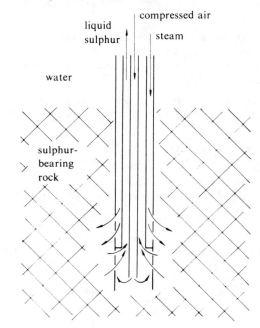

If the deposits of sulphur-bearing rock are at not too great a depth, or are under the sea-bed, then the sulphur can be mined, as (for example) in Poland.

However, in this case the sulphur must be separated from the rock and purified by flotation, melting and distillation.

The extraction of elemental sulphur has increased greatly during the last ten years from about 25 million tonnes in 1970 to about 35 million tonnes in 1979.

World production of sulphur in all forms was about 56 million tonnes in 1979.

Sulphur can be transported easily and cheaply. It is loaded in a liquid state and transported in heated containers (eg special tank ships and tank cars).

1.3.2.   Elemental sulphur from sulphur compounds

Elemental sulphur from sulphur compounds is obtained chiefly as a by-product from natural gas, petroleum and coal.

II Inorganic raw materials and large-scale products: Large-scale industrial processes

At present (1980) the amounts of sulphur which pollute the atmosphere as SO_2 during the burning of natural gas, heating oil and coal exceed the sulphur requirements of the chemical industry.

Increasing quantities of sulphur are being obtained from sulphur removal, partly because of the growing use of natural gas and petroleum for producing energy and as raw materials for chemical industry, and partly because of more stringent regulations about environmental protection.

Plants for sulphur removal in the natural gas fields are also producing considerable quantities of elemental sulphur amounting to 21 million tonnes throughout the world in 1979.

Sulphur is removed from natural gas by washing. The removal of H_2S (and of carbon dioxide, CO_2) from natural gas is effected by a wet washing process.

In order to remove sulphur from petroleum and petroleum fractions the organic sulphur compounds are first converted to H_2S by hydrogenation, and the H_2S is then absorbed.

In the Clause process hydrogen sulphide, H_2S, is converted to elemental sulphur by incomplete combustion.

Oxidation takes place on a catalyst, with controlled admission of air, so that the hydrogen in H_2S is oxidized to water vapour, H_2O, while the sulphur is not oxidized:

Hydrogen sulphide		oxygen		sulphur	water vapour

$$2\,H_2S \quad + \quad O_2 \quad \xrightarrow{\text{catalyst}} \quad 2\,S \quad +2\,H_2O$$

$$\Delta H = -443.8 \text{ kJ/mol*}$$

One normal cubic metre** of H_2S gives 1.4 kg of sulphur (or one cubic foot gives 0.0874 lb of sulphur).

1.4. Chemical properties of sulphur

At high temperatures sulphur combines with many metals and non-metals. For example, sulphur can be ignited in air and burns to give sulphur dioxide [sulphur (IV)-oxide]:

Sulphur		oxygen		sulphur dioxide

$$S \quad + \quad O_2 \quad \longrightarrow \quad SO_2;$$

$$\Delta H = -296.5 \text{ kJ/mol}$$

---

* The value of ΔH refers to the number of moles represented in the chemical equation.
**Since the volume of a gas depends strongly on the temperature and pressure, the volume at $0°C$ ($32°F$) and 1.0 bar (14.5 lb/in^2) is taken as the normal value. 1 normal cubic metre $= 1$ m^3 gas at $0°C$ and 1.0 bar.

When sulphur vapour and methane are passed over incandescent coke carbon disulphide, CS_2, is obtained:

Sulphur	methane		carbon disulphide		hydrogen sulphide
$4S$	$+$	CH_4 $\longrightarrow$	CS_2	$+$	$2H_2S$

Sulphur	carbon	carbon disulphide
$2S$	$+$ C $\longrightarrow$	CS_2

1.5. Uses of sulphur

Elemental sulphur is used for the vulcanization of rubber, in the manufacture of matches, for making carbon disulphide, for preparing pharmaceutical products and for destroying vermin.

However, of the elemental sulphur extracted or produced, almost 90% is used for the manufacture of sulphuric acid.

II Inorganic raw materials and large-scale products: Large-scale industrial processes

2. Sulphuric acid

2.1. Industrial importance

In terms of the quantity produced sulphuric acid is the most important inorganic product of chemical industry, coming ahead of ammonia, chlorine and caustic soda.

In 1979 the total production of sulphuric acid, expressed as tonnes of H_2SO_4, in the world was 135.2 million tonnes.

In 1979 sulphuric acid production, expressed as tonnes of H_2SO_4, was as follows in the countries listed below.

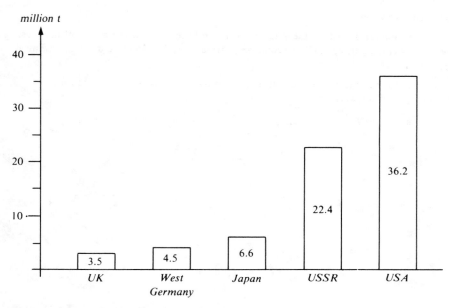

2.2. Physical properties of sulphuric acid and oleum

Pure sulphuric acid, H_2SO_4, is a colourless, transparent, oily liquid. Anydrous 100% sulphuric acid is termed the monohydrate (because $H_2SO_4 = SO_3 H_2O$). It freezes at $+10°C$ $(50°F)$ and has a density of 1.8 g/cm³ (112 lb/ft³) at room temperature.

Solutions of sulphur trioxide, SO_3, in 100% sulphuric acid are called oleum or fuming sulphuric acid. If oleum comes into contact with moist air a strongly corrosive white mist of sulphuric acid is formed.

Commercial specifications for solutions of sulphuric acid state their percentage content of H_2SO_4, while for oleum the percentage content of free dissolved SO_3 is given.

For example, 100 g (or lb) of 85% sulphuric acid consists of 85 g (or lb) H_2SO_4 and 15 g (or lb) water, while 100 g (or lb) of 65% oleum contains 65 g (or lb) SO_3 and 35 g (or lb) H_2SO_4.

It is more economical to ship oleum than sulphuric acid, since oleum has a higher percentage content of SO_3.

Similarly, oleum can absorb more water than the monohydrate, assuming that the same amount of sulphuric acid is produced in each case. There is therefore a growing trend towards the use of oleum.

Since iron is only attacked by dilute sulphuric acid, and not by concentrated sulphuric acid or oleum, steel is the most economical material for vessels used for storage and transport.

2.3.   Manufacture of sulphuric acid

2.3.1.   Summary

As shown by the scheme on p. 10, starting materials for the manufacture of sulphuric acid include elemental sulphur, sulphide ores, waste sulphates, waste sulphuric acid, and dilute sulphuric acid.

By far the most important raw materials are elemental sulphur and sulphide ores.

The scheme also shows that both elemental sulphur and sulphur combined with other elements react with atmospheric oxygen to give sulphur dioxide, SO_2.

In addition to elemental sulphur or sulphur compounds and atmospheric oxygen, water is a raw material for the manufacture of sulphuric acid.

The recovery of concentrated sulphuric acid by evaporating off water from dilute sulphuric acid constitutes an exception to the last statement.

2.3.2.   Manufacture of sulphuric acid by the double contact process

The manufacture of H_2SO_4 involves three stages:

1.   Production of SO_2

Sulphur or sulphur compounds	+	atmospheric oxygen	$\longrightarrow$	SO_2 (sulphur dioxide)

2.   Production of SO_3

SO_2	+	atmospheric oxygen	$\xrightarrow{\text{catalyst}}$	SO_3 (sulphur trioxide)

3.   Absorption of SO_3

SO_3	+	water from sulphuric acid	$\longrightarrow$	H_2SO_4 (sulphuric acid)

Figure 2. Summary of the manufacture of sulphuric acid

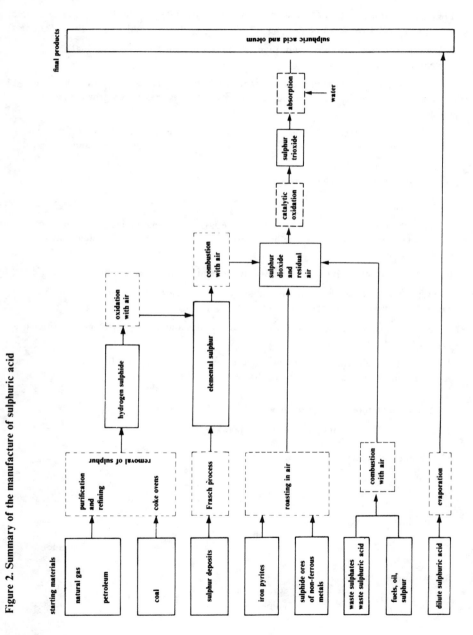

Figure 3. Double-absorption contact

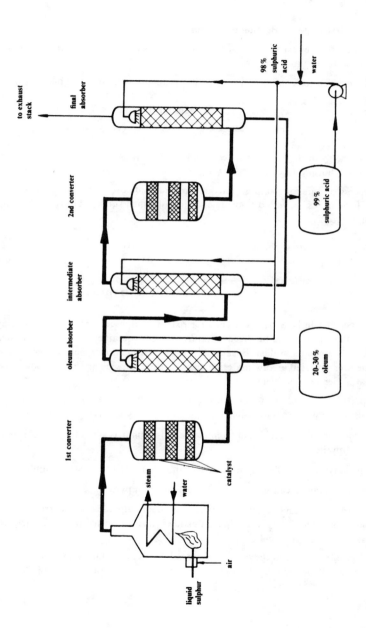

II Inorganic raw materials and large-scale products: Large-scale industrial processes

2.3.2.1. First stage: production of sulphur dioxide

This stage offers fewest problems when pure liquid sulphur can be burnt. This is done in burners similar to those used for heating oil.

The air required for combustion is dried by scrubbing with 96% sulphuric acid.

Sulphur		oxygen		sulphur dioxide

$$S \quad + \quad O_2 \quad \longrightarrow \quad SO_2; \quad \Delta H = -296.5 \text{ kJ/mol}$$

The heat energy of the exothermic reaction is used for raising steam. The combustion of one tonne of sulphur produces about 3.4 t of steam at 30 bars (435 lb/in²).

The process of obtaining sulphur dioxide from sulphide ores is known as roasting. The metal of the ore is converted to oxide and the combined sulphur burns to give sulphur dioxide:

Pyrites		atmospheric oxygen		iron (III) oxide		sulphur dioxide

$$4\,FeS_2 \quad + \quad 11\,O_2 \quad \longrightarrow \quad 2\,Fe_2O_3 \quad + \quad 8\,SO_2$$
$$\Delta H = -3310.0 \text{ kJ/mol}$$

Many types of kiln have been designed for roasting sulphides, from single-stage pyrites kilns to multi-stage, rotary and fluidizedbed kilns. The SO_2 obtained is often contaminated with considerable amounts of dust. Because of the gas purification which is necessary when sulphide ores are used as raw materials, plant costs are twice as great as when the raw material is elemental sulphur.

When pyrites is roasted in a fluidizedbed kiln 1.4 t of steam at 30 bars (435 lb/in²) is produced per tonne of pyrites; the corresponding figure for a rotary kiln is about 1 t.

The production of sulphuric acid from elemental sulphur is steadily gaining ground at the expense of pyrites.

3 t of monohydrate (100% sulphuric acid) can be obtained from 1 t of sulphur.

1.6 t of monohydrate can be obtained from 1 t of iron pyrites.

In the extraction of zinc, copper and lead from their sulphide ores the treatment of the ore necessarily produces SO_2. This sulphur dioxide is also converted to sulphur trioxide, SO_3, and finally to sulphuric acid.

Special problems arise in the case of waste sulphates and dilute sulphuric acid, which are becoming available for working up in ever increasing quantities. For example, iron (II) sulphate, which is a waste product in the manufacture of titanium dioxide, is converted to SO_2 by combustion with added sulphur:

Iron (II) sulphate		sulphur		atmospheric oxygen		iron (III) oxide		sulphur dioxide
$4 FeSO_4$	$+$	$2 S$	$+$	O_2	$\longrightarrow$	$2 Fe_2O_3$	$+$	$6 SO_2$

2.3.2.2. Second stage: oxidation of SO_2 to SO_3

The sulphur dioxide gas is oxidized further to gaseous sulphur trioxide by the oxygen remaining after the formation of SO_2:

Sulphur dioxide		atmospheric oxygen		sulphur trioxide	
$2 SO_2$	$+$	O_2	$\underset{\text{catalyst}}{\rightleftharpoons}$	$2 SO_3;$	$\Delta H = -198.0 \ kJ/mol$

This reaction only occurs at a reasonable speed at temperatures above 600°C. However, at these high temperatures the SO_3 molecule decomposes to a considerable extent into sulphur dioxide and oxygen.

This means that at 600°C (1112°F) only 70% of the SO_2 can be converted into SO_3, as indicated by the reverse arrow in the chemical equation.

However, by using a suitable catalyst the reaction can be made to proceed very rapidly at lower temperatures, at which SO_3 is stable. The manufacture of SO_3 therefore employs a catalyst containing vanadium and potassium oxides, in which the V_2O_5 is the active component. The mixture of SO_2 and air will then ignite below 500°C (932°F).

By using intermediate cooling about 95% of the SO_2 can be converted to SO_3.

2.3.2.3. Third stage: absorption of SO_3 to give sulphuric acid

For this purpose the mixed gases are passed through concentrated sulphuric acid. The water contained in the sulphuric acid reacts with the sulphur trioxide:

Sulphur trioxide		water (from conc. sulphuric acid)		sulphuric acid	
SO_3	$+$	H_2O	$\longrightarrow$	$H_2SO_4;$	$\Delta H = -132.6 \ kJ/mol$

The process of absorption is usually carried out in two successive stages, of which the first produces 20-35% oleum and the second 98% sulphuric acid.

The remaining gas, which still contains 0.5 vol.% SO_2 and 5-6 vol.% O_2, is passed over a second catalyst (hence the term "double contact process"). In this process the total yield of SO_3 exceeds 99.5%, and the problems of dealing with the effluent gas are alleviated.

While the residual gas from older installations still contained 0.2-0.5 vol.% SO_2, in the double contact process this is reduced to 100-300 ppm. Hence, in spite of increasing production, the emission of sulphur dioxide in the manufacture of sulphuric acid has been considerably reduced.

The technique of sulphuric acid manufacture is now very advanced, as is shown by the high degree of automation. For example, a modern sulphuric acid plant which uses elemental sulphur as its raw material and produces 500 t SO_3 per day requires only two workers per shift.

2.4. Properties and uses of sulphuric acid

Sulphuric acid is needed in five types of industrially important reaction:

1. H_2SO_4 reacts as a typical strong acid with metals, bases and salts;
2. sulphuric acid is used in organic chemistry for sulphonation, ie the introduction of the functional group $-SO_3H$;
3. formation of alcohol from olefins in the presence of sulphuric acid
4. sulphuric acid as the catalyst in nitration
5. H_2SO_4 combines with water.

2.4.1. Acid properties

Dilute sulphuric acid dissolves base metals as sulphates, with the evolution of hydrogen, eg

Iron		sulphuric acid		iron (II) sulphate		hydrogen
Fe	+	H_2SO_4	$\longrightarrow$	$FeSO_4$	+	H_2

The base metal lead is an exception, since a layer of insoluble lead sulphate is formed on the surface of the metal and protects it from further attack by the acid. This passivation of lead makes it possible for lead and sulphuric acid to exist in contact in the lead accumulator.

Salts of sulphuric acid, the sulphates, are readily soluble in water, with the exception of barium sulphate (heavy spar), $BaSO_4$, calcium sulphate (gypsum), $CaSO_4.2 H_2O$, and lead sulphate, $PbSO_4$.

Other examples of the acid properties of sulphuric acid are the treatment of crude phosphates to give water-soluble fertilizers:

"Phosphorite"   sulphuric acid   "superphosphate"

$$Ca_3(PO_4)_2 \quad + \quad 2\ H_2SO_4 \quad \longrightarrow \quad Ca(H_2PO_4)_2 + 2\ CaSO_4$$

and the manufacture of the fertilizer sulphate of ammonia (ammonium sulphate) from ammonia

Ammonia   sulphuric acid   ammonium sulphate

$$2\ NH_3 \quad + \quad H_2SO_4 \quad \longrightarrow \quad (NH_4)_2SO_4$$

However, since there is now a trend towards the production of sulphate-free fertilizers, sulphuric acid is often used in the fertilizer industry for making phosphoric acid (see II,6, p. 10) rather than directly:

Tertiary calcium   sulphuric   phosphoric   gypsum
   phosphate   acid   acid

$$Ca_3(PO_4)_2 \quad + \quad 3\ H_2SO_4 \xrightarrow{\ H_2O\ } 2\ H_3PO_4 \quad + \quad 3\ CaSO_4 . 2\ H_2O$$

More than half of the sulphuric acid produced is used for making water-soluble mineral fertilizers, phosphates and sulphate of ammonia.

2.4.2. Sulphonation

The introduction of the sulphonic acid group ($-SO_3H$) into organic molecules is important as a means of preparing reactive intermediates:

Benzene   sulphuric acid   benzenesulphonic   water
                                              acid

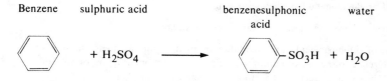

In instances where sulphonation is difficult, so that the use of H_2SO_4 gives bad yields, chlorsulphonic acid is employed. Chlorsulphonic acid is prepared from sulphur trioxide and gaseous hydrogen chloride:

$$SO_3 \quad + \quad HCl \quad \longrightarrow \quad ClSO_3H$$

The SO_3H group in the intermediate is usually split off again in subsequent syntheses.

The sulphonic acid group is however retained in the preparation of water-soluble detergents, dyestuffs and medicinal products (eg sulphonamides).

2.4.3. Alcohol production from olefins

$$\begin{array}{ccccc}
\text{olefin} & & \text{water} & & \text{alcohol} \\
R - CH = CH_2 & + & H\text{--}OH & \xrightarrow{\;H_2SO_4\;} & R\ CH_2 - CH_2 - OH
\end{array}$$

Alcohols, particularly ethanol, based on olefins industrially are produced in the USA. Sulphuric acid again acts as the catalyst.

2.4.4. Nitration

In nitration with nitrating acid, i.e. the production of nitrobenzene, sulphuric acid is used as the catalyst.

$$\begin{array}{cccc}
\text{Benzene} & \text{nitric acid} & \text{nitrobenzene} & \text{water}
\end{array}$$

This reaction is carried out in the presence of concentrated sulphuric acid. The mixture of nitric and sulphuric acids is called nitrating acid.

2.4.5. Dehydration

Concentrated sulphuric acid is highly hygroscopic, i.e. it absorbs water. Therefore it is used for drying gases (eg chlorine).

2.4.6. Review of products made from sulphuric acid

The following review shows the most important uses of sulphuric acid.

Details of the use of sulphuric acid in the manufacture of the white pigment titanium dioxide and a reference to the use of aluminium sulphate in the manufacture of aluminium silicate fillers will be found in section II.8 (Inorganic pigments and fillers).

2.4.6. Review of products made from sulphuric acid

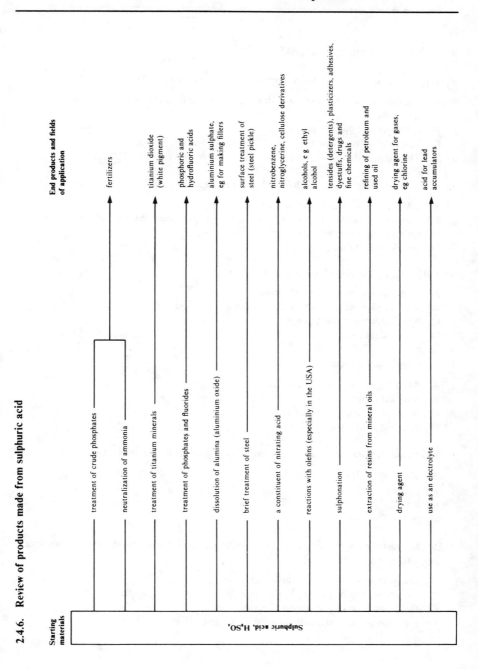

Starting materials

Sulphuric acid, H₂SO₄

- treatment of crude phosphates
- neutralization of ammonia → fertilizers
- treatment of titanium minerals → titanium dioxide (white pigment)
- treatment of phosphates and fluorides → phosphoric and hydrofluoric acids
- dissolution of alumina (aluminium oxide) → aluminium sulphate, eg for making fillers
- brief treatment of steel → surface treatment of steel (steel pickle)
- a constituent of nitrating acid → nitrobenzene, nitroglycerine, cellulose derivatives
- reactions with olefins (especially in the USA) → alcohols, e g ethyl alcohol
- sulphonation → tensides (detergents), plasticizers, adhesives, dyestuffs, drugs and fine chemicals
- extraction of resins from mineral oils → refining of petroleum and used oil
- drying agent → drying agent for gases, eg chlorine
- use as an electrolyte → acid for lead accumulators

End products and fields of application

II Inorganic raw materials and large-scale products: Large-scale industrial processes

3. Self-evaluation questions

42 Which statements on sulphur or sulphur compounds are correct?

   A   Sulphur occurs in nature only in the form of compounds owing to its high reactivity.

   B   Hydrogen sulphide is a toxic gas.

   C   The sulphur compounds in fuel oil burn to give SO_2.

   D   The biggest user of elemental sulphur is the match industry.

43 In the production of sulphur by the Frasch process the steam introduced serves to

   A   extract the sulphur from the rock containing it

   B   convert the sulphur to hydrogen sulphide, which is then obtained in the form of gas

   C   melt the sulphur out of the rock containing it

   D   purify the sulphur

44 Sulphuric acid is nowadays obtained mainly

   A   from sulphur by the Frasch process

   B   from sulphur as a by-product from natural gas and petroleum

   C   by the double contact process

   D   by the Claus process

45 Pure 100% sulphuric acid

   A   has a higher density than water

   B   is called oleum

   C   is called fuming sulphuric acid

   D   is called monohydrate

46 Which substances are used as raw materials in sulphuric acid production?

   A   elemental sulphur

   B   pyrite

   C   air

   D   water

47 In the most commonly used process for sulphuric acid manufacture, which processing stage takes place catalytically?

   A   the production of the raw materials

   B   the production of the sulphur dioxide

   C   the production of the sulphur trioxide

   D   the absorption of the sulphur trioxide to give oleum

48 The designation "double contact process" indicates that

A the catalyst employed comprises two different materials
B a processing stage is repeated with the same catalyst
C two different processing stages take place, each with a separate catalyst
D the use of this process has enabled the yield to be doubled

49 Which quantity of pure H_2SO_4 (100%) can be produced from 32 tonnes of sulphur?
To answer this question, use the table of atomic masses in section I/9, on page 5.

A 64 tonnes
B 66 tonnes
C 98 tonnes
D 100 tonnes

50 The double contact process for the production of sulphuric acid is

A a catalytic process
B an exothermic process
C a cyclic process
D a continuous process

51 Which statements about the properties and uses of sulphuric acid are correct?

A Concentrated sulphuric acid is hygroscopic, ie it binds water.
B Sulphuric acid is used primarily for the production of mineral fertilizers.
C Sulphuric acid is used in concentrated form (about 98%) in car batteries.
D Organic substances can often be made soluble by reaction with sulphuric acid.

52 Which metal is passivated by sulphuric acid?

A lead
B aluminium
C gold
D copper

53 In which compounds is sulphur hexavalent?

A SO_3
B H_2SO_4
C SO_2
D $(NH_4)_2SO_4$

II Inorganic raw materials and large-scale products: Large-scale industrial processes

54 Which two large-scale industrial processes include processing stages in which equilibrium reactions occur?

A double contact process
B ammonia synthesis
C phosphorus production
D brine electrolysis

6 Phosphorus and the Manufacture of Phosphoric Acid

1. Phosphorus

1.1. Properties

Phosphorus (chemical symbol P) does not occur in nature as the element, since it is very reactive. It must be prepared chemically from compounds containing phosphorus, the rock phosphates.

Elemental phosphorus occurs in three different forms, also called allotropes. These are yellow (or white) phosphorus, red phosphorus and black phosphorus.

The black form can only be obtained under artificial conditions.

In the industrial manufacture of phosphorus the white form is always obtained first, and can then be converted into the red.

White phosphorus is extremely reactive. For example, it ignites spontaneously in air at room temperature, and must therefore be stored under water. It melts at 44°C (111°F) and boils at 281°C (538°F). White phosphorus is very poisonous. The fatal dose for a human being is 0.1 g (0.035 oz). On standing in light it is slowly transformed into the stable red allotrope, with luminescence. Red phosphorus contains less energy and is therefore less reactive. It does not ignite in air below 250°C (482°F). Red phosphorus melts at 590°C (1094°F). It is non-poisonous.

1.2. Occurrence

Compounds of phosphorus, especially phosphates, are important constituents of animal and plant organisms, where they occur in the bones and in the material of nerves and cells.

They are therefore very necessary constituents of animal feeds and of fertilizers.

Crude phosphates occur in huge deposits, for example in the Kola peninsula in the USSR. This consists of apatite, a volcanic rock composed of calcium phosphate containing fluoride, $Ca_5(PO_4)_3F$. On weathering this turns into phosphorite, of which there are considerable deposits in the USA in Tennessee, Florida and Idaho, in North Africa in Algeria and Morocco, in Togoland, and in the Karaton mountains in the USSR.

Many iron ores contain considerable amounts of phosphates, which appear as Thomas phosphate, a valuable by-product of iron smelting.

The value of crude phosphate depends on its content of pure phosphate; this is expressed commercially as % BPL (bone phosphate of lime).

This means the calculated percentage content of pure calcium phosphate of composition $Ca_3(PO_4)_2$.

Known world resources are at present given as 12,700 million t. Of this 5200 million t (more than 40%) is in Morocco and Spanish Morocco. The USA has 2700 million t and the USSR 1500 million t, representing respectively 21% and 11% of world resources.

An explosive increase in the demand for crude phosphate has occurred during the last 40 years, parallel with the increased production of elemental phosphorus and phosphoric acid. In 1979 world extraction of phosphate rock was 127 million t, and world production of phosphorus was 686,000 t (see diagram on p. 13).

II Inorganic raw materials and large-scale products: Large-scale industrial processes

1.3. Manufacture of phosphorus by the electrothermic process

Only a few countries possess extensive production capacity for phosphorus (and phosphoric acid), namely the USA, USSR, Canada, Western Germany and the Netherlands.
The phosphorus industry in central Europe is an extreme example of geographical separation between the occurrence of raw materials and their working up. Although crude phosphate contains only about 15% combined phosphorus, it is transported over long distances.

1.3.1. Chemical principles

In the phosphates, salts of phosphoric acid (H_3PO_4), the phosphorus is combined with oxygen. This is emphasized when the formula of calcium phosphate, $Ca_3(PO_4)_2$, is written as $3\ CaO \cdot P_2O_5$. The production of phosphorus from phosphates involves two chemical reactions:

1. The combined phosphorus in P_2O_5 must be freed from oxygen, ie reduced. The reducing agent employed is carbon.
2. All accompanying substances (CaO, fluorides) must be converted into new, stable compounds. This is achieved by the addition of quartz sand and the formation of silicate slags.

The reduction of the combined phosphorus is an endothermic reaction, ie energy must be supplied continuously to keep the reaction going.

Phosphorus pentoxide		carbon		elemental phosphorus		carbon monoxide	

$$P_2O_5 \quad + \quad 5\ C \quad \longrightarrow \quad 2\ P \quad + \quad 5\ CO; \quad \Delta H = +939.3\ kJ/mol*$$

The required energy is supplied through electrodes as electrical energy, giving rise to the name **electrothermic process**.
The formation of slag can be represented by the following simplified equation:

Calcium oxide		silicon dioxide		calcium silicate

$$CaO \quad + \quad SiO_2 \longrightarrow CaSiO_3$$

The overall equation for the production of phosphorus from calcium phosphate is relatively complicated:

Calcium phosphate	silica	coke	phosphorus	carbon monoxide	slag

$$Ca_3(PO_4)_2 \ + \ 3\ SiO_2 \ + \ 5\ C \xrightarrow{\text{electrical energy}} 2\ P \ + \ 5\ CO \ + \ 3\ CaSiO_3$$

---

* The value of ΔH refers to the number of moles represented in the chemical equation

1.3.2.   Flow scheme for the manufacture of phosphorus by the electrothermic process

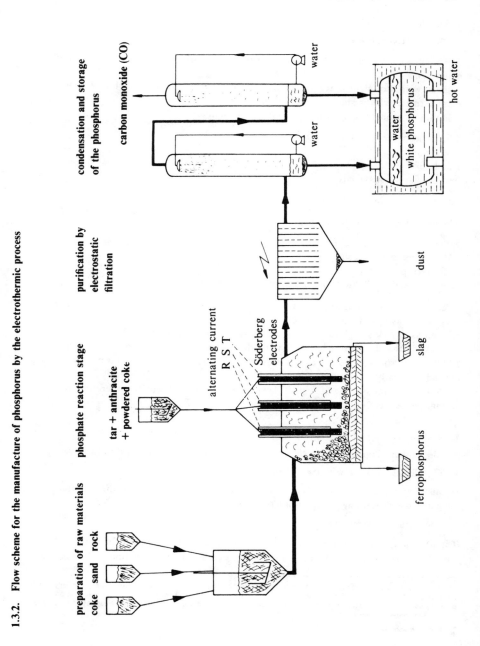

1.3.3. Material and energy balance sheet

The electrothermic production of phosphorus involves particularly large quantities of materials.

A given weight of elementary phosphorus requires twelve times as much raw materials.

A modern electrical reduction furnace requires the following quantities of raw materials, assuming phosphate rock with an average P_2O_5 content:*

8.00 t phosphate rock (31% P_2O_5)
2.80 t silica
1.25 t coke
0.05 t electrode materials

12.10 t starting materials

The following reaction products are obtained:

1.00 t elemental phosphorus
7.70 t calcium silicate slag
0.15 t ferrophosphorus
2500 normal m³ (88,000 ft³) of gas containing 85 vol. % CO

The yield of phosphorus amounts to 92–93%. The phosphorus obtained by this process is very pure, having a P-content of 99.7%.

As well as having a high turnover of material, this process is one of the most energy-intensive in the chemical industry. The production of 1 t of phosphorus requires about 13,000 kWh (kilowatt hours) of electrical energy. The price of elemental phosphorus thus depends critically on the cost of energy. The white phosphorus obtained is used predominantly for making pure phosphoric acid, the price of which is therefore also strongly dependent on energy prices.

1.3.4. Process stages**

An electrothermic plant for making white phosphorus can be divided into three sections.

1. The electric furnace with its electrodes

In this section the raw materials are fed in, reduction takes place, and the reaction products (slag and gas containing phosphorus) are removed.

2. The filtration system

The filtration system consists of electrostatic filters and arrangements for removing the dust from the filters. It serves to remove dust particles from the gas containing phosphorus.

---

*P_2O_5, phosphorus pentoxide, is the anhydride of phosphoric acid, H_3PO_4. It is common practice, especially in the fertilizer industry, to indicate the quality of a product by its P_2O_5 content. However, the P_2O_5 content is only a quantity used for convenience in calculating, and does not imply that crude phosphate or phosphate fertilizers actually contain the chemical substance P_2O_5.

** See flow scheme p.5.

3. The condensation section

The purified phosphorus vapor is transformed into a liquid in the condensation section. The liquid phosphorus at 60°C (140°F) can be conveyed further by pumping.

Notes on 1:

Lumps of phosphate rock are ground up with dry coke and sand and fed into the furnace. Reduction takes place at temperatures around 2000°C (3600°F). Slag and ferrophosphorus are obtained as by-products. Ferrophosphorus is used as a raw material in the steel industry. At the high temperatures involved both ferrophosphorus and silicate slag are liquid, and can be easily separated on account of their different densities.

Söderberg electrodes

In order to ensure continuous economic running the preparation and positioning of the electrodes must be controlled. During the reduction of the P_2O_5 the electrodes are burned away and must be continuously moved forward. This is effected by the use of Söderberg electrodes. The electrode material consists of a calcined mixture of anthracite, coke, tar and pitch in definite proportions. New casings filled with the pasty electrode material can be welded on without interrupting the working of the furnace. The electrodes are propelled hydraulically into the furnace. A Söderberg electrode has a diameter of 1.35 m (4.43 ft), an average length of 10 m (32.81 ft) and weighs about 25 t (55,000 lb).
The life-time of a reaction furnace is five to eight years.

Notes on 2:

The furnace gas, containing the phosphorus vapour, is freed from accompanying dust impurities in the filtration system. The high content of combustible carbon monoxide in the furnace gases is utilized to pre-heat the raw materials.

Notes on 3:

The condensed phosphorus is stored in steel containers, air being excluded.

1.4. Uses of phosphorus

Apart from military uses the very reactive white phosphorus is important only as an intermediate product.
A small proportion of white phosphorus is converted to red phosphorus by slow heating in the absence of air.

1.4.1. Red phosphorus

Red phosphorus is used in the match industry and for pyrotechnics. It is mixed with glass powder to form the striking surface for safety matches.
Red phosphorus has been used recently to render expanded plastic materials self-extinguishing in case of fire.

Combustion produces polyphosphoric acids, which cover the plastic with a film which prevents access of air.

1.4.2. Phosphorus pentoxide

Phosphorus pentoxide (phosphoric oxide) is formed when phosphorus burns.

$$
\underset{\substack{\text{Phosphorus} \\ 4\,P}}{} \quad + \quad \underset{\substack{\text{atmospheric} \\ \text{oxygen} \\ 5\,O_2}}{} \xrightarrow{\hspace{2cm}} \underset{\substack{\text{phosphorus pentoxide} \\ \text{(phosphoric oxide)} \\ 2\,P_2O_5\ (=P_4O_{10})}}{} ;\ \Delta H = -2984\ \text{kJ/mol}
$$

On account of its strongly hygroscopic (ie dehydrating) properties the main industrial use of phosphorus pentoxide is as a drying agent for gases.

1.4.3. Halogen compounds of phosphorus

The following halogen compounds of phosphorus are of industrial importance:
Phosphorus pentachloride, PCl_5,
Phosphorus trichloride, PCl_3,
Phosphorus oxychloride, $POCl_3$.

In order to obtain them white phosphorus is burnt in chlorine gas,

$$
\underset{\substack{\text{Phosphorus} \\ 2\,P}}{} \quad + \quad \underset{\substack{\text{chlorine} \\ 3\,Cl_2}}{} \xrightarrow{\hspace{2cm}} \underset{\substack{\text{phosphorus} \\ \text{trichloride} \\ 2\,PCl_3}}{} ;\quad \Delta H = -644.8\ \text{kJ/mol}
$$

Since phosphorus has a maximum valency of five, an excess of chlorine gives PCl_5, which can be converted into $POCl_3$ by treating with the calculated amount of water.
These three compounds are used as starting materials for preparing organic phosphorus compounds, especially esters of phosphoric acid.
These esters have become very important as plasticizers for plastic materials. They are also valuable as additives for motor fuels and lubricants.
Phosphorus pentachloride is also used as a chlorinating agent for organic compounds.

1.4.4. Phosphorus pentasulphide

The production of phosphorus pentasulphide received a great impetus from the discovery of its use as a starting material for the preparation of pesticides, fungicides and insecticides (see 2.3.2., Phosphate esters). Its reaction with various organic compounds also provides an easy route to additives for lubricants and flotation agents.

1.4.5. Pure phosphoric acid

By far the greatest proportion of white phosphorus is used to make pure phosphoric acid (also called thermal phosphoric acid).

2. Phosphoric acid

Phosphoric acid is manufactured from crude phosphates by two essentially different processes.
In the first elemental phosphorus is prepared by the electrothermic process and then converted into phosphoric acid.
In the second the crude phosphates are converted to phosphoric acid by wet treatment with acids.

2.1. Pure phosphoric acid

Phosphoric acid of high purity can easily be obtained from elemental phosphorus. The process involves two stages.

1st stage

The phosphorus is burnt to give phosphorus pentoxide by spraying liquid phosphorus into air through multiple nozzles. The fine droplets burn spontaneously to $P_4O_{10} (= 2\ P_2O_5)$

| | atmospheric | phosphorus |
| Phosphorus | oxygen | pentoxide |

$$4\ P\ +\ 5\ O_2\ \longrightarrow\ 2\ P_2O_5\ (=P_4O_{10});\quad \Delta H = -2984\ kJ/mol$$

1 t of phosphorus gives 2.3 t of phosphorus pentoxide.

2nd stage

The phosphorus pentoxide is converted to phosphoric acid by taking up water from a dilute solution of phosphoric acid, which is circulated.
Heat is again liberated.

Phosphorus pentoxide   water   phosphoric acid

$$2\ P_2O_5\ (=P_4O_{10})\ +\ 3\ H_2O\ \longrightarrow\ 4\ H_3PO_4;\qquad \Delta H = -368.6\ kJ/mol$$

The H_3PO_4 obtained by this method is sometimes termed "thermal phosphoric acid", since the acid is obtained by a process of combustion, with evolution of heat.

2.2. Phosphoric acid manufacture by the wet process

The wet process is used mainly to convert water-insoluble crude phosphates into phosphoric acid, which can then be used to make suitable soluble forms of calcium phosphate for the

II Inorganic raw materials and large-scale products: Large-scale industrial processes

fertilizer and fodder industries. The object of this is to remove impurities from the crude phosphates and finally to obtain high-grade water-soluble phosphates.

The main impurities are silicates and fluorides of calcium, aluminium, and iron.

The crude phosphate is treated with concentrated sulphuric acid, H_2SO_4, after its particle size has been reduced to less than 0.15 mm (0.000006 in).

2.2.1. Chemical principles

The chemical principle involved in this process is that phosphoric acid, an acid of medium strength, can be displaced from its salts, contained in the crude phosphate, by the strong acid sulphuric acid. The calcium is converted to calcium sulphate, which is almost insoluble, and the phosphoric acid can then be separated:

$$Ca_3(PO_4)_2 + 3\ H_2SO_4 \longrightarrow 3\ CaSO_4 + 2\ H_3PO_4$$

The complex changes which occur during the industrial process can be summarized in the following chemical equation:

Crude phosphate	sulphuric acid	water	gypsum	phos-phoric acid	hydro-gen fluoride

$$Ca_5(PO_4)_3F + 5\ H_2SO_4 + 10\ H_2O \longrightarrow 5\ CaSO_4 \cdot 2\ H_2O + 3H_3PO_4 + HF$$

The reaction is exothermic, with $\Delta H = -2959$ kJ/mol.

The hydrogen fluoride is absorbed as a silicate compound.

2.2.2. Material balance sheet and process stages

The manufacture by the wet process of phosphates containing the equivalent of 1 t P_2O_5 requires, for example:

3.1 t crude phosphate with a P_2O_5 content of 34% and

2.94 t 96% sulphuric acid.

As a by-product 5.5 t of gypsum is obtained, containing 20–30% of adhering water. The wet process can thus be divided roughly into the following stages:

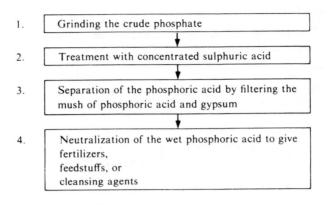

1. Grinding the crude phosphate

2. Treatment with concentrated sulphuric acid

3. Separation of the phosphoric acid by filtering the mush of phosphoric acid and gypsum

4. Neutralization of the wet phosphoric acid to give fertilizers,
feedstuffs, or
cleansing agents

2.3. Uses of phosphoric acid

2.3.1. Phosphates and polyphosphates

More than 90% of pure phosphoric acid is neutralized in order to make the following phosphates:

a. Na_3PO_4 and Na_2HPO_4, which are important constituents of mineral additives for fodder, and of tonics;

b. K_3PO_4 and K_2HPO_4, which are added to liquid cleansing agents;

c. $(NH_4)_2HPO_4$, which is required as a nutrient in the biological purification of sewage;

d. the calcium phosphates $CaHPO_4$ and $Ca(H_2PO_4)_2$ are constituents of fodder and phosphate fertilizers.

Phosphates for cleansing agents

The polyphosphate pentasodium triphosphate, $Na_5P_3O_{10}$, occupies a special position as an essential constituent of synthetic cleansing agents. About 60% of the pure phosphoric acid produced in Europe is used for this purpose. On account of their dispersant effect and the effect of surface-active substances in increasing cleansing properties, cleansing agents contain 30–50% of this condensed phosphate. The condensed phosphates (polyphosphates) convert the calcium and magnesium ions, which determine the hardness of water, into salts which remain water-soluble even when the washing is boiled. Tertiary sodium phosphate, Na_3PO_4, is used to soften water by precipitating insoluble calcium and magnesium phosphates.

Fertilizers and fodder

More than 90% of the production of phosphate ores is converted into fertilizers either by the wet process for phosphoric acid or by treatment with nitric acid. Multiple fertilizers are often prepared by treating crude phosphate with nitric acid, followed by cooling in order to crystallize out calcium nitrate tetrahydate, and finally neutralization of the solution with ammonia.

Baking powders

When mixed with sodium bicarbonate, $NaHCO_3$, sodium hydrogen phosphate is used as a baking powder. The reaction between these two salts evolves CO_2.

Acidifying agents

Phosphoric acid is used directly for the surface treatment of metals, and, in a pure form, for acidifying beverages.

Foods and cosmetics

When pure phosphoric acid is neutralized with calcium oxide, CaO, calcium hydroxide, $Ca(OH)_2$ or calcium carbonate, $CaCO_3$, calcium phosphate is obtained

Phosphoric acid calcium oxide calcium phosphate water

$$2\ H_3PO_4\quad +\quad 3\ CaO\ \longrightarrow\ Ca_3(PO_4)_2\quad +\quad 3\ H_2O$$

It is used as an additive to foodstuffs and to cosmetic preparations, particularly toothpaste. In addition to calcium phosphates, sodium phosphates are of considerable importance in food technology, for example in the cheese and meat industries and in the manufacture of pudding mixes.

Textiles

Urea phosphate is used for flame-proofing textiles. Disodium phosphate, Na_2HPO_4, is used to load silk fibres.

2.3.2. Phosphate esters

It has already been mentioned that esters of phosphoric acid are of considerable importance for making hard, brittle plastics (see IV,7, Plastics). Certain esters of phosphoric and thiophosphoric acid are effective insecticides, eg

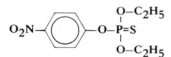 Parathion

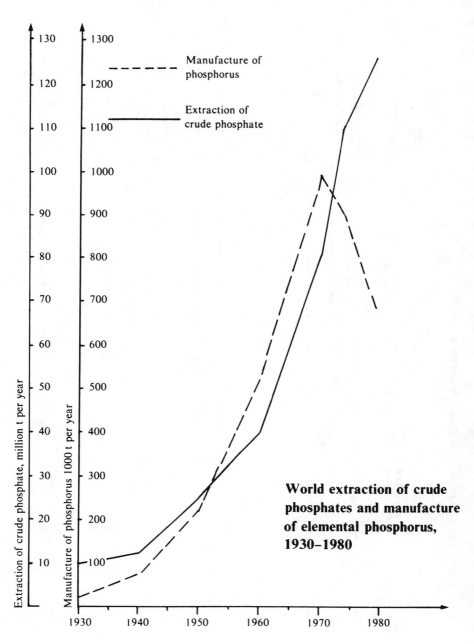

World extraction of crude phosphates and manufacture of elemental phosphorus, 1930–1980

2.3.3. Review of products derived from phosphorus and phosphoric acid

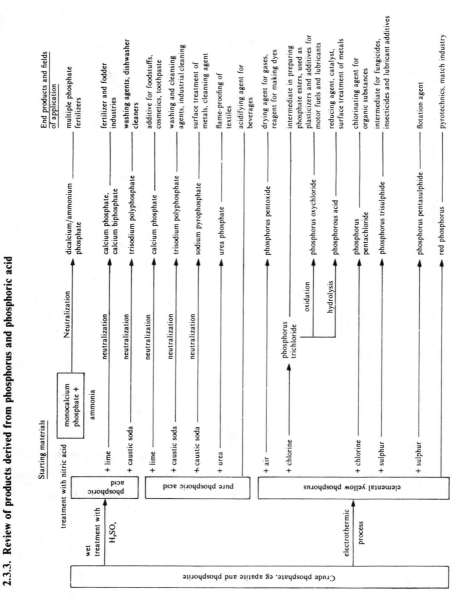

3. Self-evaluation questions

55 Which statement on phosphorus is correct?

A   Phosphorus occurs in nature only in compounds.
B   White phosphorus is self-igniting and therefore has to be stored under petroleum.
C   Red phosphorus is toxic.
D   Elemental phosphorus is a liquid at room temperature.

56 In which form is phosphorus most frequently found in nature?

A   as calcium phosphate
B   as ammonium phosphate
C   as phosphorus pentoxide
D   elemental

57 The industrial production of phosphorus is termed an electrothermal process because

A   it is an electrolytic process
B   the resulting phosphorus vapours are purified by electrofiltration
C   the development of heat is monitored electrically
D   the necessary energy is provided as electrical energy

58 In the production of phosphoric acid by the wet dissolution method the solvent used is

A   water
B   sulphuric acid
C   coke
D   steam

59 Pentasodium triphosphate is used as a

A   fertilizer component
B   detergent component
C   herbicide
D   feed additive

60 Phosphorus and phosphorus compounds are used in the production of

A   fertilizers and matches
B   toothpastes and drinks
C   baking powder and car batteries
D   drying agents and phosgene

61 Phosphorus pentoxide, P_2O_5, is

A   a fertilizer salt
B   a hygroscopic compound
C   the anhydride of phosphoric acid, H_3PO_4
D   the combustion product of phosphorus

II Inorganic raw materials and large-scale products: Large-scale industrial processes

62 Phosphorite is used in the production of

    A   fertilizers
    B   phosphoric acid
    C   phosphorus
    D   iron

7 Inorganic Oxo-compounds. Hydrogen Peroxide and Perborates

1. Definition of terms

The oxo-compounds belong to the class of complex compounds.

Complex compounds have a characteristic structure. A distinction is drawn between a single central atom and the atoms which surround this central atom, known as ligands.

For example, the structure of ammonium chloride is:

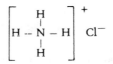

In NH_4^+, the ammonium ion, the nitrogen N is the central atom and the hydrogen atoms H are the ligands. Complex ions are often placed in square brackets.

In complex compounds the central atoms can be either metals or nonmetals, eg

Chromium	Cr	Chlorine	Cl
Manganese	Mn	Sulphur	S
Aluminium	Al	Oxygen	O
Iron	Fe	Nitrogen	N
Copper	Cu	Phosphorus	P
		Carbon	C

These central atoms can be surrounded by atoms, molecules or ions. This produces so-called compounds of higher order, or complex compounds.

> If the ligands of a complex compound consist of oxygen atoms, it is termed an oxo-compound

The following are two examples of inorganic oxo-compounds:

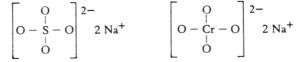

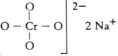

sodium sulphate   sodium chromate

2. Inorganic oxo-acids

Among the most important oxo-compounds are the oxygen-containing acids of the nonmetals and their salts (cf I.7).

The outer electron shell of oxygen has room for one more electron pair:

$$I\overline{O}$$

An oxygen atom can therefore attach itself to a central atom which possesses free electron pairs.

For example, there are four acids containing chlorine and oxygen. These are as follows, arranged in increasing order of the number of oxygen atoms in the complex ion:

1. Hypochlorous acid, $HClO$

$$H^+ \, [\, Cl - O \,]^-$$

2. Chlorous acid, $HClO_2$

$$H^+ \, [\, O - Cl - O \,]^-$$

3. Chloric acid, $HClO_3$

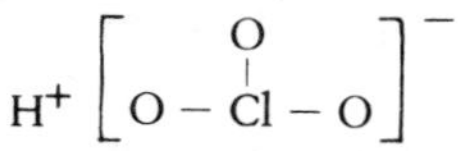

$$H^+ \begin{bmatrix} & O & \\ & | & \\ O - & Cl & - O \end{bmatrix}^-$$

4. Perchloric acid, $HClO_4$

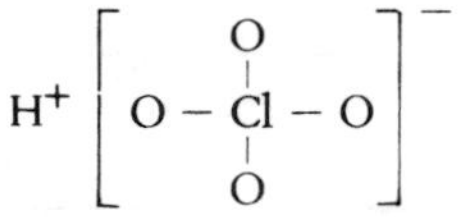

$$H^+ \begin{bmatrix} & O & \\ & | & \\ O - & Cl & - O \\ & | & \\ & O & \end{bmatrix}^-$$

The most important oxo-acid of an element is named by attaching the syllable "-ic" to the name of the element. Thus the most important oxo-acids of chlorine, sulphur and phosphorus are

$HClO_3$	chloric acid
H_2SO_4	sulphuric acid
H_3PO_4	phosphoric acid

The names of the salts of these acids end with the syllable "-ate", eg chlorate, sulphate, phosphate.

7 Inorganic oxo-compounds.
Hydrogen peroxide and perborates

2.1. Important salts of inorganic oxo-acids

2.1.1. Sodium hypochlorite and bleaching powder

Aqueous solutions of hypochlorous acid decompose slowly in the dark and very rapidly in sunlight to give hydrochloric acid and oxygen:

Hypochlorous acid	hydrochloric acid		oxygen
$2 \text{ HClO} \longrightarrow$	2 HCl	$+$	2 O

This behaviour makes hypochlorous acid one of the strongest oxidizing agents known. Atomic oxygen, which is formed initially in the decomposition, has a very intense bleaching and disinfecting effect.

The salts of hypochlororous acid, the hypochlorites, also act as very strong oxidizing agents in aqueous solution and are used to bleach paper, cellulose, textiles, etc, and also as disinfectants. Important examples are sodium hypochlorite and chloride of lime (bleaching powder).

Sodium hypochlorite can be prepared by passing chlorine into sodium hydroxide solution:

Chlorine		sodium hydroxide		sodium hypochlorite		sodium chloride		water
Cl_2	$+$	$2 \text{ NaOH} \longrightarrow$		NaOCl	$+$	NaCl	$+$	H_2O

A solution of sodium hypochlorite can be obtained directly from the products of electrolysis of sodium chloride, by allowing the chlorine produced at the anode to mix with the sodium hydroxide produced simultaneously at the cathode. The resulting solution is known as "chlorine bleach".

If chlorine is allowed to react with slaked lime (calcium hydroxide Ca(OH)_2) the product is bleaching powder, or chloride of lime:

Chlorine		calcium hydroxide		bleaching powder		water
Cl_2	$+$	$\text{Ca(OH)}_2 \longrightarrow$		CaOCl_2	$+$	H_2O

Bleaching powder is a mixed calcium salt of hydrochloric acid, HCl, and hypochlorous acid, HClO. Its importance has diminished considerably in recent years. Today chlorine is usually transported as the liquid element, and rarely as bleaching powder. The bleaching solutions required are prepared as a rule from chlorine at the point where they are needed.

The alkali metal salts of chlorous acid, HClO_2, are also strong oxidizing agents, like those of HClO. They are used mainly for bleaching textiles and cellulose.

2.1.2. Chlorates

The most important salts of chloric acid are potassium chlorate, $KClO_3$, and sodium chlorate, $NaClO_3$.

Working with the alkali metal chlorates requires great care, since in the presence of even small quantities of oxidizable material, such as sulphur, phosphorus or organic compounds, friction can cause explosions.

On heating, the alkali metal chlorates decompose readily with the evolution of oxygen, eg

Potassium chlorate	potassium chloride	oxygen
$2\ KClO_3$ $\longrightarrow$	$2\ KCl$ $+$	$3\ O_2$

Alkali metal chlorates are therefore often used as oxidizing agents in practice.
Potassium chlorate is used in large quantities in the manufacture of matches. Safety matches, used since 1848, have heads composed of potassium chlorate as an oxidizing agent together with combustible material and binder. The striking surface consists of a mixture of glass powder and red phosphorus. When the match is struck the red phosphorus catches fire and ignites the readily combustible mixture in the match head. The alkali chlorates are also used industrially in making fireworks and as weed killers.

2.1.3. Perchlorates

Perchloric acid, $HClO_4$, is one of the strongest acids known. If it comes into contact with combustible materials, such as paper and wood, its powerful oxidizing properties lead to explosive ignition. Perchloric acid produces painful wounds on the skin, which heal with difficulty. The perchlorates are the most stable oxygen compounds of chlorine. The explosives industry therefore uses ammonium perchlorate, NH_4ClO_4, which is insensitive to blows but explodes when ignited, in preference to the less stable and therefore more dangerous chlorates. The perchlorates are also used as rocket propellants and in pyrotechnics.

2.1.4. Permanganates

The permanganates are complex compounds in which the central atom is manganese

$$\left[O - \overset{\displaystyle O}{\underset{\displaystyle O}{Mn}} - O \right]^{-} Me^{+}$$

Me^+ = metal ion

The permanganate ion, MnO_4^{-}, has a characteristic purple colour.

By far the most important permanganate is potassium permanganate, $KMnO_4$, which crystallizes in deep purple prisms.

Potassium permanganate readily loses oxygen, and therefore has many applications as an oxidizing agent. For example, it is used for bleaching wool, silk and other fibres, and for decolorizing oil. It is also a disinfectant.

2.1.5.   Chromates

The central atom in the chromates is chromium, Cr:

$$
\left[\begin{array}{c} O \\ | \\ O - Cr - O \\ | \\ O \end{array} \right]^{2-} \quad Me^{2+}
$$

Chromates have some importance as tanning agents and as pigments (see II,8).

3. Peroxo-compounds

Peroxo-compounds contain oxygen atoms directly bound to one another, and are thus characterized by a —O—O— bridge. The simplest compound in this class is hydrogen peroxide, (sometimes called hydrogen superoxide):

H—O—O—H   ; H_2O_2

Both water and hydrogen peroxide are composed of the same two elements, hydrogen and oxygen, but hydrogen peroxide contains a larger proportion of oxygen. It can therefore be regarded as an oxidation product of water:

$2 H_2O + O_2 \longrightarrow 2 H_2O_2$; $\Delta H = +189.1$ kJ/mol*

Potassium peroxodisulphate is an example of a more complicated peroxo-compound:

$$
\left[\begin{array}{c} O \qquad\quad O \\ | \qquad\quad | \\ O - S - O - O - S - O \\ | \qquad\quad | \\ O \qquad\quad O \end{array} \right]^{2-} \quad 2\ K^+ \qquad K_2S_2O_8
$$

This compound also contains two oxygen atoms linked together.

---

* The value of ΔH refers to the number of moles represented in the chemical equation.

II Inorganic raw materials and large-scale products: Large-scale industrial processes

3.1. Industrial manufacture of hydrogen peroxide

The importance of hydrogen peroxide has increased considerably during the last few decades. In 1980 world production was more than 500,000 tonnes, reckoned as the 100% product.

The current world capacity is 700,000 tonnes per year. About 90% of world production employs organic processes. By far the most important of these is the anthraquinone process, which in Europe and USA accounts for about 97% of the hydrogen peroxide produced.

This process employs derivatives of anthraquinone, eg

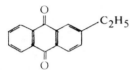

2-ethylanthraquinone

The anthraquinone derivative is dissolved in a mixture of solvents, and is converted to the corresponding anthrahydroquinone by passing in hydrogen in the presence of a catalyst. The catalyst used for hydrogenation is removed by filtration, and in a second reaction stage the solution of hydroquinone is bubbled with air.

This reconverts it to the anthraquinone derivative, and at the same time hydrogen peroxide is formed:

2-ethylanthraquinone + H$_2$ $\longrightarrow$ 2-ethylanthrahydroquinone

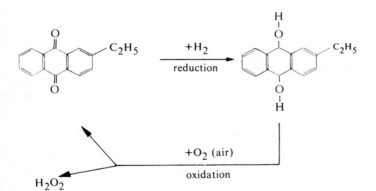

The hydrogen peroxide thus produced is extracted with water from the reaction mixture of anthraquinone + hydrogen peroxide.

The solution of anthraquinone remaining after the extraction of the hydrogen peroxide can be used afresh as starting material after drying and purification.

Figure 1. Manufacture of hydrogen peroxide (H_2O_2) by the anthraquinone process

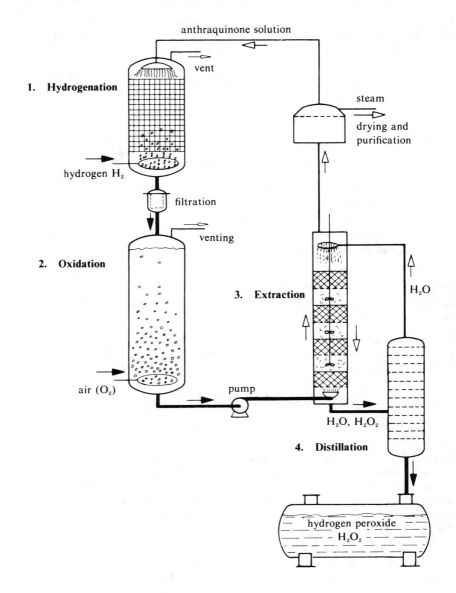

The crude hydrogen peroxide solution is purified in subsequent stages and concentrated by distilling off water.

The cyclic process (see flow scheme, p. 9) can be repeated continuously. However, a proportion of the anthraquinone derivative used is always decomposed: it then loses its ability to produce hydrogen peroxide and must be replaced.

The industrial manufacture of hydrogen peroxide by the anthraquinone process can thus be divided into four stages:

> 1. Hydrogenation of the anthraquinone derivative
>   Reduction
> 2. Bubbling with air and formation of hydrogen peroxide
>   Oxidation
> 3. Extraction of hydrogen peroxide with water
> 4. Purification and concentration by distillation.

3.2. Properties and uses of hydrogen peroxide

Pure anhydrous hydrogen peroxide is a colourless liquid. It is stable at room temperature both in the pure state and in aqueous solution.

Under normal pressure (1013 mbar or 14.5 lb/in^2) hydrogen peroxide freezes at $-0.4°C$ (31.3°F), and boils at 150°C (302°F). Its molecular weight is 34.

Hydrogen peroxide decomposes with the evolution of much heat into water and oxygen. This occurs on heating or in the presence of various materials, eg finely divided silver, platinum, manganese dioxide, dust particles, alkalis, or rough surfaces.

$$2 H_2O_2 \longrightarrow 2 H_2O + O_2; \quad \Delta H = -189.1 \text{ kJ/mol}$$

Particles of dust falling into highly concentrated hydrogen peroxide solutions can cause violent explosions.

Even traces of alkali derived from glass can cause the slow decomposition of hydrogen peroxide. For this reason concentrated solutions are stored in glass vessels coated with paraffin wax. Small quantities of various substances are often added to industrial hydrogen peroxide to improve its stability by retarding or preventing its decomposition. Substances which have this effect include phosphoric acid, barbituric acid and uric acid.
The importance of hydrogen peroxide as a disinfectant and bleaching agent depends on its marked tendency to lose oxygen.

Hydrogen peroxide is sold as an aqueous solution, either 3% or 30% (perhydrol). It is used in medicine as a disinfectant, for example as a mouth wash and for rinsing out wounds.

Hydrogen peroxide (sometimes under the name hydrogen superoxide) is often used for cosmetic purposes, for example for giving hair a blonde colour, and also for bleaching straw, feathers, ivory, sponges, glue, fats, leather, furs, etc. It is used to an increasing extent for bleaching fibres such as wool, cotton, silk and artificial fibres. Hydrogen peroxide is more and more displacing chlorine for bleaching, since it acts more rapidly and permanently and inflicts less damage on the fibres.

Pure hydrogen peroxide is a weak acid and can therefore form salts. These salts are called **peroxides**.

3.3. Metal peroxides

Important salts of hydrogen peroxide are

Sodium peroxide	Na_2O_2
Barium peroxide	BaO_2

When sodium peroxide reacts with water, hydrogen peroxide is first formed (I), which then splits off oxygen (II):

I. $2 Na_2O_2 + 4 H_2O \longrightarrow 2 H_2O_2 + 4 NaOH$

II. $2 H_2O_2 \longrightarrow 2 H_2O + O_2$

The salts of hydrogen peroxide therefore have powerful oxidizing properties, and are used for making bleaching solutions for various animal and vegetable products.

Sodium peroxide is manufactured industrially on a large scale. Metallic sodium, in rotating drums, is oxidized in two stages by blowing in air.

1.   Sodium   oxygen   sodium oxide

     $4 Na + O_2 \longrightarrow 2 Na_2O; \quad \Delta H = -832.6 \text{ kJ/mol}$

2.   Sodium oxide   oxygen   sodium peroxide

     $2 Na_2O + O_2 \longrightarrow 2 Na_2O_2$

The oxidation to sodium oxide occurs when the temperature inside the drum is about 150 to 200°C (300 to 400°F), and the sodium oxide is converted to sodium peroxide at about 350°C (660°F).

3.4. Urea peroxohydrate

In urea peroxohydrate hydrogen peroxide, HO—OH, adds on to the organic compound urea

$$\begin{array}{c} H_2N \\ \diagdown \\ \diagup \\ H_2N \end{array} C = \overline{\underline{O}} + H_2O_2 \longrightarrow CO(NH_2)_2 \cdot H_2O_2$$

This adduct is the only important organic peroxohydrate, and is sold under the names percarbamide or carbamide perhydrate. It contains about 35% hydrogen peroxide.

Urea peroxohydrate is a solid, which is stable when stored under dry and cool conditions. For this reason it is often described as "solid hydrogen peroxide".

It is used for bleaching hair and as a disinfectant in treating wounds.

3.5. Peroxoborates

The salts of boric acid are called borates. Orthoboric acid, H_3BO_3, has the following structural formula:

Borates are obtained by replacing one or more of the hydrogen atoms of orthoboric acid by metal atoms.

Metaboric acid, HBO_2, can be derived formally from orthoboric acid by the loss of a water molecule.

HO—B—O H $\xrightarrow{-H_2O}$ HO—B=O

The peroxoborates differ from the orthoborates in having attached to boron an extra oxygen atom, which can be easily split off, thus giving rise to oxidizing and bleaching properties. Like all peroxo-compounds, the peroxoborates contain —O—O— bridges.

The most important compound of this class is sodium peroxoborate, often called sodium perborate for sort. It has the following structure:

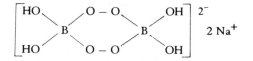

The ionic lattice of this salt contains additional molecules of water of crystallization, so that the overall formula of sodium peroxoborate is

$NaBO_2(OH)_2 \cdot 3\ H_2O$ Sodium peroxoborate trihydrate

Sodium perborate is the most stable perborate, and provides a safe way of storing hydrogen peroxide as a solid.

The perborates decompose in aqueous solution to give hydrogen peroxide, which they resemble in properties, in particular in their ability to lose oxygen readily. Perborates are therefore important as cleansing and bleaching agents for a wide variety of materials. Commercial sodium perborate contains about 10.4% active oxygen. Washing powders for textiles usually consist of a mixture of surface-active substances, soda and sodium perborate. The proportion of the last substance is often as high as 30% (cf. IV,6).

Oxygen baths are obtained by using a mixture of sodium perborate and substances which decompose the hydrogen peroxide formed on dissolution into water and oxygen.

In addition to the alkali perborates, the perborates of the alkaline earth metals have some importance. In contrast to the alkali perborates they are sparingly soluble in water and are therefore used as constituents of toothpaste and disinfecting powders.

3.5.1. Large-scale manufacture of sodium peroxoborate

The industrial preparation of sodium peroxoborate is conveniently carried out in two stages. In the first stage a solution of sodium metaborate is prepared from borax, $Na_2B_4O_7$, and caustic soda solution:

Borax		caustic soda		sodium metaborate		water
$Na_2B_4O_7$	+	$2\ NaOH$	$\longrightarrow$	$4\ NaBO_2$	+	H_2O

The solution of sodium metaborate is then converted to sodium peroxoborate by the action of hydrogen peroxide:

Sodium metaborate		hydrogen peroxide		water		sodium peroxoborate
$NaBO_2$	+	H_2O_2	+	$3\ H_2O$	$\longrightarrow$	$NaBO_2(OH)_2 \cdot 3\ H_2O$

In order to obtain sodium peroxoborate of uniform quality the second stage is carried out as a continuous process.

The apparatus for the continuous production of sodium peroxoborate consists of a vertical column through which the reaction solution is circulated by means of a pump. The liquid passes up the column to an overflow, and is then returned to the base of the column through a filter and a collecting vessel. Sodium metaborate is supplied to the overflow and hydrogen peroxide is injected at the foot of the column. The widening of the column at its upper end serves to reduce the flow rate of the liquid and thus prevents particles of peroxoborate from being swept away by the overflowing liquid.

As a precaution the circulating liquid is filtered before re-entering the column. Part of the liquid is removed at the bottom of the column, and after filtering off the crystals of peroxoborate the mother liquor is returned to circulation through the collecting vessel. The moist salt obtained is finally dried and removed.

Figure 2. Plant for the continuous production of sodium peroxoborate

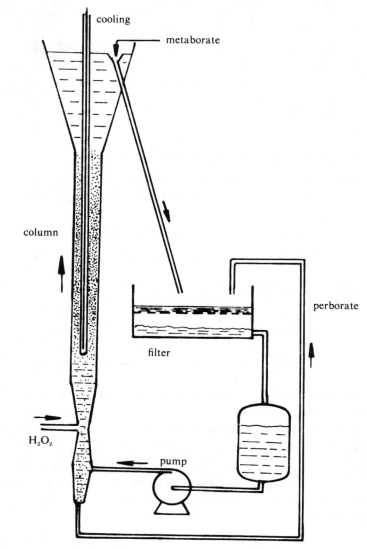

4. Self-evaluation questions

63   Which statement is applicable to all oxo compounds?

A   They readily release oxygen.
B   They can be stabilized only by the use of additives.
C   They are complex compounds whose central atoms are surrounded by oxygen atoms as ligands.
D   They are complex compounds whose central atoms are oxygen atoms.

64   Which compound is an oxo compound?

A P_2O_5
B H_2O
C H_2SO_4
D   NaOH

65   Which compounds are peroxides?

A PbO_2
B Na_2O_2
C H_3BO_3
D H_2O_2

66   Peroxo compounds are compounds that

A   contain an —O— grouping in the molecule
B   contain two oxygen atoms in the molecule
C   contain oxygen atoms that are linked to another element by a double bond
D   contain an —O—O— grouping in the molecule

67   Which statements about hydrogen peroxide are correct?

A   Hydrogen peroxide is a strong acid whose salts are called peroxides.
B   Hydrogen peroxide readily releases oxygen and therefore has a powerful oxidizing action.
C   Hydrogen peroxide is as stable a compound as H_2O.
D   Hydrogen peroxide is employed in dilute form as a disinfectant.

68 In which processing operation is hydrogen peroxide produced industrially by the anthraquinone method?

A hydrogenation
B oxidation
C extraction
D distillation

69 Sodium perborate is used for

A fertilizers
B detergents
C insecticides
D foodstuffs

8 Inorganic Pigments and Fillers

1. Inorganic pigments

There is some justification for dealing with inorganic fillers and pigments in the same chapter. Not only are pigments and fillers often manufactured together, but also the dividing line between them is fluid.

Thus in paper making, silicates and silicic acids are used as fillers, since they increase the volume of the paper and prevent the printing ink from going through, and also as pigments, since they improve the whiteness and opaqueness of the paper.

Carbon black is another example of the use of the same substance as a reinforcing filler (in rubber mixes) and also as a pigment (in printers' ink). In general the dividing line between pigments and fillers is drawn according to the following principles:

1. pigments are more finely divided then fillers;
2. pigments are more expensive than fillers;
3. white pigments have a higher refractive index than white fillers.

1.1. Definition of terms

The common term for all substances used to impart colour is colorants. Colorants can be divided into pigments and dyestuffs.

The fundamental difference between dyestuffs and pigments lies in their solubility. Dyestuffs are soluble in water or organic solvents, and are mainly used in aqueous solution, for example for dyeing textiles. Pigments, on the other hand, are insoluble substances. They are used in cases where solutions are unsuitable, for example for colouring plastics and paints, and for printing on paper.

The small size of pigment particles is necessary in order to obtain maximum colouring power. Pigments are therefore pulverized in ball mills or between rollers and are then incorporated in the material to be coloured.

> Pigments are tinted or neutral colorants.* Pigments are finely divided powders which are practically insoluble in solvents and binding agents.

The insolubility of pigments in water, organic solvents, plasticizers and binders means that they cannot be dissolved out of the coloured material and cannot run or smudge.

On a chemical basis a distinction is made between inorganic and organic pigments.

---

* Black and white are termed neutral colours.

Organic colorants are compounds with complicated structures (see IV,9, colorants). In contrast, inorganic colorants are chemically very simple.

Most inorganic pigments are oxides of metals, such as titanium dioxide, TiO_2, iron (III) oxide, Fe_2O_3, and zinc oxide, ZnO. Compared with organic colorants, the different chemical constitution of the inorganic pigments is responsible for their greater density, the fact that they are in general less sensitive to light, and their usually much greater stability to heat.

Thus it is not possible to use an organic pigment for an enamel which is to be fused on to a metal surface.

Organic and inorganic pigments can be used either side by side or in combination. In the latter case use is made of the more brilliant colours of the organic pigments.

1.2. Uses and industrial importance

Naturally occurring iron oxide minerals have been used in prehistoric times for cave paintings.

Even in antiquity the inorganic pigments white lead and red lead were made. However, pigments were not made in industrial quantities until the 19th century.

Present day demands for inorganic pigments could not be met, either in quantity or in quality, by naturally occurring minerals.

Inorganic pigments are therefore made predominantly by synthetic methods. These products possess higher purity, more uniform quality, more intense colour, and special properties in use which are lacking in the naturally-occurring minerals.

A special use of coloured and neutral inorganic pigments is in the colouring of paints and of synthetic materials (plastics). They are also used for colouring concrete, ceramics, glass, enamels and printing inks. The uses of white pigments extend from stove enamels through white paper and white shirts made of synthetic fibres to zebra-crossing markings.

In terms of both quantity and value white pigments are by far the most important group of inorganic pigments.

1. About half of the world production of inorganic pigments (excluding carbon blacks) refers to the white pigment titanium dioxide (see p. 11).

   The world capacity for this most important pigment almost doubled between 1960 and 1970, reaching 2.17 million t per year. Up to 1979 it had risen to more than 2.6 million t per year. About 40% of world capacity for TiO_2 is in Western Europe, 27% in USA.

2. After titanium dioxide zinc oxide is the next most important pigment, contributing about 20% of world production of inorganic pigments (excluding carbon blacks).

   This white pigment is particularly important in Eastern Europe for use in paints. Less than 15% of world production of zinc oxide (0.77 million t in 1969) is in Western

Europe. In the western world zinc oxide is used mainly as an activator and accelerator for the vulcanization of rubber, and only to a lesser extent as a pigment for paints and in the glass and ceramic industries.

Zinc oxide is also used in the production of duplicating paper, since some varieties of zinc oxide exhibit electrostatic phenomena when illuminated.

3. Synthetic iron oxide pigments (see p. 14) contribute about 10% of world production of inorganic pigments (excluding carbon blacks).

Synthetic and naturally occurring iron oxide pigments are the most important coloured pigments. Between 1960 and 1971 the annual production of coloured and black iron oxide pigments increased from about 300,000 t to about 500,000 t. About 30% of this amount refers to naturally occurring iron oxide, which, however, is mainly used when it is not necessary to maintain exactly the same tint, as for example in the building industry.

About half of the world production of synthetic iron oxides occurs in West Germany.

4. The following compounds contribute together about 10% of world production of inorganic pigments (excluding carbon blacks):

Lead pigments	Lead dioxide	PbO
	Red lead	Pb_3O_4
	White lead	$2\ PbCO_3.Pb(OH)_2$
Lithopones	(see p. 10)	

The production of white lithopones is falling off throughout the world, with the exception of the USSR.

Of the lead pigments the production figures for black lead dioxide are by far the highest. However, it is used predominantly for making lead accumulators, and not as a pigment.

Red lead is a constituent of undercoat paints used to protect against corrosion.

White lead is used nowadays only as a specialized product.

5. The following coloured pigments contribute together less than 5% of world production:

Chromate pigments (lead chromate, zinc chromate, barium chromate)
Chromium oxide
Molybdate orange
Cadmium yellow (cadmium sulphide)
Cadmium selenide
Blue iron pigments (eg Prussian Blue)
Ultramarine

These coloured pigments are used to a considerable extent for colouring paints and plastics and in colour printing.

The inorganic pigments are much cheaper than organic dyestuffs and pigments on account of their simple nature.

In contrast to the rapid and research-intensive developments in the field of organic colorants and the rapid changes in the range of coloured textiles available, only a few inorganic compounds have been recently developed as pigments. In the field of inorganic pigments attention has been paid primarily to improving the properties of existing pigments.

1.3. Chemical and physical properties of pigments

Inorganic pigments should as far as possible exhibit no chemical reactivity. They should be stable and should remain unchanged. Certain pigments used for anti-corrosion paints actually improve their protective action: they are known as active pigments.

The excellent resistance of inorganic pigments towards light, weather and heat is a sign of their chemical stability.

A few pigments which contain water of crystallization, for example green chromium oxide hydrate, may lose this water on heating and hence change their tint.

Inorganic pigments do not react with inorganic compounds or mixtures, such as water, air, cement, lime, asbestos cement, concrete and glass. They also ought not to react with organic compounds such as solvents, linseed oil, synthetic resins, plasticizers and plastics, in which they may be embedded. The yellowing of paints, accompanied by a chalky appearance, indicates decomposition of the binder, which may be caused by the pigment.

If there are very stringent requirements for chemical inertness the surface of the pigment particles may be treated with completely inactive substances. This is done, for example, with some special titanium dioxide pigments.

The following physical properties of pigments are of primary importance:
1. high hiding power, and good brightening or colouring power;
2. for white pigments, a refractive index as high as possible;
3. as insoluble as possible in binders (eg plastics) and solvents (eg water);
4. particles of small and uniform size;
5. particles which can be easily dispersed in the binder.

1.4. The optical properties of pigmented paints, plastics, concrete, etc

1.4.1. Colour

The colour or neutral shade of a non-luminescent material, for example pigmented plastic or paper, is known as the body colour.

The body colour depends on the kind of light (sunlight, filament electric bulbs or fluorescent tubes) with which the object is illuminated.

White sunlight is a mixture of the spectral colours which may be seen separately in the rainbow. A body which appears black in sunlight is one which absorbs all these colours. Carbon black is almost completely black: it absorbs about 95% of the incident radiant energy.

The radiant energy absorbed can heat a coloured object, or may bring about chemical reactions in it.

Violet light, and the invisible ultraviolet rays, are particularly rich in energy.

If the reflected light lacks certain colours, then the object will appear red, green, etc, in sunlight, depending upon which colours remain in the reflected light. If a body reflects not only red light, but also part of the white light, then we perceive the colour as pink. Similarly, white and blue gave pale blue, white and yellow give cream, etc. The smaller the amount of light reflected by a body, the darker will its colour appear. In this way red can turn into orange or dark brown, and similarly for other colours.

If all the colours of sunlight are reflected to the same extent the colour is perceived as white. Polished silver surfaces reflect about 90% of the intensity of each spectral colour.

> Coloured objects absorb part of the incident light. The body colour depends on the mixture of colours which is reflected

Surfaces which are not highly polished, but are more or less uneven do not produce specular reflection, but diffuse reflection in different directions.

1.4.2.   Refraction, refractive index

When light falls on a body part of it is reflected at the surface, while another portion penetrates into the body. This latter portion is refracted, ie it alters its direction. The greater this refraction, the greater is the refractive index of the material in question (see Table on p. 8).

1.4.3.   Hiding power, colouring and brightening power

Since pigments are insoluble colorants pigment materials contain finely divided (dispersed) particles of pigment. The light which has penetrated the material is diffusely reflected by these small particles and repeatedly refracted, and hence scattered. At the same time the light is weakened by absorption.

The combination of these factors prevents the light from penetrating deeply into the pigmented object. In this way the background of a varnish is hidden. and a transparent varnish is rendered opaque by pigmentation.

In a film of paint of given thickness a pigment of greater **hiding power** does not need to be used at such a high concentration in order to render the paint opaque, so that it is impossible to tell whether the background is black or white.

Similarly, a transparent synthetic fibre can be rendered opaque and matt by the use of a white pigment.

Table: Refractive indices of binders, fillers and white pigments

1. **Air**		.0
2. **Binders**		
Water		1.3
Urea-formaldehyde resin		1.6
3. **Fillers**		
Class	Substance	
Silicic acid	Silica	1.45
	Kieselguhr (diatomaceous earth)	1.45
Silicates	Aluminium silicate	1.6
	Kaolin (china clay)	1.6
	Calcium silicate	1.5
	Sodium aluminium silicate	1.5
Carbonates	Chalk	1.6
	Precipitated calcium carbonate	1.6
Oxides	Alumina	1.6
Sulphates	Barium sulphate (blanc fixe)	1.65
4. **White pigments**		
Class	Substance	
Sulphides	Zinc sulphide	2.4
Oxides	Zinc oxide	2.0
	Titanium dioxide (anatase)	2.55
	Titanium dioxide (rutile)	2.75

The following two factors are decisive in determining the hiding power of a pigment.

1. Its refractive index: the higher the refractive index the greater the hiding power.
2. The size and shape of its particles: they must be spherical and as small as possible.

In carbon black pigments the diameter of the primary particles is between 15 and 30 nm.* In carbon black used as a filler for rubber the particle diameter is about 1000 times as great.

For titanium dioxide the optimum particle size is between 400 and 500 nm.*

In order to obtain the maximum **colouring power** (or, for white pigments, the maximum **brightening power**) the optimum particle size must be preserved in the binder. If the pigment particles agglomerate to some extent in the binder to form so-called secondary structures, then both the colouring power and the hiding power will fall.

The maximum colour intensity is thus obtained when each particle of the pigment is surrounded by binder. Many pigments are given a special surface treatment with the object of making them more "binder-compatible" and thus reducing their tendency to agglomerate. Such pigments can be more easily dispersed in the binder.

> When the particles have their optimum size and shape the hiding power increases with increasing refractive index.

Compared with the pigments, the refractive indices of fillers differ only slightly from those of the appropriate binders. If the binder and the filler have the same refractive index, then the filler is just as invisible in the binder as water in water. The widespread fashion for transparent shoe-soles is made possible by using a filler with the same refractive index as rubber.

Under special conditions, however, such fillers can act as white pigments. The explanation is as follows.

Just as transparent ice can be converted by crushing into white, opaque snow, paper treated with fillers can be white and opaque. This effect differs from that obtained with white pigments: the filler brightens the paper and renders it more opaque because the open structure of the paper entraps air, thus creating many air/fibre and air/filler surfaces at which the light is reflected diffusely.

In the same way, the white and opaque appearance of crushed ice is due to diffuse reflection from the newly formed air/ice interfaces.

Wet snow is less white than dry snow. This is partly because the water masks the irregularly shaped surfaces, and partly because there is only a small difference between the refractive indices of ice and water.

---

* 1 nm (nanometre) = 1 millionth of a millimetre = 3.94 hundred-millionth of an inch.

II Inorganic raw materials and large-scale products: Large-scale industrial processes

For the same reason filled white paper becomes grey if it is impregnated with paraffin wax or oil. Similarly, concrete blended with chalk looks white when dry, but becomes much darker when wet, since the refractive index of chalk does not differ much from that of water (cf. Table, p. 8).

On the other hand, because of the high refractive index of titanium dioxide, paper containing this pigment remains white even when impregnated with liquids, and the white markings on roads are still readily visible on rainy days.

1.4.4. Luminescent pigments

Luminescent pigments (luminophores) constitute a special class. When they are illuminated, or irradiated with energy in other ways, they shine either for very short periods (fluorescence) or for a longer time (phosphorescence).

The most important representative of the luminescent pigments is zinc sulphide, ZnS. Its phosphorescence is caused by the incorporation of very small concentrations of copper ions, Cu^{2+}, in the ionic lattice of ZnS. Inorganic luminophores are used for marking entrances and exits, for rendering electron beams visible (for example on the screens of television sets) and in fluorescent tubes for transforming ultraviolet radiaton into visible light.

1.5. The chemical preparation of important pigments

1.5.1. Precipitation from aqueous solution

Most inorganic pigments, including the most important ones, are prepared by precipitation reactions. These are chemical reactions in which the crude pigment is deposited as a crystalline solid by the addition of one solution to another.

One of the two solutions is frequently prepared by dissolving a metal (eg zinc, lead, cadmium) in an acid (eg hydrochloric acid) or a mixture of acids (eg sulphuric and nitric acids). These solutions must often be carefully purified, so as to prevent the tint of the pigment being modified by foreign metal ions (eg Cu^{2+}). The second solution contains the anion (eg S^{2-} from sodium sulphide) which combines with the metal cation to give the insoluble crude pigment.

The conditions of precipitation must be controlled so as to produce pigment particles of the required size. Important factors are the temperature and the pH, whether the two solutions are run in simultaneously or consecutively, and the speed of the mixer.

Precipitation of lithopones (barium sulphate/zinc sulphide)

$$BaS \quad + \quad ZnSO_4 \quad \longrightarrow \quad BaSO_4\downarrow \quad + \quad ZnS\downarrow$$

Barium sulphide	zinc sulphate		white		white

The value of the lithopone is increased by a high content of zinc sulphide.

Precipitation of chrome yellow (lead chromate)

$$Pb(NO_3)_2 \quad + \quad Na_2CrO_4 \quad \longrightarrow \quad PbCrO_4 \downarrow \quad + \quad 2\,NaNO_3$$

Lead nitrate   sodium chromate   yellow

Other chromate pigments, such as zinc chromate (zinc yellow) and barium chromate (barium yellow) are also prepared by precipitation.

Precipitation of cadmium yellow (cadmium sulphide)

$$CdCl_2 \quad + \quad Na_2S \quad \longrightarrow \quad CdS \downarrow \quad + \quad 2\,NaCl$$

Cadmium   sodium sulphide   yellow
chloride

After precipitation the pigments are separated, washed and dried. In some instances, for example for the chromate pigments and cadmium yellow, it is then only necessary to grind the pigment in order to separate crystals which have grown together.

Often, however, the crude pigment first precipitated has a particle size or crystal structure which differ widely from the optimum characteristics. This is the case for titanium dioxide, iron oxide, lithopones, cadmium red, etc. These crude pigments are processed further by ignition. During this heat treatment the constituent cations and anions of the pigment may under some circumstances re-arrange their positions in space; ie the crystal structure may change.

1.5.1.1.   Preparation of titanium dioxide

Titanium dioxide pigments are manufactured by two competing processes.

In the USA titanium dioxide is made on a large scale by the **chloride process**. In this process the raw materials (see below) are treated with chlorine.

$$2\,TiO_2 \quad + \quad 3C \quad + \quad 4\,Cl_2 \quad \longrightarrow \quad 2\,TiCl_4 \quad + \quad CO_2 \quad + \quad 2\,CO$$

The pure titanium tetrachloride, $TiCl_4$, is then converted into titanium dioxide of high purity, the chlorine being recovered at the same time:

$$TiCl_4 \quad + \quad O_2 \quad \longrightarrow \quad TiO_2 \quad + \quad 2\,Cl_2$$

This process has the advantage that there are no by-products to dispose of.

Outside the USA titanium dioxide is produced predominantly by the **sulphate process**.

II Inorganic raw materials and large-scale products: Large-scale industrial processes

The raw material employed is the mineral ilmenite, which is digested with sulphuric acid. Ilmenite contains 30% to 55% of TiO_2 by weight, according to the extent to which it is contaminated with foreign material, the so-called gangue.

Ilmenite is the most abundant and hence the most important raw material for the production of titanium dioxide.

In the treatment with sulphuric acid each kilogram (2.2 lb) of iron combined in the ilmenite requires 1.8 kg (4.0 lb) of 100% sulphuric acid to react with it.

In the sulphate process some difficulties arise in the budget for sulphuric acid and in the treatment of waste materials (ferrous sulphate and dilute sulphuric acid contaminated with ferrous sulphate).

This treatment is necessary in order to avoid pollution of the environment.

A second raw material for the manufacture of titanium dioxide pigments is the rarer and more expensive mineral rutile, which consists of more or less pure TiO_2. Rutile is also extracted from sands which may contain only 0.1% of rutile.

In the chloride process for titanium dioxide pigments the raw material employed is almost exclusively rutile.

The **manufacture of titanium dioxide by the sulphate process** can be divided into four stages:

1. dissolution of ilmenite and reduction of ferric ions;
2. removal of the gangue and part of the iron (II) sulphate (ferrous sulphate);
3. precipitation of titanium dioxide;
4. subsequent treatment of the titanium dioxide.

1. Dissolution and reduction

The dissolution of ilmenite produces iron (III) sulphate (ferric sulphate), which would cause difficulties in the precipitation of the titanium dioxide by forming insoluble red-brown iron (III) hydroxide (ferric hydroxide).

$$4\ FeTiO_3\ +\ 14\ H_2SO_4\ \longrightarrow\ 4\ Ti(SO_4)_2\ +\ 2\ Fe_2(SO_4)_3\ +\ 14\ H_2O$$

| Ilmenite | sulphuric acid | | titanium sulphate | ferric sulphate | water |

By adding the calculated quantity of scrap iron the Fe^{3+} of the ferric sulphate is reduced to Fe^{2+} of iron (II) sulphate:

$$Fe_2(SO_4)_3\ +\ Fe\ \longrightarrow\ 3\ FeSO_4$$

| Ferric sulphate | iron | iron (II) sulphate |

Figure 1. Flow scheme for the manufacture of titanium dioxide by the sulphate process

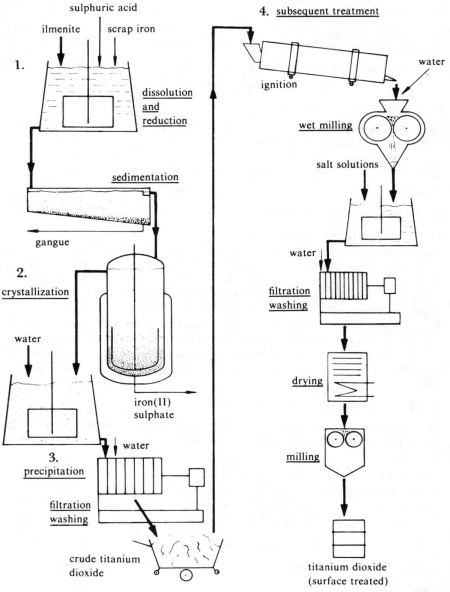

2. Removal of the gangue and part of the ferrous sulphate

Since the gangue is not attacked by concentrated sulphuric acid it can be separated from the dissolved ilmenite in settling tanks.

The dissolution reaction is exothermic, so that the solution becomes hot. On cooling, part of the ferrous sulphate formed separates out as the hydrated salt $FeSO_4 \cdot 7\ H_2O$.

This by-product can be used to recover sulphuric acid (see II,5, Manufacture of sulphuric acid).

3. Precipitation of titanium dioxide

On diluting the strongly acid solution, the titanium sulphate is hydrolyzed by the added water. Hydrated titanium dioxide is precipitated.

$$Ti(SO_4)_2\ +\ 2\ H_2O\ \longrightarrow\ TiO_2 . aq\downarrow\ +\ 2\ H_2SO_4$$
$$\text{white}$$

The precipitation is aided by adding crystal nuclei.

4. Subsequent treatment of titanium dioxide

The crude pigment is filtered off from the dilute sulphuric acid (containing ferrous sulphate) and is carefully washed. It is then ignited (calcined) in a rotary tube furnace. At temperatures above $1000°C$ ($1800°F$) the anatase crystal structure is converted into the more closely packed rutile structure. Titanium dioxide is more useful as a pigment in the rutile form, which has the higher refractive index (see Table, p. 8).

The crystallization process in the rotary tube furnace produces particles which are too large, so the product is then ground in a mill. This is usually followed by chemical treatment, in which hydroxides and oxides of aluminium, silicon and zinc are deposited on the surface of the pigment particles. This increases the stability of the pigment to light and weather.

1.5.1.2. Preparation of iron oxide pigments

The colour spectrum of the various iron oxide pigments ranges from bright yellow through orange, red and brown to black. Their optical effectiveness is very high in proportion to their cost.

They can be manufactured from cheap starting materials, for example the dilute acid (containing ferrous sulphate) from the treatment of ilmenite, and the pickling solutions used in the iron and steel industry. The addition of scrap iron to these solutions reacts with most of the excess acid and more iron goes into solution:

$$Fe\ +\ H_2SO_4\ \longrightarrow\ FeSO_4\ +\ H_2$$

| Scrap iron | sulphuric acid | iron (II) sulphate | hydrogen |

On neutralization, for example with caustic soda, iron (II) hydroxide is precipitated.

$$FeSO_4 \quad + \quad 2NaOH \quad \longrightarrow \quad Fe(OH)_2 \downarrow \quad + \quad Na_2SO_4$$

| Iron (II) sulphate | sodium hydroxide | iron (II) hydroxide | sodium sulphate |

By passing in controlled amounts of air a part or all of the $Fe(OH)_2$ is oxidized to iron (III) hydroxide, $Fe(OH)_3$, which is also insoluble.

After filtering off and drying and/or ignition iron oxide pigments can be obtained in three basic colours:

yellow	FeO(OH)	basic iron (III) oxide
red	Fe_2O_3	iron (III) oxide
black	$FeO \cdot Fe_2O_3$	(Fe_3O_4) iron (II,III) oxide

Another important process for the production of iron oxide pigments makes use of the fact that iron oxides are precipitated when nitrobenzene is reduced to aniline by iron filings in acid solution. By adding suitable metal salts valuable pigments can be obtained.

By washing and filtering off followed by drying or ignition, pigments can be obtained which are bright yellow, red, bluish red, brown or black.

Hard iron oxide pigments are used as ingredients of polishes, and ferromagnetic iron oxides are used in magnetic tapes for storing information.

1.5.2.   Other methods of preparing pigments

In the gaseous state: at high temperatures metal vapours can react with oxygen to give pigments, eg

$$2\,Zn \quad + \quad O_2 \quad \longrightarrow \quad 2\,ZnO$$

zinc oxide
(zinc white)

or chlorides of metals can be converted to oxides in an oxyhydrogen flame, eg

$$\underbrace{6\,H_2 + 3\,O_2}_{\text{Oxyhydrogen}} \quad + \quad \underset{\substack{\text{aluminium}\\\text{chloride}}}{4\,AlCl_3} \quad \longrightarrow \quad \underset{\substack{\text{aluminium}\\\text{oxide}}}{2\,Al_2O_3} \quad + \quad 12\,HCl$$

In melts: for example, ores containing cobalt can be fused with quartz sand and potassium carbonate (potash) to give a dark blue pigment which is used for colouring ceramics, glass and enamels.

II Inorganic raw materials and large-scale products: Large-scale industrial processes

2. Inorganic fillers

2.1. Definition of terms

Fillers are substances which are added to numerous products, for example tablets, cosmetic preparations, soaps, washing and cleansing powders, printing and artists' colours, adhesives, PVC floor coverings, plastic sheeting, moulded artificial resins, vulcanized rubber mixes, facing plaster, textiles, paper and carpets.

With the exception of carbon black, sawdust and shredded paper or textiles, all important fillers are white inorganic powders.

Like pigments, fillers are insoluble in binders and solvents.
Nowadays fillers rarely serve merely as cheap make-weights or diluents. They are predominantly auxiliary substances which improve the quality of a product. In this sense the term **active fillers** may be used.

Active fillers can improve the useful properties of a product, or may even introduce new ones.

2.2. Uses and industrial importance

Natural and artificial rubbers first acquired their supreme industrial importance through the use of carbon black as a filler. Apart from carbon black, the strength and elasticity of vulcanized rubber mixes is often improved considerably by white fillers such as calcium silicate, sodium aluminium silicate and highly dispersed silicic acids.

Fillers are used in paper to influence a large number of industrially important properties. They decrease the transparency of thin paper, increase its whiteness (see II,8, p. 7) and produce smooth surfaces suitable for printing, on which the printing ink is retained without coming through; they considerably increase the volume of the paper, and under some circumstances may retain pigments and cheap fillers so that the paper does not become powdery on printing: they cause computer tape to unroll smoothly and silently, and they regulate the combustibility of cigarette paper.

The uses of fillers in the textile industry resemble those in the paper industry.

In paints, printing inks and artists' colours fillers serve to thicken the product. They improve the covering power and help to prevent sedimentation and coagulation of the pigment particles.

When added in larger quantities they are used to produce a matt finish. The choice between heavy fillers such as barium sulphate and light ones such as precipitated silicates may depend on whether the paint is to be sold by weight or volume.

Suitable fillers are also used as thickening agents in toothpaste, creams, lotions, ointments, plastic pastes and dispersions, adhesives, primers, polishes and shoe creams.

Fillers are also added to powders and granules which have a tendency to cake. Fillers which are physiologically harmless are added in small quantities to table salt and to food concentrates such as instant coffee and soup and sauce powders.

Many fillers also find uses in the chemical laboratory as absorbents for vapours, liquids and solids. They are used as drying agents and for separating mixtures.

An interesting use of modified fillers is in the purification of effluents. They have the property of absorbing from the effluent several times their own weight of oil or solvents.

Fillers are also of some importance as carriers for active substances such as catalysts and insecticides.

Just as in the case of pigments, fillers can be divided into those which occur naturally and those which are produced synthetically, sometimes known as extenders. The following fillers occur naturally:

1. heavy spar or barytes — $BaSO_4$ (barium sulphate);
2. chalk — $CaCO_3$ (calcium carbonate);
3. kieselguhr or diatomaceous earth — SiO_2 (silica) containing water;
4. kaolin and clay — calcium aluminium silicates;
5. china clay — specially pure kaolin;
6. powdered asbestos, quartz, slate and rock.

The following fillers are produced synthetically:

1. carbon black — C (carbon)
   carbon black is mainly used as a filler and small particles are used as the finely divided carbon black pigment;
2. silicic acids — $(SiO_2)_n \cdot \times H_2O$
   and modified silicic acids containing oxides such as CaO, MgO, Al_2O_3 and BaO.
3. blanc fixe — $BaSO_4$ (barium sulphate);
4. satin white — a mixture of $CaSO_4$ and Al_2O_3;
5. calcium silicate — $CaSiO_3$ containing water;
6. aluminium silicate — $Al_2O_3 \cdot SiO_2$ containing water;
7. hydrated alumina — $Al_2O_3 \cdot H_2O$;
8. precipitated calcium carbonate — $CaCO_3$.

With the exception of carbon black, clay and powdered asbestos, slate and rock, all the above fillers are white solids.

2.3.   Physical properties

2.3.1.   External and internal surfaces, thixotropy

All fillers are finely divided solids. Their most prominent common property is their large
surface.

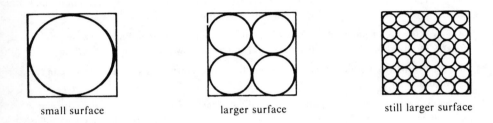

| small surface | larger surface | still larger surface |

The greater the degree of subdivision of a substance, the greater is its external surface.

The primary particles of a filler usually have diameters between 10 and 100 nm.* The specific
surface of a filler lies between 10 and 400 m² per gram (340 and 13,600 yd² per ounce),
depending on whether the particles have smooth, rounded or fissured surfaces.

> Fillers with particularly large external surfaces are
> particularly active.

Active fillers include carbon blacks (soots), kieselguhr, precipitated and pyrogenic silicic
acids, and precipitated calcium and aluminium silicate.

The large external surface allows the forces of attraction between the binder and the filler
to come fully into play. At the same time the particles of the filler also adhere together to
some extent, forming a secondary structure which holds the binder together in a kind of
scaffolding. In this way the mechanical strength of "filled" elastomers such as rubber or
silicone rubber is increased.

> The forces of attraction between the binder and the particles of filler
> and between the filler particles themselves are responsible for the
> reinforcing effect of active fillers

* See footnote on page 9 of this section.

A particularly interesting example is the effect of highly active silicic acids in producing "non-drip" paints. These paints are **thixotropic**, ie they will flow under the pressure of a paint brush or a roller, but when the pressure is released they immediately become solid again.

Thixotropy is particularly desirable in paints applied by dipping, since it prevents the paint from running off and dripping.

Thixotropic properties arise in the following way.

The particles of the filler form a framework or scaffolding in the binder.

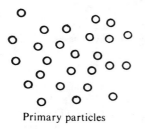  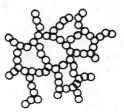

Primary particles   Secondary structure
                                                              in the binder

When a force is applied this framework is destroyed. When the shearing force is removed the framework forms again.

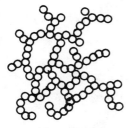

shearing force

at rest

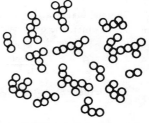

non-drip paint, due to   liquid paint, due to
structural framework   destruction of framework

Other fillers, for example the gels formed by silicic acid and metal oxides, are characterized by a particularly large **internal surface**. The particles of these fillers are strongly fissured, for example they may have a capillary structure. They can therefore easily absorb moisture or other liquids, and can thus improve the fluidity of powders, or can act as thickeners by taking up solvents.

Fillers with a large internal surface are particularly important as adsorbents.

2.3.2. Optical properties

It has already been mentioned that silicate fillers and silicic acids can act as pigments in paper owing to light scattering (see II,8, 1.4.3. p. 7).

2.4. The chemical preparation of important fillers

2.4.1. Carbon blacks

Carbon blacks represent a pure and finely divided form of carbon with a small and variable content of hydrogen and oxygen.

They are produced when gaseous, liquid or solid hydrocarbons burn with an insufficient supply of oxygen, so that combustion is incomplete and the flame is smoky. These **combustion blacks** are active fillers. They are particularly important as reinforcing fillers in the rubber industry.

1. **Furnace black** is made from residual oil and tar oil, which are rich in aromatic hydrocarbons, especially naphthalene and higher aromatics. The oil is sprayed into the flames of the furnace, where it is cracked into carbon black and hydrogen. Furnace blacks have been produced recently which are so finely divided that they can be used as pigments.
2. **Channel black** is obtained by the incomplete combustion of gaseous or vaporized hydrocarbons. It is the ideal carbon black for use as a pigment.
3. **Lampblack** is less important than the two preceding types of black.
4. **Thermal blacks** are obtained in the complete absence of oxygen. Gaseous hydrocarbons such as methane or acetylene are decomposed thermally:

$$CH_4 \xrightarrow{\text{high temperatures}} C \quad + \quad 2\,H_2$$

$$\text{Methane} \qquad\qquad \text{carbon black} \qquad \text{hydrogen}$$

Thermal blacks are not so finely divided as those produced by combustion, but are good conductors of electricity. They are therefore used for making electrodes. A well known example is the carbon electrodes of dry batteries.

The inactive thermal blacks are used for special purposes, for example vulcanized products which need to conduct electricity.

2.4.2. Precipitation of silicates and silicic acid from aqueous solution

In this process two solutions are mixed, and react to give the insoluble filler, which is precipitated. This method is known as **wet precipitation**. One of the two solutions to be mixed is always a solution of water-glass (sodium silicate). It is made by fusing together finely ground quartz sand and soda, followed by dissolution of the melt in water.

The second solution required to produce the filler may be either

a)   an acid, for the preparation of precipitated silicic acid:

$$Na_2O \cdot (SiO_2)_3 \ + \ 2\,HCl \ \xrightarrow{\ H_2O\ } \ 3\,SiO_2 \cdot x\,H_2O \ + \ 2\,NaCl$$

| Water-glass solution | hydrochloric acid | polysilicic acid |

or

b)   an acid salt solution, for example of aluminium sulphate or calcium chloride, for the preparation of precipitated silicates:

$$Na_2O \cdot (SiO_2)_3 + 3\,CaCl_2 \ \xrightarrow{\ H_2O\ } \ 3\,CaSiO_3 \cdot H_2O \ + \ 2\,NaCl \ + \ 4\,HCl$$

| Water-glass solution | calcium chloride | calcium silicate |

Primary particles of the required size can be obtained by choosing suitable conditions for precipitation. The necessary secondary structure is formed by the agglomeration of several primary particles.

Precipitated silicic acids consist of porous particles, and therefore possess an internal surface.

Precipitated silicic acids having a specially large internal surface (eg 90% internal and 10% external surface) are chosen for use as adsorbents.

2.4.3.   Pyrogenic silicic acids

Pyrogenic silicic acids, obtained in flames, are particularly pure products consisting of more than 98% SiO_2.*

They form very small spherical particles with a non-porous surface, and therefore have a high activity.

They are prepared by the reaction of the vapour of silicon tetrachloride in an oxyhydrogen flame. The combustion of hydrogen with oxygen produces water vapour, which converts silicon tetrachloride into silica. Gaseous hydrogen chloride is formed as a by-product.

$$\underbrace{2\,H_2 + O_2}_{\text{Oxyhydrogen}} \ \longrightarrow \ \underset{\text{water vapour}}{2\,H_2O}$$

---

*   Although SiO_2 (silicon dioxide, or silica) is actually the anhydride of silicic acid, the terms "silica" and "silicic acid" often occur interchangeably in common usage.

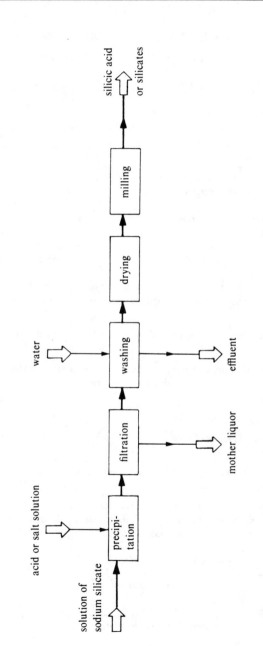

Figure 2. Flow scheme for the preparation of silicic acids and silicates by wet precipitation

$$\text{SiCl}_4 \quad + 2\,\text{H}_2\text{O} \quad \xrightarrow{\;1000°\text{C}\;} \quad \text{SiO}_2 \quad + 4\,\text{HCl}$$

silicon
tetrachloride   silicon dioxide

Since $SiCl_4$ reacts with H_2O in a flame the process is known as "flame hydrolysis".

Just as in the case of carbon black the spherical primary particles of the silicic acid formed join together into chains and bands, which separate out as flakes from the combustion gases.

Subsequent treatment with hot moist air is used to remove residual hydrogen chloride, HCl, which adheres to the extensive surface of the particles.

Figure 3. Flow scheme for the flame hydrolysis of silicon tetrachloride

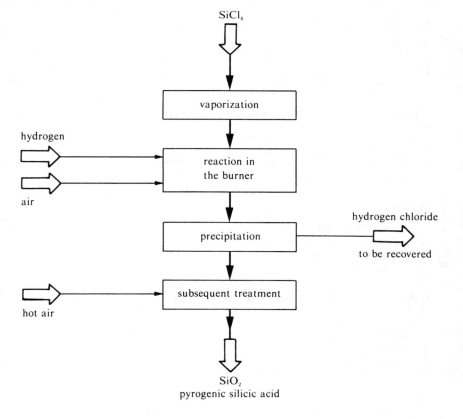

3. Self-evaluation questions

70   Pigments are

    A   good colorants only when in a fine state of dispersion
    B   water-soluble colorants
    C   virtually insoluble substances
    D   identical to dyes

71   The most important inorganic pigments in terms of quantity are

    A   iron oxide pigments
    B   lithopones
    C   titanium dioxide pigments
    D   zinc oxide pigments

72   Which compounds are white pigments?

    A   iron oxide
    B   zinc oxide
    C   titanium dioxide
    D   lead chromate

73   Which factors govern the hiding power of a pigment dispersion?

    A   the refractive index of the pigment
    B   the shape of the pigment particles
    C   the pigment particle size range
    D   the thickness of the dispersion layer

74   Fillers are

    A   more finely dispersed than pigments
    B   soluble in binders and solvents
    C   generally cheaper than pigments
    D   additives

75   Which fillers occur in nature and therefore do not need to be manufactured
synthetically?

    A   chalk
    B   aluminium silicate
    C   carbon black
    D   kieselguhr

76 Which of the following are used on a large scale both as a filler and as a pigment?

A   iron oxides
B   carbon blacks
C   titanium dioxides
D   rock flour

77 Which products contain fillers?

A   paper
B   creams
C   paints
D   rubber

9 Nuclear Chemistry

1. Theoretical principles

1.1. Elements and isotopes

Every atom contains an atomic nucleus. Outside the nucleus there are only negative elementary particles, the extranuclear electrons (see 1.5).

The nucleus is composed of positively charged elementary particles, **protons**, and electrically neutral elementary particles, **neutrons**.

The elementary particles which make up the nucleus are given the common name of nucleons.

The number of protons in an atomic nucleus is always equal to the number of extranuclear electrons.

> Number of protons in the atomic nucleus =
> number of extranuclear electrons

The protons and neutrons in the nucleus are held together in an extremely small space. They contribute the mass of the atom. The masses of the proton and the neutron are almost the same.

> An atomic nucleus is composed of elementary
> particles, protons and neutrons

The following symbols are used as abbreviations for electrons, protons and neutrons:

electrons e^-
protons p^+
neutrons n

The number of mass units contained in these elementary particles, or in atomic nuclei composed of them, is indicated by a number at the top left-hand side of the corresponding symbol.

> The neutron contains 1 mass unit; it is therefore written as ^1n.
> The proton also contains 1 mass unit, and is hence written as ^1p.
> The nucleus of a hydrogen atom also has 1 mass unit; hence ^1H.
> The nucleus of a helium atom contains 4 mass units; hence ^4He.

The number of units of positive charge carried by an elementary particle, or by an atomic nucleus composed of elementary particles is indicated by an additional subscript on the left-hand side of the symbol.

> A neutron has zero charge, and is therefore written ^1_0n.
> A proton has 1 unit of positive charge, and is therefore written ^1_1p.

The nucleus of a hydrogen atom has one unit of positive charge, and is therefore written $^1_1$H.

A helium nucleus has two units of positive charge, and is therefore written $^4_2$He.

> The superscript on the left of the symbol for an element gives the number of units of mass in its nucleus.
> The subscript on the left gives the number of units of positive charge on the nucleus.

The concept of an element

All the atoms of a given element have the same number of protons in their nuclei.

The nucleus of hydrogen	contains 1 proton
The nucleus of helium	contains 2 protons
The nucleus of carbon	contains 6 protons
The nucleus of oxygen	contains 8 protons
The nucleus of uranium	contains 92 protons

> An element is a substance consisting of atoms which all contain the same number of protons in their nuclei

Isotopes

The nucleus of a given element always contains the same number of protons, but the number of neutrons may vary.

> A given element always has the same number of protons in its nucleus, but the number of neutrons can vary.

Atoms which have the same number of protons but different numbers of neutrons are called isotopes.

For example, there are three isotopes of hydrogen:

Number of protons	+	number of neutrons	=	number of mass units	Name	Symbol
1	+	0	=	1	Hydrogen	$^1_1$H
1	+	1	=	2	Deuterium	$^2_1$D
1	+	2	=	3	Tritium	$^3_1$T

Some isotopes, for example tritium, have radioactive properties, ie they emit high-energy radiation.

By far the most abundant isotope of hydrogen is $^1_1$H. It consists of only one proton, and the symbol $^1_1$H is often used in place of $^1_1$p.

Monoisotopic elements

Elements in which the atoms always contain the same number of neutrons are termed monoisotopic.

There are 21 naturally occurring monoisotopic elements, including sodium and aluminium. Naturally occurring sodium consists only of atoms with 11 protons and 12 neutrons.

1.2. Nuclear processes

A nuclear process means the modification of an atomic nucleus by the loss, gain or alteration of some of its constituents.

If a proton is added to a lithium nucleus, which consists of 3 protons and 4 neutrons, two helium nuclei are produced.

$$^7_3 \text{Li} + ^1_1 \text{H} \longrightarrow ^4_2 \text{He} + ^4_2 \text{He}$$

The element lithium has been transmuted into the element helium by a nuclear process. Energy is liberated, so that this nuclear process is an exothermic reaction.

> Nuclear chemistry is concerned with the transmutation of elements by modifying the atomic nucleus
> The products of nuclear processes are new elements or isotopes.

1.3. Carrying out nuclear transformations

The element to be transformed is bombarded with particles in specially designed equipment. These particles may be neutrons, protons, α-particles, or other light nuclei (α-particles consist of helium nuclei $^4_2$He).

The "missiles" can be obtained from the disintegration of radioactive elements, or from other sources. Thus Sir Ernest Rutherford, the founder of nuclear chemistry, used a disintegration product of thorium as an α-particle gun in his first transmutation of an element. He bombarded nitrogen and obtained oxygen and protons.

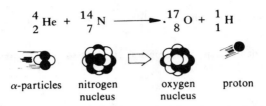

$$_2^4\text{He} + \, _7^{14}\text{N} \longrightarrow \cdot \, _8^{17}\text{O} + \, _1^1\text{H}$$

| α-particles | nitrogen nucleus | | oxygen nucleus | proton |

Particles of higher energy can be obtained by accelerating them, ie giving them a higher velocity. The apparatus required for this purpose is known as a particle accelerator.

1.4. Energy transformations

The energy changes associated with nuclear processes are around ten million times as great (for the same quantity of material) as those for chemical processes. These enormous amounts of energy are stored in the nucleus as binding energy. This is necessary in order to hold together the protons, which have like charges, and become more difficult the larger the number of protons in the nucleus. If a nucleus disintegrates into smaller fragments part of the binding energy is no longer required and is given up to the surroundings.

For example, when uranium disintegrates according to the equation

| Neutron | | uranium | | krypton | | barium | | neutrons |

$$_0^1\text{n} + \, _{92}^{235}\text{U} \longrightarrow \, _{36}^{89}\text{Kr} + \, _{56}^{144}\text{Ba} + 3\, _0^1\text{n}$$

the amount of energy liberated is about 16.75 billion kJ/mol (15.88 billion Btu/mol) of uranium atoms.

1.5. The stability of atomic nuclei, radioactivity

1.5.1. Natural and artificial radioactivity

When protons and neutrons combine to form an atomic nucleus it is found that the relative numbers of protons and neutrons must be accurately adjusted if the nucleus is to be stable. For small nuclei the optimum ratio is 1:1; for example helium contains 2 protons and 2 neutrons. As the number of protons increases the number of neutrons needed to hold the nucleus together rises, and in the largest naturally occurring nuclei the neutron/proton ratio is about 1:1.59. Uranium 238 contains 146 neutrons and 92 protons, so that (number of neutrons)/(number of protons) = 146/92 = 1.59.

When the number of protons in the nucleus exceeds a certain value, even the optimum ratio is insufficient to ensure stability. The nuclei disintegrate spontaneously, with a definite average lifetime, into smaller nuclei, with the emission of radiation. This phenomenon is known as **natural radioactivity**.

If the optimum proton:neutron ratio is altered by bombarding the nucleus with neutrons or α-particles the nucleus may again disintegrate with the emission of high-energy radiation. This case is known as **artificial radioactivity**.

1.5.2. The law of radioactive decay

The characteristic quantity for radioactive disintegration is the **half-life**. This is the time required for the disintegration of half of any arbitrary number of atoms originally present.

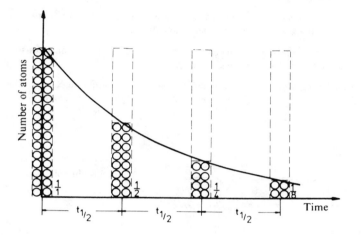

The half-lives of individual radioactive elements or isotopes vary greatly. This is shown by the following comparison:

Half-lives of some isotopes

Nucleus disintegrating			Nucleus formed			Half-life = $t_{1/2}$	
$^{232}_{90}$	Th	(thorium)	$^{228}_{86}$	Rn	(radon)	$1.41 \cdot 10^{10}$	years
$^{238}_{92}$	U	(uranium)	$^{234}_{90}$	Th	(thorium)	$4.51 \cdot 10^{9}$	years
$^{210}_{82}$	Pb	(lead)	$^{210}_{83}$	Bi	(bismuth)	19.4	years
$^{60}_{27}$	Co	(cobalt)	$^{60}_{28}$	Ni	(nickel)	5.24	years
$^{90}_{38}$	Sr	(strontium)	$^{90}_{39}$	Y	(yttrium)	28	years
$^{140}_{56}$	Ba	(barium)	$^{140}_{57}$	La	(lanthanum)	12.8	days
$^{214}_{82}$	Pb	(lead)	$^{214}_{83}$	Bi	(bismuth)	28.4	minutes
$^{213}_{83}$	Bi	(bismuth)	$^{209}_{81}$	Tl	(thallium)	$4.2 \cdot 10^{-6}$	seconds

1.5.3. Nuclear radiation

If nuclear radiation is passed between two parallel metal plates bearing positive and negative charges respectively, the rays are split into three parts. The nuclear radiation is thus not homogeneous.

One part of the beam is deflected towards the plate bearing a negative charge. This part consists of positively charged helium nuclei, also called α-rays.

Another part of the radiation is deflected towards the positive pole. This consists of negatively charged particles, electrons, called β-rays.

The undeflected rays consist of electromagnetic radiation, resembling X-rays. They are called γ-rays. Since the α- and β-rays consist of particles they have been called corpuscular rays.

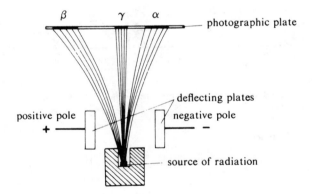

Diagram showing the deflection of nuclear radiation

1.5.4. The penetrating power of nuclear radiation

The penetrating power of the different kinds of radiation varies greatly, and depends on their energy. Since the α- and β-rays consist of material particles they are very quickly slowed down by matter, losing their kinetic energy. They therefore cause serious damage if they penetrate into the tissues and cells of the body.

The γ-rays consist purely of electromagnetic energy, and therefore have great penetrating power. Penetrating power is expressed in terms of the "half-value thickness", which is the thickness of the layer required to reduce the original radiation intensity by half.

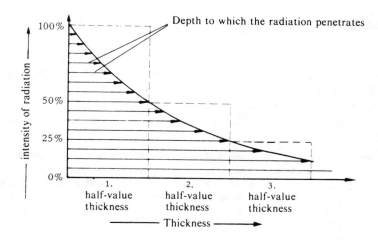

Explanation of the diagram

The half-value thickness is taken as 2 cm (0.79 in). In the first 2 cm (0.79 in) 50% of the incident radiation is absorbed and 50% is transmitted. In the second half-value thickness half of the remaining 50% (25% of the original intensity) is again absorbed.

In each half-value thickness the intensity at the beginning of the layer is reduced to one half. Strictly speaking, zero intensity is never reached. For example, the intensity of hard (ie high-energy) β-radiation is halved by a 1.5 mm (0.059 in) layer of water or plastic, while hard γ-radiation requires 1.5 cm (0.59 in) of lead, α-rays are almost completely absorbed by a sheet of paper, since their particles are relatively "very large".

1.5.5. Detection of nuclear radiation

Nuclear radiation is not perceived by the human senses. The simplest method of detection is by means of a fluorescent zinc sulphide (ZnS) screen. When a particle hits the screen it produces a small flash of light, which can be observed under a microscope. Photographic plates are also affected by nuclear radiation, as well as by visible light.

Modern procedures record the number of particles emitted by means of electronic devices, for example Geiger counters.

The extensive use of radiochemical methods of investigation is due to the extremely high sensitivity of present-day radiation detectors: 10^{-12} to 10^{-18} mols of a radioactive substance, ie down to 1 million radioactive nuclei, can be detected as a matter of routine.

II Inorganic raw materials and large-scale products: Large-scale industrial processes

1.6. Artificial nuclear transformations

1.6.1. Nuclear fission

The best known example of nuclear fission is that of uranium, which, inter alia, forms the basis of the generation of energy in nuclear power plants.

When uranium is bombarded with slow (thermal) neutrons, the uranium nucleus is split. The uranium $^{235}_{92}U$ is broken down into barium and krypton, and a further three neutrons are emitted.

If the newly formed neutrons are slowed down (for example by means of graphite) they also become thermal neutrons and are able to split uranium nuclei. If an emitted neutron can on average cause more than one nuclear fission, the multiplication factor is said to be greater than unity. The number of fission neutrons then increases like an avalanche and a chain reaction is set up. A chain reaction is always a possibility if the mass of the fissionable material is large enough.

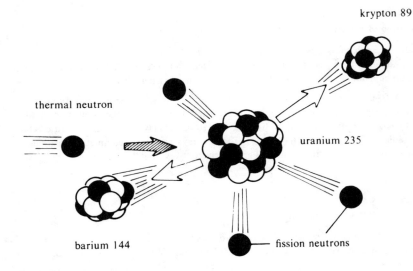

krypton 89

thermal neutron

uranium 235

barium 144

fission neutrons

If the multiplication factor is only slightly greater than unity the fission process can be easily regulated by "mopping up" some of the neutrons. In this way a controlled nuclear fission can be produced, as in a nuclear reactor.

The interior of a nuclear reactor thus contains a large number of free neutrons. Nuclear reactors are used as neutron sources for the production of industrial quantities of radioactive isotopes.

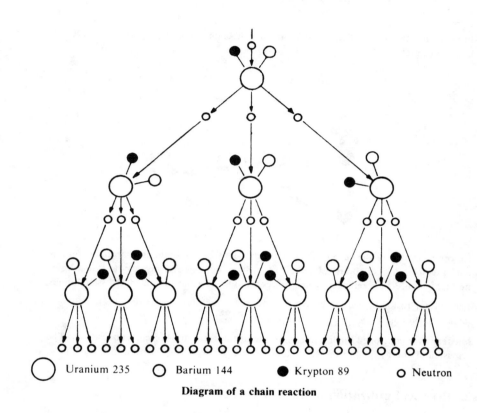

| | Uranium 235 | ○ | Barium 144 | ● | Krypton 89 | ○ | Neutron |

Diagram of a chain reaction

If the multiplication factor is considerably greater than unity the chain reaction acquires an explosive character. This produces an uncontrolled nuclear reaction, which, as is well known, is used in the atom bomb.

1.6.2.   Nuclear fusion

The combination of two small nuclei to give a larger one is called nuclear fusion. Like nuclear fission, this process liberates large amounts of energy.

> Energy is liberated by the fission of nuclei containing more than 60 nucleons. If the number of nucleons is less than 60, energy is liberated by nuclear fusion.

II Inorganic raw materials and large-scale products: Large-scale industrial processes

For example, the fusion of two deuterium nuclei liberates about 2.1×10^8 kJ/mol (200 billion Btu/mol).

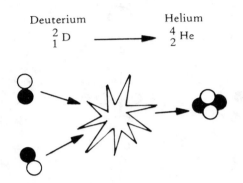

However, because of the large forces of repulsion between the similarly charged deuterium nuclei temperatures of about 100 million °C (180 million °F) are necessary to bring about nuclear fusion. At present such temperatures can only be attained in nuclear fission bombs.

The nuclear fusion of deuterium to give helium has therefore so far only been achieved in the hydrogen bomb.

Mastery over controlled nuclear fusion is one of the great aims of current research. The provision of energy for future generations depends critically on the attainment of this goal.

2. Practical utilization of nuclear energy

At present most electrical energy is still obtained by burning coal, petroleum or natural gas. However, during the last few decades petroleum and natural gas have become some of the most important sources of raw materials for organic reactions, including the production of foodstuffs. Since these raw materials cannot be replaced from other sources, and since the reserves of petroleum and natural gas are not inexhaustible, in the long term they must be reserved for chemical purposes.

The importance of the production of energy from nuclear fuels is therefore increasing steadily.

The share of nuclear energy in the electricity supply

This sketch depicts the percentage share of nuclear energy in the electricity generation of several western industrialized nations. It demonstrates that it will no longer be possible to dispense with electricity from nuclear power in the industrialized nations. The proportions already contributed by this source are too large.

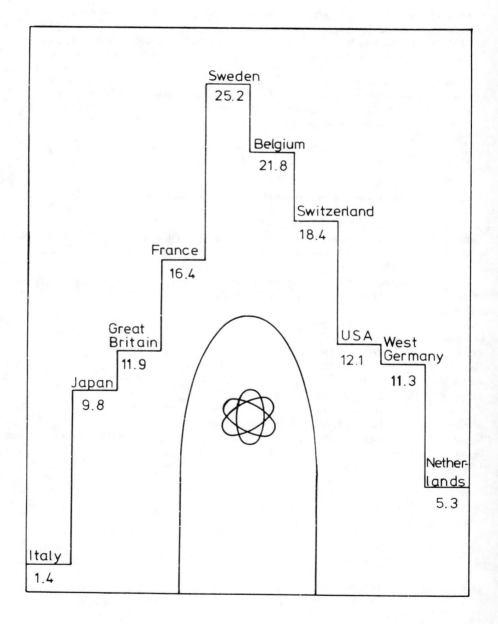

Sweden 25.2
Belgium 21.8
Switzerland 18.4
France 16.4
Great Britain 11.9
Japan 9.8
USA 12.1
West Germany 11.3
Netherlands 5.3
Italy 1.4

II Inorganic raw materials and large-scale products: Large-scale industrial processes

Up until 1980, 233 nuclear power stations with a net output of 132,588 MW(elec) had been commissioned throughout the world. The USA comes first with a net output of 52,477 MW(elec), followed by Japan with 15,007 MW(elec), the USSR with 14,245 MW(elec) and France with 11,001 MW(elec). Western Germany with 8620 MW(elec) only comes fifth.

A further 337 nuclear power stations with a net output of 317,581 MW(elec) are in the course of construction or have been ordered throughout the world. The following table provides a survey of the nuclear power stations of the countries which have made the greatest progress in the utilization of nuclear energy.

The nuclear power stations of the world, broken down by country
(position as at mid-1980)

Country	In operation, under construction and on order		Of which in operation		Of which under construction		Of which on order	
	Number	MW(elec)	Number	MW(elec)	Number	MW(elec)	Number	MW(elec)
West Germany	29	25,062	14	8,620	11	11,931	4	4,511
East Germany	13	5,120	4	1,340	5	2,100	4	1,680
France	58	52,588	18	11,001	29	29,095	11	12,650
United Kingdom	29	15,512	18	8,118	10	6,194	1	1,200
Japan	34	23,888	24	15,007	6	5,087	4	3,794
Canada	26	15,572	11	5,516	10	6,256	5	3,800
Sweden	12	9,460	6	3,740	6	5,720	-	-
Soviet Union	60	40,505	34	14,245	12	11,260	14	15,000
Spain	17	14,058	3	1,073	12	11,178	2	1,927
USA	199	193,436	72	52,477	94	103,109	33	37,850

2.1. Mode of operation of a nuclear power plant

The production of electrical energy from nuclear energy is carried out in nuclear power plants.

This process involves five stages.

Stage 1

The nuclear energy is converted into heat energy by nuclear fission in a nuclear reactor, and is transferred to a heat transfer agent, for example carbon dioxide gas (primary heat transfer agent).

The primary heat transfer agent becomes radioactive by direct contact with the nuclear fuel. It passes through a closed circuit, so that the radioactivity does not reach the surroundings.

Stage 2

The carbon dioxide gas (the primary heat transfer agents) gives up its heat to water (the secondary heat transfer agent) in a heat exchanger. The water is vaporized and leaves the heat exchanger as high-pressure superheated steam.

Stage 3

The steam drives a turbine, thus converting the heat energy into mechanical energy (kinetic energy). This lowers the temperature and pressure of the steam.

Stage 4

The turbine drives a generator, on the same principle as a bicycle dynamo. In the generator the kinetic energy is converted into low-voltage electrical energy.

Stage 5

The electrical energy is transformed from a low to a high voltage and is fed into the electricity grid.

Diagram of a nuclear power plant

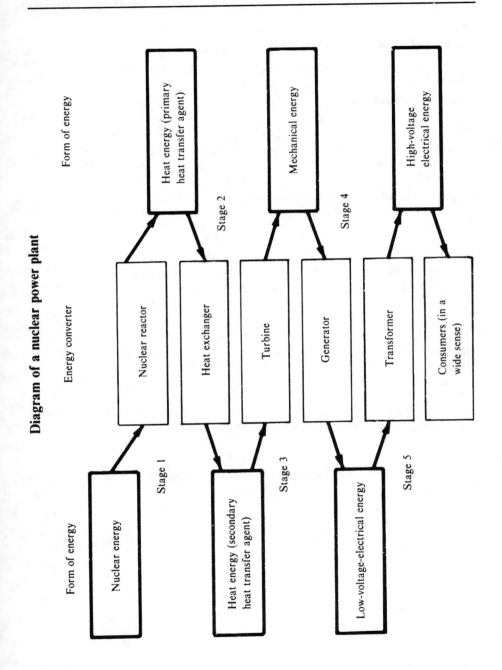

Form of energy — Nuclear energy

Energy converter — Nuclear reactor

Form of energy — Heat energy (primary heat transfer agent)

Stage 1

Heat energy (secondary heat transfer agent)

Heat exchanger

Stage 2

Stage 3

Turbine

Mechanical energy

Low-voltage-electrical energy

Generator

Stage 4

Stage 5

Transformer

High-voltage electrical energy

Consumers (in a wide sense)

2.2. Construction of a nuclear reactor

The centre of a nuclear power plant is the nuclear reactor.

The nuclear fuel is in the form of long rods, called fuel elements, which are distributed equidistantly in a block of graphite (the moderator).

The purpose of the graphic block is to slow down the neutrons produced by nuclear fission (fission neutrons), since only slow neutrons of low energy are able to cause further nuclear fission. Control rods are built into the reactor in order to regulate its output.

The heat developed in the fuel elements is removed by a cooling agent, for example carbon dioxide gas, which flows continuously round the elements. The reactor is surrounded by a steel casing and by a thick layer of concrete, which serve as radiation shields. Extensive measuring, regulating and safety devices prevent any leaks of radioactive material, and make the nuclear reactor an extremely safe installation which is easy to maintain.

Isolated failures in nuclear power plants have shown, that both perfect technology and perfect maintenance are of paramount importance.

A development in reactor design which has some promise for the future is the breeder reactor, also called the "fast breeder", which uses plutonium Pu 239 as the fissile material. The fast neutrons emitted by the fission process convert cheap uranium 238 into plutonium, so that fresh fissile material is continuously produced - hence the term "breeder".

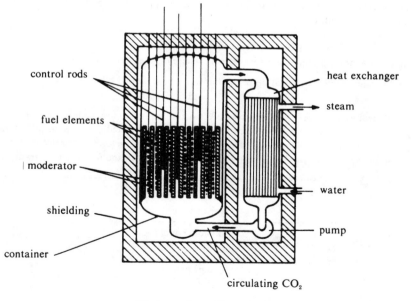

control rods

fuel elements

moderator

shielding

container

heat exchanger

steam

water

pump

circulating CO_2

Diagram of a nuclear reactor

II Inorganic raw materials and large-scale products: Large-scale industrial processes

3. Practical application of nuclear fission products

The products of nuclear fission are used widely in science and industry as **radiation sources** and **tracer isotopes**.

3.1. Radioactive labelling, tracer isotopes

The replacement of an atom in a molecule by a radioactive atom of the same element does not affect the chemical or physical properties of the molecule. However, the molecule is now provided with a "transmitter", which emits radiation which can be detected.

In this way it is possible to "locate" the molecule and to determine its situation. This procedure is called radioactive labelling or isotopic labelling, and the isotopes used are termed tracer isotopes, since they serve to trace the position of the labelled molecule. If the labelled atoms are incorporated into other molecules as a result of chemical reaction, these molecules will then "transmit" and hence reveal the presence in them of the atoms in question.

3.2. Radioactive substances in medicine

Medicine employs substances with radioactive labelling for testing recently developed drugs and for diagnosis. In investigating a drug the following questions can be answered:
1. How quickly does the active substance become distributed in the body?
2. Where is it concentrated?
3. How is it eliminated?
4. How long does it remain in the body?
5. What degradation products are formed, and where do they remain?

Information concerning concentration in the body can be obtained by means of autoradiography.

In autoradiography a radioactve biological object is brought into close contact with a photographic film. In regions where the radioactivity is concentrated the photographic film is blackened by the action of radiation. The developed film thus gives a picture of the distribution of the radioactive material in the tissue, as shown in the following diagram:

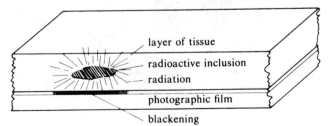

The principle of autoradiography

In medical diagnosis radioactively labelled substances (trace elements, vitamins, hormones, proteins) are introduced into the body in order to obtain information about the functioning of individual organs from the rate at which these substances migrate.

For example, the thyroid hormone thyroxin contains a large proportion of iodine.

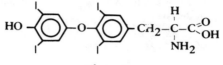

Thyroxin

This hormone is produced continuously in the thyroid gland and passes into the blood stream. If the body is supplied with radioactive iodine instead of ordinary iodine this radioactive iodine is incorporated into the thyroxin. By measuring the radioactivity of the blood protein, information can be obtained about the functioning of the thyroid gland. By comparing with "normal" values it is easy to detect either abnormally low or abnormally high thyroid activity. If the amount of iodine taken up exceeds the normal value, this indicates iodine deficiency in the body, while a decreased iodine uptake suggests malfunctioning of the thyroid.

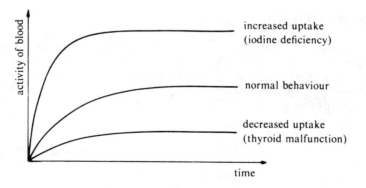

3.3.   Radioactive substances in research

3.3.1.   Elucidation of chemical reaction mechanisms

Isotopic labelling has become an indispensable aid in the investigation of chemical and biological processes. What is the principle underlying this method? If we add to an element a small proportion of one of its isotopes, not present in the naturally occurring mixture of

---

isotopes, then this isotope will take place proportionately in all physical and chemical processes which the element undergoes. This can be followed particularly easily with a radoactive isotope which emits radiation.

The added isotope is called a tracer. It is added in extremely small amounts, since the detection of nuclear radiation is highly sensitive. For example, 10^{-11} g (ie one hundred thousand millionth part of a gram or 330 thousand billionth part of an ounce) of the radioactive isotope $^{14}_{6}C$ can be detected.

The following isotopes are of particular importance:

Isotope	Half-life	Radiation
C 14	5568 years	β
H 3	12.26 years	β
P 32	14.2 days	β
I 131	8 days	β, γ
Na 24	15 hours	β, γ
S 35	88 days	β
Co 60	5.2 years	β, γ
Sr 90	53 days	β

The use of tracers in chemistry will now be illustrated by two examples.

The isomerization of n-butane

n-butane isomerizes to i-butane in the gas phase in the presence of a catalyst. If one of the terminal methyl groups is labelled with C 14, on analysing the isotope distribution in i-butane and unchanged n-butane after isomerization radioactivity is found to be present not only in the methyl groups, but also in the tertiary carbon atom of the i-butane and in the methylene group of the n-butane.

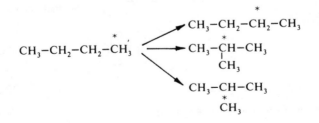

Hence in addition to the isomerization reaction, n-butane must undergo a re-arrangement in which a methyl group and a methylene group change places. This reaction is only detectable by means of isotopic labelling.

3.3.2.   Industrial research techniques

This refers to investigations which can be carried out on industrial equipment without interrupting its operation. If substances emitting γ-radiation are used it is possible to follow and to measure the movement of the radioactive substance even through thick walls, since γ-rays have high penetrating power and can pass unhindered through the walls of the container.

Example:

In the production of plastics several charges of polymer powder are mixed in a silo by a blast of air, so as to obtain a uniform product. In order to determine how long a period of mixing is required, about 1 kg (2.2 lb) of the material is radioactively labelled and is added to the contents of the silo at the top. After the mixing has been started samples are taken from time to time from the top and the bottom. As mixing progresses the activity at the top of the silo decreases, and that at the bottom increases. Equal activity at the top and the bottom indicates that mixing is complete.

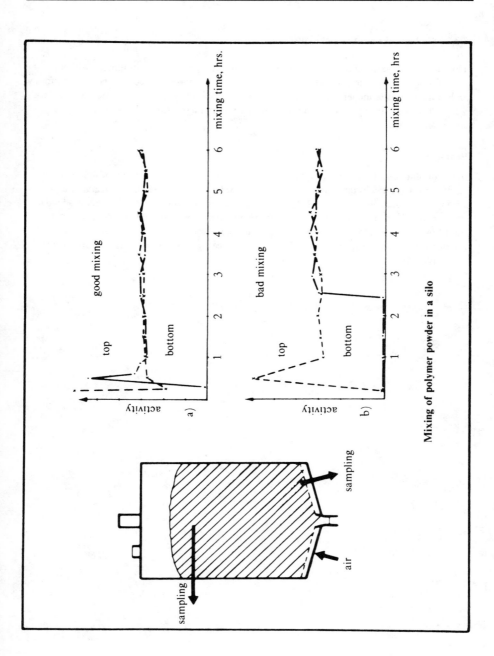

Mixing of polymer powder in a silo

3.3.3. Measurements of wear

It is not usually possible to make an accurate estimate of the extent of wear or deterioration of machines and industrial installations without stopping, completely emptying and dismantling the equipment. The incorporation of radioactive material makes it possible to carry out continuous checks and hence to judge with greater certainty the correct time for replacing worn-out parts.

For example, in order to monitor the deterioration of blast furnace linings radioactive sources of cobalt 60 ($^{60}_{27}$Co) are built into the brickwork, and their activity measured outside the casing.

If some of the cobalt has been removed by the wearing out of the lining the activity outside it will decrease. From the diminution in activity it is possible to estimate the thickness of the lining (after allowing for the natural nuclear decay), and hence the time at which it needs to be renewed.

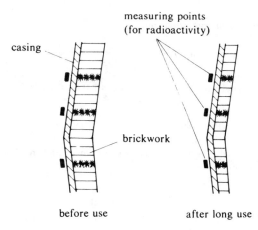

before use after long use

4. Application of nuclear radiation

4.1. Medical diagnosis and therapy

Radioisotopes are used as sources for X-ray diagnosis. Small portable sources containing thallium 170 ($^{170}_{81}$Tl) as the radiation source are now available and can be used to take X-ray pictures at the scene of an accident. In medical therapy nuclear radiation is used for treating tumours and cancerous tissues, since unhealthy or rapidly growing tissues are harmed or destroyed by nuclear radiation to a greater extent than healthy tissues. Radium, which has been used exclusively for many years, is replaced nowadays by inexpensive isotopes such as cobalt 60 ($^{60}_{27}$Co).

4.2. Sterilization

High doses of radiation kill bacteria. Medical equipment, clothing, containers, etc, can therefore be sterilized with γ-rays.

Treatment of foodstuffs with radiation has similar effects. As well as killing bacteria it prevents sprouting (for example of potatoes), both of which lead to better keeping qualities.

4.3. Non-destructive testing of materials

X-rays have been used for many years for the non-destructive testing of materials. X-ray equipment is expensive to buy and has high running costs. The use of radioactive sources has the advantage that no energy supply is required, and the necessary apparatus is portable and easy to handle.

Disadvantage factors are the relatively long exposure times required and the need for radiation protection.

The following isotopes are used as radiation sources:

$^{170}_{69}$ Tm (thulium) for steel up to 6 mm (0.24 in) thick

$^{192}_{77}$ Ir (iridium) for steel between 6 and 50 mm (0.24 and 2.0 in) thick

$^{60}_{27}$ Co (cobalt) for steel between 50 and 500 mm (2.0 and 20 in) thick

The object to be tested is placed between the source of radiation and the detector (photographic film or counter).

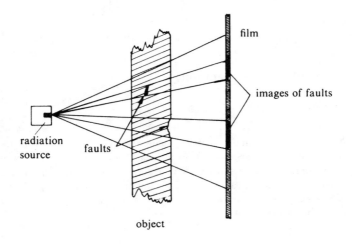

If the material contains a foreign body the radiation will be less attenuated at this point, and the underlying part of the photographic film will be blackened to a greater extent.

4.4. Light sources

The radiation emitted during nuclear decay can be converted directly into visible light by using luminescent substances.

The choice of a radioisotope for this purpose must take into account the following points:

1. the nucleus must not emit any high-energy γ-radiation;
2. the disintegration products should be as harmless as possible;
3. the luminescent substance should be destroyed as little as possible by the radiation;
4. the half-life should be as long as possible;
5. the nuclide should be simple to handle;
6. the nuclide should be inexpensive.

These conditions are satisfied by the following three isotopes:

$${}^{3}_{1}\text{H}\ \text{(tritium)}\qquad t_{1/2} = 12.26\ \text{years}$$

$${}^{85}_{36}\text{Kr}\ \text{(krypton)}\qquad t_{1/2} = 10.3\ \ \text{years}$$

$${}^{147}_{61}\text{Pm}\,\text{(promethium)}\qquad t_{1/2} =\ \ 2.65\ \text{years}$$

The above isotopes are all β-emitters, ie they emit electrons. The proportion of γ-radiation is less than 1%.

The highest light efficiency (20 to 30%) is obtained by using ZnS (zinc sulphide) as the phosphor. One application is in the light buoy.

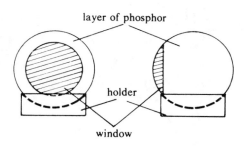

layer of phosphor

holder

window

The lead-glass sphere has a radius of 2 cm (0.79 in) and a window of area 6 cm² (0.93 in²). It is filled with radioactive krypton. The layer of phosphor is on the inner surface of the hollow sphere. Since this kind of light source cannot be switched on and off, it is turned to "dark" by rotating the window in the holder.

4.5. Radio-isotopes in preparative chemistry

The energy of nuclear radiation is between a thousand and ten million times greater than the binding energy between atoms in a molecule.

It is therefore not surprising that the action of nuclear radiation on chemical compounds produces changes in their molecules.

The primary effect is almost always to split the molecule up into fragments (radicals). These radicals can then react with one another to produce a wide variety of secondary products.

Compound ——→ cleavage ——→ radicals ——→ reaction ——→ new compounds

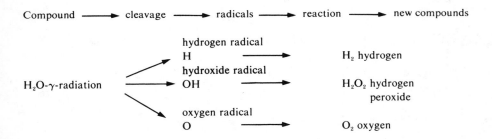

Since radicals play an important part in the formation of high polymers (plastics), radionuclides have been introduced to an increasing extent for promoting polymerization and cross-linking.

Example:

Ethylene is a hydrocarbon which has a double bond in its molecule;

Ethylene

The effect of high-energy radiation is to open up the double bond with the formation of a radical

$$
\begin{array}{cc}
\text{H} & \text{H} \\
| & | \\
\cdot\,\text{C} - \text{C}\,\cdot \\
| & | \\
\text{H} & \text{H}
\end{array}
\quad \text{Ethylene radical}
$$

This radical is very reactive. It can add on to both ends of an ethylene molecule

$$
\begin{array}{cc}
\text{H} & \text{H} \\
| & | \\
\text{C} - \text{C} \\
| & | \\
\text{H} & \text{H}
\end{array}
\left[
\begin{array}{cc}
\text{H} & \text{H} \\
| & | \\
\text{C} - \text{C} \\
| & | \\
\text{H} & \text{H}
\end{array}
\right]
\begin{array}{cc}
\text{H} & \text{H} \\
| & | \\
\text{C} - \text{C} \\
| & | \\
\text{H} & \text{H}
\end{array}
$$

original ethylene radical

The terminal free valencies continue to add on further ethylene molecules until two radicals meet each other. This leads to interruption of the growth of the chain.

The gas ethylene has been converted into the solid plastic polyethylene.

5. Self-evaluation questions

78   Which statements concerning the term "element" are correct?

A   An element is a basic chemical substance.
B   An element consists of atoms with identical numbers of protons.
C   An element is a pure element if it consists only of atoms whose number of protons corresponds to the number of neutrons.
D   In the atoms of an element the number of nucleons and the number of protons are always identical.

79   The term "isotopes" connotes

A   the types of atom of an element
B   only the atoms of pure elements
C   atoms with the same number of protons but differing numbers of neutrons
D   atoms with the same number of neutrons but differing numbers of protons

80   The number of isotopes of the element hydrogen occurring in nature is

A   3
B   2
C   1
D   0

81   Nuclear chemistry is concerned with

A   the transformations of elements
B   changing the electronic shells of atoms
C   changing the nuclei of atoms
D   creating compounds from elements

82   The term "artificial radioactivity" connotes

A   the build-up of new atomic nuclei by bombarding existing nuclei
B   the disintegration of atomic nuclei by bombarding them with neutrons or α-particles
C   non-induced disintegration of the atomic nuclei of naturally-occurring elements
D   any disintegration of atomic nuclei in which energy is released

83 How are radioactive rays deflected?

A α β γ
B β γ α
C γ α β
D α γ β

photographic plate

deflector plates
positive pole   negative pole
+ ──────────   ─

radiation source

84 α-rays are

A very fast-flying helium nuclei
B very fast-flying electrons
C electromagnetic rays
D corpuscular rays

85 Which radioactive rays possess the greatest penetration power?

A α-rays
B β-rays
C γ-rays

86 How can radioactive rays be detected?

A by exposure of photographic plates
B by rendering them visible on a zinc sulphide fluorescent screen
C by Geiger-counters
D by magnifying them under an optical microscope

87 Nuclear fusion occurs when

A energy is required in a radiochemical process
B a uranium-type atomic bomb explodes
C a hydrogen bomb explodes
D protons in the nucleus of an atom are united to form neutrons

II Inorganic raw materials and large-scale products: Large-scale industrial processes

88 A radiochemical chain reaction occurs if, in the decomposition of atomic nuclei,

A a sufficiently large mass of the fissile material is present
B an average of less than one neutron is released per disintegrating atomic nucleus
C an average of one neutron is released per disintegrating atomic nucleus
D an average of more than one neutron is released per disintegrating atomic nucleus

89 The half-life of a radioactive element is the time

A after which half the fissile material is used up in a nuclear-chemical chain reaction
B after which the radiation intensity of a naturally disintegrating element has declined to half the initial intensity
C during which exactly half the number of atoms initially present have disintegrated
D within which the radiaton intensity penetrating a protective layer of lead is reduced by half

90 A nuclear reaction has two heat transfer systems,

A because this prevents the radioactivity from escaping outwards
B because this improves heat transfer
C because the primary heat transfer agent becomes radioactive as a result of direct contact with the nuclear fuel
D because this enables the thermal energy produced to be utilized better

10 Answers to the Self-evaluation Questions on Sections 1–9

1.	A, B	2.	A, B
3.	D	4.	B, C
5.	C	6.	A
7.	C		

8.	B, C	9.	A, B, C
10.	B, C	11.	D
12.	A	13.	A
14.	A, B	15.	A, C, D
16.	B, D	17.	C, D
18.	B, C, D	19.	B
20.	A, C, D	21.	A, C

22.	A, B	23.	A, B, C, D
24.	A, C	25.	B
26.	D	27.	A, B
28.	B, C, D	29.	A
30.	A, D	31.	C

32.	D	33.	A
34.	A, C, D	35.	B
36.	D	37.	C, D
38.	A, C, D	39.	B, C, D
40.	A	41.	A

42.	B, C	43.	C
44.	B, C	45.	A, D
46.	A, B, C, D	47.	C
48.	B	49.	C
50.	A, B, D	51.	A, B, D
52.	A	53.	A, B, D
54.	A		

55.	A	56.	A
57.	D	58.	B
59.	B	60.	A, B
61.	B, C, D	62.	A, B, C

63.	C	64.	C
65.	B, D	66.	D
67.	B, D	68.	B
69.	B		

70.	A, C	71.	D
72.	B, C	73.	A, B, C, D
74.	C, D	75.	A, D
76.	B	77.	A, B, C, D

78.	A, B	79.	A, C
80.	A	81.	A, C
82.	B	83.	B
84.	A, D	85.	C
86.	A, B, C	87.	C
88.	A, D	89.	B, C
90.	A, C		

III
Organic Chemistry Classification and Nomenclature

1 General, Classification, Nomenclature

1. Composition of organic compounds, hydrocarbons

Organic chemistry is the chemistry of carbon compounds. In organic compounds **carbon** is combined with only a **small number of other elements**. In order of their abundance these are: hydrogen, oxygen, nitrogen, halogens (especially chlorine), sulphur, phosphorus; metals and other elements occur less frequently.

A large number of organic compounds contain only carbon and hydrogen and are called **hydrocarbons**. Petroleum and natural gas, which are the most important raw materials for the synthesis of organic compounds, consist mainly of hydrocarbons.

Although organic compounds contain only a small number of elements, there exists a very large number of organic compounds, about 1 million, compared with only about 100000 inorganic compounds. This is due to the special bonding characteristics of the carbon atom.

2. Basic chain and ring structures

In contrast to almost all other elements, an unlimited number of C-atoms can combine with one another. This leads to compounds with basic frameworks which are **chains, rings**, or three-dimensional networks.

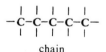

chain

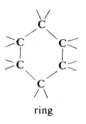

ring

3. Single, double and triple bonds

Another peculiarity of C-atoms is that they can be linked together by single, double or triple bonds:

$$\equiv C-C\equiv \qquad \geq C=C\leq \qquad -C\equiv C-$$

Since carbon is always tetravalent these units can combine with further carbon atoms, or with atoms of other elements.

Single bonds between C-atoms (C-C bonds) are stable at ordinary temperatures, as are C-H bonds. This is because these bonds are non-polar (see Polar bonds, I, 5.5.3).
C=C and C≡C bonds are less stable. In double and triple bonds the extra bonds are

III Organic chemistry classification and nomenclature

considerably less stable and are easily broken. This means that such compounds are considerably more reactive than the more inert substances containing C-C and C-H bonds.

The reactivity of C-compounds is further increased if they contain in addition to hydrogen other elements such as chlorine, oxygen or nitrogen. This is due to the fact that bonds with these elements are polar.

4. Functional groups

The insertion into a hydrocarbon of halogens, or other reactive atoms or groups of atoms, alters its properties, in particular its chemical behaviour. One then speaks of **functional groups**, for example $-Cl$, $-OH$, $-SO_3H$, $-NH_2$.

5. Saturated and unsaturated hydrocarbons

A hydrocarbon is said to be saturated if it contains only C-C bonds, so that all the remaining bonds of the carbon atoms are combined with H-atoms.

On the other hand, if on account of the presence of a multiple bond between C-atoms the C-atoms can combine with only one or two H-atoms, then the hydrocarbon is said to be unsaturated, since it contains fewer hydrogen atoms than the maximum number corresponding to a saturated hydrocarbon.

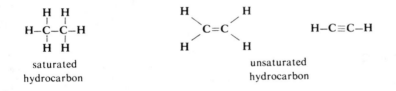

saturated hydrocarbon

unsaturated hydrocarbon

Unsaturated compounds are very reactive. By adding on hydrogen atoms or other atoms or groups of atoms the multiple bond is destroyed and a saturated compound is formed. This makes it possible to introduce functional groups and thus to create new possibilities of reaction for the compound.

6. Aromatic systems

In addition to single and multiple bonds there exists a type of bonding known as **aromatic**. All aromatic compounds are based on a symmetrical cyclic framework. The atoms in the ring are not united by normal single or double bonds. The bonds are evened out, producing a ring system in which all the bonds are equivalent.

Aromatic compounds are thus not unsaturated in the usual sense of the word, but have rings which are much more stable.

The most important aromatic compound is benzene.

7. Principles for the classification of organic compounds

The enormous number of C-compounds can be uniquely arranged, classified and named. This is based on the following principles:

a. whether the basic framework of the molecule is a chain or a ring;
b. the nature and number of bonds in the basic framework;
c. the nature and number of functional groups.

8. Aliphatic and alicyclic compounds

The basic framework of an organic molecule consists of a chain or ring of C-atoms. The C-chain can be either straight (continuous), or can be branched one or more times:

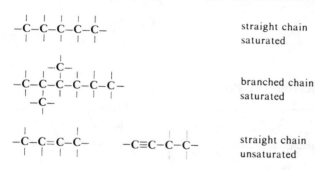

straight chain
saturated

branched chain
saturated

straight chain
unsaturated

Saturated and unsaturated C-compounds having a chain structure are called acyclic or **aliphatic.***

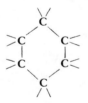

cyclic
saturated

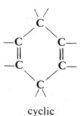

cyclic
unsaturated

Cyclic saturated and unsaturated compounds are termed **alicyclic** compounds.

In the simplest case the remaining bonds are all joined to hydrogen atoms, and the compound is then a hydrocarbon.

---

* From the Greek **aleiphar**, fat: a group of substances having chain structures.

9. Carbocyclic and heterocyclic compounds

If its ring consists only of C-atoms a compound is called **carbocyclic** or **isocyclic**.*

The most important carbocyclic compounds are the aromatic compounds, derived from benzene.

If the cyclic framework of a molecule contains not only C-atoms, but also one or more hetero-atoms,** such as O, N or S, the compound is called **heterocyclic**.

carbocyclic   heterocyclic

10. Principles of nomenclature, IUPAC rules

The **nomenclature** of organic compounds is governed by the following principles:

The name of a compound is referred to that of the hydrocarbon on which it is based. Compounds which contain elements other than C and H are regarded as derivatives of hydrocarbons.

Systematic nomenclature is now internationally agreed. It is the result of the 1892 Geneva Conference on Nomenclature and the modern **IUPAC rules***** worked out by international commissions.

In practice trivial names are often used alongside the systematic names.

---

* From the Greek **isos**, the same.
** From the Greek **heteros**, different.
*** IUPAC = International Union of Pure and Applied Chemistry.

11. Self-evaluation questions

1   Organic chemistry is

   A   the chemistry of non-metals
   B   the chemistry of organisms
   C   the chemistry of carbon compounds
   D   the chemistry of hydrocarbons only

2   Which group contains elements occurring frequently in organic compounds?
Choose just one answer!

   A   O — Cl — H — N
   B   H — S — Fe — O
   C   F — P — N — H
   D   C — O — H — N

3   The fact that there are so many organic compounds is due to

   A   the frequent occurrence of carbon in comparison with other elements
   B   the especially low reactivity of carbon
   C   the chain-formation and ring-formation of carbon atoms with one another
   D   the especially high reactivity of carbon

4   Which hydrocarbons are saturated?

A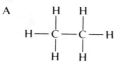

B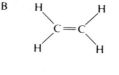

C   H — C≡C — H

D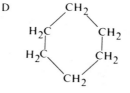

5 Which compounds are carbocyclic?

A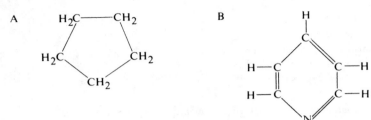

B

C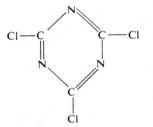

D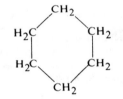

2   Saturated Hydrocarbons

1. Methane, tetrahedral models, rational formulae, homologous series

In saturated hydrocarbons the carbon atoms are joined together by single bonds. The simplest saturated hydrocarbon is **methane**, CH_4, in which a single carbon atom is bound to four H-atoms.

A good three-dimensional picture of the CH_4 molecule is given by the **tetrahedral model**, in which the four H-atoms are at the vertices of a regular tetrahedron, the centre of which is occupied by the C-atom.

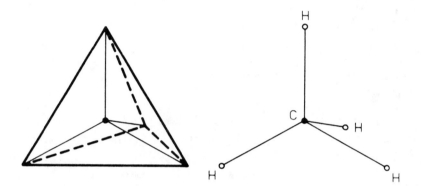

The symmetrical model shows that all four C-H bonds are equivalent. In a given reaction it is therefore immaterial which of the four H-atoms is replaced by a univalent atom or a univalent group of atoms. In each case the same compound will be obtained:

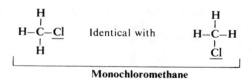

Monochloromethane

III Organic chemistry classification and nomenclature

All saturated hydrocarbons can be derived from methane by replacing one or more H-atoms by a C-atom or a chain of C-atoms:

$$
\begin{array}{ccc}
\text{H} & & \\
| & & \\
\text{H--C--H} & \text{CH}_4 & \text{CH}_4 \\
| & & \\
\text{H} & & \\
\end{array}
$$

$$
\begin{array}{ccc}
\text{H H} & & \\
|\ \ | & & \\
\text{H--C--C--H} & \text{CH}_3\text{--CH}_3 & \text{C}_2\text{H}_6 \\
|\ \ | & & \\
\text{H H} & & \\
\end{array}
$$

$$
\begin{array}{ccc}
\text{H H H} & & \\
|\ \ |\ \ | & & \\
\text{H--C--C--C--H} & \text{CH}_3\text{--CH}_2\text{--CH}_3 & \text{C}_3\text{H}_8 \\
|\ \ |\ \ | & & \\
\text{H H H} & & \\
\end{array}
$$

$$
\begin{array}{ccc}
\text{H H H H} & & \\
|\ \ |\ \ |\ \ | & & \\
\text{H--C--C--C--C--H} & \text{CH}_3\text{--(CH)}_2)_2\text{--CH}_3 & \text{C}_4\text{H}_{10} \\
|\ \ |\ \ |\ \ | & & \\
\text{H H H H} & & \\
\end{array}
$$

| structural | rational | molecular |
| formulae | formulae | formulae |

The rational formula is a simplified version of the structural formula. The constitution of the compound can be deduced from the rational formula, but only rarely from the molecular formula.

If the hydrocarbons are arranged in order of increasing number of carbon atoms an **homologous series** is obtained. In an homologous series each member differs from the preceding one by the addition of the group $-\text{CH}_2-$

2. Alkanes (paraffins)

The state of aggregation of the saturated chain hydrocarbons depends on the length of the chain.*

Homologous series of alkanes		State of aggregation
CH_4	Methane	
C_2H_6	Ethane	gaseous
C_3H_8	Propane	
C_4H_{10}	Butane	
C_5H_{12}	Pentane	
C_6H_{14}	Hexane	
C_7H_{16}	Heptane	
C_8H_{18}	Octane	liquid
C_9H_{20}	Nonane	
$C_{10}H_{22}$	Decane	
$C_{11}H_{24}$	Undecane	
.		
.		
$C_{16}H_{34}$	Hexadecane	
$C_{17}H_{36}$	Heptadecane	solid
$C_{18}H_{38}$	Octadecane	
.		
.		

All the alkanes have names formed by adding the syllable -ane to the Greek numeral, except for the first four alkanes, which have trivial names with the same ending.
The general molecular formula of the alkanes is C_nH_{2n+2} where n is the number of C-atoms in the molecule.

---

* The Table gives the state of aggregation for straight-chain alkanes at room temperature and normal pressure (20°C and 1013 mbar).

III Organic chemistry classification and nomenclature

3. Structural isomerism

In the homologous series of the alkanes the member having four C-atoms can have two possible structures for the carbon skeleton:

and

straight chain   branched chain

For the next homologue there are three possible structures:

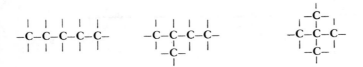

Compounds which have the same molecular formula but different structures are called **isomers***, and the phenomenon is called **isomerism**.
If, as in the example given, the phenomenon is due to branching of the carbon chain, it is called **structural isomerism.**

4. Rules for naming isomers

According to earlier nomenclature, if there is a single branching at the second C-atom of a C-chain, this is indicated by the prefix iso- (i-). A double branching at the second C-atoms is shown by the prefix neo-. The isomer with a straight chain then has the prefix n- (normal).
The two isomers of butane are thus named as follows:

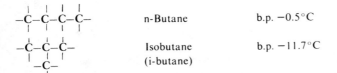

n-Butane   b.p. $-0.5\,°C$

Isobutane   b.p. $-11.7\,°C$
(i-butane)

The number of possible isomers increases rapidly with the number of C-atoms:

C_1	2 isomers	C_7	9 isomers
C_5	3 isomers	C_{10}	75 isomers
C_6	5 isomers	C_{12}	355 isomers

---

* From the Greek **isos**, the same, **meros**, a part; ie containing equal quantities of the same elements.

The modern nomenclature for isomers is based on the following rule:

The name of the longest hydrocarbon chain is used as the basis, and the isomers are named as derivatives of these hydrocarbons. The side chains must then be named separately.
In order to specify the point of branching the carbon atoms in the chain are numbered. The numbering of the C-atoms in the longest chain begins at the end which is nearest to the point of branching.

Of the three isomers of pentane,

has four C-atoms in its longest chain. The CH_3 group is attached to the second C-atom. This group is called methyl, so that the pentane isomer shown is called 2-methylbutane. The names of the three isomers of pentane (C_5H_{12}) are thus:

$-\overset{\|}{\underset{\|}{C}}-\overset{\|}{\underset{\|}{C}}-\overset{\|}{\underset{\|}{C}}-\overset{\|}{\underset{\|}{C}}-\overset{\|}{\underset{\|}{C}}-$	n-Pentane	b.p. $36\,°C$
$-\overset{\|}{\underset{\|}{C}}-\overset{\|}{\underset{\|}{C}}-\overset{\|}{\underset{\|}{C}}-\overset{\|}{\underset{\|}{C}}-$ $\quad -\overset{\|}{\underset{\|}{C}}-$	2-Methylbutane (isopentane)	b.p. $28\,°C$
$\quad -\overset{\|}{\underset{\|}{C}}-$ $-\overset{\|}{\underset{\|}{C}}-\overset{\|}{\underset{\|}{C}}-\overset{\|}{\underset{\|}{C}}-$ $\quad -\overset{\|}{\underset{\|}{C}}-$	2,2-Dimethylpropane (neopentane)	b.p. $9.4\,°C$

Branched hydrocarbons have lower boiling points than unbranched ones, as shown by the above examples.
The five isomers of hexane (C_6H_{14}) have the following carbon skeletons and names:

1 2 3 4 5 6
C–C–C–C–C–C   n-Hexane

1 2 3 4 5
C–C–C–C–C   2-Methylpentane
$\quad |$
CH_3 (isohexane)

$\quad CH_3$
$\quad |$
1 2 3 4
C–C–C–C   2,2-Dimethylbutane
$\quad |$
$\quad CH_3$

```
1  2  3  4  5
C–C–C–C–C          3-Methylpentane
      |
      CH3
```

```
1  2  3  4
C–C–C–C            2,3-Dimethylbutane
|    |
H3C  CH3
```

5. Univalent radicals, alkyl (R-)

The **alkyl radicals** are derived from the alkanes. They contain one H-atom less than the corresponding alkane. They are named by replacing the ending -ane in the name of the alkane by -yl:

CH_3-	Methyl group	
C_2H_5-	Ethyl group	CH_3-CH_2-

C_3H_7-	n-Propyl
	or
	isopropyl
		$CH_3-CH_2-CH_2-$
		
		CH_3-CH-
		     $\vert$
		     CH_3

C_4H_9-	n-Butyl
	or
	isobutyl
		$CH_3-CH_2-CH_2-CH_2-$
		
		$CH_3-CH-CH_2-$
		     $\vert$
		     CH_3

The univalent radicals derived from the pentanes are called pentyl, isopentyl and neopentyl, or sometimes amyl and isoamyl.
The general abbreviation for a univalent alkyl radical is R-.

6. Divalent radicals from alkanes

The **divalent radicals** of alkanes contain two fewer H-atoms than the corresponding alkane. According to an older but still common nomenclature they are named by substituting the ending -ylene for -ane in the name of the alkane, provided that the two free valencies (and hence sites for combination) are on two different C-atoms. (Methylene is an exception to this rule.)

$-CH_2-$ Methylene radical

eg in

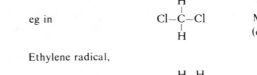

Methylene chloride
(dichloromethane)

$-C_2H_4-$ Ethylene radical,

eg in

Cl–C–C–Cl

Ethylene chloride
(1,2-dichloroethane)

If on the other hand both the free valencies are on the same carbon atom the ending -ylidene is used.

$CH_3-\overset{|}{\underset{|}{C}}H$ Ethylidene radical

eg in

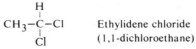

Ethylidene chloride
(1,1-dichloroethane)

In modern nomenclature both the position of the Cl-atoms and the point of branching are indicated by numbers.

7. Cycloalkanes

If a chain of C-atoms joined by single bonds is closed to form a ring we obtain alicyclic (cycloaliphatic) compounds. Hydrocarbons belonging to this class are called **cycloalkanes** (alternatively cycloparaffins or naphthenes). The two most important cycloalkanes are:

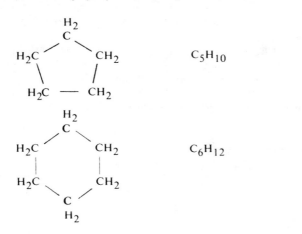

C_5H_{10} Cyclopentane

C_6H_{12} Cyclohexane

III Organic chemistry classification and nomenclature

Their general molecular formula is C_nH_{2n}. Their names follow those of the alkanes, with the addition of the prefix cyclo-.

In order to avoid confusion with aromatic compounds the abbreviated formulae for the cycloalkanes are sometimes written with an H in the ring to emphasize their saturated character.

C_6H_{12} Cyclohexane

A cycloalkane with a side chain is

$C_5H_9-CH_3$ Methylcyclopentane

The univalent radicals are called

C_5H_9- Cyclopentyl-

$C_6H_{11}-$ Cyclohexyl-

The general abbreviation R- is also used for these radicals.

8. Occurrence, uses and reactions of saturated hydrocarbons. Substitution

Natural gas and petroleum consist mainly of alkanes and cycloalkanes in proportions which depends on their origin. In addition to their uses as sources of energy (heating gas, heating oil, engine fuel) and as lubricants, natural gas and petroleum are acquiring a rapidly growing importance as chemical raw materials (see Petroleum and natural gas as raw materials, IV, 1).

The saturated hydrocarbons (alkanes and cycloalkanes) form the least reactive class of hydrocarbons. Under suitable conditions one or more of their hydrogen atoms can be replaced by other atoms or groups of atoms. This process of **substitution** produces new compounds, the derivatives of the hydrocarbons.

Substitution can be used to introduce functional groups. For example in chlorination one or more H-atoms are replaced by Cl:

Cl_2 + $H-CH_3$ $\longrightarrow$ $Cl-CH_3$ +   HCl

Chlorine   methane   (mono)chloromethane   hydrogen

                       = methyl chloride   chloride

9. Self-evaluation questions

6 Two adjacent members of a homologous series of compounds always differ from each other

   A in having different structural formulae but the same empirical formula
   B by one carbon atom
   C by one CH_2 group
   D in their physical properties

7 Alkanes are

   A hydrocarbons
   B unsaturated compounds
   C constituents of natural gas and petroleum
   D relatively low-reactivity compounds

8 The compound

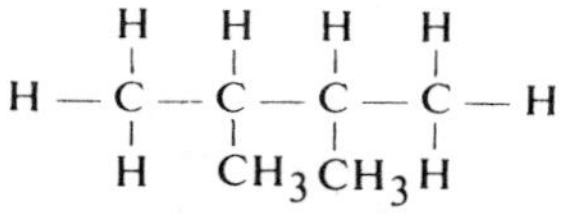

$$H - \overset{\overset{\displaystyle H}{|}}{\underset{\underset{\displaystyle H}{|}}{C}} - \overset{\overset{\displaystyle H}{|}}{\underset{\underset{\displaystyle CH_3}{|}}{C}} - \overset{\overset{\displaystyle H}{|}}{\underset{\underset{\displaystyle CH_3}{|}}{C}} - \overset{\overset{\displaystyle H}{|}}{\underset{\underset{\displaystyle H}{|}}{C}} - H \text{ is}$$

   A an isomer of hexane
   B identical to the compound

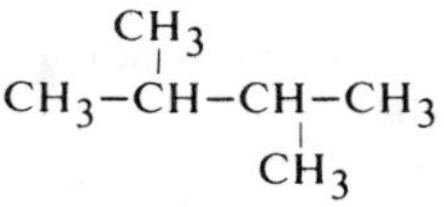

$$CH_3-\overset{\overset{\displaystyle CH_3}{|}}{C}H-\overset{\underset{\displaystyle CH_3}{|}}{C}H-CH_3$$

   C 2,3-dimethylbutane
   D a derivative of hexane

9 Which two of the following compounds are isomers?

   A $CH_3-CH_2-\overset{\underset{\displaystyle CH_3}{|}}{C}H-CH_3$ B $CH_3-CH_2-\overset{\underset{\displaystyle Cl}{|}}{C}H-CH_3$

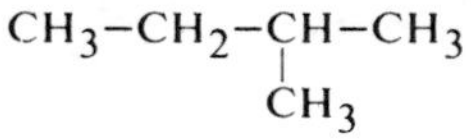

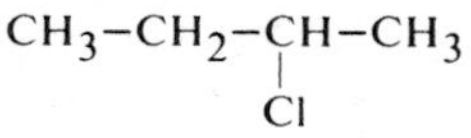

   C $CH_3-CH_2-CH_2-CH_2-CH_3$ D $CH_3-\overset{\underset{\displaystyle CH_3}{|}}{C}H-CH_3$

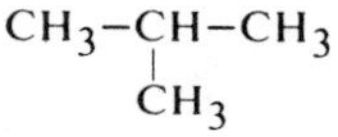

III Organic chemistry classification and nomenclature

10 Light petroleum contains pentanes. Give the structural formulae of the three isomeric pentanes.

11 Cycloalkanes are

A main constituents of petroleum
B compounds with the general empirical formula C_nH_{2n+2}
C low-reactivity compounds that take part only in addition reactions with other substances
D aromatic hydrocarbons

12 Which formulae represents 1,1-dichloroethane?

A CH_2Cl-CH_2Cl B $CHCl_2-CH_3$
C CH_3-CHCl_2 D $CCl_2=CH_2$

13 In the formula

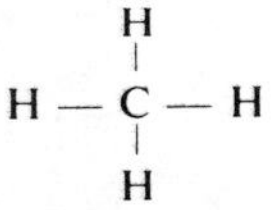

$$H - \overset{\displaystyle H}{\underset{\displaystyle H}{\overset{|}{\underset{|}{C}}}} - H$$

each valency stroke represents

A an electron
B the force of attraction between two particles with opposite charges
C a cloud of many electrons
D a joint pair of electrons

3 Unsaturated Hydrocarbons

1. Alkenes (olefins)*, homologous series

In unsaturated hydrocarbons two or more C-atoms are joined by double or triple bonds.
Most of them also contain single bonds.

Acyclic (chain) unsaturated hydrocarbons with **double bonds** are called **alkenes** or **olefins**.
The first member of the homologous series with only one $C = C$ bond has two C-atoms.
The first few members have two commonly used names, the first according to the IUPAC
rules and the second the name generally used in industry.**

C_2H_4	$CH_2=CH_2$	Ethene or ethylene
C_3H_6	$CH_2=CH-CH_3$	Propene or propylene
C_4H_8	$CH_2=CH-CH_2-CH_3$	1-Butene or butylene
C_5H_{10}	$CH_2=CH-CH_2-CH_2-CH_3$	1-Pentene
C_6H_{12}	$CH_2=CH-(CH_2)_3-CH_3$	1-Hexene

Their general molecular formula is C_nH_{2n}.

The names of all alkenes end in -ene.

C_2 to C_4 alkenes are gases at room temperature, C_5 to C_{16} liquids, and C_{17} onwards solids.

2. Nomenclature of the alkenes

Isomerism occurs from butene (C_4H_8) onwards, depending on the branching of the C-
skeleton and the position of the double bond. **Nomenclature** follows the same rules as for
the alkanes, but it is necessary in addition to specify the position of the double bond.

The longest chain gives the name, which ends in -ene. The number preceding the name
denotes the carbon atom which precedes the double bond. The numbering of the C-atoms
begins at the end which is nearest to
1. a double bond, or
2. a branch point.

III Organic chemistry classification and nomenclature

The three butenes are shown below:

$$\overset{1}{C}H_2=\overset{2}{C}H-\overset{3}{C}H_2-\overset{4}{C}H_3 \qquad \text{1-Butene}$$

$$\overset{1}{C}H_3-\overset{2}{C}H=\overset{3}{C}H-\overset{4}{C}H_3 \qquad \text{2-Butene}$$

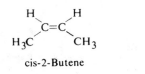

Isobutene
or isobutylene
or methylpropene

3. Cis-trans isomerism

The double bond leads to a new kind of isomerism. The two structural formulae

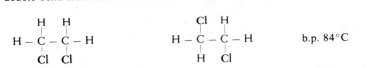

b.p. 84°C

represent the same molecule, 1.2-dichloroethane, since the two halves of the molecule can rotate freely about the C-C bond.

By contrast, the C = C bond is rigid. The two formulae

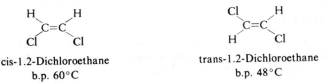

cis-1.2-Dichloroethane
b.p. 60°C

trans-1.2-Dichloroethane
b.p. 48°C

therefore represent two different molecules.

In the cis-form* the two chlorine atoms lie on the same side of the molecule. In the trans-form* they are diagonally opposite one another. This kind of spatial isomerism is called **cis-trans isomerism**. Another example is

cis-2-Butene

trans-2-Butene

---

* cis = on this side
  trans = on the other side.

4. Alkene radicals

The following names are used for some important **alkene radicals**:

$CH_2 = CH-$ Vinyl, eg in $CH_2 = CH-Cl$ vinyl chloride
$CH_2 = CH-CH_2-$ Allyl, eg in $CH_2 = CH-CH_2 = Cl$ allyl chloride

The abbreviation R-, eg R-Cl, is again used for these univalent radicals.

A divalent radical is

$CH_2 = C$ Vinylidene   eg in $CH_2 = CCl_2$ vinylidene chloride

5. Dienes, Trienes

There are also hydrocarbons containing two or more double bonds. The number of double bonds is indicated by inserting the numerical prefixes di-, tri-, tetra-, etc, before the final syllable -ene; hence **diene, triene**, etc. The position of the double bonds is shown in the usual way by means of numbers.

$$\overset{1}{C}H_2 = \overset{2}{C} = \overset{3}{C}H - \overset{4}{C}H_3$$ 1.2-Butadiene

$$\overset{1}{C}H_2 = \overset{2}{C}H - \overset{3}{C}H = \overset{4}{C}H_2$$ 1.3-Butadiene

$$\overset{1}{C}H_2 = \overset{2}{\underset{CH_3}{C}} - \overset{3}{C}H = \overset{4}{C}H_2$$ 2-Methyl-1.2-butadiene (isoprene)

$$\overset{1}{C}H_2 = \overset{2}{C}H - \overset{3}{C}H = \overset{4}{C}H - \overset{5}{C}H = \overset{6}{C}H_2$$ 1.3.5-Hexatriene

6. Conjugated, cumulative and isolated double bonds

The arrangement of double bonds in a molecule is classified in the following way:
Conjugated double bonds: alternating double and single bonds, eg in 1.3.5-hexatriene
Cumulative double bonds: double bonds adjacent to one another, eg in 1.2-butadiene
Isolated double bonds: at least two single bonds between each pair of double bonds.
A large number of conjugated double bonds in a molecule may cause it to be coloured (see Colorants, IV,9).

7. Cycloalkenes (cyclo-olefins)

Like the alkanes, alkenes may also exist as rings. They are called **cycloalkenes** (or cyclo-olefins):

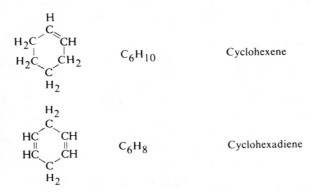

C_6H_{10} Cyclohexene

C_6H_8 Cyclohexadiene

The following large unsaturated rings with olefinic properties have recently become important (hydrogen atoms have been omitted):

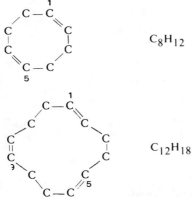

C_8H_{12} 1.5-Cyclo-octadiene

$C_{12}H_{18}$ 1.5.9-Cyclododecatriene

Both of these have symmetrically disposed double bonds and can be synthesized from butadiene.

8. Alkynes (acetylenes)

Unsaturated hydrocarbons containing **triple bonds** are called **alkynes** or acetylenes (after the first member of the series). They therefore have two alternative names:

C_2H_2	$CH \equiv CH$	Ethyne
		or acetylene
C_3H_4	$HC \equiv C-CH_3$	Propyne
		or methylacetylene
C_4H_6	$CH \equiv C-CH_2-CH_3$	1-Butyne
	and	or ethylacetylene
	$CH_3-C \equiv C-CH_3$	2-Butyne
		or dimethylacetylene

The names of alkynes are recognized by the final syllable **-yne**.

9. Nomenclature and radicals of the alkynes

The **nomenclature** of the alkynes follows rules analogous to those given for the alkenes. Univalent **radicals** of the alkynes (denoted in general by R-) are the following:

| $CH \equiv C-$ | Ethynyl |
| $CH \equiv C-CH_2-$ | Propargyl |

10. Occurrence, uses and reactions of unsaturated hydrocarbons; addition, polymerization, monomers, polymers, initiators, mixed polymers (copolymers)

In contrast to the saturated hydrocarbons, unsaturated hydrocarbons constitute only a very small percentage of natural gas and petroleum. The large quantities which are required are obtained by the chemical modification of the above raw materials in oil refineries (see Petroleum and natural gas as raw materials, IV,1).

Unsaturated hydrocarbons — alkenes, cycloalkenes and alkynes — are reactive compounds. Because of their marked tendency to react they constitute readily convertible **raw materials** for a large number of **syntheses** carried out on a large scale industrially (see Hydrogenation (methanol and oxo syntheses) IV,3; Plastics, synthetic resins, synthetic rubber, IV,7; Colorants, IV,9).

The $C = C$ and $C \equiv C$ bonds are the reactive sites (see I,5), which undergo important reactions, especially additions, including polymerizations.

Addition means the adding on of molecules to unsaturated hydrocarbons. One such molecule is added on to each double bond in the unsaturated molecule. The double bond is opened up, so that addition can take place at each of the two C-atoms originally connected by a double bond.

III Organic chemistry classification and nomenclature

Molecules which add on readily include hydrogen, halogens, hydrogen halides, water and oxygen:

$$H_2C = CH_2 \quad + \quad Cl-Cl \quad \longrightarrow \quad \begin{matrix} H_2C-CH_2 \\ | \quad | \\ Cl \quad Cl \end{matrix}$$

| Ethene | chlorine | 1.2-dichloroethane (ethylene chloride) |

$$H_2C = CH_2 \quad + \quad H-Cl \quad \longrightarrow \quad H_3C-CH_2Cl$$

| Ethene | hydrogen chloride | chloroethane (ethyl chloride) |

$$H_2C = CH_2 \quad + \quad H-OH \quad \longrightarrow \quad H_3C-CH_2-OH$$

| Ethene | water | ethanol (ethyl alcohol) |

Addition to C = C bonds produces saturated compounds without any other products.

For the alkynes, if only one molecule is added to each C = C bond, only one bond will be broken and the product will be a molecule containing a C = C bond:

$$H-C \equiv C-H \quad + \quad H-H \quad \longrightarrow \quad H_2C = CH_2$$

| Ethyne (acetylene) | hydrogen | ethene (ethylene) |

$$H-C \equiv C-H \quad + \quad H-Cl \quad \longrightarrow \quad H_2C = CH-Cl$$

| Ethyne | hydrogen chloride | chloroethene (vinyl chloride) |

Polymerization is a special kind of addition. Unsaturated molecules are not only able to add on molecules of a different kind, but can add on to one another. An example of this is the addition of ethylene to itself:

$$\overline{CH_2{=}CH_2} \; + \; \overline{CH_2{=}CH_2} \; + \; \overline{CH_2{=}CH_2}$$
$$\longrightarrow \; -CH_2-CH_2-CH_2-CH_2-CH_2-CH_2-$$

or in a simplified form:

$$n \, CH_2 = CH_2 \quad \longrightarrow \quad +CH_2{-}CH_2{+}_n$$

| **Ethene (ethylene)** | **polyethylene (PE)** |
| **monomer** | **polymer** |

The substance consisting of single unsaturated molecules is called the **monomer**, and the macromolecular substance produced the **polymer**. In the example given gaseous ethylene is converted to solid polyethylene, consisting of thread-like macromolecules. In industry polymerization is started by means of **initiators**, which open up the double bond.
Mixed polymers (copolymers) are formed by the polymerization of mixtures of different monomers (see Plastics IV,7).

11. Self-evaluation questions

14 Alkenes are compounds that

    A   possess the general empirical formula C_nH_{2n-2}
    B   are the only unsaturated hydrocarbons
    C   possess a double bond between two carbon atoms in the molecule
    D   occur only in unbranched form

15 Which pair of compounds are cis-trans isomers?

A
$$\begin{array}{cc} H & Cl \\ | & | \\ C = C \\ | & | \\ H & Cl \end{array} \quad \text{and} \quad \begin{array}{cc} Cl & H \\ | & | \\ C = C \\ | & | \\ Cl & H \end{array}$$

B
$$\begin{array}{cc} CH_3 & H \\ | & | \\ C = C \\ | & | \\ H & CH_3 \end{array} \quad \text{and} \quad \begin{array}{cc} H & CH_3 \\ | & | \\ C = C \\ | & | \\ CH_3 & H \end{array}$$

C
$$\begin{array}{cc} Cl & Cl \\ | & | \\ C = C \\ | & | \\ H & H \end{array} \quad \text{and} \quad \begin{array}{cc} H & H \\ | & | \\ C = C \\ | & | \\ Cl & Cl \end{array}$$

D
$$\begin{array}{cc} F & H \\ | & | \\ C = C \\ | & | \\ H & F \end{array} \quad \text{and} \quad \begin{array}{cc} F & F \\ | & | \\ C = C \\ | & | \\ H & H \end{array}$$

16 Which compounds possess conjugated double bonds in the molecule?

    A $CH_2 = CH-CH = CH-CH = CH_2$
    B $CH_2 = CH-CH = CH_2$
    C $CH_3-CH = CH-CH_3$
    D $CH_3-CH = C = CH_2$

17 Which of the following compounds can be polymerized?

    A   Ethylene
    B   Chloroethane
    C   Vinyl chloride
    D   Benzene

III Organic chemistry classification and nomenclature

18 Insert crosses in the correct spaces in the following scheme.

	saturated	unsaturated	isocyclic	heterocyclic
Furan				
$CH_3(CH_2)_4CH_3$ n-Hexane				
$H_2C = CH-CH_3$ Propene				
Ethylene oxide				
Cyclohexane				

19 Which of the following compounds are hydrocarbons?

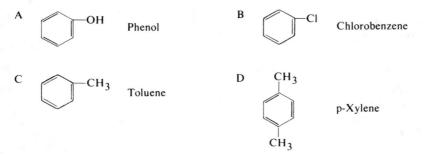

A   —OH   Phenol

B   —Cl   Chlorobenzene

C   —CH_3 Toluene

D CH_3 ... CH_3 p-Xylene

20 Which compounds are named correctly?

A $CH_3-CH=CH_2$ 2-propene
B $CH_2=CH-CH=CH_2$ 1.4-butadiene
C $CH_2=CH-\underset{\underset{CH_3}{|}}{C}=CH_2$ 2-methyl-1.3-butadiene

D

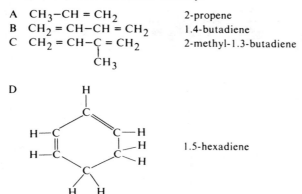

     1.5-hexadiene

21 The term "addition" connotes

A   the reaction of unsaturated single molecules to form macromolecules
B   the addition of molecules to alkenes and alkines, accompanied by a cleavage of the macro bond
C   the addition of molecules to unsaturated compounds, accompanied by a cleavage of relatively small molecules (such as HCl, H_2O)
D   also polymerization

22 A feature of 2-pentene is the fact that its molecule

A   contains two double bonds
B   is substituted at the second carbon atom
C   contains five carbon atoms
D   possesses a double bond between the second and third carbon atoms

23 Fill out the following Table:

Name of radical	Corresponding hydrocarbon	Formula of radical
Methyl		
Ethyl		
Vinyl		
Propenyl		

III Organic chemistry classification and nomenclature

24 Which type of reaction is represented by the following reaction diagram?

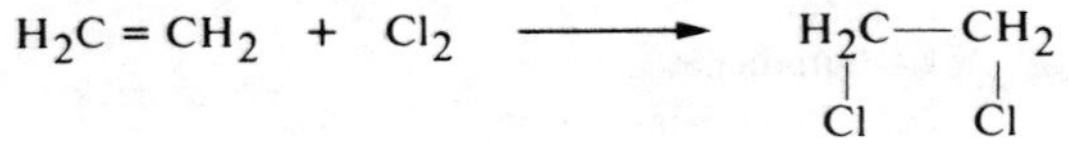

$$H_2C = CH_2 \ + \ Cl_2 \ \longrightarrow \ \underset{\underset{Cl}{|} \quad \underset{Cl}{|}}{H_2C - CH_2}$$

A   substitution
B   polymerization
C   elimination
D   addition

4 Aromatic Hydrocarbons

1. Benzene, structure

Aromatic compounds (arenes) have as their basic structure an aromatic system in which all the bonds are equivalent. The most important parent substance of aromatic compounds is **benzene**, having the molecular formula C_6H_6.

The usual structural formula for benzene does not express the equivalence of the bonds.

or the simplified form

In order to represent the fact that the benzene ring contains neither single nor double bonds the following symbol is also used:

2. Reactions of benzene, substitution reactions, aromatic character

The chemical **reactions of benzene** resemble in no way those of the unsaturated cycloalkenes. The typical reactions of benzene do not consist of additions to the double bond, but in the replacement of hydrogen by other atoms or groups of atoms. The aromatic structure of the ring remains unchanged by these **substitution reactions**.

The fact that the H-atoms on the benzene ring are readily substituted without any change in either the carbon skeleton or the bond system of the ring is referred to as **aromatic character**. The benzene nucleus is thus preserved in substitution reactions.

There are four important substitution reactions, which will be described for the parent substance benzene.

2.1. Chlorination

In **chlorinations** H-atoms on the ring are replaced by Cl-atoms:

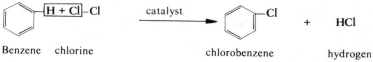

| Benzene | chlorine | | chlorobenzene | | hydrogen chloride |

III Organic chemistry classification
and nomenclature

Since only one H-atom is replaced by a Cl-atom, this is termed monochlorination. In dichlorination dichlorobenzene is produced.

2.2. Sulphonation

In **sulphonation** H-atoms on the ring are replaced by SO_3H groups (sulphonic acid groups):

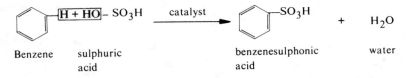

| Benzene | sulphuric acid | | benzenesulphonic acid | water |

Benzenedisulphonic acids are formed when two H-atoms are replaced by SO_3H groups.

2.3. Nitration

In **nitrations** H-atoms on the ring are replaced by NO_2 groups (nitro groups):

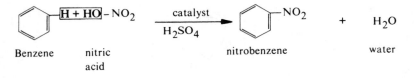

| Benzene | nitric acid | | nitrobenzene | water |

On disubstitution dinitrobenzene is formed.

2.4. Alkylation

Alkylation, the introduction of alkyl groups, can also be effected by a substitution reaction:

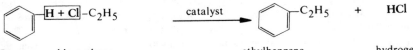

| Benzene | chloroethane | | ethylbenzene | hydrogen chloride |

On a large industrial scale a more important alkylation reaction is the addition of benzene to unsaturated hydrocarbons:

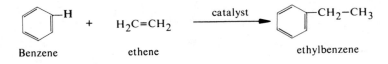

| Benzene | ethene | | ethylbenzene |

3. Mononuclear aromatic hydrocarbons

The alkylbenzenes are derived from benzene. They are mononuclear aromatic hydrocarbons.

3.1. Alkylbenzenes

The following are important representatives of the **alkylbenzenes**:

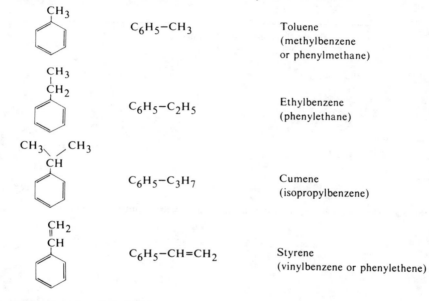

$C_6H_5-CH_3$ — Toluene (methylbenzene or phenylmethane)

$C_6H_5-C_2H_5$ — Ethylbenzene (phenylethane)

$C_6H_5-C_3H_7$ — Cumene (isopropylbenzene)

$C_6H_5-CH=CH_2$ — Styrene (vinylbenzene or phenylethene)

The C_6H_5 group is called phenyl.

3.2. Positional isomerism

Positional isomerism arises when more than one of the H-atoms of benzene is substituted. It is then necessary to specify which of the C-atoms in the ring bear the substituents. When two H-atoms are substituted three isomeric derivatives can be formed:

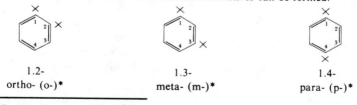

1.2-
ortho- (o-)*

1.3-
meta- (m-)*

1.4-
para- (p-)*

* Greek *orthos* = straight, correct; hence by analogy; next to
  meta = after; hence by analogy; further away
  para = beyond; hence by analogy; opposite

III Organic chemistry classification and nomenclature

The positions are numbered in a clockwise direction.
There are therefore three dimethylbenzenes: 1. o-Xylene

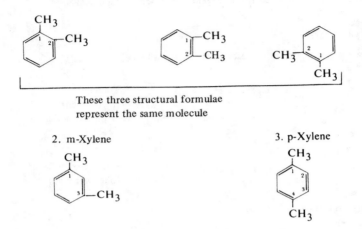

These three structural formulae
represent the same molecule

2. m-Xylene

3. p-Xylene

If there are three substituents on the benzene ring they are numbered analogously:

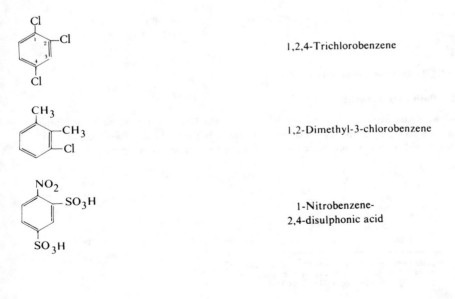

1,2,4-Trichlorobenzene

1,2-Dimethyl-3-chlorobenzene

1-Nitrobenzene-
2,4-disulphonic acid

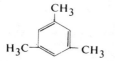

1,3,5-Trimethylbenzene

3.3. Nucleus and side chain, nuclear and side-chain chlorination

In alkylbenzene molecules a distinction is made between the **nucleus** and the **side chain**. In substitution reactions H-atoms from either the nucleus or the side chain can be replaced, leading to different products.

One must therefore distinguish between nuclear chlorination and side-chain chlorination.

Nuclear chlorination:

   + Cl_2 $\longrightarrow$    +   HCl

Methylbenzene (toluene)   1-methyl-2-chlorobenzene
                                            (o-chlorotoluene)

Side-chain chlorination:

   + Cl_2 $\longrightarrow$ +   HCl

Methylbenzene   Phenylmethyl chloride
(toluene)   (benzyl chloride)

If a side chain contains double bonds, these can lead to polymerization, leaving the nucleus unchanged:

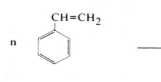 $\longrightarrow$

Styrene   polystyrene
(vinylbenzene)

3.4. Univalent aromatic radicals, aryl, Ar-

The names of radicals derived from aromatic hydrocarbons vary, depending on whether the H-atom has been removed from the benzene nucleus or from the side chain.
The univalent radicals have the ending -yl:

C_6H_5 radical
phenyl radical

C_6H_5-CH_2 radical
phenylmethyl radical
or benzyl radical

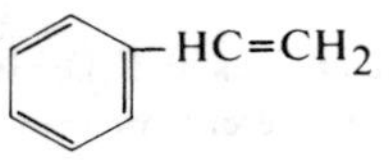

Phenylethene
(vinylbenzene, styrene)

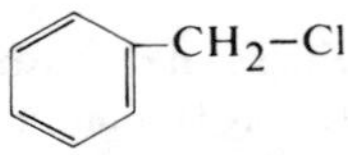

Phenylmethyl chloride
(benzyl chloride)

Univalent aromatic radicals in which one H-atom has been removed from the nucleus are called **aryls**: the general abbreviation is **Ar-**.

3.5. Divalent aromatic radicals

Divalent radicals have the ending -ylene for substitution in the nucleus, and -ylidene for substitution in the side chain:

C_6H_4 radical
phenylene radical

C_6H_5-CH radical
phenylmethylidene
or benzal radical

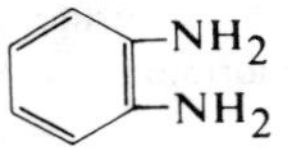

o-Phenylenediamine
(1,2-diaminobenzene)

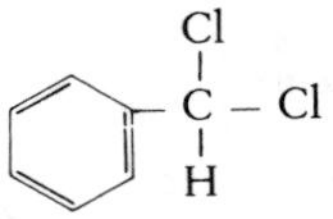

Phenylmethylidene chloride
(benzal chloride)

4. Polynuclear aromatic hydrocarbons

In addition to the mononuclear compounds there are also **polynuclear aromatic hydrocarbons,** which are divided into two groups.

4.1. Singly linked rings

The first group contains rings which are linked by **single bonds** or through **C-atoms**:

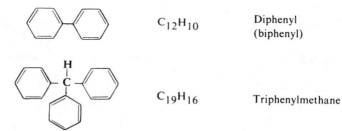

$C_{12}H_{10}$ Diphenyl (biphenyl)

$C_{19}H_{16}$ Triphenylmethane

In triphenylmethane three of the H-atoms of methane have been replaced by phenyl groups.

4.2. Condensed ring systems

In the second group of polynuclear aromatic systems two or more rings are joined together by shared C-atoms. They are called **condensed ring systems**:

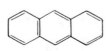

$C_{10}H_8$ $C_{14}H_{10}$ $C_{14}H_{10}$

Naphthalene   Anthracene   Phenanthrene

III Organic chemistry classification and nomenclature

These hydrocarbons are also aromatic.

The C-atoms of the naphthalene nucleus are distinguished either by numbers or by Greek letters:

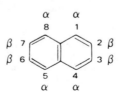

$$\begin{array}{cc} \alpha & \alpha \\ 8 & 1 \end{array}$$

A naphthalene derivative containing one functional group can exist in two isomeric forms:

α-Naphthol   β-Naphthol

The univalent radical of naphthalene, $C_{10}H_7$, is called either α-naphthyl or β-naphthyl

α-Naphthylamine

The divalent radical of naphthalene, $C_{10}H_6$, is called naphthylene

1,8-Naphthylenediamine

The C-atoms of the trinuclear ring systems are numbered as follows:

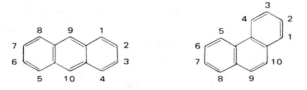

5. Occurrence and uses of aromatic hydrocarbons

Petroleum contains only small amounts of aromatic hydrocarbons. They can be produced from saturated hydrocarbons by dehydrogenation (see Petroleum and natural gas as raw materials, IV,1).

Benzene is also obtained in the manufacture of coke from coal (see Coal as a raw material, IV,2).

Aromatic hydrocarbons are among the most important **raw materials** for chemical **syntheses**. In particular, numerous intermediates are obtained by the chlorination, sulphonation, nitration and alkylation of benzene.

III Organic chemistry classification and nomenclature

6. Self-evaluation questions

25 Which formula expresses the identical nature of the C-C linkages in the benzene molecule?

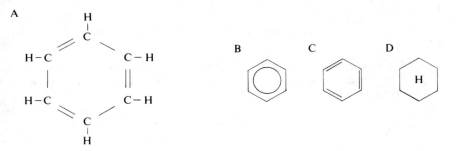

A

B   C   D

26 Which compound belongs to the condensed ring systems?

A

benzene

B

diphenyl

C

naphthalene

D

triphenylmethane

27 Naphthalene has the empirical formula

A $C_{10}H_6$
B $C_{12}H_8$
C $C_{14}H_{10}$
D $C_{10}H_8$

28 The typical features of benzene include:

A its aromatic character

B the possibility of introducing substitutes for the linked H atoms without destroying the benzene ring

C the ability to take part in addition reactions in which the double bonds in the ring are split open

D its low toxicity

29 Which two reactions with benzene are possible?

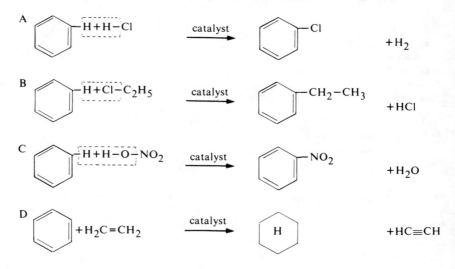

30 Which compounds are correctly named?

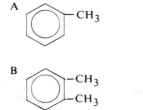

A   phenylmethane

B   m-xylene

III Organic chemistry classification and nomenclature

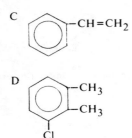

C   —CH=CH$_2$ vinylbenzene

D   —CH$_3$ —CH$_3$ Cl   1,2,3-dimethylchlorobenzene

31 How many isomeric dichlorobenzenes are there?

A   three
B   six
C   nine
D   twelve

32 What is the following compound called?

A   3,4,6 trichlorobenzene
B   2,4,5 trichlorobenzene
C   1,2,4 trichlorobenzene
D   1,4,5 trichlorobenzene

33 Which two reactions are possible in the chlorination of methyl benzene?

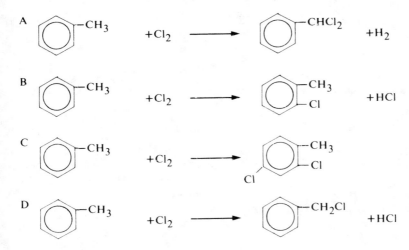

5 Heterocyclic Compounds

1. Hetero-atoms, mononuclear and condensed heterocycles

Ring systems which contain atoms other than C (**hetero-atoms**), especially N, O and S, are called **heterocyclic compounds**. The term also includes systems in which heterocyclic compounds are condensed with (for example) benzene nuclei. Out of the large number of existing heterocycles we shall consider here only the most important 5- and 6-membered rings and a few **condensed heterocycles**.

As regards nomenclature, extensive use is still made of established trivial names. Rational formulae and molecular formulae are not useful.

2. Five-membered rings

The following are simple **five-membered rings**:

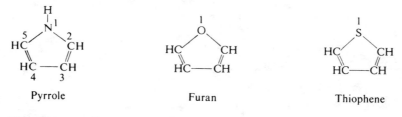

| Pyrrole | Furan | Thiophene |

The numbering of the ring atoms begins with the hetero-atom.
Pyrrole, furan and thiophene exhibit aromatic character (see III, 4,2).
A completely hydrogenated furan is

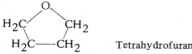

Tetrahydrofuran (THF) b.p. 66°C (151°F)

The following heterocycles have two hetero-atoms in the ring:

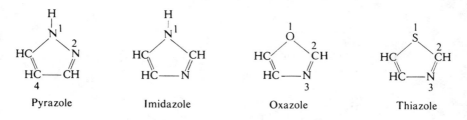

| Pyrazole | Imidazole | Oxazole | Thiazole |

The numbering begins with one hetero-atom and continues in the direction of the second hetero-atom.

3. Six-membered rings

The most important simple six-membered ring containing one hetero-atom is:

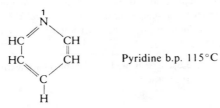

Pyridine b.p. 115°C

Pyridine possesses aromatic character.
The three isomeric 6-membered rings with two N-atoms are:

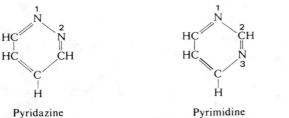

Pyridazine   Pyrimidine   Pyrazine

4. Condensed heterocyclic ring systems

The following are important condensed heterocyclic ring systems:

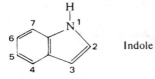

Indole

and the two isomers

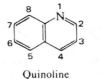   and

Quinoline   Isoquinoline

Compounds with three condensed rings and one hetero-atom include:

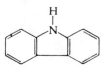

Carbazole

Acridine

5. Modern nomenclature

The hetero-atom is denoted by the prefix and the ring size by the characteristic ending.

The prefixes for the hetero-atoms are as follows:

Element	Prefix	Element	Prefix
O	Oxa-	N	Aza-
S	Thia-	P	Phospha-
Se	Selena-	As	Arsa-
Te	Tellura-	Sb	Stiba-
Si	Sila-	Bi	Bisma-

The final a of the prefix is dropped when it is combined with the ending.
The endings denoting the ring size are:

Number of atoms in the ring	Rings containing N		Rings without N	
	unsaturated	saturated	unsaturated	saturated
3	-irine	-iridine	-irene	-irane
4	-etine	-etidine	-etene	-etane
5	-ole	-olidine	-ole	-olane
6	-ine	*	-ine	-ane
7	-epine	*	-epine	-epane

* In these cases saturation is expressed by inserting the prefix parhydro- before the ending of the original unsaturated compound.

III Organic chemistry classification and nomenclature

Examples:

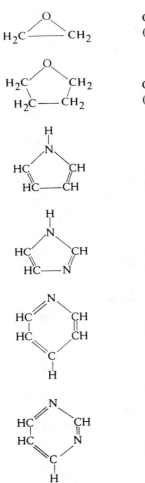

Oxirane
(ethylene oxide)

Oxolane
(tetrahydrofuran)

Azole
(pyrrole)

1,3-Diazole
(imidazole)

Azine
(pyridine)

1,3-Diazine
(pyrimidine)

6. Occurrence, uses and reactions of heterocyclic compounds

Many heterocyclic ring systems constitute the basis of numerous substances in the animal and plant kingdoms. Complicated substances such as haemoglobin, chlorophyll and many alkaloids (nitrogenous plant bases) are derivatives of pyrrole. Other alkaloids are based on pyridine (nicotine), quinoline (quinine) or isoquinoline (the opium alkaloids). Indigo is a derivative of indole.

Because of the presence of the hetero-atoms heterocyclic compounds are not hydrocarbons. They often possess aromatic properties and hence undergo the same kind of substitution reactions as the benzene hydrocarbons. They are used for making pharmaceutical preparations and colorants.

Tetrahydrofuran (THF) is used as a solvent. Ethylene oxide is an extremely reactive intermediate which is used for a number of syntheses (see Oxidation, IV,4).

III Organic chemistry classification and nomenclature

7. Self-evaluation questions

34 How many hetero atoms are contained in thiazole?

A   none
B   two
C   three
D   five

35 In pyrazine the hetero atoms are in the

A   ortho position
B   meta position
C   para position
D   1,4 position

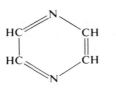

36 Carbazole has the empirical formula

A $C_{13}H_{10}N$
B $C_{12}H_{12}N$
C $C_{13}H_9N$
D $C_{12}H_9N$

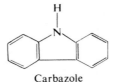

Carbazole

37 Hetero atoms in compounds are indicated by customary prefixes. Which two combinations of element and prefix are correct?

A   Oxygen . . . Oxa
B   Hydrogen . . . Perhydro
C   Phosphorus . . . Aza
D   Sulphur . . . Thia

6 Derivatives of Hydrocarbons

III Organic chemistry classification and nomenclature

1. Singly and multiply substituted hydrocarbons, summary

A large number of derivatives are based on the hydrocarbons. If only one H-atom is replaced by a functional group a **singly substituted hydrocarbon** is obtained. In a **multiply substituted hydrocarbon** several H-atoms have been replaced.

The following table gives a summary of the most important functional groups. A typical aliphatic and/or aromatic compound is shown for each group of compounds.

Functional group	Examples		Name of the group of compounds
	Aliphatic compound	Aromatic compound	
$-Cl$	C_2H_5Cl	⬡$-Cl$	Chlorohydrocarbons
$-OH$	CH_3OH	⬡$-OH$	Alcohols ⎱ hydroxy Phenols ⎰ compounds
$-O-$	$H_5C_2-O-C_2H_5$		Ethers
$-SO_3H$		⬡$-SO_3H$	Sulphonic acids
$-NO_2$		⬡$-NO_2$	Nitro compounds
$-NH_2$	$C_2H_5-NH_2$	⬡$-NH_2$	Amines
$-CHO$	CH_3-CHO	⬡$-CHO$	Aldehydes
$-\underset{O}{\overset{\|}{C}}-$	$CH_3-CO-CH_3$		Ketones
$-COOH$	CH_3-COOH	⬡$-COOH$	Carboxylic acids

2. Halogen derivatives of hydrocarbons

These derivatives can contain either one or several halogen atoms as substituents in the molecule. If they contain only one halogen atom they correspond to the types R-Hal, Ar-Hal or Ar-$(CH_2)_n$-Hal.

III Organic chemistry classification and nomenclature

2.1. Monohalogen derivatives, primary, secondary and tertiary carbon atoms

The following are important monohalogen derivatives of aliphatic hydrocarbons:

H_3C-Cl
(Mono)chloromethane
(methyl chloride)

H_3C-CH_2-Cl
(Mono)chloroethane
(ethyl chloride)

$CH_3-CH_2-\underset{\underset{Cl}{|}}{CH_2}$
1-Chloropropane
(n-propyl chloride)

$CH_3-\underset{\underset{Cl}{|}}{CH}-CH_3$
2-Chloropropane
(i-propyl chloride)

The older nomenclature, still in use, takes into account whether the C-atom bearing the functional group is attached to one, two or three other C-atoms. The C-atom is then described as a **primary, secondary or tertiary C-atom** respectively

$CH_3-CH_2-CH_2-\underset{\underset{Cl}{|}}{CH_2}$
1-Chlorobutane
(n-butyl chloride)
or prim. butyl chloride)

$CH_3-CH_2-\underset{\underset{Cl}{|}}{CH}-CH_3$
2-Chlorobutane
(sec. butyl chloride)

$CH_3-\underset{\underset{Cl}{|}}{\overset{\overset{CH_3}{|}}{C}}-CH_3$
2-Methyl-2-chloropropane
(tert. butyl chloride)

An example of an unsaturated chlorohydrocarbon is

$H_2C=CH-Cl$
Chloroethene
(vinyl chloride)

Examples of aromatic chlorohydrocarbons are

 —Cl   Chlorobenzene

 —CH_2-Cl Phenylchloromethane
(phenyl methyl chloride
or benzyl chloride)

 —CH_3
—Cl   1-Methyl-2-chlorobenzene
(o-chlorotoluene)

2.2. Multiply halogenated hydrocarbons

If more than one H-atom in the molecule of a hydrocarbon is replaced by halogen atoms the result is a **multiply halogenated hydrocarbon**:

$$H-\underset{\underset{Cl}{|}}{\overset{\overset{H}{|}}{C}}-Cl \qquad CH_2Cl_2 \qquad \text{Dichloromethane}$$
Dichloromethane
(methylene chloride)

$$H-\underset{\underset{Cl}{|}}{\overset{\overset{Cl}{|}}{C}}-Cl \qquad CHCl_3 \qquad$$
Trichloromethane
(chloroform)

$$Cl-\underset{\underset{Cl}{|}}{\overset{\overset{Cl}{|}}{C}}-Cl \qquad CCl_4 \qquad$$
Tetrachloromethane
(carbon tetrachloride)

$$\underset{\underset{Cl}{|}}{CH_2}-\underset{\underset{Cl}{|}}{CH_2}$$
1,2-Dichloroethane
(ethylene chloride)

$$CH_2=\underset{\underset{Cl}{|}}{C}-Cl$$
1,1-Dichloroethene
(vinylidene chloride)

$$\underset{\underset{Cl}{|}}{CH}=\underset{\underset{Cl}{|}}{C}-Cl$$
Trichloroethene

$C_6H_6Cl_6$ Hexachlorocyclohexane

1,4-Dichlorobenzene
(p-dichlorobenzene)

III Organic chemistry classification and nomenclature

A given molecule may also contain atoms of more than one kind of halogen. The most important compounds of this kind are the **chlorofluoromethanes**, for example

$$\begin{array}{c} Cl \\ | \\ H-C-F \\ | \\ F \end{array}$$ Chlorodifluoromethane (FC 22)

$$\begin{array}{c} Cl \\ | \\ F-C-Cl \\ | \\ Cl \end{array}$$ Trichlorofluoromethane (FC 11)

2.3. Reactions and uses

The introduction of one or more halogen atoms into a hydrocarbon molecule offers fresh possibilities for the transformation of these compounds in subsequent reactions. The halogen introduced can be replaced by other functional groups or by moieties of other molecules. The halogenated aromatic hydrocarbons exhibit particularly marked reactivity. Of these the most important industrial intermediate for syntheses (see IV,5) is chlorobenzene, which is made from chlorine and benzene:

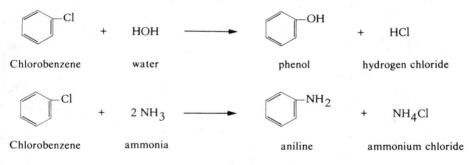

| Chlorobenzene | water | | phenol | hydrogen chloride |

| Chlorobenzene | ammonia | | aniline | ammonium chloride |

The reactivity of halogenated aliphatic hydrocarbons is considerably lower, and they are therefore less suitable for subsequent reactions. Their importance lies in their use as non-inflammable solvents, especially for fats, for example in the extraction of oils from seeds and the degreasing of metal surfaces. Thus chlorinated hydrocarbons such as trichloroethene and methylene chloride are used as end-products. The chlorofluoromethanes are also important only as end-products. They are used as propellants for aerosols (sprays) and for the manufacture of foamed products. They are also used as refrigerants of low toxicity in refrigerators and air conditioners.

Under suitable conditions the chlorohydrocarbons can also be used as starting materials for preparing other compounds, among which the **metallo-organic compounds** are particularly important:

C_2H_5Br	+	Mg	⟶	C_2H_5MgBr
Ethyl bromide		magnesium		ethyl magnesium bromide

Because of the presence of the magnesium these compounds are considerably more reactive than the chlorohydrocarbons. The organo-aluminium compounds constitute another industrially important group of substances. Among these diethylaluminium chloride $(C_2H_5)_2AlCl$, is of prime importance as a catalyst for polymerizations.

In unsaturated chlorohydrocarbons such as $H_2C=CHCl$ the double bond is the most reactive part of the molecule. This chlorohydrocarbon therefore possesses the typical properties of an unsaturated compound. It is readily polymerized:

n $H_2C=CH$ ⟶ $\left[CH_2-CH\right]_n$
      | |
      Cl   Cl

Chloroethene
(vinyl chloride)

polyvinyl
chloride (PVC)

3. Hydroxy and polyhydroxy compounds

If an organic compound contains a hydroxy group (OH) as a functional group it is called a **hydroxy compound**, or a **polyhydroxy compound** if the molecule contains more than one hydroxy group.

3.1. Monohydric alcohols, primary, secondary and tertiary alcohols, alcohol radicals

If the hydroxy group is combined with an aliphatic or alicyclic radical or with the side chain of an aromatic radical, the resulting compound is called a **monohydric alcohol** and has the general formula R-CH or $Ar-(CH_2)_n-OH$.

These compounds can be named in three different ways:
a. The ending -ol is added to the name of the hydrocarbon (the terminal e being omitted) and the position of the OH group is indicated by a numerical prefix. General class names are: alkanols, alkenols, alkynols.

b. The name of the hydrocarbon radical is followed by the word **alcohol**.

c. In large molecules with a complex structure the prefix **hydroxy-** is used. The position of the OH-group in the molecule is indicated by a number preceding the prefix.

III Organic chemistry classification and nomenclature

Alkanols

CH_3-OH

Methanol
(methyl alcohol)

CH_3-CH_2-OH

Ethanol
(ethyl alcohol)

There are two isomers of propanol:

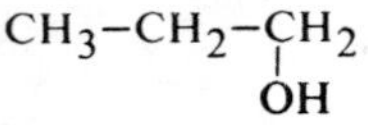

$CH_3-CH_2-\underset{\underset{OH}{|}}{CH_2}$

1-Propanol
(n-propyl alcohol)

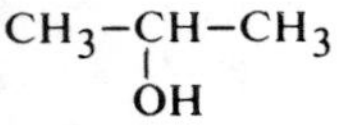

$CH_3-\underset{\underset{OH}{|}}{CH}-CH_3$

2-Propanol
(i-propyl alcohol)

The four isomeric butanols are as follows:

$CH_3-CH_2-CH_2-CH_2-OH$

1-Butanol
(n-butyl alcohol)
or prim. butyl alcohol)

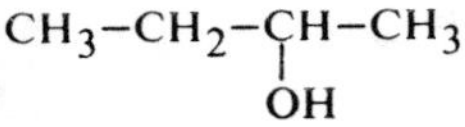

$CH_3-CH_2-\underset{\underset{OH}{|}}{CH}-CH_3$

2-Butanol
(sec. butyl alcohol)

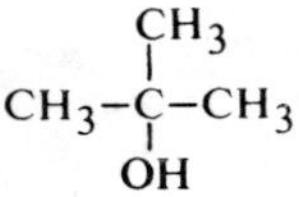

$CH_3-\underset{\underset{OH}{|}}{\overset{\overset{CH_3}{|}}{C}}-CH_3$

2-Methyl-2-propanol
(tert. butyl alcohol)

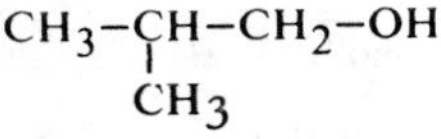

$CH_3-\underset{\underset{CH_3}{|}}{CH}-CH_2-OH$

2-Methyl-1-propanol
(i-butyl alcohol)

$CH_3-(CH_2)_3-CH_2-OH$

1-Pentanol
(1-hydroxypentane
or n-amyl alcohol)

A distinction is made between **primary, secondary and tertiary alcohols**, as shown in the above examples. In a primary alcohol the OH-group is bound to a terminal C-atom, which is directly joined to only one other C-atom, and is therefore a primary C-atom. In secondary and tertiary alcohols the OH-group is linked to a carbon atom which is bound directly to two or three other C-atoms respectively.

Summary:

General formula	OH-group linked to	Class of alcohol
$R-CH_2-OH$	prim. C-atom	prim. alcohol
$R-CH-R$ $\quad\ \,$OH	sec. C-atom	sec. alcohol
$\quad\ \,$R $R-C-OH$ $\quad\ \,$R	tert. C-atom	tert. alcohol

Univalent **alcohol radicals** which combine through their C-atoms are denoted by the following prefixes when they occur in large molecules:

$-CH_2-OH$ Hydroxymethyl-
                                          (methylol-)

$-CH_2-CH_2-OH$ Hydroxyethyl-
                                          (ethylol-)

fcr example

$HO-H_2C$ ⟨ H ⟩ CH_2-OH 1,4-Dimethylolcyclohexane

Univalent alcohol radicals which combine through their oxygen atoms are named by adding a syllable -oxy to the (abbreviated) name of the hydrocarbon radical:

CH_3-O- Methoxy

C_2H_5-O- Ethoxy

Alicyclic alcohol

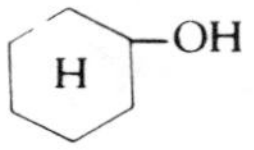

⬡—OH $\quad$ $C_6H_{11}-OH$ Cyclohexanol
(H)   (cyclohexyl alcohol)

III Organic chemistry classification and nomenclature

Aromatic alcohols

In aromatic alcohols the functional OH-group is not linked directly to the aromatic ring system, but to the "aliphatic" C-atoms of the side chain:

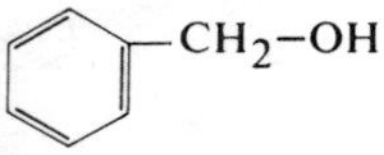

$-CH_2-OH$ (benzene ring)

Phenylmethanol
(benzyl alcohol)

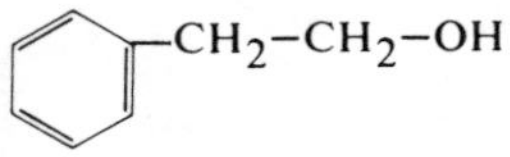

$-CH_2-CH_2-OH$ (benzene ring)

Phenylethanol
(phenylethyl alcohol)

Alkenols

$CH_2=CH-CH_2-OH$

Propene-3-ol
(allyl alcohol)

Alkynols

$CH\equiv C-CH_2-OH$

Propyne-3-ol
(propargyl alcohol)

$$CH\equiv C-\underset{\underset{OH}{|}}{\overset{\overset{CH_3}{|}}{C}}-CH_2-CH_3$$

3-Methyl-1-pentyne-3-ol

3.2. Polyhydric alcohols

The molecules of polyhydroxy compounds contain more than one OH-group. The hydroxy groups are always on different C-atoms.

If the OH-groups are linked to an aliphatic or alicyclic radical the compound is a **polyhydric alcohol**.

The number of OH-groups is indicated by inserting the syllables di-, tri, etc, before the ending -ol.

The **dihydric alcohols (diols)** are also called **glycols**, after the first member of the homologous series:

CH_2-OH
$|$
CH_2-OH

Ethanediol
(ethylene glycol,
or glycol)

$$\begin{array}{l} CH_2OH \\ | \\ CHOH \\ | \\ CH_3 \end{array}$$

1,2-Propanediol
(1,2-propylene
glycol)

$$\begin{array}{l} CH_2-C\equiv C-CH_2 \\ | \qquad\qquad | \\ OH \qquad\quad OH \end{array}$$

2-Butyne-1,4-diol

The most important **trihydric alcohol** is

$$\begin{array}{l} H_2C-OH \\ | \\ H-C-OH \\ | \\ H_2C-OH \end{array}$$

Propanetriol
(glycerine
or glycerol)

Alcohols with four or more hydroxy groups have the ending -ite or itol:

Tetrites (erythrites)
Pentites
Hexites

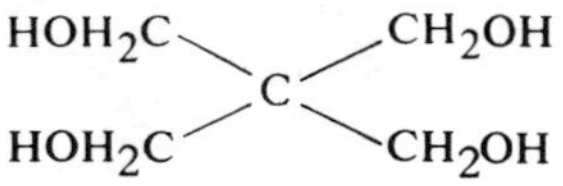

$$\begin{array}{l} HOH_2C \qquad\quad CH_2OH \\ \qquad\qquad \diagdown C \diagup \\ HOH_2C \qquad\quad CH_2OH \end{array}$$

Pentaerythrite
(pentaerythritol)

3.3. Reactions and uses

Alcohols are used as end products for the following purposes:

Solvents	eg methanol, butanol
Extractants	eg isopropanol
Potable spirits	only ethanol, obtained by alcoholic fermentation
Anti-freeze for car radiators	eg ethylene glycol
Cosmetic preparations	eg glycerol (glycerine)

Alcohols are used mainly as intermediates in the following reactions:

a) Formation of **alcoholates** (alkoxides):

$$2\ CH_3OH \quad + \quad 2\ Na \quad \longrightarrow \quad 2\ CH_3ONa \quad + \quad H_2 \uparrow$$

Methanol	metallic sodium	Na methylate (methoxide)	hydrogen

Alcoholates are salts and are readily soluble in water.

III Organic chemistry classification and nomenclature

b) Formation of **ethers** by eliminating water, either between two separate alcohol molecules (intermolecularly):

$$C_2H_5 \boxed{OH \ + \ H} OC_2H_5 \longrightarrow C_2H_5-O-C_2H_5 \ + \ H_2O$$

Ethanol   ethanol   diethyl ether   water

or internally within a polyhydric alcohol (intramolecularly), which produces cyclic ethers:

$$\begin{array}{c} CH_2-CH-CH_3 \\ \ \ | \ \ \ \ \ | \\ O\boxed{H \ \ OH} \end{array} \longrightarrow \begin{array}{c} H_2C-CH-CH_3 \\ \diagdown \diagup \\ O \end{array} \ + \ H_2O$$

1,2-Propylene glycol   propylene oxide   water

c) Formation of **esters** by reaction with organic or inorganic acids:

$$CH_3-CO\boxed{OH \ + \ H}O-C_2H_5 \ \rightleftharpoons \ CH_3-COO-C_2H_5 \ + \ H_2O$$

                                            acetic acid ethyl ester
Acetic acid   ethanol   (ethyl acetate)   water

$$O_2S\begin{array}{l} \boxed{OH \ \ \ \ \ H}OCH_3 \\ \ \ \ \ \ + \\ \boxed{OH \ \ \ \ \ H}OCH_3 \end{array} \rightleftharpoons O_2S\begin{array}{l} \diagup O-CH_3 \\ \diagdown O-CH_3 \end{array} \ + \ 2H_2O$$

                                   sulphuric acid
                                   dimethyl ester
Sulphuric   (dimethyl sulphate
    acid   methanol $(CH_3)_2SO_4)$ water

d) Formation of **aldehydes** and **carboxylic acids**, or of **ketones**, by oxidation (dehydrogenation):

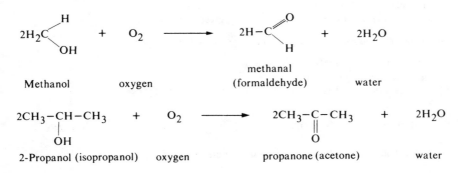

$$2H_2C\begin{array}{l} \diagup H \\ \diagdown OH \end{array} \ + \ O_2 \longrightarrow 2H-C\begin{array}{l} \diagup\!\!\diagup O \\ \diagdown H \end{array} \ + \ 2H_2O$$

                                       methanal
Methanol   oxygen   (formaldehyde)   water

$$2CH_3-CH-CH_3 \ + \ O_2 \longrightarrow 2CH_3-\underset{O}{\overset{||}{C}}-CH_3 \ + \ 2H_2O$$
$$\ \ \ \ \ \ \ | \\ \ \ \ \ \ OH$$

2-Propanol (isopropanol)   oxygen   propanone (acetone)   water

The oxidation of primary alcohols gives aldehydes, which can be further oxidized to carboxylic acids. Secondary alcohols give ketones on oxidation. Tertiary alcohols cannot be oxidized.

3.4. Monohydric phenols

Another group of hydroxy compounds contains one or more OH-groups attached to the nucleus of an aromatic hydrocarbon. If a single OH-group is present on a benzene ring the compound is a **monohydric phenol** of the type Ar-OH. The simplest hydroxy compound of benzene is

     C_6H_5-OH Phenol
(hydroxybenzene)

The univalent radical C_6H_5-O- is called phenoxy.

The three isomeric monohydroxy derivatives of toluene are the cresols:

1-Methyl-2-   1-Methyl-3-   1-Methyl-4-
hydroxybenzene   hydroxybenzene   hydorxybenzene
(o-cresol)   (m-cresol)   (p-cresol)

The monohydroxynaphthalenes are:

OH

1-Naphthol
(α-naphthol)

OH

2-Naphthol
(β-naphthol)

III   Organic chemistry classification and nomenclature

3.5. Polyhydric phenols

Important examples of **polyhydric phenols**, with several OH-groups on the benzene ring, are provided by the isomeric dihydroxybenzenes:

o-Dihydroxy-
benzene
(pyrocatechol)

m-Dihydroxy-
benzene
(resorcinol)

OH

OH

p-Dihydroxy-
benzene
(hydroquinone)

1,2,3-Trihydroxybenzene
(pyrogallol)

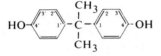

4,4'-Dihydroxydiphenyl-
dimethylmethane
(bisphenol A)

3.6. Reactions and uses

A soap solution of a mixture of cresols is used directly as a disinfectant under the name of Lysol.

In industrial chemistry phenols, cresols and naphthols are used as starting materials or intermediates for the following reactions:

a)   Formation of **phenolates**
Phenols are weak acids and therefore react with alkalis to give water-soluble phenolate salts:

Phenol

sodium
hydroxide

Na-phenolate

water

b) Formation of **phenol ethers**

Phenol   methanol   methyl phenyl ether   water
                                        (anisole)

c) Formation of **phenol-formaldehyde resins**, which are used in considerable quantities for moulded thermo-setting plastics.

d) **Quinones** are produced by the oxidation (dehydrogenation) of dihydroxy derivatives of benzene, naphthalene and anthracene:

    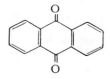

p-Benzoquinone
(p-quinone)   1,4-Naphthaquinone   9,10-Anthraquinone

Quinones and their derivatives are intermediates in the synthesis of colorants.

4. Ethers

If two hydrocarbon radicals are linked through an O-atom the resulting molecule is called an **ether**. The hydrocarbon radicals may be either alkyl or aryl, giving ethers of the types R-O-R, Ar-O-R and Ar-O-Ar.

They are named by the word ether preceded by the names of the radicals.

4.1. Simple and mixed ethers, polyethers, cyclic ethers, epoxides

If the two radicals are the same the molecule is symmetrical and is called a **simple ether**, for example

$C_2H_5-O-C_2H_5$ Diethyl ether ("ether")

[diphenyl ether structure]   Diphenyl ether

If the two radicals are different, the substance is called a **mixed ether**:

CH₃–O–⟨phenyl⟩ Methyl phenyl ether

$CH_2=CH-O-C_2H_5$ Vinyl ethyl ether

Another group of ethers comprises the glycol ethers:

$HOH_2C-CH_2-O-CH_2-CH_2OH$ Diglycol (diethylene glycol)

$HOH_2C-CH_2-O-CH_2-CH_2-O-CH_2-CH_2OH$ Triglycol (triethylene glycol)

CH_2-O-CH_3
CH_2-OH Methylglycol
(ethylene glycol monomethyl ether)

–O–CH₂–CH₂–OH Phenylglycol (ethylene glycol monophenyl ether)

If a compound contains several ether bridges it is called a **polyether**, for example the polyglycols.

$HOH_2C(-CH_2-O-CH_2)_n -CH_2OH$

The following are examples of **cyclic ethers**:

 or more simply ⟨O, H⟩

Tetrahydrofuran (THF)

H_2C⟨O⟩CH_2
H_2C ⟨O⟩ CH_2 1,4-Dioxan

Cyclic ethers in which the oxygen forms part of a three-membered ring are called **epoxides** (oxiranes):

H_2C—CH_2 over O Ethylene oxide (oxirane)

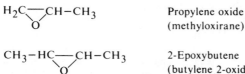

H_2C—CH–CH_3 (with O bridge)	Propylene oxide (methyloxirane)
CH_3–HC—CH–CH_3 (with O bridge)	2-Epoxybutene (butylene 2-oxide)
H_2C—CH–CH_2–Cl (with O bridge)	Epichlorhydrin

4.2. Reactions and uses

Apart from being inflammable most ethers are unreactive and hence stable. On the other hand, cyclic ethers containing a three-membered ring are extremely reactive.

The stable ethers are used as solvents, as hydraulic liquids (eg brake fluid), or, if they have a high boiling point, as heat transfer agents.

The characteristic reactions of epoxides are addition or polymerization with cleavage of the three-membered ring:

$$H_2C\text{—}CH_2 \text{ (with O bridge)} + H_2O \longrightarrow \underset{\underset{OH \quad OH}{|\quad\;|}}{H_2C\text{—}CH_2}$$

Ethylene oxide   water   ethanediol (ethylene glycol)

$$n\; H_2C\text{—}CH\text{–}CH_3 \text{ (with O bridge)} \longrightarrow \big[O\text{–}CH_2\text{–}CH_2\text{ –}CH_2\big]_n$$

Propylene oxide   polypropylene oxide (PPO)

5. Thiols and organic sulphides

Organic compounds of the type R-SH are called **thiols** or **thioalcohols** or **mercaptans**. Thiols are named by adding the ending -thiol to the name of the hydrocarbon, or the word mercaptan after the name of the radical:

C_2H_5-SH   Ethanethiol (ethyl mercaptan)

If two hydrocarbon radicals are linked through an S-atom, R-S-R, the compound is called an **organic sulphide** or thioether:

CH_3–S–C_2H_5 Methyl ethyl sulphide

Alcohol radicals can also be linked by an S-bridge to give a sulphide:

$$HOH_2C-CH_2-S-CH_2-CH_2OH \qquad \text{Thiodiglycol}$$

Long-chain and cyclic thiols and sulphides are present in small concentrations in crude petroleum (often up to 2% by weight), and give it an unpleasant smell. Before petroleum products are burnt or used in chemical syntheses these sulphur compounds must be removed, since they are catalyst poisons, and contribute to the pollution of the environment, because when petroleum products are used for heating the products of combustion contain sulphur dioxide (SO_2).

(Desulphurization has become an important source of sulphur.)

6. Sulphonic acids

Sulphonic acids contain the functional group $-SO_3H$. sometimes also called the sulpho-group.

They are named by adding -sulphonic acid to the name of the hydrocarbon.

Of the aliphatic acids of the type $R\text{-}SO_3H$ only a mixture of long-chain acids containing 10 to 20 C-atoms is of any industrial importance. Its salts constitute the active principle of some completely synthetic washing agents:

$$CH_3-(CH_2)_{10}-CH_2-SO_3H \qquad \text{Dodecanesulphonic acid.}$$

The following are aromatic sulphonic acids of the type $Ar\text{-}SO_3H$:

$-SO_3H$ Benzenesulphonic acid

SO_3H α-Naphthalenesulphonic acid

$C_{12}H_{25}-\!\!\!\!\!\text{—}SO_3Na$ Sodium dodecylbenzenesulphonate (the most important detergent)

In the last formula the way in which the SO_3Na group is attached to the benzene nucleus signifies that this group can be linked to any of the C-atoms in the ring.

A phenolsulphonic acid is:

Phenol-2,4-disulphonic acid

The introduction of the SO₃H group (see IV, 6.1), followed by neutralization with caustic soda, produces the sodium salts of sulphonic acids, sodium sulphonates. These salts are readily soluble in water, so that sulphonation and neutralization can be used to render organic compounds water-soluble. This procedure is used in the synthesis of colorants, pharmaceuticals and detergents.

7. Nitro compounds

The functional group $-NO_2$ is called nitro when the N-atom is linked directly to a C-atom. Of the **nitro-compounds**, only the aromatic derivatives of the **type AR-NO₂** are of industrial importance. They are obtained by nitration (see IV, 6.4).

They are always named by placing the prefix nitro- before the name of the hydrocarbon:

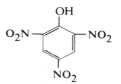

Nitrobenzene   1,5-Dinitronaphthalene

The following are some nitro compounds containing other substituents in the benzene ring:

1-Hydroxy-2-nitrobenzene
(o-nitrophenol)

1-Hydroxy-2,4,6-trinitrobenzene
(2,4,6-trinitrophenol, or
picric acid)

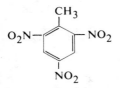

1-Methyl-2,4,6-trinitrobenzene
or trinitrotoluene (TNT, an
explosive)

Industrially the most important nitro compound is nitrobenzene, which can be readily
reduced to aniline and is thus an intermediate for a large number of colorants.

8. Amines

The functional group $-NH_2$ is called the amino group. As with the nitro group, the N-
atom is directly linked to a C-atom. The group name for these compounds is **amines**.
Compounds of the type **R-NH₂** or **Ar-NH₂** are the simplest amines.

8.1.   Primary, secondary and tertiary amines; simple, mixed and cyclic amines

In order to obtain a clear representation of the large number of different amines they are
regarded as derivatives of ammonia, NH_3. Of the three H-atoms in NH_3 either one, two or
all three can be replaced by organic radicals.

This leads to amines of the following types:

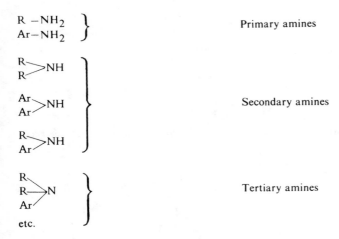

For amines the terms primary, secondary and tertiary refer to the N-atoms, and not to the
C-atom, as is usually the case. In primary amines the N-atom of the NH_2 group is linked
directly to only one C-atom, in secondary amines to two, and in tertiary to three C-atoms.

In the nomenclature of amines the ending -amine is added to the names of the hydrocarbon radicals which are linked to the N-atom. In molecules with a complicated structure the prefix amino- is used for the NH_2 group.

If a molecule contains several amino groups it is termed a diamine, triamine, polyamine, etc.

Aliphatic amines

CH_3-NH_2 (mono) Methylamine

$(CH_3)_2NH$ Dimethylamine

$(CH_3)_3N$ Trimethylamine

$(C_2H_5)_3N$ Triethylamine

$\begin{array}{c} CH_3 \\ \diagdown \\ C_2H_5 \diagup \end{array} NH$ Methylethylamine

$CH_3(CH_2)_{14}-NH_2$ Pentadecylamine

$NH_2-CH_2-CH_2-NH_2$ Ethylenediamine
(1,2-diaminoethane)

$NH_2-(CH_2)_6-NH_2$ Hexamethylenediamine
(1,6-diaminohexane)

Alicyclic amine

Cyclohexylamine

Aromatic amines

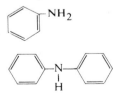

Aniline
(phenylamine, aminobenzene)

Diphenylamine

III Organic chemistry classification and nomenclature

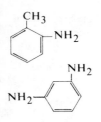 o-Aminotoluene
(o-toluidine)

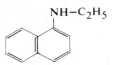 m-Phenylenediamine

Mixed amines (alkylarylamines)

 N.N-Dimethylaniline

The terminology N.N- indicates that both the methyl groups are attached to the same N-atom.

 N-Ethyl-α-naphthylamine

Cyclic amines

$$H_2C \underset{\underset{H}{|}}{\overset{CH_2}{N}} $$ Ethyleneimine

Morpholine

8.2. Aminoalcohols, aminophenols, aminosulphonic acids, nitranilines

The following are examples of compounds which contain other functional groups in addition to amino groups:

Aminoalcohols

$HO-CH_2-CH_2-NH_2$ (mono)Ethanolamine

$$HO-CH_2-CH_2 \diagdown$$
$$\qquad\qquad\qquad NH \qquad\qquad \text{Diethanolamine}$$
$$HO-CH_2-CH_2 \diagup$$

$$HO-CH_2-CH_2 \diagdown$$
$$\qquad\qquad\qquad N-CH_2-CH_2-OH \quad \text{Triethanolamine}$$
$$HO-CH_2-CH_2 \diagup$$

$$C_2H_5 \diagdown$$
$$\qquad\quad N-CH_2-CH_2-OH \qquad \text{Diethylethanolamine}$$
$$C_2H_5 \diagup$$

$-NH-CH_2-CH_2OH$ N-Hydroxyethylaniline

Aminophenol

m-Aminophenol

Aminobenzenesulphonic acids

NH₂ ... SO₃H

Orthanilic acid
(o-aminobenzenesulphonic acid)

NH₂ ... SO₃H

Metanilic acid
(m-aminobenzenesulphonic acid)

III Organic chemistry classification and nomenclature

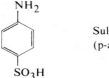

Sulphanilic acid
(p-aminobenzenesulphonic acid)

Nitroanilines (nitranilines)

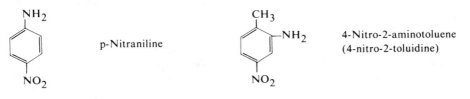

p-Nitraniline

4-Nitro-2-aminotoluene
(4-nitro-2-toluidine)

8.3. Reactions and uses

Amines, aminoalcohols, aminophenols and aminosulphonic acids are important intermediates for syntheses, eg of pharmaceutical products, additives for mineral oils and pesticides. Di- and triethanolamine are also used to make toilet preparations, lubricants, polishes, amino-soaps, industrial emulsifiers and wetting agents. The amines of long-chain hydrocarbons are used as flotation agents in the treatment of ores (see IV,6).

Amines have a basic reaction. Aliphatic amines are strong bases, while aromatic amines are weak ones. The following are important reactions of amines:

a. Salt formation
   Strong acids give water-soluble ammonium salts:

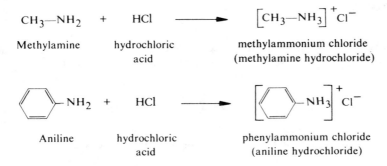

If the three H-atoms of NH_3 are replaced by alkyl radicals by reaction with alkyl halides (alkylation), tertiary amines are obtained. These can be alkylated further to give **quaternary ammonium compounds**:

| Trimethylamine | methyl iodide | tetramethylammonium iodide |

b. The alkylation of tertiary aliphatic amines gives quaternary ammonium compounds (see under a. above).

The industrially important alkylation of primary aromatic amines leads to N-substituted anilines:

| Aniline | methyl iodide | N,N-dimethylaniline | hydrogen iodide |

c. Conversion to acid amides (see III, 6.13.3):

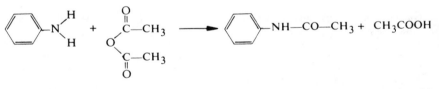

| Aniline | acetic anhydride | acetanilide | acetic acid |

d. Preparation of isocyanates:
primary amines react with phosgene ($COCl_2$) to give isocyanates:

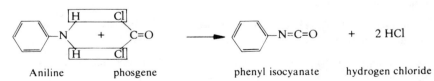

| Aniline | phosgene | phenyl isocyanate | hydrogen chloride |

The di-isocyanates obtained from diamines by this method are the starting materials for polyurethanes (see IV,7).

III Organic chemistry classification and nomenclature

e. Diazotization and coupling

The action of nitrous acid (HNO_2), prepared from hydrochloric acid (HCl) and sodium nitrite ($NaNO_2$), introduces a further nitrogen atom into primary aromatic amines. This reaction is called **diazotization**. The products are **diazonium* salts**, in which two N-atoms linked to one another are attached to one C-atom in the ring:

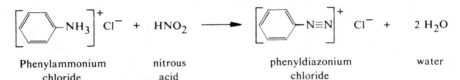

| Phenylammonium chloride | nitrous acid | phenyldiazonium chloride | water |

Phenols or primary or secondary aromatic amines can be coupled to the terminal N-atom of the diazonium group. This **coupling** produces **azo compounds****. The two N-atoms of the azo group, $-N=N-$, form a bridge between the two molecules which are coupled together:

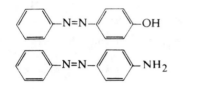

p-Hydroxyazobenzene

p-Aminoazobenzene

Numerous azo compounds are used as colorants (see IV,9). In these colorants the parent substance is

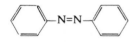

Azobenzene

9. Aldehydes

Aldehydes contain the functional group $-CHO$, which has the structure $-C\overset{H}{\underset{O}{\diagup\diagdown}}$

A distinction is made between aliphatic aldehydes of the type **R-CHO** and aromatic aldehydes of the type **Ar-CHO**.

Aldehydes can be formed by the oxidation (dehydrogenation) of primary alcohols (see III, 6.3.3); this is the origin of the name aldehyde: **al**cohol **dehyd**rogenatus.

---

*From the French **diazote**, signifying two N-atoms attached to one ring.
From the French **azote, signifying 1 N-atom per ring.

Aldehydes are names as follows:

a. Starting with the C-atom of the aldehyde group, the longest carbon chain in the molecule is found. The ending -al is added to the name of this hydrocarbon. Numbering begins at the C-atom of the aldehyde group.
b. Simple aldehydes have trivial names, mostly ending in -aldehyde.

9.1. Simple aldehydes

The following are examples of simple aldehydes:

Alkanals

H—CHO
Methanal
(formaldehyde)

CH_3—CHO
Ethanal
(acetaldehyde)

C_2H_5—CHO
Propanal
(propionaldehyde)

$CH_3-CH_2-CH_2-CHO$
n-Butanal
(butyraldehyde)

CH_3
   ＼
      CH—CHO
   ／
CH_3
i-Butanal
(isobutyraldehyde)

Alkenals

CH_2=CH—CHO
Propenal
(acrolein)

CH_3—CH=CH—CHO
2-Butenal
(crotonaldehyde)

Alkynal

CH≡C–CHO
Propynal
(propargylaldehyde)

Aromatic aldehyde

Benzaldehyde

III Organic chemistry classification and nomenclature

9.2. Dialdehydes, hydroxyaldehydes

An example of a dialdehyde is

CHO
|
CHO

Ethanedial
(glyoxal)

Among aldehydes containing other functional groups are the **hydroxyaldehydes**, which contain both aldehyde and alcohol groups. They are also called **aldols** (**ald**ehyde-alcoh**ols**).

CHO
|
CH$_2$OH

Hydroxyethanal
(glycoaldehyde)

CHO
|
CHOH
|
CH$_2$OH

2,3-Dihydroxypropanal
(glyceraldehyde)

H$_3$C—CH—CH$_2$—CHO
      |
     OH

3-Hydroxybutanal
(acetaldol)

A similar molecular structure occurs in a group of sugars, called **aldoses** because they contain an aldehyde group. Their most important representative is:

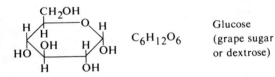

Glucose
(grape sugar
or dextrose)

$C_6H_{12}O_6$

9.3. Reactions and uses

Aldehydes are extremely reactive compounds. The reactive site in the molecule is the CHO group, $-C{<}_O^H$ with its double bend between C and O. The group $-\overset{|}{\underset{\parallel O}{C}}-$ is called the carbonyl group.

a. Reduction:

$$CH_3-C{<}_O^H \quad + \quad H_2 \quad \longrightarrow \quad CH_3-CH_2OH$$

Ethanal
(acetaldehyde)

hydrogen

ethanal
(ethyl alcohol)

b. Oxidation:

$$2\ CH_3\!-\!C\!\!\stackrel{H}{\underset{O}{\diagdown}} \quad + \quad O_2 \quad \longrightarrow \quad 2\ CH_3\!-\!C\!\!\stackrel{OH}{\underset{O}{\diagup}}$$

| Ethanal | oxygen | ethanoic acid |
| (acetaldehyde) | | (acetic acid) |

c. Addition:

| Ethanal | hydrogen | ethanal |
| (acetaldehyde) | cyanide | cyanhydrin |

d. Polymerization

Several aldehyde molecules (three, four, or many) can add on to one another to form polymers, of which the long-chain macro-molecules are important as plastics:

$$n \; \underset{H}{\overset{H}{C}}\!\!=\!\!O \quad \longrightarrow \quad \left[\underset{H}{\overset{H}{C}}\!-\!O\right]_n$$

Formaldehyde   polyformaldehyde

e. The formation of synthetic resins by the reaction of formaldehyde with phenol or urea or melamine (see IV,4).

Aldehydes are very important as intermediates, but are used directly in only small quantities.

10. Ketones

The functional group of **ketones** is the carbonyl group $C=O$, also called the oxo group or the keto group. The C-atom or the carbonyl group is linked to two other C-atoms. A distinction is made between simple and mixed ketones, according to whether the two hydrocarbon radicals are the same or different. They are of the following types:

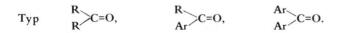

Typ $\underset{R}{\overset{R}{\diagdown}}C\!\!=\!\!O,$ $\underset{Ar}{\overset{R}{\diagdown}}C\!\!=\!\!O,$ $\underset{Ar}{\overset{Ar}{\diagdown}}C\!\!=\!\!O.$

III Organic chemistry classification and nomenclature

Several methods can be used in the nomenclature of ketones:

a. The longest chain of carbon atoms which contains the carbonyl group gives the name of the parent hydrocarbon. This is followed by the ending -one, with a numerical prefix to indicate the position of the carbonyl group. The numbering starts at the end which is nearest to the carbonyl group.

b. The names of the two hydrocarbon radicals linked to the carbonyl group are given, followed by the word ketone.

c. For ketones with complex structures the prefixes keto- and oxo- are used, with numbers to indicate their positions.

d. Simple and industrially important ketones have trivial names in addition.

10.1. Simple and mixed ketones

The following are simple ketones:

$$CH_3 \text{---} \underset{\underset{O}{\|}}{C} \text{---} CH_3$$

Propanone
(dimethyl ketone or acetone)

$$C_2H_5 \text{---} \underset{\underset{O}{\|}}{C} \text{---} C_2H_5$$

3-Pentanone
(diethyl ketone or 3-oxopentane)

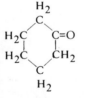 or more simply

Cyclohexanone

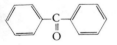

Benzophenone
(diphenyl ketone)

The following are mixed ketones:

$$CH_2 \text{---} \underset{\underset{O}{\|}}{C} \text{---} C_2H_5$$

Butanone
(methyl ethyl ketone)

$$\overset{3}{CH_3} \text{---} \overset{2}{CO} \text{---} \overset{1}{CH} = CH_2$$

1-Buten-3-one
(methyl vinyl ketone)

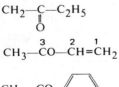

Acetophenone
(methyl phenyl ketone)

10.2. Diketones, hydroxyketones, ketoaldehydes

The following are **diketones**:

$$H_3C-\underset{\underset{O}{\|}}{C}-\underset{\underset{O}{\|}}{C}-CH_3$$

Butadione
(diacetyl)

$$H_3C-\underset{\underset{O}{\|}}{C}-CH_2-\underset{\underset{O}{\|}}{C}-CH_3$$

2,4-Pentadione
(acetylacetone)

The **hydroxyketones** are examples of ketones containing other functional groups. They contain both keto and alcohol groups, and are also called keto-alcohols.

$$\begin{array}{l} CH_2OH \\ | \\ C=O \\ | \\ CH_2OH \end{array}$$

Dihydroxyacetone

This type of structure is shared by a group of sugars called **ketoses** on account of their keto-groups. Their most important representative is:

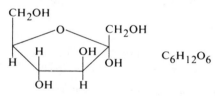

$C_6H_{12}O_6$

Fructose
(levulose)

An example of a **keto-aldehyde** is

$$H_3C-\underset{\underset{O}{\|}}{C}-CHO$$

Propanal-2-ono
(2-oxopropanal or methylglyoxal)

10.3. Reactions and uses

The simple ketones acetone and methyl ethyl ketone are used in large quantities as solvents. Chemical reactions with ketones as starting materials are of less importance.
Ketones can be hydrogenated to secondary alcohols:

$$CH_3-\underset{\underset{O}{\|}}{C}-CH_3 \quad + \quad H_2 \quad \longrightarrow \quad CH_3-\underset{\underset{OH}{|}}{\overset{\overset{H}{|}}{C}}-CH_3$$

Dimethyl ketone (acetone)　　　hydrogen　　　　　2-propanol (isopropanol)

Conversely, acetone can be prepared by the oxidation of isopropanol.

III Organic chemistry classification and nomenclature

11. Carbohydrates

The term **carbohydrate** or saccharide is used to describe a group of organic compounds which mostly contain only H- and O-atoms in addition to C-atoms. Most carbohydrates have the molecular formula $(CH_2O)_n$, which accounts for their name (carbon hydrates).

Carbohydrates occur in large quantities as sugar, starch and cellulose, particularly in trees and plants. Together with proteins and fats, carbohydrates constitute the third important group of our foodstuffs.

Glucose and fructose (see III, 6.9.2 and III, 6.10.2) are simple sugars with six C-atoms per molecule, and are called **monosaccharides.**

Ribose is a monosaccharide with 5 carbon atoms per molecule. It is an important building block in the molecular chain of the nucleic acids. The nucleic acids are the carriers of genetic information in the cells.

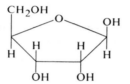

The molecules of cane sugar and beet sugar are called **disaccharides**. They contain twelve C-atoms, and can be split up into one molecule of glucose and one molecule of fructose.

The **polysaccharides** are carbohydrates of high molecular weight, of which the most important examples are starch and cellulose. Cellulose is the structural material of plants; starch constitutes the energy reserve in seeds (eg grain), roots (eg carrots) and tubers (eg potatoes).

Di- and polysaccharides which have been broken down into monosaccharides can be further split up by fermentation in the presence of enzymes (organic catalysis occurring in malt and yeast). This last process produces ethyl alcohol and CO_2 gas, and it is therefore known as **alcoholic fermentation**:

Breakdown of polysaccharides:

$$(C_6H_{10}O_5)_n \quad + \quad n\ H_2O \quad \xrightarrow{\text{catalyst}} \quad n\ C_6H_{12}O_6$$

| Potato starch, corn starch | water | | glucose |

Alcoholic fermentation

$$C_6H_{12}O_6 \xrightarrow{\text{catalyst}} \quad 2\ C_2H_5OH \quad + \quad 2\ CO_2$$

Glucose   alcohol   carbon dioxide

Alcoholic drinks are produced by fermentation.

12. Carboxylic acids

The functional group of the carboxylic acids is the carboxyl group -COOH. This group consists of a carbonyl group and a hydroxyl group: $-C\overset{\displaystyle O}{\underset{\displaystyle OH}{\big\langle}}$

12.1. Monocarboxylic acids

If the molecule of the carboxylic acid contains only one carboxyl group it is called a **monocarboxylic acid** of the type R-COOH or Ar-COOH.

There are several systems of nomenclature:

a. The longest C-chain starting with the C-atom of the carboxyl group is found. The name of the acid is then given by adding -oic acid to the name of the corresponding hydrocarbon. Numbering starts with the C-atom of the carboxyl group.

b. The ending -carboxylic acid is added to the name of the hydrocarbon to which the carboxyl group is attached. In this nomenclature the C-atom of the carboxyl group is not included in the name of the hydrocarbon.

c. For molecules with a complex structure the carboxyl group can be shown using the prefix carboxy- with a number to indicate its position.

d. Trivial names can be used for naturally-occurring carboxylic acids.

Alkanoic acids (saturated monocarboxylic acids)

H−COOH   Methanoic acid
(formic acid)

CH_3-COOH Ethanoic acid
(methanecarboxylic acid
or acetic acid)

III Organic chemistry classification and nomenclature

C_2H_5—COOH

Propanoic acid
(ethanecarboxylic acid
or propionic acid)

CH_3–CH_2–CH_2–COOH

n-Butanoic acid
(propanecarboxylic acid
or n-butyric acid)

C_4H_9—COOH

Valeric acid

C_5H_{11}—COOH

Caproic acid

The alkanoic acids from C_1 to C_9 are liquids. From C_1 to C_3 they have a pungent odour, and from C_4 to C_9 an unpleasant rancid smell. From C_{10} onwards they are solids, and they become more odourless as the chain length increases. The higher acids of this homologous series are known as fatty acids because they can be obtained from naturally occurring fats and oils.

C_{12}: $C_{11}H_{23}$—COOH   Lauric acid

C_{16}: $C_{15}H_{31}$—COOH   Palmitic acid

C_{18}: $C_{17}H_{35}$—COOH   Stearic acid

Another aliphatic monocarboxylic acid is

—CH_2—COOH   Phenylacetic acid

Alkenoic acids (unsaturated monocarboxylic acids)

C_3: CH_2=CH—COOH

Propenoic acid
(acrylic acid

C_4: CH_2=CH—CH_2—COOH

3-Butenoic acid
(vinylacetic acid)

CH_2=C—COOH
          |
         CH_3

Methylpropenoic acid
(meth(yl)acrylic acid)

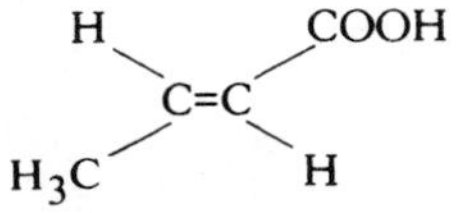

trans-2-Butenoic acid
(crotonic acid)

Another unsaturated monocarboxylic acid is

—CH=CH—COOH

3-Phenylpropenoic acid
(cinnamic acid)

Animal and vegetable fats and oils contain not only saturated but also some unsaturated carboxylic acids. The most important are the following C_{18} acids:

CH_3—$(CH_2)_7$—CH=CH—$(CH_2)_7$—COOH Oleic acid

CH_3—$(CH_2)_4$—CH=CH—CH_2—CH=CH—$(CH_2)_7$—COOH Linoleic acid

CH_3–CH_2–CH=CH–CH_2–CH=CH–CH_2–CH=CH–$(CH_2)_7$–COOH Linolenic acid

Linoleic and linolenic acids are used industrially as drying oils (linseed oil varnish).

Aromatic monocarboxylic acid

—COOH

Benzenecarboxylic acid
(benzoic acid)

Carboxylic acid of a heterocyclic compound

—COOH

Pyridine-3-carboxylic acid
(nicotinic acid)

12.2. Dicarboxylic acids

Dicarboxylic acids contain two carboxyl groups in their molecules.

Saturated dicarboxylic acids

$COOH$
$COOH$

Ethanedioic acid
(oxalic acid)

H_2C <COOH <COOH

Propanedioic acid
(methanedicarboxylic acid
or malonic acid)

CH_2–COOH
CH_2–COOH

Butanedioic acid
(ethanedicarboxylic acid
or succinic acid)

III Organic chemistry classification and nomenclature

CH₂—CH₂—COOH
|
CH₂–CH₂–COOH

Hexanedioic acid
(butane-1, 4-dicarboxylic acid
or adipic acid)

Of these the most important industrially is adipic acid, which is used as a starting material for the manufacture of polyamides (see IV, 8.3).

Unsaturated dicarboxylic acids

H–C–COOH
‖
H–C–COOH

cis-Ethenedicarboxylic acid
(maleic acid)

H–C–COOH
‖
HOOC–C–H

trans-Ethenedicarboxylic acid
(fumaric acid)

These acids are important for the production of hardening synthetic resins.

Aromatic dicarboxylic acids

—COOH
—COOH

o-Benzenedicarboxylic acid
(phthalic acid)

COOH

COOH

p-Benzenedicarboxylic acid
(terephthalic acid)

These two isomeric acids are both of great industrial importance, phthalic acid for making plasticizers, and terephthalic acid as a starting material for manufacturing polyesters (synthetic fibres and varnishes).

12.3. Reactions and uses, acyl groups

a. Acid properties
Aqueous solution of carboxylic acids are weakly acidic compared with solutions of the inorganic acids HCl, H_2SO_4, HNO_3.

They form salts with alkalis:

CH₃—COOH   +   HONa $\longrightarrow$ CH₃–COONa   +   H₂O

Acetic acid   sodium hydroxide   sodium acetate   water

The following are important carboxylic acids and their sodium salts:

H—COOH	Methanoic acid (formic acid)	H—COONa	Sodium formate
CH_3—COOH	Acetic acid	CH_3—COONa	Sodium acetate
C_2H_5—COOH	Propionic acid	C_2H_5—COONa	Sodium propionate
$C_{17}H_{35}$—COOH	Stearic acid	$C_{17}H_{35}$—COONa	Sodium stearate
⬡—COOH	Benzoic acid	⬡—COONa	Sodium benzoate

Formic acid is used for precipitating natural rubber from latex.

Acetic acid is used in the textile industry as a weak acid for adjusting the pH of dye baths.

Benzoic acid and sodium benzoate are important preservatives.

The sodium and potassium salts of fatty acids are constituents of soaps and detergents (see IV, 6).

b. Far more important is the use of carboxylic acids as starting materials for making their derivatives. A distinction can be made between reactions on the carboxyl group and substitution in hydrocarbon radical. The importance of these derivatives justifies treatment in separate sections.

(A) Substitutions in the carboxyl group:

(1) Esters $R-C\overset{O}{\underset{OR}{\diagdown}}$

(2) Carboxylic acid halides $R-C\overset{O}{\underset{Hal}{\diagdown}}$

(3) Carboxylic acid amides $R-C\overset{O}{\underset{NH_2}{\diagdown}}$

(4) Carboxylic acid anhydrides $R-C\overset{O}{\underset{O-\underset{\underset{O}{\|}}{C}-R}{\diagdown}}$

(5) Nitriles $R-C\equiv N$

With the exception of the nitriles all these derivatives contain the **acyl group** $R-C\overset{O}{\diagdown}$ or $Ar-C\overset{O}{\diagdown}$

(B) Substitutions in the hydrocarbon radical:

(1) Halogenated carboxylic acids

(2) Hydroxycarboxylic acids

(3) Aminocarboxylic acids

In these derivatives the COOH group remains unchanged in the molecule.

13. Derivatives of carboxylic acids (A)

The first group of carboxylic acid derivatives (A) are formed by replacing the OH in the carboxyl group by another functional group or radical. Except for the nitriles this leaves the univalent acyl groups unchanged. The acyl groups are named after the corresponding acids:

$$H-C{\overset{\nearrow O}{\underset{\searrow}{}}}$$ Formyl

$$CH_3-C{\overset{\nearrow O}{\underset{\searrow}{}}}$$ Acetyl

$$C_2H_5-C{\overset{\nearrow O}{\underset{\searrow}{}}}$$ Propionyl

$$C_6H_5-C{\overset{\nearrow O}{\underset{\searrow}{}}}$$ Benzoyl

13.1.   Esters, esterification, condensation, polycondensation; waxes, fats and oils; soaps, saponification, polyesters

Esters are formed from acids and alcohols with the elimination of water:

| Acid | + | alcohol | $\xrightleftharpoons[\text{ester hydrolysis}]{\text{esterification}}$ | ester | + | water |

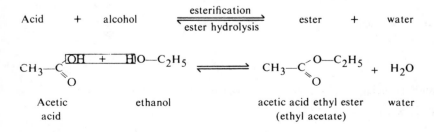

| Acetic acid | ethanol | | acetic acid ethyl ester (ethyl acetate) | water |

Esterifications are equilibrium reactions (see I, 3.12). Esterifications belong to the class of condensation reactions.

Condensation means the reaction of two like or unlike molecules to form a larger molecule with the elimination of small molecules such as H_2O or HCl.

If a large number of condensations produce macromolecules the process is called **polycondensation**.

The esters of the carboxylic acids are extremely important.

The esters of acetic acid play a predominant part as solvents for resins, adhesives and varnishes.

$CH_3—\underset{O}{\overset{\|}{C}}—O—C_2H_5$ Acetic acid ethyl ester
(ethyl acetate, or
"acetic ester")

$CH_3—\underset{O}{\overset{\|}{C}}—O—C_4H_9$ Acetic acid butyl ester
(butyl acetate)

These and other similar esters of acetic acid have a fruity odour and are used in fruit essences and aromas.

Esters of higher fatty acids and higher monohydric alcohols occur as waxes, for example:

$C_{15}H_{31}—\underset{O}{\overset{\|}{C}}-O-C_{30}H_{61}$ Palmitic acid myricyl ester

a constituent of beeswax.

Esters of the higher fatty acids with the trihydric alcohol glycerol are found in animal and vegetable **fats and oils**. The three OH-groups of the glycerol are usually esterified with molecules of different fatty acids:

$CH_2—O—C\overset{O}{\diagdown}C_{15}H_{31}$
$CH\ —O—C\overset{O}{\diagdown}C_{17}H_{35}$
$CH_2—O—C\overset{O}{\diagdown}C_{17}H_{33}$ Tri-ester of glycerol
with palmitic, stearic
and oleic acids

Saturated and unsaturated fatty acids predominate respectively in solid fats and liquid oils.

The most important reaction of the fats and oils is their decomposition into fatty acids and alcohols, which can be effected either by steam or by boiling with aqueous alkalis. If the decomposition is carried out with pure H_2O the free fatty acids are obtained together with

III Organic chemistry classification and nomenclature

glycerol. Decomposition by means of alkalis, on the other hand, yields the alkali metal salts of the fatty acids, the **soaps**. For this reason this type of ester decomposition is called **saponification***

Fat + **caustic soda** ⟶ **soap** + **glycerol**

The following are the most important **esters of dicarboxylic acids**:

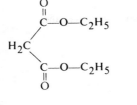

Propanedioic acid diethyl ester (ethyl malonate, or "malonic ester")

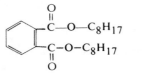

o-Phthalic acid dioctyl ester (dioctyl phthalate, or DOP) a plasticizer for plastics

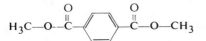

Terephthalic acid dimethyl ester (dimethyl terephthalate, or DMT)

Polyesters of maleic and of terephthalic acid are of industrial importance as synthetic resins and fibres (see IV, 7 and 8).

In addition to esters of the carboxylic acids there are some industrially important **esters of inorganic acids:**

$$O \underset{O}{\overset{}{=}} S \overset{O-CH_3}{\underset{O-CH_3}{}}$$

Sulphuric acid dimethyl ester (dimethyl sulphate), a methylating agent

$$CH_2-O-NO_2$$
$$CH\ -O-NO_2$$
$$CH_2-O-NO_2$$

(Tri)nitric acid glyceryl ester (glycerol trinitrate; incorrectly described as nitroglycerine), an explosive

---

* From the Latin **sapo** = soap.

$$O=P\overset{\displaystyle O-C_8H_{17}}{\underset{\displaystyle O-C_8H_{17}}{-O-C_8H_{17}}}$$

Phosphoric acid trioctyl ester (trioctyl phosphate), a plasticizer

13.2. Carboxylic acid halides

The carboxylic acid chlorides are the most important compounds among the **carboxylic acid halides**. The OH of the COOH group is replaced by Cl:

$$CH_3\overset{\displaystyle}{\underset{\displaystyle O}{-C}}-Cl$$

Acetyl chloride

Benzoyl chloride

13.3. Carboxylic acid amides

In the **carboxylic acid amides** the OH of the COOH group has been replaced by NH₂:

$$H\overset{\displaystyle}{\underset{\displaystyle O}{-C}}-NH_2$$

Methanoic acid amide (formic acid amide, or formamide)

A derivative of formamide is

$$H\overset{\displaystyle}{\underset{\displaystyle O}{-C}}-N\overset{\displaystyle CH_3}{\underset{\displaystyle CH_3}{<}}$$

N,N-dimethylformamide

the solvent for synthetic polyacrylonitrile fibre.

$$CH_3\overset{\displaystyle}{\underset{\displaystyle O}{-C}}-NH_2$$

Ethanoic acid amide (acetic acid amide, or acetamide)

A derivative is

$$CH_3\overset{\displaystyle}{\underset{\displaystyle O}{-C}}-NH-\!\!\!\left\langle\!\!\!\bigcirc\!\!\!\right\rangle\!\!\!-O-C_2H_5$$ Phenacetin

a constituent of febrifuges (antipyretics).

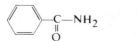

Benzoic acid amide (benzamide)

III Organic chemistry classification and nomenclature

13.4. Carboxylic acid anhydrides

Acid anhydrides are formed by eliminating H_2O from two carboxyl groups. The following are the most important acid anhydrides:

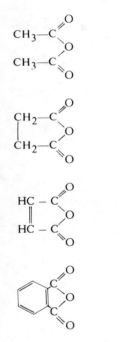

Acetic acid anhydride (acetic anhydride)

Butanedioic acid anhydride (succinic anhydride)

Butenedioic acid anhydride (maleic anhydride)

o-Benzenedicarboxylic acid anhydride
(phthalic anhydride)

13.5. Nitriles

The functional group of the **nitriles** is the univalent nitrile group $C\equiv N$, which is linked to a hydrocarbon radical. The nitriles are also described as alkyl cyanides R-CN or aryl cyanides Ar-CN.

CH_3—CN

Ethanoic acid nitrile (acetonitrile or methyl cyanide)

H_2C=CH—CN

Propenoic acid nitrile (acrylonitrile or vinyl cyanide)

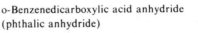

1,6-Hexanedioic acid dinitrile (adiponitrile)

$$\text{[benzene ring]} \begin{array}{l} -CN \\ -CN \end{array}$$

o-Phthalic acid dinitrile (o-phthalodinitrile or phthalonitrile)

Nitriles can be converted into the corresponding carboxylic acids by reaction with water, NH_3 being split off:

$$CH_3\text{—}CN \;+\; 2\,H_2O \;\longrightarrow\; CH_3\text{—}C\overset{\displaystyle O}{\underset{\displaystyle OH}{}} \;+\; NH_3$$

Acetonitrile   water   acetic acid   ammonia

14. Derivatives of carboxylic acids (B)

Another group of carboxylic acid derivatives (B) is derived by replacing H-atoms in the hydrocarbon radical of the carboxylic acid, the carboxyl group COOH remaining unchanged. The position of substituents in the hydrocarbon radical can be indicated in two different ways:

a) The C-atoms of the longest hydrocarbon radical are numbered starting with the C-atom of the carboxyl group.

$$\begin{array}{cccccc} 6 & 5 & 4 & 3 & 2 & 1 \\ C-&C-&C-&C-&C-&COOH \end{array}$$

b) In older nomenclature the C-atoms are indicated by Greek letters, beginning with the C-atom next to the carboxyl group. The terminal C-atom of the hydrocarbon radical is sometimes designated by ω independent of the length of the chain.

$$\begin{array}{cccccc} C-&C-&C-&C-&C-&COOH \\ \epsilon & \delta & \gamma & \beta & \alpha & \\ (\omega) & & & & & \end{array}$$

c) In aromatic carboxylic acids the position of substituents on the aromatic ring is indicated in the usual way (see III, 4.3.2 and III, 4.4.2).

14.1 Halogenated carboxylic acids

The most important compounds in this group are the chlorocarboxylic acids:

$$\underset{\displaystyle Cl}{H_2C}\text{—}COOH$$

Monochloroacetic acid

Cl$_3$C—COOH   Trichloroacetic acid

CH$_3$—CH—COOH   2-Chloropropanoic acid
         | (α-chloropropionic acid)
         Cl

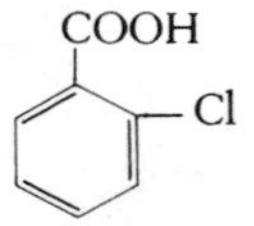

COOH
    — Cl   2-Chlorobenzenecarboxylic acid (2-
                                 chlorobenzoic acid or o-chlorobenzoic
                                 acid)

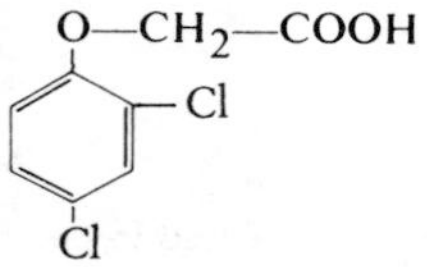

O—CH$_2$—COOH
    — Cl   2,4-Dichlorophenoxyacetic acid (2,4
                                 D), a constituent of weed-killers
Cl

14.2. Hydroxycarboxylic acids, lactones

If an H-atom in the hydrocarbon radical is replaced by a hydroxyl group, OH, the compound is a **hydroxycarboxylic** acid:

H$_2$C—COOH   Hydroxyethanoic acid (hydroxyacetic
    | acid or glycollic acid)
    OH

CH$_3$—CH—COOH   2-Hydroxypropanoic acid (α-
         | hydroxypropionic acid or lactic acid)
         OH

                                 The salts of lactic acid are called lactates.

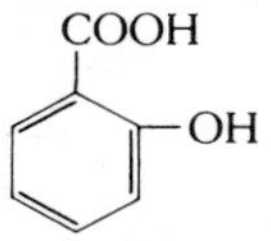

COOH
    —OH   2-Hydroxybenzenecarboxylic acid
                                 (o-hydroxybenzoic acid or salicylic
                                 acid)

In this group of compounds the relative positions of the OH and COOH groups is of particular importance. While α-hydroxy acids are stable, β-hydroxy acids have a tendency to eliminate water intramolecularly, giving unsaturated carboxylic acids. The elimination of water is also possible for γ, δ and higher hydroxy acids. In this case the product is an inner ester with a ring structure, known as a **lactone**:

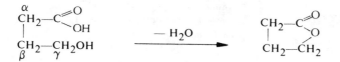

γ-Hydroxybutyric acid γ-butyrolactone

The following are hydroxy derivatives of polybasic carboxylic acids:

COOH
|
CHOH
|
CHOH
|
COOH

2,3-Dihydroxybutanedioic acid
(dihydroxysuccinic acid or tartaric
acid)

The salts are called tartrates:
potassium hydrogen tartrate is tartar

COOH
|
CH₂
|
HO—C—COOH
|
CH₂
|
COOH

3-Hydroxy-3-carboxypentanedioic acid
(2-hydroxypropane-1,2,3-tricarboxylic
acid or citric acid)

3,4,5-Trihydroxybenzoic acid (gallic
acid),
a constituent of tannins

14.3. Aminocarboxylic acids, lactams, polyamides, proteins

If the substituent in the hydrocarbon radical is an amino group, NH₂, the compound is an **aminocarboxylic acid**, often described more briefly as an amino-acid:

H₂C—COOH
|
NH₂

Aminoethanoic acid (aminoacetic acid,
glycocoll, or glycine)

III Organic chemistry classification and nomenclature

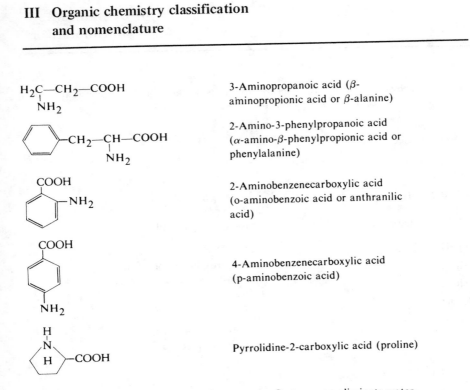

$$H_2C—CH_2—COOH$$
$$| \\ NH_2$$

3-Aminopropanoic acid (β-aminopropionic acid or β-alanine)

$$\text{—}CH_2—CH—COOH$$
$$| \\ NH_2$$

2-Amino-3-phenylpropanoic acid (α-amino-β-phenylpropionic acid or phenylalanine)

COOH
—NH$_2$

2-Aminobenzenecarboxylic acid (o-aminobenzoic acid or anthranilic acid)

COOH

NH$_2$

4-Aminobenzenecarboxylic acid (p-aminobenzoic acid)

—COOH

Pyrrolidine-2-carboxylic acid (proline)

Aliphatic amino-acids which contain four or more C-atoms can eliminate water intramolecularly to form cyclic compounds known as **lactams**. The following reaction is of industrial importance:

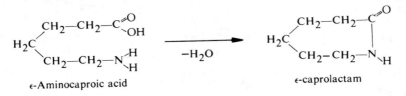

ϵ-Aminocaproic acid ϵ-caprolactam

ϵ-Caprolactam can be converted by polycondensation into the synthetic fibre **Polyamide 6** (see IV, 8).

The **proteins** have structures resembling the polyamides. Both are macromolecular substances. The macromolecules are formed by the elimination of water between COOH and NH$_2$ groups. The linking group

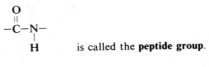

is called the **peptide group**.

Section of a Polyamide-6 molecule:

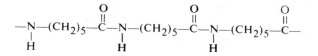

Section of a protein molecule:

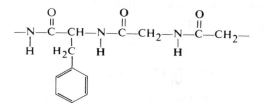

In polyamides each unit of the macromolecule consists of at least six C-atoms (including the C-atom of the peptide group). In contrast to the polyamides the macromolecules of the proteins are built of like or unlike α-amino-acids. The only β-amino-acid which occurs is β-alanine. In this case the units therefore contain only two or at the most three C-atoms.

The term **peptide** is used to describe compounds which contain roughly three to ten amino-acids. Only high-molecular compounds made up of amino-acids are called **proteins**.

Proteides (or conjugated proteins) contain in addition to amino-acids other substances such as phosphoric acid, carbohydrates and nucleic acids. The nucleus and the plasma of living cells consist of substances of this type.

Amino-acids are necessary for building up the body proteins of human beings and animals. A distinction is made between **essential amino-acids**, which have to be supplied to the body in food, and **non-essential amino-acids**, which the body can synthesize itself.

15. Derivatives of sulphonic acids

Just as for the carboxylic acids there are two types of sulphonic acid derivatives, depending on whether substitution has occurred in the sulphonic acid group SO_3H (better written as SO_2OH) or in the hydrocarbon radical.

The following are examples of replacement of the OH in the sulpho group:

$\text{C}_6\text{H}_5\text{—SO}_2\text{—Cl}$ Benzenesulphonic acid chloride
                                             (benzenesulphonyl chloride)

III   Organic chemistry classification and nomenclature

⬡—SO₂NH₂

Benzenesulphonic acid amide
(benzenesulphonamide)

CH₃—⬡—SO₂—O—C₂H₅

p-Toluenesulphonic acid ethyl ester

Examples of sulphonic acids substituted in the hydrocarbon radical are sodium dodecylbenzenesulphonate (see III, 6.6) and sulphanilic acid (see III, 6.8.2).

A derivative of sulphanilic acid is

p-Aminobenzenesulphonic acid amide
(sulphanilamide),
an important bactericide

16. Derivatives of carbonic acid

The **derivatives of carbonic acid** have structures based on the compound H_2CO_3,

$$O = C \underset{OH}{\overset{OH}{\diagdown}}$$

The anhydride of H_2CO_3 (carbonic acid) is CO_2 (carbon dioxide), which is often inaccurately referred to as carbonic acid. Although carbonic acid itself is not stable, its derivatives can be prepared. Some of them are very important industrially.

In the important derivatives one or both of the OH groups in the structural formula have been replaced.

The **halides** of carbonic acid

$$O = C \underset{Cl}{\overset{Cl}{\diagdown}}$$ $COCl_2$

Phosgene
(carbonic acid dichloride)

are of considerable importance in the synthesis of isocyanates, polycarbonates and colorants (see Plastics and colorants, IV, 7 and 9).

The **esters** of carbonic acid

$$O=C\begin{cases} O-C_2H_5 \\ O-C_2H_5 \end{cases}$$

Carbonic acid diethyl ester
(diethyl carbonate)

are used as solvents and plasticizers.

The **monoamide** of carbonic acid is

$$O=C\begin{cases} NH_2 \\ OH \end{cases}$$

Carbamic acid

Its salts, the **carbamates**

$$O=C\begin{cases} NH_2 \\ ONH_4 \end{cases}$$

Ammonium carbamate

and esters, the **urethanes**

$$O=C\begin{cases} NH_2 \\ O-C_2H_5 \end{cases}$$

Ethyl urethane

can be prepared.

The **diamide** of carbonic acid is

$$O=C\begin{cases} NH_2 \\ NH_2 \end{cases}$$

$CO(NH_2)_2$ Urea

which is the end product of protein metabolism in human beings and mammals. Urea is a high-grade fertilizer (see II, 4). It is also a starting material for the manufacture of aminoplastics (see IV, 7).

Urea can be converted to **melamine**, which is also a component of aminoplastics:

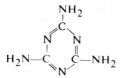

Melamine

III Organic chemistry classification and nomenclature

17. Self-evaluation questions

38 Which of the following compounds is the solvent trichlorethylene?

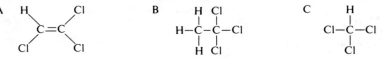

39 Give the rational formulae for the missing products.

Reduction product		Oxidation product
	$CH_3-C\overset{O}{\underset{H}{<}}$	
	$H_3C-CH-OH$ $\quad\quad\quad CH_3$	

40 The functional group -COOH is found in

   A   carboxylic acids
   B   ketones
   C   aldehydes
   D   alcohols

41 Which compound is an aromatic alcohol?

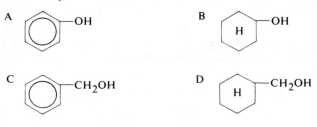

42 Give the formulae of the following:

   A   2-Chlorobutane
   B   3-Nitrotoluene
   C   1-Bromo-4-nitrobenzene

43 Which of the following compounds are primary amines?

A 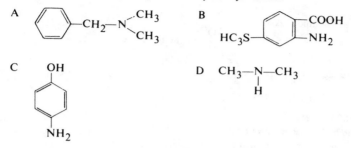 B

C   D CH$_3$—N—CH$_3$
                             |
                             H

44 Name the following:

A 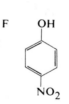 B

C CH$_3$—OH D CH$_3$—C—CH$_3$
                      ‖
                      O

E HC≡CH F

45 Glycerin is

A a water-soluble alcohol
B a trivalent alcohol
C a product of fat cleavage
D toxic

46 which compounds are ethers?

A H$_2$C—CH$_2$ B
        \ /
         O

C C$_2$H$_5$—O—CH$_3$ D CH$_3$—O—O—CH$_3$

III Organic chemistry classification and nomenclature

47 Aldehydes are

  A highly reactive compounds

  B compounds containing the group $-C\underset{H}{\overset{O}{\diagup}}$ in the molecule

  C important intermediates

  D oxidation products of primary alcohols

48 Carbohydrates are compounds that

  A occur naturally in fats

  B belong to the hydrocarbons group

  C usually have the general empirical formula $(CH_2O)_n$

  D yield drinkable alcohol by fermentation

49 Which statements about terephthalic acid are correct?

  A It is a dicarboxylic acid

  B Its formula is $\longrightarrow$

  C It is of great importance in the production of polyesters

  D It is highly corrosive like sulphuric acid

50 What are fatty acids?

  A unsaturated dicarboxylic acids

  B long-chain alkanoic acids

  C aromatic carboxylic acids

  D aromatic dicarboxylic acids

51 Which substances are esters?

  A beeswax

  B soaps

  C edible fats

  D lubricating oils

52 Which of the sections of the macromolecule illustrated below is/are termed a polypeptide group?

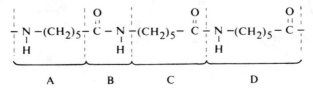

53 Which types of reaction do the alkenes undergo?

A   substitution reactions
B   addition reactions
C   condensation reactions
D   elimination reactions

54 Which of the following occur as by-products in the chlorination of methane?

A   chlorine and hydrogen
B   hydrogen chloride
C   hydrogen
D   water

55 The reaction of a carboxylic acid with an alcohol is

A   a neutralization reaction
B   esterification
C   an equilibrium reaction
D   dehydrogenation

56 In which contexts do the following compounds play a part?
Which of the statements are correct?

A   formaldehyde . . . . plastics
B   benzoic acid . . . . preservatives
C   glycol . . . . production of fats and oils
D   vinyl chloride . . . . production of polystyrene

57 The functional group -OH is found in

A   hydroxycarboxylic acids
B   phenols
C   alkanols
D   carbohydrates

58 Aniline   — NH$_2$ is

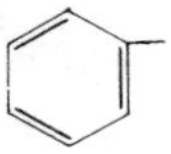

A   a substituted hydrocarbon
B   an aromatic compound
C   a heterocyclic compound
D   an amide

III Organic chemistry classification and nomenclature

59 Complete the following structural formulae:

A   COOH

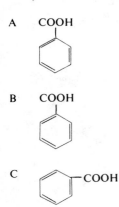

o-Benzenedicarboxylic acid (phthalic acid)

B   COOH

p-Benzenedicarboxylic acid
(terephthalic acid)

C   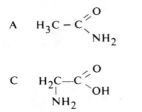—COOH

o-Hydroxybenzoic acid (salicylic acid)

60 Which of the following compounds is an amino-acid?

A $H_3C - C \underset{NH_2}{\overset{O}{\diagup}}$

B   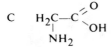 NH$_2$

C $H_2C - C \underset{NH_2}{\overset{O}{\diagup}} OH$

7 Biopolymers

1. Glucose

Important biopolymers for the formation of living organisms and for safeguarding their existence are the carbohydrates, proteins and fats. The main building block for carbohydrates and cellulose is glucose.

By means of photosynthesis, it is continuously produced in the plant cells from water and carbon dioxide with the aid of chlorophyll as biocatalyst and through the supply of solar energy.

carbon dioxide	water		glucose	water	oxygen

$$6\ CO_2 \quad + \quad 12\ H_2O \xrightarrow[\text{chlorophyll}]{h \cdot v} C_6H_6(OH)_6 \quad + \quad 6\ H_2O \quad + \quad 6\ O_2$$

$$\Delta H = +2820 \text{ kJ/mol*}$$

Ring structure of the glucose molecule

With the formation of oxygen bridging bonds and the elimination of water are formed the polymer products starch, cellulose or glycogen, the depot carbohydrate in the liver.

The synthesis of protein substances takes place through the linkage of amino acids. They are the building blocks of the proteins and thus of life itself.

Up to now, 26 naturally-occurring amino acids have been discovered, 8 of which are essential. Essential amino acids cannot be synthesized by the human organism and must therefore be furnished in the diet.

The characteristic feature of an amino acid is the amino group. In the amino acids occurring in nature, they are bound in the α-position to the carboxyl group.

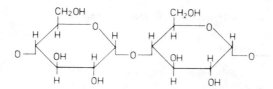

Section from the structural formula of a starch molecule

*The value of ΔH refers to the number of moles represented in the chemical equation.

III Organic chemistry classification and nomenclature

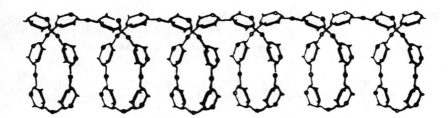

Schematized helical structure of a section of an amylose molecule

2. Amino acids

The amino acid with the simplest structure is glycine, an α-aminoacetic acid.

$$
\begin{array}{c}
H \\
| \\
H - C - COOH; \\
| \\
NH_2
\end{array}
\qquad\qquad
\begin{array}{c}
H \\
| \\
R - C - COOH \\
| \\
NH_2
\end{array}
$$

The peptide bonds are formed through a reaction between an amino group and carboxyl group with elimination of water:

$$
H_2N-CH_2-\underset{\underset{O}{\|}}{C}-OH + H-\underset{\overset{H}{|}}{N}-CH_2-COOH \xrightarrow{-H_2O} H_2N-CH_2-\underset{\underset{O}{\|}}{C}-\underset{\overset{H}{|}}{N}-CH_2-COOH
$$

The amide group $\left[-\underset{\underset{O}{\|}}{C}-\underset{\overset{H}{|}}{N}- \right]$ is frequently described as a peptide bond.

By varying the sequence, i.e. changing the order of sequence of the amino acids in a protein chain and modifying the three-dimensional structure of these chains it is possible to synthesize a large number of different protein molecules.

3. Amino bases

The nucleic acids are synthesized from ribose, phosphoric acid and the amino bases. The ribose and phosphoric acid molecules are linked to one another via ester bonds. They produce a threadlike polymer chain structure.

The amino bases are linked with the ribose molecules in a systematic sequence. The linkage takes place through a carbon-nitrogen bond, i.e. the amino bases are bound to the ribose molecules by N-glycoside linkage.

A ribose molecule, a phosphoric acid molecule and an amino base form a nucleotide building block. The nucleric acids are made up from a large number of nucleotide building blocks to form a polyester macromolecule:

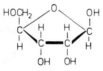

Ribose structure

Phosphoric acid structure

Five amino bases are of decisive importance as building molecules of the nucleic acids. These are the three amino bases of the pyrimidine structure cytosine, thymine and uracil:

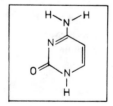

C = cytosine,

T = thymine,

U = uracil

The other two amino bases possess the purine structure

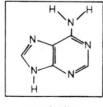

A = adenine

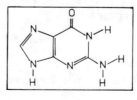

G = guanine

A section of a nucleic acid structure is illustrated by the simplified diagram shown below:

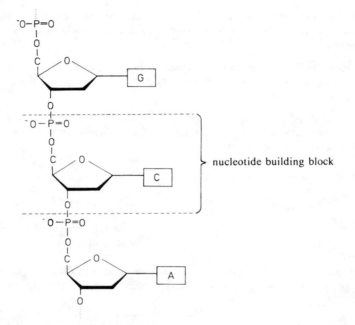

nucleotide building block

Nucleic acids are the carriers of genetic information. As polymeric chain molecules they encode the information in the sequence of their nucleotides and thus of the amino bases. The amino bases are arranged like letters or symbols in a programming language.

4. Acetic acid

Acetic acid, which contains two carbon atoms, functions as the building block of fatty acids.

The linkage reactions necessary for the biosynthesis of fatty acids take place between activated acetic acid molecules. These are acetyl radicals connected with the coenzyme A via a sulphur bridge bond.

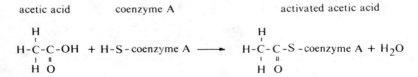

acetic acid   coenzyme A   activated acetic acid

The acetic acid building block with its two carbon atoms is the reason why the fatty acids occurring in nature always contain an even number of carbon atoms.

5. Isoprene

In addition to glucose, amino acids and acetic acid as building blocks of carbohydrates, proteins and fatty acids, mention should be made of isoprene and porphine as building elements of natural substances.

Isoprene, 2-methyl-1,3-butadiene,

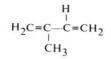

can be regarded as a building block of natural rubber:

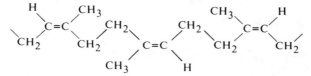

Section of the structural formula of natural rubber

However, it is not only encountered as a building unit in natural rubber, but also in the fat-soluble vitamins A, axerophthol
     D, calciferol
     E, tocopherol and
     K, phylloquinone.

These vitamins consist of derivatives of partly cyclized isoprenoid polymers. The isoprene unit can be regarded as the building block on which all these are based. See figs 1 and 2 on which the models of structural formulae of vitamins A, D_3, E and K are illustrated.

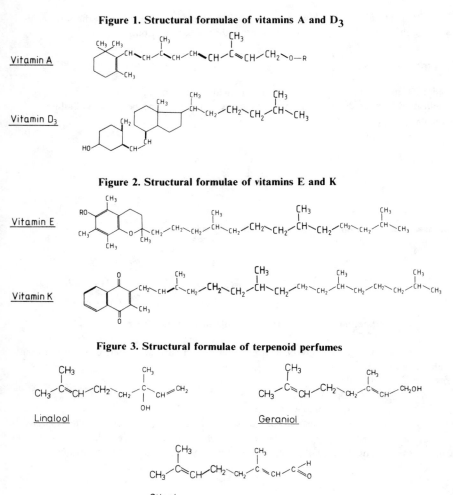

Figure 1. Structural formulae of vitamins A and D₃

Vitamin A

Vitamin D₃

Figure 2. Structural formulae of vitamins E and K

Vitamin E

Vitamin K

Figure 3. Structural formulae of terpenoid perfumes

Linalool

Geraniol

Citral

The carbon skeletons of almost all terpenoid perfumes are made up from the basic isoprene unit (see fig. 3).

Mention will be made only of the best known, such as

> linalool
> geraniol and
> citral.

6. Porphin

The porphin pigments include the biologically important colouring components: i.e.
   blood, hemin
   leaf green, chlorophyll*
   vitamin B_{12}, cyanocobalamin.

As their basic structure, all contain porphine, which is synthesized from 4 pyrrol rings linked to each other by 4 methine bridges.

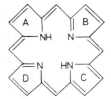

Model of the molecular structure of the porphin ring

Figure 4. Structural formula of chlorophyll a and b

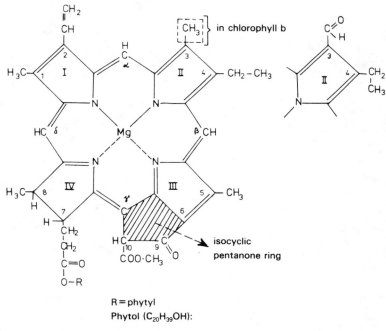

R = phytyl
Phytol ($C_{20}H_{39}OH$):

---

* chloros (Greek) = yellow-green
  phyllum (Latin) = leaf

Figure 5. Structural formula of hemin

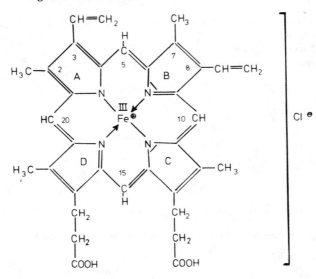

Figure 6. Structural formula of vitamin B$_{12}$

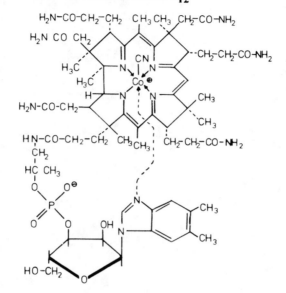

The synthetically produced phthalocyanines are likewise based on the structure of the porphin ring. The four methine groups are replaced by nitrogen bridges, $-N=$, and a benzene ring is condensed at each pyrrol nucleus. The metals that can combine to form a complex are copper, nickel, etc.

Figure 7. Structural formula of copper phthalocyanine

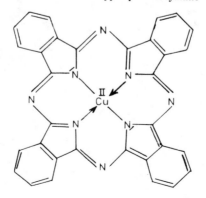

8 Answers to the Self-evaluation Questions on Sections 1–6

1 C   2 D   3 C   4 A,D   5 A,D

6 C,D   7 A,C,D   8 A,B,C   9 A,C

10 A

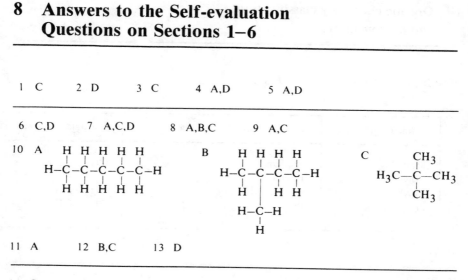

11 A   12 B,C   13 D

14 C   15 D   16 A,B

17 A,C The ending -ene indicates it. Vinyl is the radical $CH_2=CH-$

18

	saturated	unsaturated	isocyclic	heterocyclic
Furan		x		x
$CH_3(CH_2)_4CH_3$ n-Hexane	x			
$H_2C=CH-CH_3$ Propene		x		
Ethylene oxide	x			x
Cyclohexane	x		x	

III Organic chemistry classification and nomenclature

19 B,D   20 C   21 A,B,D   22 C,D

23	Name of radical	Corresponding hydrocarbon	Formula of radical
	Methyl	Methane	CH_3-
	Ethyl	Ethane	C_2H_5-
	Vinyl	Ethene (ethylene)	$CH_2=CH-$
	Propenyl	Propene (propylene)	$CH_2=CH-CH_2-$

24 D

25 B   26 C   27 D   28 A,B   29 B,C

30 A,C   31 A   32 C   33 B,D

34 B   35 C,D   36 D   37 A,D

38 A

39	Reduction product		Oxidation produkt
	C_2H_5OH	$CH_3-C{\overset{O}{\underset{H}{\diagdown}}}$	CH_3COOH
		$H_3C-\underset{CH_3}{\overset{\mid}{CH}}-OH$	$H_3C-\underset{O}{\overset{\parallel}{C}}-CH_3$

40 A   41 C

42 A $CH_3-\underset{Cl}{\overset{\mid}{CH}}-CH_2-CH_3$

B (benzene ring with CH_3 and $-NO_2$)

C (benzene ring with Br and NO_2)

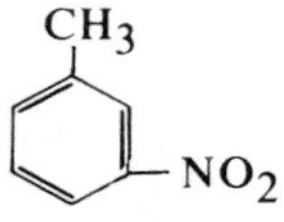

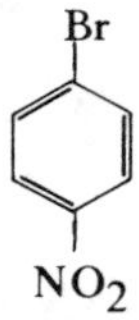

43 B, C   44 A Chlorobenzene
                  B Phenylethene or Vinylbenzene or styrene
                  C Methanol or methyl alcohol
                  D Dimethyl ketone or acetone or propanone
                  E Acetylene or ethyne
                  F p-Nitrophenol or l-hydroxy-4-nitrobenzene

45 A,B,C   46 A,B,C   47 A,B,C,D   48 C,D   49 A,C

50 B   51 A,C   52 B   53 B   54 B

55 B,C   56 B   57 A,B,C,D   58 A,B

59 A   B 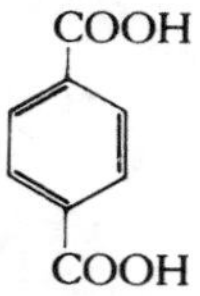   C

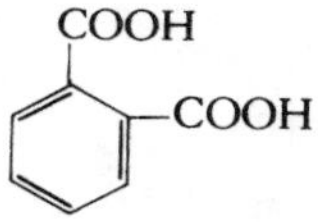

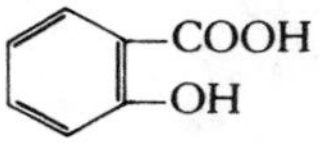

60 C

III/8-3

IV
Organic Raw Materials and Large-Scale Products: Industrial Processes

1 Crude Oil and Natural Gas as Raw Materials

1. Petrochemistry and hydrocarbons

Petrochemistry is an important sector of the chemical industry.

> The term petrochemistry (more correctly petroleum chemistry) covers all the industrial processes and chemical syntheses for which hydrocarbons form the starting materials.

The main sources of these hydrocarbons are crude oil and natural gas (in both cases the crude product).
Hydrocarbons or mixtures of hydrocarbons are isolated or produced from these two raw materials and also from coal.

Intermediates and end products are then obtained by synthetic processes from these primary products.

The following stages in processing are closely connected:

1. the raw materials for petrochemistry are crude oil, natural gas and coal;
2. primary products are obtained from the raw materials, and
3. are converted into intermediates and end products.

In order to understand why hydrocarbons are so important for organic syntheses, it is necessary to take a look at the structure of these compounds. Hydrocarbons (for details see III, 1, 2, 3 and 4) consist of molecules in the shape of chains or rings, the basic framework of which is formed by C-atoms linked to one another.

Saturated hydrocarbons with a molecular structure consisting of a large number of recurring — CH_2 — groups, methylene groups as they are called, form the main constituent of crude petroleum.

Examples are shown below:

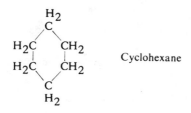

Cyclohexane

$$H_3C–CH_2–CH_2–CH_2–CH_2–CH_2–CH_2–CH_2–CH_2–CH_3 \qquad \text{Decane}$$

IV Organic raw materials and large-scale products: Industrial processes

The main constituent of natural gas is methane, CH_4, also a saturated hydrocarbon.

$$
\begin{array}{c}
\quad\ H \\
\quad\ | \\
H - C - H \\
\quad\ | \\
\quad\ H
\end{array}
$$

Since methane is a much lighter molecule than the larger and more inert hydrocarbon molecules that occur in petroleum, it is a gas at room temperature.

In petrochemistry, reactive molecules are produced from these relatively unreactive, saturated hydrocarbons. There are two possible routes for this:

1. either by splitting off hydrogen, which leads to the formation of molecules with $C = C$ carbon double bonds or $C \equiv C$ carbon triple bonds, or aromatic hydrocarbons, such as

   benzene,

2. or by substitution, ie replacing H atoms by reactive atoms or groups of atoms.

The starting material for both methods is hydrocarbons, ie compounds in which hydrogen is linked to carbon. These hydrocarbons are then converted into reactive primary products and intermediates.

Against this, a coal-based chemical industry must develop processes for converting carbon into hydrocarbons. These processes require additional operations and incur corresponding costs. Until the beginning of the 1950s, hydrocarbons were produced in Germany by this route by way of coal hydrogenation (see IV, 3, page 3). The hydrogenation of coal can be represented by the following overall equation:

Coal		Steam		Methylene group		carbon dioxide
3 C	+	$2\,H_2O$	$\longrightarrow$	$2\,(-CH_2-)$	+	CO_2
36 g		36 g		28 g		44 g

$$\Delta H = +119 \text{ kJ/mol* } (500\,^\circ C)$$

This means that 36 g of carbon (atomic weight of $C = 12$) yield only 28 g of methylene groups.

So 1.29 t of pure carbon must be used to obtain one tonne of methylene groups.

As shown in the equation, 1/3 of the carbon used is not hydrogenated to give hydrocarbons, but is lost as carbon dioxide.

---

* The value of ΔH refers to the number of moles represented in the chemical equation.

It stands to reason that hydrocarbons produced by coal hydrogenation are relatively expensive (see the Fischer-Tropsch synthesis in the chapter "Coal as a raw material", IV, 2, page 23).

In crude oil and also in natural gas these methylene groups are present in the form of chain or ring compounds.

This is the real reason why crude oil and natural gas have become important as a basis for petrochemistry.

> The hydrogen content of the hydrocarbons in crude oil and natural gas constitutes the real value of these materials as a raw material for petrochemical processes.

The ease with which petrochemical raw materials and secondary products can be transported over large distances through pipelines constitutes a further advantage.

2. Petrochemical raw materials

2.1. Crude oil

2.1.1. Chemical composition

Some hundreds of different compounds can be detected in crude oil. These range from very small, volatile constituents up to very large molecules, which remain in the tar when crude oil is distilled. Irrespective of the site where it is found crude oil consists almost entirely of hydrocarbons. In addition, it contains varying proportions of compounds containing sulphur, oxygen or nitrogen.

The removal of the sulphur-containing organic compounds, in particular, encounters difficulties. The processing of crude oil in the refinery is determined by its composition.

A distinction must be drawn between:

1. **the paraffinic crude oils;**
   these contain more than 50% of saturated hydrocarbons with a straight or branched chain (alkanes or paraffins);

2. **naphthenic crude oils;**
   these contain predominantly alicyclic hydrocarbons, in particular derivatives of cyclopentane and cyclohexane, and aromatic hydrocarbons; examples are crude oil from Arkansas, California, Venezuela, Baku and Rumania;

3. and crude oils with a mixed composition;

these are composed of paraffinic and naphthenic constituents; examples are petroleum from the Middle East and Africa.

The proportions of gasoline and oil and tar fractions vary greatly within the three groups mentioned.

2.1.2. Occurrence

A distinction is drawn between proved, probable and possible oil fields.

The oil-bearing potentialities of large areas of the surface of the earth have not yet been investigated. The presence of further reserves can, however, be presumed from the evidence of geological strata. These are termed possible reserves.

The known reserves of crude oil in the world were stated in 1979 at 87.3 billion tonnes. Well over half are located in the Middle East (Saudi Arabia, Kuwait, Iran, Ira Abu Dhabi etc.). With 65 million t, the Federal Republic of Germany accounts for 0.08% of crude oil deposits.

57.6%	of the present known deposits are accounted for by the Middle East,
11.1%	by the USSR,
6.7%	by Latin America (Venezuela, Mexico, Argentine),
5.5%	by North Africa (Libya, Algeria, Tunisia, Egypt),
4.4%	by the USA (Texas, Louisiana, Alaska),
3.7%	by Western Europe (Britain, Norway etc.),
3.4%	by the remainder of Africa (Nigeria etc.),
3.1%	by the People's Republic of China,
3.1%	by the other Far East countries,
0.9%	by Canada
0.5%	by the remaining countries.

100.0%

Western Europe's most recent and richest source of oil is the North Sea. It belongs to the national territories of Britain, the Netherlands, Denmark and Norway. Worldwide, 3.25 billion t of crude oil were pumped from wells in 1979.

Figure 1: World crude oil reserves in million t. Position in 1979: total reserves 87,250

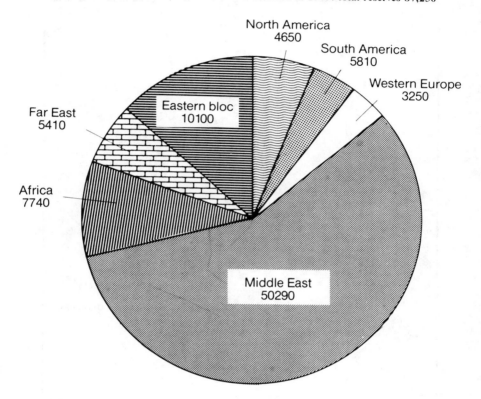

Figure 2:   Oil and gas fields in the North Sea

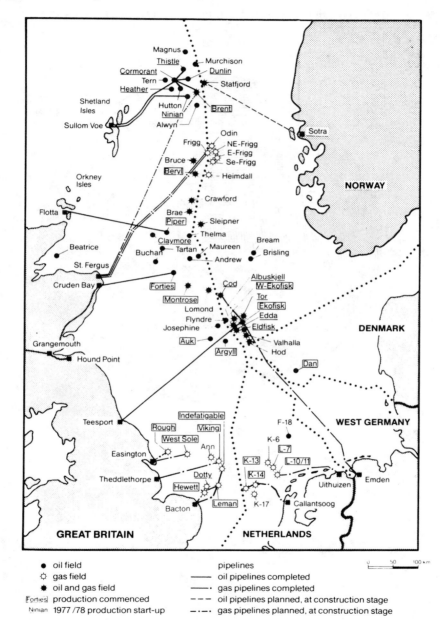

● oil field	pipelines
○ gas field	—— oil pipelines completed
✳ oil and gas field	——· gas pipelines completed
Forties production commenced	– – – oil pipelines planned, at construction stage
Ninian 1977 /78 production start-up	–·–· gas pipelines planned, at construction stage

2.1.3. Extraction

The crude oil reserves are not lodged in large subterranean lakes, but are absorbed in porous sandstone and limestone rocks.

When a bore-hole is sunk, the oil diffuses slowly through the rock to this bore-hole and is thus brought to the surface.

The rate of diffusion depends on:
a) the capillary nature of the rocks;
b) the viscosity of the petroleum;
c) the gas pressure or water pressure in the oil field.

The oil must, therefore, be extracted at an optimum rate, in order to prevent the oil capillaries from being broken, which would disturb the flow of oil or interrupt it permanently, by blocking up the pores in the rock.

It is therefore not possible to shut down oil-wells and to start extraction again as often as one would like. This is why it is necessary for oil-producing countries to supply and to sell crude oil even in times of political crisis.

At the present time there are 700,000 bore-holes extracting oil in the world, of which 625,000 are in the United States.

The greatest depth was reached in 1967 in the USA: 7260 m (23,820 ft) in West Texas.

Total world production of crude oil has been as follows:

1960	1051 million t
1970	2336 million t
1974	2870 million t
1978	3095 million t
1979	3252 million t

Production output has more than trebled in the last 20 years.

2.1.4. Transport

Crude oil is transported by means of tankers and in pipelines. The capacity of a pipeline depends on:

a) the inside diameter of the pipe (12 to 48 inches)
b) the power employed at the pumping stations;
c) the flow properties of the petroleum.

In some cases the pipes are heated in order to make the oil less viscous.

Some pipelines are of continental dimensions. Comecon pipelines in the USSR run for more than 2000 km (1240 miles) from the Volga-Ural region to the refinery and to the chemical plant at Schwedt/Oder in Eastern Germany.

IV Organic raw materials and large-scale products: Industrial processes

Branch lines run from this pipeline to Leuna in East Germany and through Slovakia (the Bratislava refinery) to North Bohemia (Leutensdorf) and Hungary.

In Saudi Arabia the Trans-Arabian pipeline, built in 1950, runs for 1750 km (1088 miles) to Saida.

For important oil pipelines in Western Europe and Northern America, see Figures 4 and 5 on pages 14, 15

2.2. Natural gas

2.2.1. Chemical composition

The composition of natural gas is very variable, depending on its origin.

It always contains methane CH_4 and can also contain the hydrocarbons ethane C_2H_6, propane C_3H_8 and butane C_4H_{10}.

In addition, it contains the following in varying proportions:

hydrogen	H_2
nitrogen	N_2
carbon dioxide	CO_2
hydrogen sulphide	H_2S
helium	He.

The H_2S must be removed from natural gas before it can be used for heating or as a raw material for the manufacture of primary products (see V, 3).

2.2.2 Occurrence

Natural gas occurs frequently as gas bubbles near to oil fields. Natural gas is also dissolved in crude oil. It can occur in the surroundings of coal deposits as a product of subsequent carbonization.

The proved reserves of natural gas in the world are estimated at around 74,730 billion cubic metres.* This quantity is approximately equivalent to two thirds of the calorific value of the world's crude oil deposits. Of this total, the eastern bloc countries account for almost 40%, followed by the countries in the Middle East with natural gas reserves of over 20,700 billion m^3.

5.5% of world reserves of natural gas are in Western Europe, the major proportion being located in the Netherlands and Britain. The Netherlands now occupy first place among Western European producers of natural gas.

---

* The volume of gaseous substances depends on the pressure and the temperature. 1 standard cubic metre is 1 m^3 of gas at $0°C$ and 1.013 bars pressure. One billion m^3 equals approximately 35 billion ft^3.

The Federal Republic of Germany has natural gas deposits with proved reserves of 190 billion m³. 75% of this volume is located in the area between Weser and Ems. World production of natural gas in 1979 amounted to almost 1,500 billion m³

Figure 3: **World reserves of natural gas in billion m³ Position in 1979: total reserves 74,370.**

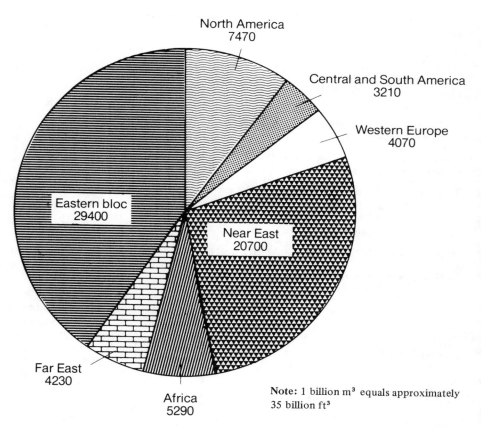

Note: 1 billion m³ equals approximately 35 billion ft³

2.2.3. Extraction

Boring is carried out for natural gas by methods similar to those for crude oil.

Extraction of natural gas in the world in 1979 was about 1,618 billion m³.

IV Organic raw materials and large-scale products: Industrial processes

Natural gas is mainly used in power stations for power generation and in households. Purified crude gas is used in industry (blast furnaces and the melting of glass) as a means of transferring energy.

Natural gas is also used to an increasing extent as a raw material for the petrochemical industry.

2.3. Grid for transporting crude oil, natural gas and petrochemical products

The transport of large quantities of liquids and gases by pipelines is a very economical process.

In the context of the petrochemical industry a distinction should be made, in principle, between the following:
a) a crude oil pipeline;
b) an ethylene pipeline;
c) a pipeline for finished products;
d) a gas pipeline (natural gas or petroleum gas);
e) a long distance gas pipeline.

The purpose of the long distance gas pipeline is to transport coke-oven gas, refinery gas, cracked gases and mixtures of gases.

This pipeline grid provides a reliable means of exchange for the materials produced by the petroleum industry, the petrochemical industry and gas works in cities.

2.3.1. Crude oil pipelines in Western Europe and North America

At present the crude oil grid in the world has a total length of 1.8 million km (1.1 million miles).

Of these pipelines, the following may be mentioned:

Name	Countries	Starting point and terminal of pipeline	Length (km)	(miles)
Mid Valley	USA	Texas-Ohio	1600	994
Trans Alska Pipeline	USA	Prudhoe Bay-Valdez	1280	795
Interprovincial Pipe Line	Canada/USA/ Canada	Edmonton-Toronto	3100	1926
North Peruvian pipeline	Peru	Concordia-Bayovar	850	528
SONATRACH	Algeria	Hassi Messaoud-Arzew	850	528
		Mesdar-Skikda	750	466
Orient pipeline	Iraq/Turkey	Kirkuk-Dörtyol	981	610
Tapline	Saudi Arabia/Jordan/ Syria/Lebanon	Abqaiq-Sidon	1715	1066
NWO (North-West oil pipeline)	Fed. Rep. of Germany	Wilhelmshaven-Wesseling	384	239

TAL (Transalpine oil pipeline)	Italy/Austria/Fed. Rep. of Germany	Trieste-Ingolstadt	464	288
		Ingolstadt-Karlsruhe	286	178
SPLSE (Sociéte du Pipeline Sud-Européen)	France/Fed. Rep. of Germany/ Switzerland	Labéra-Karlsruhe	782	486
		Lavéra-Strassburg	714	444
COMECONpipeline	USSR-Eastern Europe	Mosyr-(via Brest/Plock)- Halle	1150	715
Transsiberian oil pipeline	USSR	Tuimasy-Angarsk	3700	2300
		Tuimasy-Leningrad	1800	1200

The crude oil pipelines start from important ports handling ocean freight and run to refineries situated inland, the locations of which are selected because of important consumer industries, such as the chemical industry, for the products of the refinery.

2.3.2. The link between the chemical industry and refineries

In the early days of the petrochemical industry, the raw materials available were the by-products obtained in the refining of crude oil for the fuel and power industries (motor fuels and fuel oil).

Refinery capacities in 1979 amounted in
the USA to	889.5 million t
Great Britain to	122.6 million t
the FRG to	153.9 million t
Western Europe to	975.6 million t.

Every change in the production structure in the fuel and power industries therefore had an effect of some kind on the range of by-products available. See the flow-sheet in Figure 4 on page 16 entitled "From petrochemical raw materials to chemical end-products".

As the demand increased for primary products such as ethylene, acetylene, propylene and benzene, especially as raw materials for plastics and synthetic fibres, the parallel products obtained at the refineries no longer afforded a satisfactory supply of raw materials.

When planning new refineries, therefore, the oil companies and the chemical industry united to form vertical combines.

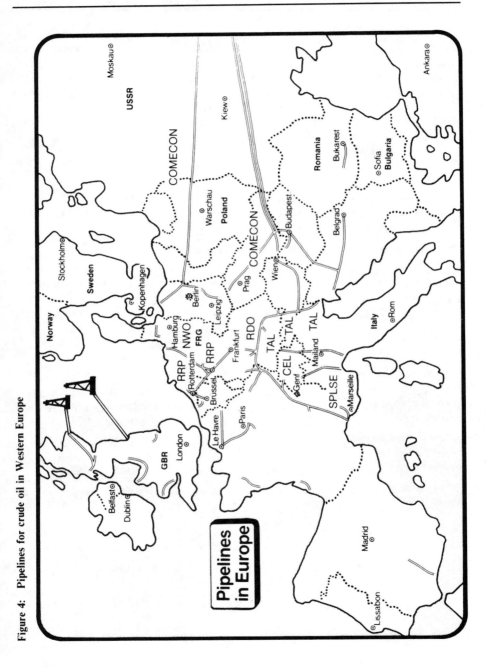

Figure 4:   Pipelines for crude oil in Western Europe

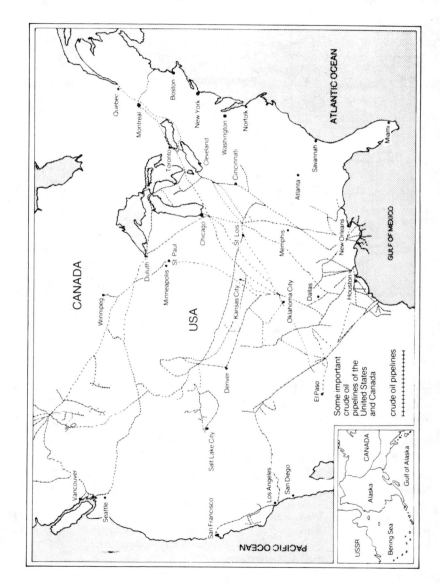

Figure 5: Pipelines for crude oil in North America

IV Organic raw materials and large-scale products: Industrial processes

Figure 6: From petrochemical raw materials to chemical end-products

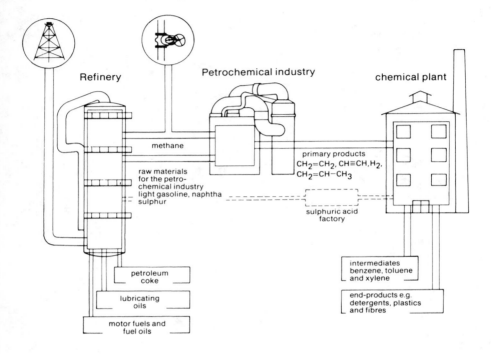

2.3.3. Pipelines for finished products

Refinery products such as petrol, diesel oil, fuel oil, middle distillates, chemical intermediates and hydrogen are conveyed in these pipelines.

Ethylene, a product obtained by cracking special naphtha fractions of crude oil, is transported by pipeline to centres of the chemical industry for further processing

Of these pipelines, the following may be mentioned:

Name	Countries	starting point and terminal of pipeline	Length (km)	(miles)
Little Big Inch	USA	Texas-New York	2385	1482
Plantation	USA	Louisiana-North Carolina	2025	1258
RMR Rhine-Main pipeline)	Netherlands/Fed. Rep. of Germany	Rotterdam-Ludwigshafen	665	413
Trapil	France	Gonfreville-Paris	185	115
UKOP	Great Britain	Thames-Merseyside	390	242

3. Petrochemical processes and unit processes

3.1. Distillation

Distillation is the most important separation process employed for processing crude oil. The difference in boiling point between the individual constituents is utilized to separate the crude oil into its components.

Another separation process makes use of the different solubilities of the hydrocarbons.

It is characteristic of refineries that they are not housed in large factory buildings; instead, the furnace, the reactors, columns and pipelines stand in the open.

This is necessary for safety reasons, since the products concerned, whether gases or liquids, are all readily combustible. A refinery is monitored and controlled from central control points by a few skilled workers by means of instrumentation and control systems.

The crude oil is generally first distilled under atmospheric conditions, that is to say under normal pressure. The distillation residue can then be fractionated further under vacuum.

Figure 7 on page 18 shows the flow-sheet of distillation under atmospheric pressure. This enables all the components of the crude oil boiling below 350°C (662°F) to be separated. The distillation of crude oil is effected by means of continuously-operated fractionation columns, fitted with bubble trays. The function of the bubble trays is to increase the degree of separation obtained.

The distillation process can be subdivided into the following stages:

1. Preheating the crude oil

The crude oil is preheated to about 300°C (572°F) in a tube furnace, a system of heated tubes, and is injected into the fractionating column one-third of the way from the bottom.

The material to be distilled thus brings with it a large part of the heat required for vaporization.

2. The process taking place in the rectifying section or enrichment section of the column.

The part of the crude oil that vaporizes at the temperature of injection rises upwards in the enrichment section of the column. The other, non-volatile fraction flows downwards over the bubble trays of the so-called stripping zone.

3. The process taking place in the stripping zone.

Hot steam is additionally blown in below the lowest tray in the stripping zone; the extra energy thus provided causes a further part of the less volatile components of the crude oil to vaporize.

Figure 7:   Flow-sheet of distillation

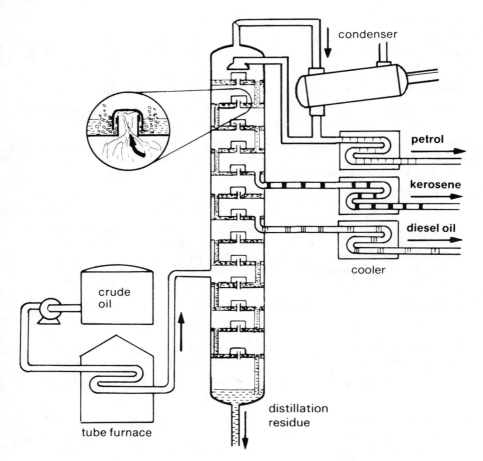

condenser

petrol

kerosene

diesel oil

cooler

crude
oil

distillation
residue

tube furnace

4. Drawing off the distillates.

The distillates are withdrawn from the fractionating column in a series of cuts with accurately defined boiling ranges. The distillates are withdrawn from the fractionating column at various heights, depending on their boiling ranges. A stationary temperature gradient is set up in the column from the stripping zone to the head of the column.

a) The cut with the lowest boiling range (in this example 40 to 175°C, or 104 to 347°F), called straight-run gasoline, is taken off at the head of the column.

b) At the foot of the column the distillation residue flows out at a temperature of about 375°C (707°F). It is passed through a heat exchanger in order to transfer its high content of heat energy into the crude oil flowing in to replace it.

The distillation residue can then be separated into different components by further fractional distillation under vacuum.

The residues from this vacuum distillation are used as heavy fuel oil or as bitumen for road construction.

c) The cuts between the straight-run gasoline at the head of the column and the distillation residue are withdrawn from the column in the form of sidestreams. They are passed through small steam-heated stripping columns, mounted on the outside of the fractionating column.

The function of the steam-heated stripping columns is to remove the lower-boiling fractions from a cut with a specified boiling range. The various cuts are always contaminated with lower-boiling fractions on the bubble trays, since all these fractions have first to pass through the lower trays before they can reach the head or the upper section of the column.

These lower-boiling fractions are recycled to the main column.

5. Distillation capacity.

A fractionating column is designed to have a capacity enabling it to distil about 250 t of crude oil, or 300 m³ (10,500 ft³) per hour. These quantities correspond to six to seven train loads per day, each of 40 to 50 tank wagons. This example alone shows that it is just not possible to transport crude oil by rail or road, but only by pipeline.

3.1.1. Crude oil fractions

The fractions into which crude oil is separated are characterized by their boiling points. The lower the boiling point, the shorter the carbon chains of the hydrocarbons.

The following main fractions are collected:
1. gas, consisting of hydrogen and methane;
2. LPG, liquefied petroleum gas, consisting of hydrocarbons with a chain length of C_2 to C_4;
3. naphtha: boiling range 30 to 100°C (86 to 212°F);
4. heavy naphtha: boiling range 100 to 200°C (212 to 392°F);
5. kerosene: boiling range 200 to 250°C (392 to 482°F);

6. gas oil (diesel oil and fuel oil): boiling range 250 to 350°C (482 to 662°F);
7. lubricating oil: boiling range 350 to 500°C (662 to 932°F);
8. bitumen: boiling range above 500°C (932°F).

Even today the major part of crude oil is used for fuel and power. Some 53% of refinery output is used for heating alone. To this must be added the high proportion of motor spirit. Up to now only 5–7% is used as a raw material for the chemical industry.

In the long term, the supply of fuel and power will have to be switched over to use nuclear energy, and perhaps also solar energy, in order to ensure supplies of crude oil as a raw material for valuable chemical products and for the production of protein by biochemical synthesis for human nourishment.

3.1.2. Gasoline

The fractions used as gasoline must meet very specific requirements. In order to achieve a high anti-knock rating, these fractions should contain a proportion of branched paraffins and aromatics. The octane number is a measure of the anti-knock rating.

This number indicates the percentage of isooctane in an isooctane/n-heptane mixture. The octane number of 100% pure isooctane has been taken as 100 and that of n-heptane as 0.

Isooctane — 2,2,4-trimethypentane

$$
\begin{array}{ccccccccc}
 & H & & CH_3 & H & & H & & H \\
 & | & & | & | & & | & & | \\
H - & C & - & C & - & C & - & C & - & C - H \\
 & | & & | & | & & | & & | \\
 & H & & CH_3 & H & & CH_3 & & H \\
\end{array}
$$

n-Heptane

$$
\begin{array}{ccccccccccccc}
 & H & & H & & H & & H & & H & & H & & H \\
 & | & & | & & | & & | & & | & & | & & | \\
H - & C & - & C & - & C & - & C & - & C & - & C & - & C & - H \\
 & | & & | & & | & & | & & | & & | & & | \\
 & H & & H & & H & & H & & H & & H & & H \\
\end{array}
$$

The cetane number is an analogous way of defining the ignition properties of a diesel fuel relative to a mixture of cetane (n-hexadecane) and α-methylnaphthalene. The cetane number of cetane has been taken as 100 and that of α-methylnaphthalene as 0.

n-Hexadecane $\qquad$ $H_3C - (CH_2)_{14} - CH_3$

α-Methylnaphthalene

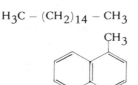

For Otto engines the octane number should be between 80 and 100.

In contrast with motor spirit, diesel fuel consists of straight-chain paraffins. Its cetane number should be less than 45.

Similarly, fuels for jet propulsion should contain unbranched paraffins as far as possible, otherwise crystallization can take place readily as a result of cooling when flying at great altitudes.

3.2. Chemical reactions

3.2.1. Petrochemical primary products

The following is a list of important primary products produced from crude oil (and in some cases also from natural gas) as a raw material:

hydrogen	H_2
methane	CH_4
ethylene	$H_2C = CH_2$
acetylene	$HC \equiv CH$
propylene	$H_2C = CH - CH_3$
1-butene	$H_2C = CH - CH_2 - CH_3$
isobutene	$H_2C = \underset{\underset{CH_3}{\vert}}{C} - CH_3$

1,3-butadiene $H_2C = CH - CH = CH_2$

benzene

toluene

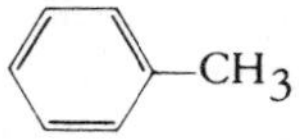

o-xylene

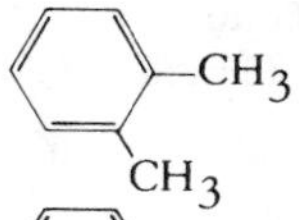

p-xylene

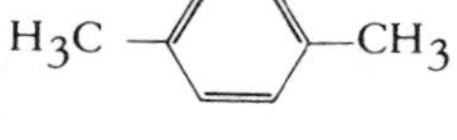

IV Organic raw materials and large-scale products: Industrial processes

These primary products of the petrochemical industry undergo further conversion to give plastics, synthetic rubber, synthetic resins, synthetic fibres, detergents, pharmaceuticals, colorants and so on. Hydrogen is used for the manufacture on a large industrial scale of ammonia, NH_3.

3.2.2. Thermal and catalytic cracking

Cracking means the splitting of long-chain hydrocarbons into two or more short-chain hydrocarbons by breaking the carbon-carbon bonds.

Depending on the particular end products desired, cracking can be carried out at varying high temperatures under pressure and in the presence of catalysts.

The possible routes it can take are shown in the diagram below.

Taking n-butane as an example:

$$
\begin{array}{l}
\text{Butane} \\
CH_3{-}CH_2{-}CH_2{-}CH_3 \xrightarrow{435^\circ C}
\end{array}
\left\{
\begin{array}{ll}
\underset{\text{Ethylene + Ethane}}{H_2C{=}CH_2 + H_3C{-}CH_3} & 38\,\% \\[2em]
\underset{\text{Propylene + Methane}}{CH_3{-}CH{=}CH_2 + CH_4} & 50\,\% \\[2em]
\underset{\text{Butylene}}{CH_3{-}CH_2{-}CH{=}CH_2} + \underset{\text{+ Dimethylethylene}}{H_3C{-}CH{=}CH{-}CH_3} + \underset{\text{+ Hydrogen}}{H_2} & 12\,\%
\end{array}
\right.
$$

In the course of time certain principles have been established for the cracking reaction:

1. at low cracking temperatures, the molecule breaks in the middle;
2. as the temperature increases, multiple cracking takes place, the larger fraction being olefinic (that is to say unsaturated), and the smaller being paraffinic;
3. as the chain length increases, cleavage of the carbon-carbon bond takes prior place over dehydrogenation; dehydrogenation always leads to a greater olefinic content;
4. the temperature affects the position at which the break in the carbon chain takes place: at about 400°C (752°F) the chains break preferentially in the middle; as the temperature increases, the point of cracking travels more towards the end of the hydrocarbons;
5. at low temperatures, about 400 to 600°C (752 to 1112°F), the cleavage of large molecules predominates, with subsequent insomerization, cyclization and dehydrogenation reactions.

Medium temperatures between 600 and 1000°C (1112 to 1832°F) produce mainly small, unsaturated hydrocarbon molecules.

The main products obtained are:
    ethylene,
    propylene, and
    butylene.

Temperatures above $1000°C$ ($1832°F$) yield mainly acetylene and ethylene. This is the case with high temperature pyrolysis.

    The rule is that the cleavage products become smaller and their hydrogen content decreases as the temperature increases.

3.2.3. High temperature pyrolysis (HTP)

High temperature pyrolysis is a special form of cracking.

This process makes it possible to crack light naphtha fractions, consisting of saturated aliphatic hydrocarbons with a boiling range between $25°C$ ($77°F$) and $130°C$ ($266°F$), into unsaturated hydrocarbons, especially ethylene, at temperatures of about $2700°C$ ($4892°F$).

This cleavage reaction is endothermic, that is to say it consumes energy. The energy required for cleavage must be supplied to the reaction system. On the other hand, the reaction products, such as ethylene and acetylene, are not stable in this high temperature range. The residence time must therefore be kept very short and the cracked gas must then be chilled to temperatures below $500°C$ ($932°F$). The actual duration of the reaction, which is identical with the residence time in the reaction zone, is less than 3×10^{-3} seconds (three thousandths of a second).

After chilling, the next phase is the removal of unwanted coproduced gases. The products are obtained 99.9% pure.

The ratio in which the various products, for example ethylene, are formed can be controlled by the cracking temperature.

HPT is a continuous process in which the unreacted hydrocarbons are recycled. The operating time is 12 to 18 months. Only then is it necessary to shut down the plant for inspection (see Figure 8, page 24).

Carrying out the process:

High temperature pyrolysis can be subdivided into four stages:
1. producing the high cracking temperature in the furnace;
2. the cracking reaction in the reaction tube;
3. chilling the products of cracking;
4. separating the products of cracking and purifying them from by-products.

1. Producing the cracking temperature
The by-products formed together with the unsaturated hydrocarbons consist essentially of methane, carbon monoxide and hydrogen. They are fed to the combustion chamber as combustion gases and, when well mixed with oxygen, are used to produce the heat energy

at temperatures of about 2700°C (4892°F). In addition, steam is blown in, as required, to protect the furnace walls and to control the flame temperature.

2. The cracking reaction
The light naphtha fractions, that is to say the hydrocarbons to be cracked, are injected into the hot gases at the end of the combustion chamber or at the head of the reaction tube, where the flame has burnt itself out. The cracking reaction starts at once.

3. Chilling the cracked products
After flowing through the reaction tube the temperatures have fallen from 2700°C (4892°F) to about 1200°C (2192°F) and, at the end of this tube, the cracked gases are chilled with oil or water, or, as the specialist would say, quenched.

After quenching, the cracked product is cooled to below 500°C (932°F) and the cracking reaction is thus stopped. Ethylene and acetylene are stable at this temperature.

The heat taken up in the quenching oil is utilized for generating steam in an indirect heat exchanger.

4. Separating and purifying the cracked products
After chilling, the cracked products are washed with oil in order to remove the heavy hydrocarbons and are cooled to 40°C (104°F) in the so-called final gas coolers. The last residues of gasoline and the water are condensed out in the final gas cooler.

Figure 8: The manufacture of ethylene and acetylene

1. generating heat and   2. compression   3. 1st purification stage;   2nd purification stage;   4. low-temperature
   the cracking reaction CO_2 removal   washing with   distillation
                                                                        light naphtha

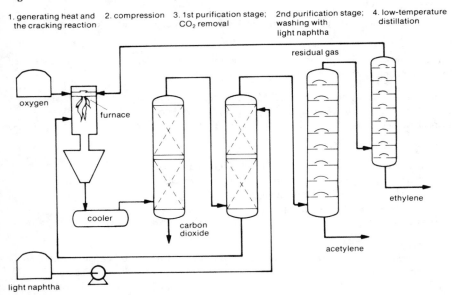

The cracked gas is compressed to about 16 bars (203 lb/in^2) and the CO_2 is washed out, using a solution of potassium methyl taurinate as absorbent. All the remaining unwanted constituents, such as C_3 hydrocarbons and higher, are removed at 30°C (86°F) with the feed naphtha and are recycled to the reaction tube for cracking, to be converted into ethylene or acetylene in the second or third pass.

After this preliminary purification, the cracked gas, now consisting only of acetylene, ethylene, methane, hydrogen and carbon monoxide, is passed into the acetylene recovery unit and washed at −30°C (−22°F) with acetone.

The acetylene extracted is 99.9% pure.

When the gas has been freed from acetylene, it is compressed further to 30 bars and is liquefied in three stages (−45°C, −104°C and −200°C; −89°F, −155°F and −328°F) and is fractionally distilled. This low-temperature distillation enables the residual cracked gas to be separated into 99.9% pure ethylene and a residual gas fraction, consisting of carbon monoxide, methane and hydrogen.

Utilizing the residual gases

The greater part of this residual gas fraction is generally burnt with oxygen in the combustion chamber and is thus used to supply the heat required in the cracking reaction. A fairly small excess fraction of the residual gas is withdrawn from the plant and used to generate steam.

Frequently, part of the residual gases is also used as synthesis gas. The methane is fed to chlorination for the production of chloromethanes. Methane and hydrogen are also fed to ammonia synthesis as raw materials (see II,3).

The carbon monoxide produced and the hydrogen are also required in the oxo synthesis (see IV,3).

Figure 9: The temperature profile in the furnace-reactor system

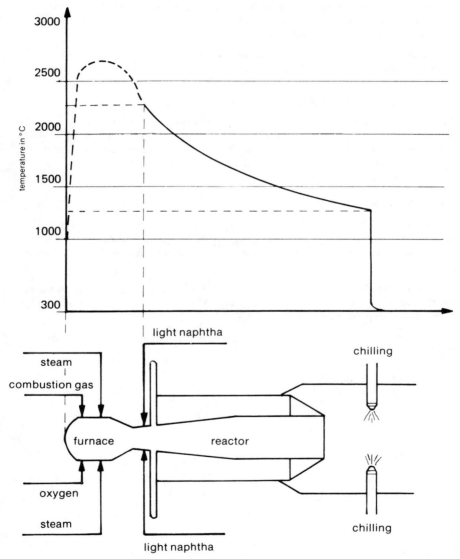

3.2.4. Dehydrogenation

Depending on the product desired, dehydrogenation, ie the removal of hydrogen from the hydrocarbons, is carried out not only by the application of heat but also in the presence of catalysts.

The dehydrogenation always produces unsaturated hydrocarbons and hydrogen.

The following is an example of thermal dehydrogenation:

$$H_3C - CH_3 \longrightarrow H_2C = CH_2 \quad + \quad H_2$$

Ethane   Ethylene   Hydrogen

The following is an example of catalytic dehydrogenation

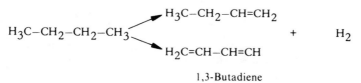

Butylene

$$H_3C-CH_2-CH_2-CH_3 \begin{cases} H_3C-CH_2-CH=CH_2 \\ H_2C=CH-CH=CH \end{cases} + \quad H_2$$

1,3-Butadiene

3.2.5. Reforming and platforming

By reforming we mean a special form of cracking under medium pressures. The application of heat is of very short duration, amounting to about 10 to 20 seconds. This process is further accelerated by means of catalysts. Platinum deposited on bauxite, Al_2O_3, is often used as the catalyst (so-called platforming). Reforming initiates rearrangements leading to highly branched hydrocarbons and thus produces high-octane petrols.

Isomerization reactions

$$H_3C-CH_2-CH_2-CH_2-CH_3 \longrightarrow H_3C-\underset{\underset{CH_3}{|}}{CH}-CH_2-CH_3$$

n-Pentane   Isopentane

and also dehydrogenation by cyclization are promoted. Ring compounds are produced from chain hydrocarbons.

IV Organic raw materials and large-scale products: Industrial processes

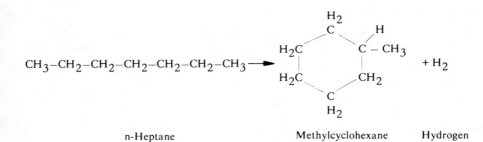

CH$_3$–CH$_2$–CH$_2$–CH$_2$–CH$_2$–CH$_2$–CH$_3$ $\longrightarrow$ + H$_2$

n-Heptane   Methylcyclohexane   Hydrogen

3.2.6. Aromatization

Aromatization is a special form of the reforming process. Toluene is obtained in this way from methylcyclohexane:

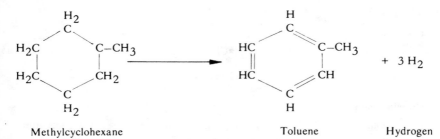

       + 3 H$_2$

Methylcyclohexane   Toluene   Hydrogen

A similar course is taken by an aromatization process that starts from n-hexane and produces benzene via cyclohexane:

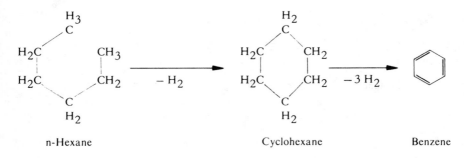

n-Hexane   Cyclohexane   Benzene

Aromatization of isobutylene gives p-xylene.

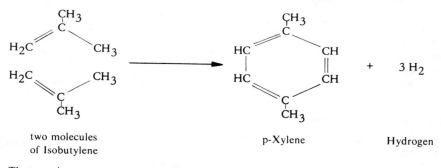

two molecules   p-Xylene   Hydrogen
of Isobutylene

The most important secondary reactions used in industry to synthesize end products directly, or through further intermediates, from the petrochemical primary products are shown in IV, 3 to 9.

IV Organic raw materials and large-scale products: Industrial processes

4. Self-evaluation questions

1 Which substances are petrochemical raw materials?

    A coal
    B natural gas
    C benzene
    D ethylene

2 Which are important constituents of crude oil?

    A elemental nitrogen
    B saturated hydrocarbons
    C elemental hydrogen
    D aromatic hydrocarbons

3 Which statements on petrochemistry are correct?

    A Petrochemistry is concerned with all industrial processes for which hydrocarbons are starting materials.
    B In petrochemistry low-reactivity, saturated hydrocarbons are produced from reactive molecules.
    C The hydrocarbons manufactured in the petrochemical industry by the hydrogenation of coal are inexpensive in comparison with the hydrocarbons deriving from crude oil and natural gas
    D Petroleum and natural gas are important in petrochemical processes owing to the high hydrogen content of the hydrocarbons.

4 In which three areas of the world are large quantities of crude oil and/or natural gas extracted?

    A Japan
    B Iraq
    C USSR
    D Saudi Arabia
    E Alaska
    F Morocco

5 The quantity of crude oil produced by a borehole in a certain time depends on

    A the viscosity of the crude oil
    B the size of the oil field
    C the size of the borehole
    D the nature of the rock in which the oil is located

6 Which statements on oil production are correct?

A   With skilled operation, boreholes can be opened and closed at will, like water taps.
B   Oilfields remain productive only if production is continuous.
C   The oil reserves currently known 1981 will last at least another 50 years at the present rate of extraction.
D   Most oil is obtained from boreholes below water level.

7 Which substances are main or subsidiary constituents of crude natural gas?

A   propane
B   methane
C   benzene
D   hydrogen

8 Which statements on natural gas are correct?

A   Natural gas always contains methane.
B   A troublesome constituent of natural gas is hydrogen sulphide.
C   Natural gas frequently occurs in the vicinity of oilfields.
D   Natural gas is often a dissolved constituent of petroleum.

9 The petrochemical processes and basic operation include

A   production of oil
B   distillation
C   desulphurization
D   cracking

10 Which fractions occur at the extraction points 1, 2 and 3 on the oil distillation column?

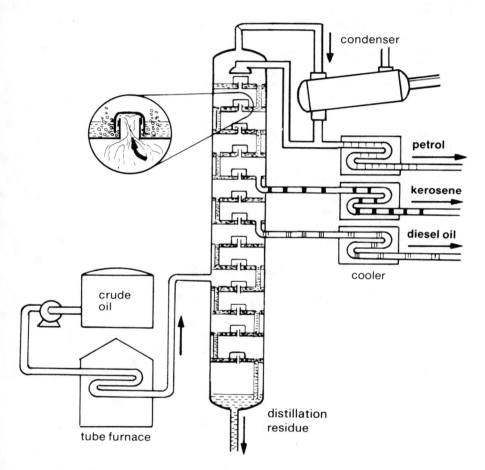

A 1 paraffin oil (kerosene), 2 gasoline, 3 diesel oil
B 1 gasoline, 2 diesel oil, 3 paraffin oil (kerosene)
C 1 paraffin oil (kerosene), 2 diesel oil, 3 gasoline
D 1 gasoline, 2 paraffin oil (kerosene), 3 diesel oil

11 Which statements on crude oil are correct?

A The majority of crude oil is employed in the chemical industry as a raw material.
B Crude oil is extracted from the oilfield with compressed air.
C In Western Europe and the USA the vast majority of electricity is produced in power stations fired with oil or natural gas.
D The great majority of oil reserves currently known are in western Asia.

12 The octane rating of gasoline provides information on

A its percentage proportion of octane
B its anti-knock properties
C its ignition properties
D its percentage content of iso-octane

13 Which substances are petrochemical primary chemicals?

A hydrogen
B nitrogen
C petroleum
D ethylene

14 The reaction

$$CH_3-CH_2-CH_2-CH_2-CH_2-CH_2-CH_3 \longrightarrow \underset{\bigcirc}{}-CH_3 + 4\,H_2$$

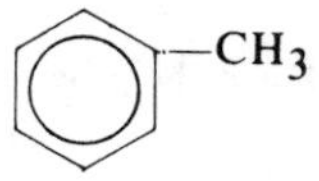

is a process of

A dehydrogenation
B aromatization
C cyclization
D isomerization

15 Cracking processes are marked by the fact that

A short-chain fractions are created from long-chain unsaturated hydrocarbons
B at relatively low temperatures (about 400 to 600°C) isomerization and cyclization frequently occur after the cleavage of large molecules
C larger cleavage products form as the temperature rises
D cleavage reactions predominate at relatively high temperatures

16 In the distillation of crude oil, separation into the individual constituents is governed by their

A density
B boiling point
C reactivity
D combustibility

IV Organic raw materials and large-scale products: Industrial processes

17 The essential value of hydrocarbons — in comparison with coal — resides in their

    A   carbon content
    B   hydrogen content
    C   chain-like structure
    D   cyclic structure

18 The anti-knock property of gasoline can be improved by the addition of

    A   alcohol
    B   tetraethyl lead
    C   branched paraffins
    D   aromatics

19 The main products of high-temperature pyrolysis (HTP) are

    A   ethylene and acetylene
    B   light petroleum spirit and hydrogen
    C   methane and ethane
    D   polyethylene and polypropylene

20 Which are the gaseous substances often added during cracking?

    A   hydrogen
    B   oxygen
    C   nitrogen
    D   steam

2 Coal as a Raw Material

1. The formation of coal

Coal is a brown to black "mineral", which is found in the ground. The collective term coals covers the various varieties, anthracite, hard coal and brown coal. Peat is the preliminary stage in the formation of coal.

Fig. 1:   Rotary bucket excavator for open-cast mining of coal*

*Acknowledgment: from the slides series of the Rheinische Braunkohlenwerke AG, Köln

Table 1: Geologic Time Chart

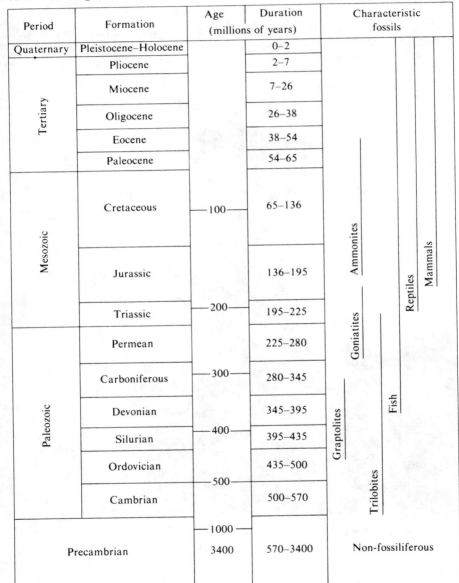

Period	Formation	Age (millions of years)	Duration (millions of years)	Characteristic fossils
Quaternary	Pleistocene–Holocene		0–2	
Tertiary	Pliocene		2–7	
Tertiary	Miocene		7–26	
Tertiary	Oligocene		26–38	
Tertiary	Eocene		38–54	
Tertiary	Paleocene		54–65	
Mesozoic	Cretaceous	100	65–136	
Mesozoic	Jurassic		136–195	Ammonites, Goniatites, Graptolites, Trilobites, Fish, Reptiles, Mammals
Mesozoic	Triassic	200	195–225	
Paleozoic	Permean		225–280	
Paleozoic	Carboniferous	300	280–345	
Paleozoic	Devonian		345–395	
Paleozoic	Silurian	400	395–435	
Paleozoic	Ordovician		435–500	
Paleozoic	Cambrian	500	500–570	
	Precambrian	1000 / 3400	570–3400	Non-fossiliferous

Hard coal is older geologically (estimated to be 300 to 400 million years old). It is deposited deep in the ground and it is therefore harder. Brown coal is 40 to 60 million years old and in some cases is found only a few metres below the earth's surface. Hard coal is now accessible at depths of 1200 m (3900 ft) and is excavated at these depths by mining at the coal-face. In the extraction of brown coal, the soil overburden is first removed; the brown coal can then be excavated directly (open-cast mining).

Hard coal and brown coal also differ in their composition (see page 11).

Coals are formed exclusively from plants. The hard coal forest in primeval times consisted of lepidodendra, horse-tails, ferns and club-moss plants of gigantic dimensions. Brown coals were formed later from plants such as are still found to some extent in the tropics. The forests were destroyed as the result of flooding, drifting and dislocations of the earth's crust and have been covered by earth. The vegetable matter has undergone a complete chemical transformation in the absence of air. The constituents of the wood, cellulose, lignin, resins and waxes, are thus no longer detectable in coal. New growth of plants developed on the superimposed soil strata and these were again destroyed by natural disasters. In this way, over long periods of time, there were formed seams of coal, interrupted by soil strata. In the formation of coal, carbonization as it is called, the successive stages were peat, brown coal, hard coal and finally anthracite.

In the USA and the Republic of South Africa, the hard coal deposits are so close to the surface that they can be worked by open-cast methods. In Europe, however, hard coal is mined underground at depths of over 3000 m (9800 ft).

2. The economic importance of coal

2.1. Coal reserves and coal mining

World coal reserves are estimated at present at 7778 billion t of hard coal and 4699 billion t of brown coal.

As a result of the exhaustion in measurable time of deposits of crude oil and natural gas and the increase in prices stemming from this, output of coal will increase considerably during the coming years. Based on the annual consumption in 1980 the reserves of hard coal known at present would last for nearly another two thousand years.

Figure 2a: Distribution of the geological reserves of hard coal, world reserves
             7,778 billion t.

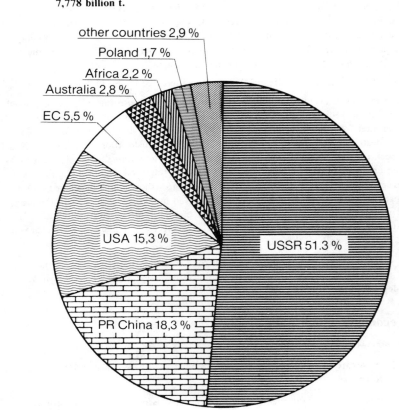

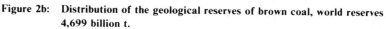

Figure 2b: Distribution of the geological reserves of brown coal, world reserves
4,699 billion t.

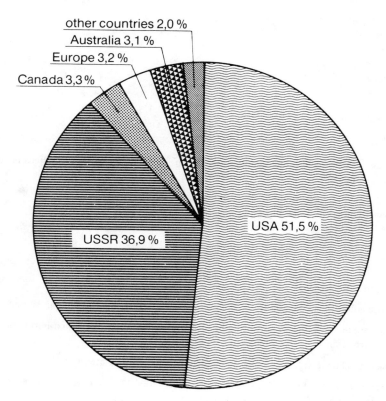

IV Organic raw materials and large-scale products: Industrial processes

Table 2a: Hard coal production in a number of important countries in millions of tonnes in 1978.

Federal Republic of Germany	90.1
Great Britain	121.7
Poland	192.6
USSR	557.5
PR China	600.0
USA	574.2
Africa	96.4
Australia	81.7
World	2632.2

Table 2b: Brown coal production in a number of important countries in millions of tonnes in 1978.

Federal Republic of Germany	123.6
German Democratic Republic	253.3
USSR	159.6
USA	26.2
World	908.6

2.2. Utilization of coal

Coal is combustible. The quality of fuels is assessed on their heat-producing value.

By heat-producing value we mean the heat of combusion of the fuel when burnt completely to give carbon dioxide, CO_2, and water, H_2O.

The heat-producing value is quoted in coal equivalents, abbreviated to CE.

1 CE is the quantity of heat liberated when 1 kilogram (kg) (2.2 lb) of hard coal of a defined quality is burnt completely to give carbon dioxide; it corresponds to 29,288 kJ

Carbon	Oxygen	Carbon dioxide

$$C \; + \; O_2 \longrightarrow CO_2 \quad ; \; \Delta H = -393.7 \text{ kJ/mol*}$$

4.19 kilojoule (3.97 Btu) is the quantity of heat required to raise the temperature of 1 kg (2.2 lb) of water by one degree centigrade (1.8 degrees Fahrenheit).

The heat-producing value depends on the C and H content of the coal or fuel.

The heat-producing value quoted for dry, ash-free hard coal is about 33,500 kJ/kg (14,400 Btu/lb), or 1.185 CE, while for brown coal the figure is 25,000 kJ/kg (10,760 Btu/lb) or 0.86 CE.

---

* The value of ΔH refers to the number of moles represented in the chemical equation.

The heat-producing value of brown coal is thus on average 25% lower than that of hard coal.

Originally, coal was used only for combustion to produce heat and other forms of energy (electricity or steam). When the coke obtainable from coal became accepted as the most economic reducing agent in the manufacture of iron in blast furnaces by reducing oxidic iron ores, coal chemistry developed at the same time, because the products obtained together with the coke, gases and tar, had to be utilized. Coal chemistry also includes processes in which synthesis gases and motor fuels are obtained from coal or coke.

Schedule 1:

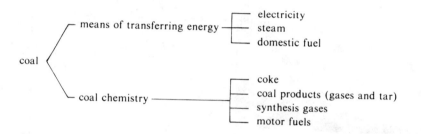

As sources of energy, oil products and natural gas have come into competition with coal. The structural change in the use of energy is shown by the fact that as late as 1950 about 90% of energy supplies were based on coal and this proportion declined to 43% by 1968. Since the first oil crisis in 1973, many countries are to an increased extent returning to the use of coal as chemical feedstock and primary energy source.

IV Organic raw materials and large-scale products: Industrial processes

Figure 3: Share of primary energy sources in world energy consumption*

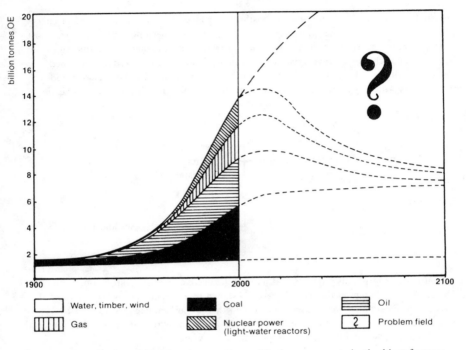

Water, timber, wind	Coal		Oil
Gas	Nuclear power (light-water reactors)		Problem field

Although the production of iron has increased steadily, improvements in the blast furnace process have continually reduced the quantity of coke required per tonne of iron. In the smelting of iron ores, coal is used as a reducing agent. The following chemical reaction takes place in the hottest zone at about 1600°C (2900°F):

$$Fe_2O_3 \quad + \quad 3\,C \quad \longrightarrow \quad 2\,Fe \quad + \quad 3\,CO$$

Iron oxide   carbon   iron   carbon monoxide

New techniques of iron manufacture have made it possible to replace coke as a reducing agent by, for example, methane from natural gas deposits or crude oil.

Coal chemistry is challenged by petrochemistry. The most important primary compound in organic chemistry, ethylene, cannot be produced profitably, or in the quantity required, from coal as a raw material. Aromatic compounds, a further group of organic compounds

* On the ordinate are plotted the energy values in oil units, 1 oil-tonne equivalent = 1.45 coal-tonne equivalents = 10,150 kcal = 42,498 kJ = 40,321 Btu.

of industrial importance, which were formerly accessible only from coal products, are now obtained mainly from crude oil products. These two examples show that the economic importance of petrochemistry at present far exceeds that of coal chemistry (see IV,1). We must wait and see what course future developments will take.

3. Coal chemistry

3.1. Composition of the various kinds of coal

Besides carbon, the major constituent, coal contains the following elements: C, H, O, N and S.

Table 3:   The quantitative chemical composition of various kinds of coal

Type of coal	% C	% H	% O	% N	% S
Peat	up to 60	5–8	30–45	1–4	0.1–2
Brown coal	60–75	5–8	20–30	0.5–1.5	0.5–2
Hard coal	75–90	2–5	10–2	up to 1	0,5–2
Anthracite	>90	<2	<2	<1	0.5–2

Carbonization, the transition from peat to hard coal, has been associated with an increase in the C content and a decrease in the H and O contents.

Converting the above percentages into the relative numbers of atoms of the elements gives an indication of the compounds present in various kinds of coal.

A high-grade hard coal, rich coal, contains relatively.

1,000 C-atoms

680 H-atoms

32 O-atoms

18 N-atoms

4 S-atoms

In organic chemistry such a high proportion of C-atoms relative to H-atoms is only found in aromatic compounds.

			C	: H
For example:	benzene	C_6H_6	1	: 1
	naphthalene	$C_{10}H_8$	1.25	: 1
	anthracene	$C_{14}H_{10}$	1.4	: 1
	pyrene	$C_{16}H_{10}$	1.6	: 1
	coronene	$C_{24}H_{12}$	2	: 1

IV Organic raw materials and large-scale products: Industrial processes

The two-dimensional structural formulae clearly show these relationships and indicate the graphite structure of the coal.

Condensed aromatic hydrocarbons and their carbon/hydrogen relationship

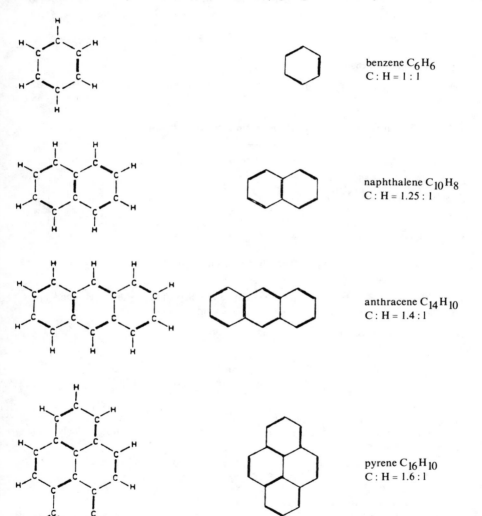

benzene C_6H_6
C : H = 1 : 1

naphthalene $C_{10}H_8$
C : H = 1.25 : 1

anthracene $C_{14}H_{10}$
C : H = 1.4 : 1

pyrene $C_{16}H_{10}$
C : H = 1.6 : 1

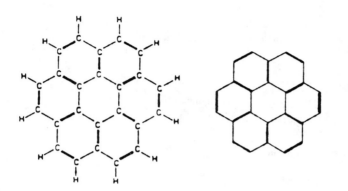

coronene $C_{24}H_{12}$
$C : H = 2 : 1$

Analyses carried out by physico-chemical methods have shown that 75% of the carbon is in the form of aromatic compounds and 25% in the form of aliphatic and alicyclic compounds.

Coal thus consists predominantly of aromatic hydrocarbons. Very large molecules are present, since a value of 3,000 is regarded as probable for the average molar mass. A certain proportion of alkylated aromatics is also present, since CH_3 side-chains have also been identified, for example toluene

Half the oxygen belongs to phenolic OH groups, for example phenol

Image of phenol structure with —OH

The other half can be assigned to carbonyl groups

$$R-\overset{\|}{\underset{O}{C}}-H \, ,$$

carboxyl groups $$R-\overset{\|}{\underset{O}{C}}-OH$$

and ether groups $$R-\overset{H}{\underset{H}{C}}-O-\overset{H}{\underset{H}{C}}-R$$

IV Organic raw materials and large-scale products: Industrial processes

Sulphur and nitrogen are linked in heterocyclic structures in very large molecules. Thiophen may be mentioned as an example of a simple sulphur-containing hetero-aromatic compound:

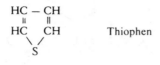
Thiophen

3.2. Conversion of coal by chemical means

3.2.1. Coal as a chemical raw material

Coal chemistry is concerned with utilization of the products obtained in degasifying, gasifying and liquefying coal (coal products), their manufacture in a pure state, improving their quality and their use in chemical syntheses. Coal or coke can also be employed as such as a reactant in chemical syntheses. Because its major constituent is carbon, coal is a valuable source of raw materials for organic products. The inorganic constituents contained in coal, particularly the sulphur, are also utilized for chemical purposes.

In USA after the First World War and in Europe after the Second World War, crude oil replaced coal as the most important source of raw materials for organic chemical compounds. Although the basic structures of aliphatic and aromatic hydrocarbon compounds are present in coal, they are all deficient in hydrogen, that is to say they do not have the –CH$_2$– fragments that form the structure of saturated hydrocarbon compounds. This is the serious disadvantage of coal compared with crude oil as a source of raw materials for the chemical industry.

A chemical industry based on coal must first hydrogenate the carbon (see IV,1, page 4).

However, in spite of these disadvantages, it has become evident recently that coal is once more a source of considerable interest for chemists. The polymer structure present in coal is of particular interest for the production of certain materials by means of specific chemical conversion processes.

In addition, coal is acquiring an increasing importance, beyond the stage of research projects, as a starting material for the microbiological production of protein.

The objective of making coal once again a product of interest to chemistry means utilizing and to some extent retaining the basic chemical structure of coal in order to convert it into a liquid, or at least soluble, intermediate product that can be processed by a chemical refinery.

In principle, three different coal conversion processes can be distinguished:

1. degasification or coking;
2. gasification;
3. liquefaction by hydrogenation.

3.2.2. Degasification or coking of coal

Coal is heated externally at high temperatures in closed ovens, in the absence of oxygen. Products volatile at the reaction temperature escape, leaving coke. This thermal up-grading can be divided into three temperature zones:

500°C to 600°C: (932°F to 1112°F)	low-temperature coking or carbonization. This is the main process employed for brown coal, but is not very important industrially.
700°C to 800°C: (1260°F to 1472°F)	medium-temperature coking. Seldom used in industry.
1000°C to 1200°C: (1832°F to 2192°F)	high-temperature coking. This is the main process employed for hard coal; this process provides the coke required as a chemical raw material for the manufacture of iron, and coal products.

The coking of hard coal at 1000°C to 1200°C (1832°F to 2192°F) produces compounds that were already present in the coal or compounds formed by high temperature re-arrangement and cleavage reactions.

The volatile substances are collected, cooled and subjected to a complicated process of separation and purification.

1 t of coal gives 314 m³ (11,090 ft³) or 320 kg (706 lb) of volatile products.

The composition of the latter is:

> 64% by volume or 56% by weight of crude gas
> 11% by volume or 12.5% by weight of tar vapours
> 25% by volume or 31.5% by weight of water vapour
>
> 100% by volume   100% by weight

The three products of coking: crude gas, coking liquor and tar are worked up separately to obtain the coal products they contain.

IV Organic raw materials and large-scale products: Industrial processes

Table 4: The composition of crude gas

H_2	54% by volume
CO	6% by volume
CO_2	3% by volume
N_2	7% by volume
CH_4	27% by volume
H_2S	8 g/m³ (499 lb/million ft³)
Other hydrocarbons	3% by volume
Benzene, toluene and xylenes	35 g/m³ (2184 lb/million ft³)
Naphthalene	10 g/m³ (399 lb/million ft³)

The vapours of the aromatic compounds are extracted from the crude gas in "benzol wash oil". After being separated from the wash oil, the liquid mixture of aromatic substances must be subjected to intensive purification and freed from S and N compounds. Subsequent distillation gives the following compounds:

Table 5: The composition of the mixture of aromatic compounds

Boiling range	Fraction	% by volume
up to 65°C (149°F)	fore-runnings	5
65°C to 93°C (149°F to 199°F)	benzene	63
93°C to 123°C (199°F to 253°F)	toluene	17
123°C to 150°C (253°F to 302°F)	xylenes	8
above 150°C (302°F)	residue	7

The distillation residue contains naphthalene.

In order to remove hydrogen sulphide, H_2S, the gas is passed over a contact mass spread out on a lattice framework, which oxidizes the H_2S and retains the sulphur formed. The sulphur is extracted from the contact mass.

The greater part of the naphthalene present in the crude gas has already been removed when washing the aromatic fraction. But the residual naphthalene must also be removed from the gas since its subsequent deposition in pipelines gives rise to a risk of blocking. Naphthalene is removed from the gas with the aid of naphthalene wash oil (obtained in the tar distillation).

Coking liquor contains 3 g (0.10 oz) of phenols and homologues and also 8 to 15 g (0.26 to 0.50 oz) of ammonia per litre. These dissolved compounds must be removed from the effluent, firstly because they are poisonous and secondly because they are valuable products for further processing.

Phenols are extracted from the liquor. After removing the extraction solvent by evaporation, a mixture of phenols of the following composition is obtained:

Phenol	52% by weight
o-Cresol	9% by weight
m-Cresol and p-cresol	16% by weight
Cresol/xylenol mixture	8% by weight
Higher phenols	15% by weight

Ammonia, combined as ammonium carbonate, can be removed from the effluent by adding milk of lime, $Ca(OH)_2$, and heating:

$$(NH_4)_2CO_3 \quad + \quad Ca(OH)_2 \longrightarrow \quad Ca\,CO_3 \quad + \quad 2\,H_2O + 2\,NH_3$$

Ammonium carbonate milk of lime

            calcium carbonate water ammonia

3.2.2.1. Coal tar

Coal tar is a lustrous, black, viscous liquid. Its density at room temperature is 1.10 g/cm³ to 1.26 g/cm³ (68.7 to 78.6 lb/ft³).

The tar is an extremely complex mixture of mainly aromatic hydrocarbons and also O, N and S compounds, most of which are also aromatic. Its average molecular mass can vary between 100 and 190 g/mol (0.22 and 0.42 lb/mol). It is estimated that about 10,000 compounds are present in the tar; so far it has been possible to identify 480 of them. Coal tar used to be the sole source of organic intermediate products, which constituted the raw material for the first synthetic dyestuffs and drugs. This is the historical reason why, even today, organic dyestuffs are listed in statistics as coal tar dyestuffs.

The following compounds, which are present to the extent of less than 1% by weight, are also obtained on an industrial scale:

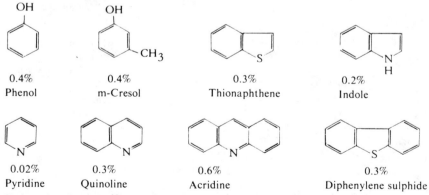

0.4%	0.4%	0.3%	0.2%
Phenol	m-Cresol	Thionaphthene	Indole
0.02%	0.3%	0.6%	0.3%
Pyridine	Quinoline	Acridine	Diphenylene sulphide

Coal tar is split up into its components by fractional distillation. The isolation of coal products has been put on an economic basis by confining distillation to only a few large central plants operating on a continuous basis.

Schedule 2:   The products of coal tar distillation

up to 180°C (356°F)	Light oil less than 3%	Mixture of aromatic compounds	Benzene, toluene and xylenes
		Crude bases	Pyridine and picoline
		Indene and coumarone	Synthetic resins
up to 210°C (410°F)	Carbolic oil less than 3%	Mixture of phenols	Phenol, cresols and xylenols
		N bases	Quinoline and isoquinoline
up to 230°C (446°F)	Naphthalene oil 10 to 12%	Naphthalene	Pure naphthalene
		Naphthalene oil	Fuel oil, benzol wash oil and impregnating oil
up to 290°C (554°F)	Wash oil 7 to 8%	Mixture consisting of	Methylnaphthalenes, diphenyl, indole, acenaphthene, fluorene and other products
up to 400°C (752°F)	Anthracene oil 20 to 28%	Crude anthracene	Pure anthracene
		Anthracene oil	Anthracene 40%
			Phenanthrene 35% Carbazole 20%
			Fluoranthrene Pyrene Naphthalene wash oil Impregnating oil Soot oil
above 400°C (752°F)	Pitch 50 to 55%	Road tar Roofing felt tar Proofing materials for buildings Briquetting pitch Electrode pitch (Söderberg-electrodes)	

Depending on their boiling points, the individual fractions contain further compounds, which can be isolated by crystallization, extraction or other separation processes.

3.2.2.2. Utilization of coal products

Coke oven gas, the constituents of which are hydrogen, methane and carbon monoxide, can be used not only as a heating gas, but also for the manufacture of synthesis gas. The separation of the small quantities of ethylene and propylene in coke oven gas is not economic.

Sulphur is converted into sulphuric acid (see II,5), which is neutralized with the **ammonia** also produced to give **ammonium sulphate,** a fertilizer (see II,4).

The aromatic hydrocarbons **benzene, toluene** and the **xylenes** are obtained not only in purifying the gas but also in the tar distillation. After purification, the mixture can be added to motor fuels to increase their octane number. These products are also produced in a pure state, however. Benzene is a solvent and an intermediate for the synthesis of chlorobenzene, nitrobenzene, benzenesulphonic acids and alkylbenzenes. o-Xylene is important industrially for the manufacture of phthalic anhydride and p-xylene for the manufacture of terephthalic acid. However, all these hydrocarbons are also obtained, to an increasing extent, from crude oil fractions (reformates).

Coumarone and **indene,** which, being benzene homologues, occur with the latter, are polymerized to give thermoplastic coumarone-indene resins.

Pyridine and its homologues are used in organic synthesis to produce herbicides, disinfectants, pharmaceuticals and vitamin preparations. The annual production of pyridine bases, from coal products alone, in West Germany is about 1,000 t.

Phenol and its methyl homologues, the cresols and xylenols, are very important industrially. This is why phenol is also produced synthetically. Coke-oven phenol, extracted from tar and coke-oven effluents, forms only about 10% (13,000 t/year) of total production in West Germany.

IV Organic raw materials and large-scale products: Industrial processes

Schedule 3: The reaction products of phenol

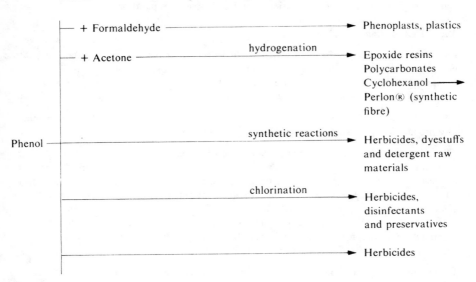

Naphthalene is the most important coal product in terms of quantity and is of great industrial importance. A few years ago, coal chemistry was the sole source of supply for naphthalene; nowadays appreciable quantities are obtained from reformed oil fractions.

Schedule 4: The reaction products of naphthalene

PA is nowadays manufactured half from naphthalene and half from o-xylene.

World production of anthracene is of the order of 30,000 t/year. Of this total, about half is produced in the Federal Republic of Germany.

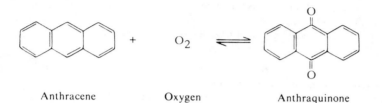

| Anthracene | Oxygen | Anthraquinone |

This is the basis of the anthraquinone dyestuffs, which are particularly important for dyeing synthetic fibres (see IV,9).

3.2.2.3. Economic problems concerning the coking of coal

During the last 15 years, world production of coal tar fluctuated by approx. 16 million tonnes per annum.

Since the main market for coke is for smelting iron ores, the future of coal tar production depends on the demand for coke in the iron and steel industry.

According to present forecasts, world consumption of coke will rise from around 400 million tonnes in 1978 to about 600 million tonnes per annum in 1990. The annual worldwide quantity of tar obtained would rise from the present 16 million to 24 million t. In Western Europe, too, a renewed increase in coke consumption is again expected during the coming years, following the drop in coke production to about 100 million t/a.

About 45 kg (99 lb) of tar are produced per tonne (2205 lb) of coke. The naphthalene content of tar is about 10% and most of it will be extracted.

3.2.2.4. Coke as a reducing agent

Coke is the most economic reducing agent for industrial processes. Depending on the reaction temperature, an atom of carbon can link itself to one or two O atoms.

Up to 900°C (1652°F), the following reaction predominates:

$$C + O_2 \longrightarrow CO_2 \qquad \Delta H = -395 \text{ kJ/mol}$$
$$\text{carbon dioxide}$$

Above 900°C (1652°F), the following reaction predominates:

$$2\,C + O_2 \longrightarrow 2\,CO \qquad \Delta H = -113 \text{ kJ/mol}$$
$$\text{carbon monoxide}$$

In most industrial reduction reactions with coke, carbon monoxide is formed predominantly. There are two reasons for this:

1. In order to achieve reaction rates (conversion in a given time) of an economic speed, the reactions must be carried out at high temperatures. Since less heat energy is liberated in the formation of CO than in the formation of CO_2, the reaction leading to CO is favoured.

2. The reduction reactions are always carried out with an excess of carbon, that is to say the quantity of carbon available is so large that the oxygen is removed almost completely by way of the compound CO, which is poorer in oxygen.

3.2.3.   Gasification of coal

3.2.3.1. Water-gas and producer gas

By gasification we mean the complete conversion of the solid substance of coal to gaseous compounds. While in the coking process only degasification of the coal takes place and coke is left as the residue, in gasification the coal is converted completely into gases, the only residue being mineral constituents, ie inorganic compounds, in the form of ash or clinker.

It is preferable to employ coke in gasification so as to produce gas mixtures that are more uniform and not contaminated by tarry constituents.

Gasification of coke is carried out by passing air and steam alternately in measured quantities over incandescent coke. The products then formed are:

> **producer gas,** main constituents N_2 and CO,
> and
> **water-gas,** main constituents CO and H_2.

These are used, individually or as a mixture, as synthesis gases for the manufacture of methanol or in the oxo reaction.

The manufacture of producer gas:

$$4\,N_2 + O_2 \quad + \quad 2\,C \quad \longrightarrow \quad 4\,N_2 + \ 2\,CO \qquad\qquad \Delta H = -226\ kJ/mol$$

$$\underbrace{}_{\text{Air}} \qquad\qquad \underset{\text{coke}}{2\,C} \qquad\qquad \underbrace{}_{\text{producer gas}}$$

The heat liberated brings the coke, which at the start was only at a red heat, up to a white heat. The supply of air is then interrupted and steam is passed over the coke, water-gas being formed:

$$H_2O \quad + \quad C \quad \longrightarrow \quad H_2 + CO \qquad\qquad \Delta H = +136\ kJ/mol$$

$$\underset{\text{Water}}{H_2O} \qquad\qquad \underset{\text{coke}}{C} \qquad\qquad \underbrace{}_{\text{water-gas}}$$

This reaction consumes heat, so that the white-hot coke is cooled. The coke is heated up again by restarting the supply of air and the formation of producer gas.

Alternate operation, blowing the coke hot by means of air and blowing it cold by means of steam, generates producer gas and water-gas alternately. Both gas mixtures must be freed from impurities, particularly sulphur compounds.

3.2.3.2. The use of producer gas and water-gas as synthesis gases

After adjustment to the correct composition, water-gas can be employed in the following syntheses:

a) The methanol synthesis:

$$CO \quad + \quad 2 H_2 \quad \longrightarrow \quad CH_3OH \qquad \text{(See IV.3)}$$

Carbon monoxide   hydrogen   methanol

$$\Delta H = -92 \text{ kJ/mol}$$

b) The oxo synthesis:

$$C_3H_6 \quad + \quad CO \quad + \quad H_2 \quad \longrightarrow \quad C_3H_7 - C\overset{\displaystyle H}{\underset{\displaystyle O}{}}$$

Propylene   carbon monoxide   hydrogen   butyraldehyde

$$\Delta H = 118 - 147 \text{ kJ/mol}$$

c) The Fischer–Tropsch synthesis:

$$n \, CO \quad + \quad (2n+1) \, H_2 \quad \longrightarrow \quad C_nH_{2n+2} \quad + \quad n \, H_2O$$

Carbon monoxide   hydrogen   saturated   water
                                                    hydrocarbons

The mixture of liquid saturated alkanes C_nH_{2n+2} can be used as a motor fuel.

Coal gasification plants are now in operation in India, Thailand, South Africa and Zambia, in Turkey, Greece and Jugoslavia, as well as in the Comecon area in Bulgaria, in the GDR and the CSR. Three other gasification plants in South Africa, China and Brazil are either under construction or have been ordered.

IV Organic raw materials and large-scale products: Industrial processes

After mixing, water-gas and producer gas are important as a synthesis gas for:

d) The ammonia synthesis:

$$N_2 \quad + \quad 3\,H_2 \quad \longrightarrow \quad 2\,NH_3 \quad ; \quad \Delta H = -92.1 \text{ kJ/mol (see II,3)}.$$

Nitrogen   hydrogen   ammonia

In the mixture of gases containing components N_2, H_2 and CO, the carbon monoxide, CO, must be removed and the H_2 content increased. Producer gas and water-gas are therefore passed together with steam over a catalyst containing Fe_2O_3, when CO and H_2O react with one another as shown below:

$$CO \quad + H_2O \quad \longrightarrow \quad CO_2 \quad + \quad H_2 \quad ; \quad \Delta H = -42 \text{ kJ/mol}$$

Carbon monoxide   water   carbon   hydrogen
                                             dioxide

Carbon dioxide, CO_2, is washed out of the gas mixture with water under pressure, thus providing a synthesis gas consisting of N_2 and an increasing content of H_2.

3.2.3.3. Economic problems

Since the oil shortage, coal gasification is increasing in importance. The USA and the Federal Republic of Germany are working intensively on joint research projects. South Africa, which has virtually no oil reserves, but instead possesses enormous reserves of easily workable coal, has already put into operation a number of large-scale plants for coal gasification.

The production of synthesis gas from liquid or gaseous hydrocarbons (oil or natural gas) is a simpler process from the point of view of chemical engineering, and the capital and operating costs are lower for this process than for synthesis gas plants using coke as a raw material (see II,1). It is considerably cheaper to transport oil and natural gas through pipelines than to transport coal. In coke plants purification of the gas is more expensive and, in addition, ash and clinker have to be removed.

Hydrocarbons obtained from oil and natural gas contain more hydrogen than those obtained from coal, so that the syntheses gases produced have a higher content of hydrogen.

3.2.4. Liquefaction of coal (coal hydrogenation)

Solid coal is converted into liquid oil products by hydrogenation. This reaction can be carried out in the presence of catalysts at temperatures of $500°C$ ($932°F$) and at a hydrogen pressure of less than 245 bars ($3550\ lb/in^2$). The reaction takes place in two stages:

1. the high-molecular compounds in the coal undergo cleavage;
2. the cleavage products are hydrogenated by hydrogen.

Increasing the hydrogen content and the cleavage of the high-molecular compounds into smaller molecules produces mixtures of saturated, liquid hydrocarbons with a boiling range of 30 to $250°C$ (86 to $432°F$) which are suitable for motor fuels.

The catalytic high-pressure hydrogenation of coal by the Bergius process was carried out in Germany during the Second World War on a large industrial scale.

In the USA, and recently also in West Germany, experiments are being carried out in small plants to continue the development of coal hydrogenation on an industrial scale, in some cases using new processes. The only country in which engine fuels are being manufactured from coal on an industrial scale is the Republic of South Africa.

Large-scale production of coal-based fuel is planned at the Sasol II and Sasol III plants of the South African Coal Oil and Gas Corp. (Sasol) in Secunda, 70 kilometers (43 miles) south of Johannesburg. It is estimated that the two plants together will have a production capacity of around 6 million tonnes of coal oil per annum.

In the USA, where high-grade coal is available from easily mined seams, the hydrogenation of coal could be completely economic and competitive. The realization that crude oil reserves will be exhausted in the foreseeable future has lent an added interest to the hydrogenation of coal.

4. Summary of the various ways in which coal is processed

Coal

Degasification

Coking 1,200 °C

Coke oven gas
- Synthesis gas — — — — Ammonia, methanol and oxo syntheses
- Sulphur — — — — Fertilizers (ammonium sulphate)
- Ammonia
- Benzene, toluene and xylenes — — — — Solvents, blending with petrol, chlorobenzene, nitrobenzene, alkylbenzenes, PA, terephthalic acid
- Naphthalene — — — — PA, naphthols and solvents

Tar
- Naphthalene — — — — see above
- Benzene, toluene and xylenes — — — — see above
- Phenol, cresols and xylenols — — — — Plastics, in herbicides, dyestuffs and detergents
- N-heterocyclic compounds — — — — Organic syntheses
- Anthracene — — — — Anthraquinone dyestuffs
- Pitch — — — — Road tar, protection of buildings and electrode pitch

Coke
- Reducing agent — — — — Manufacture of pig iron, phosphorus and calcium carbide

Gasification

Producer gas $CO = N_2$
Water-gas $CO + H_2$
} Synthesis gases — — — — Ammonia, methanol and oxo syntheses

Liquefaction
Hydrogenation

Mixture of saturated hydrocarbons — Fuels for carburettor and diesel engines

5. Self-evaluation questions

21   Which statements on coal are correct?

A Peat represents, the preliminary stage in the formation of lignite and coal.
B Coal is derived from vegetable and animal deposits.
C Hard coal is the final stage in its creation process.
D Coke and benzene are two coal products.

22   The heat-producing value is

A a measure of the quality of a fuel
B a value for heat of combustion
C governed by the carbon and hydrogen content of the fuel
D higher in coal than in lignite

23   In the coal-forming process the

A carbon content increases
B hydrogen content increases
C sulphur content increases
D oxygen content declines

24   Coking is

A a chemical decomposition reaction of coal
B a form of coal combustion
C a form of coal degasification
D the heating of coal in the absence of oxygen

25   Which statements about coal tar are correct?

A It is produced in the coking of coal.
B It contains about 10,000 different compounds.
C It is decomposed by extraction.
D It consists predominantly of aliphatic compounds.

26   Phenol and naphthalene are obtained from coal tar.
Which reaction products can in turn be obtained from these two substances?

A detergent raw materials
B plastics
C dyestuffs
D benzene

IV Organic raw materials and large-scale products: Industrial processes

27 In which industrial-scale process does coke play a part?

A pig iron production
B sulphuric acid production
C chlorine production
D nitric acid production

28 The production of producer gas is:

A a coke gasification process
B a process for producing a mixture of hydrogen and carbon monoxide
C an endothermic process
D associated with the production of CO_2

29 If a mixture of steam and combustion air enriched with oxygen is passed over lignite or coke, mixtures of gases are obtained. These are purified and act as starting materials for the production of

A ammonia
B methanol
C benzene
D hydrogen peroxide

30 Which product cannot be manufactured from carbon monoxide as shown?

A $CO + 2H_2 \longrightarrow CH_3OH$, methanol
B $CO + Cl_2 \longrightarrow COCl_2$, phosgene
C $CO + 2H_2O \longrightarrow CH_3OH + O_2$, methanol and oxygen

3 Hydrogenation (Methanol and Oxo Syntheses)

IV Organic raw materials and large-scale products: Industrial processes

1. Hydrogenation reactions

Hydrogenation is generally understood as the reaction of chemical compounds with hydrogen. Hydrogenations are always reduction reactions (see I, 6, page 13).

Some reactions have become important on a large industrial scale. These are:
1. hydrogenation of nitrogen, which produces ammonia (see II, 3);
2. hydrogenation of coal;
3. hydrogenation of fats, known as the hardening of fats;
4. hydrogenation of carbon monoxide, which produces methanol;
5. hydrogenation of carbon monoxide in the presence of unsaturated hydrocarbons, which is known as the oxo reaction.

Until the end of the Second World War the **hydrogenation of coal** had a very special importance in Germany. Its task was to supply the key products for organic chemistry from coal, which is available in the country in adequate quantities. Coal chemistry, as it is called, developed processes for isolating and synthesizing hydrocarbons in particular from coal. Now that the price of crude oil has increased so much, not least as a result of the foreseeable exhaustion of crude oil reserves, the importance of coal as a supplier of energy and as a raw material has increased considerably. Research projects to cope with this situation are being given special priority.

The hydrogenation of coal involves reactions between hard coal and brown coal or peat and wood on the one hand and hydrogen or hydrogen-containing compounds, such as water, on the other hand. These processes are carried out in the presence of catalysts, at elevated temperatures and under pressure and the reactions they are intended to promote can be summarized by the following overall equation:

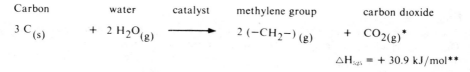

Carbon water catalyst methylene group carbon dioxide

$$3\ C_{(s)} + 2\ H_2O_{(g)} \longrightarrow 2\ (-CH_2-)_{(g)} + CO_{2(g)}{}^*$$

$$\triangle H_{525} = +30.9\ kJ/mol{}^{**}$$

The hydrogenation of fats makes it possible to convert cheap vegetable and animal fats containing unsaturated fatty acids into odourless and tasteless, hard fats, ie fats with higher melting points are obtained. These hardened fats are the starting materials for the manufacture of margarine and are raw materials for the manufacture of soap. The hydrogenation of fats is carried out between 125°C and 200°C (257°F and 392°F). The hydrogen pressure is 5 to 7 bars (72.5 to 102 lb/in²).

---

* It means (s) = solid; (l) = liquid; (g) = gas.
** The value of △H refers to the numbers of moles represented in the chemical equation.

IV Organic raw materials and large-scale products: Industrial processes

The following reaction takes place:

Unsaturated fatty acid

hydrogen

$$CH_3-(CH_2)_x-\underset{\underset{H}{|}}{C}=\underset{\underset{H}{|}}{C}-(CH_2)_x-\underset{\underset{O}{\|}}{C}-OR \ + \ H_2 \quad \xrightarrow{\text{catalyst}}$$

saturated fatty acid

$$CH_3-(CH_2)_x-CH_2-CH_2-(CH_2)_x-\underset{\underset{O}{\|}}{C}-OR \ ; \ \triangle H= \ -150.7 \ kJ/mol$$

It is common to all hydrogenation reactions that they are carried out at elevated temperatures under pressure and exclusively in the presence of catalysts. They are all reactions that produce heat (exothermic).

2. The methanol synthesis

2.1. Physical and chemical properties of methanol

Methanol is a colourless liquid with a typical alcoholic odour. It absorbs water readily from the moisture of the air (hygroscopic).

The chemical formula of methanol is $H-\underset{\underset{H}{|}}{\overset{\overset{H}{|}}{C}}-OH$ and its molar mass is 32 g/mol (32 lb/mol).

The characteristic features of this compound are the OH group and the methyl group. Organic compounds containing the OH group as a functional group are classified as alcohols. Another (older) name for methanol is, therefore, methyl alcohol. Because of its OH group and its short aliphatic hydrocarbon radical, the methyl group, methanol is closely related to water and, for example, is soluble in the latter in all proportions.

$$H_3C-OH \quad \text{Methyl alcohol}$$

$$H-OH \quad \text{Water}$$

However, the methyl group is responsible for the fact that methanol is also miscible with most organic solvents.

Methanol burns with a pale blue flame, giving water and carbon dioxide in accordance with the following equation:

| Methyl alcohol | atmospheric oxygen | | water | carbon dioxide |

$$2\ CH_3OH_{(l)} \quad + \quad 3\ O_{2\ (g)} \quad \longrightarrow \quad 4\ H_2O_{(l)} \quad + \quad 2\ CO_{2(g)}$$

$$\triangle H = -1457\ kJ/mol$$

Its flash point is 11°C (51.8°F).
The explosive limits in air are 5.5 to 44% by volume of methanol.
Its boiling point at 1.013 bars pressure is 64.7°C (148°F), its melting point is −97.9°C (−144.2°F) and its density at 20°C (68°F) is 0.7915 g/ml (49.4 lb/ft³).

Methyl alcohol is very poisonous. In contrast with ethyl alcohol, C_2H_5OH, it is only partially broken down in the body.

Formic acid, H−COOH, and formaldehyde $H-\underset{\underset{O}{\|}}{C}-H$ are formed.

Formaldehyde precipitates or hardens protein. The oxidation processes in the body are thus blocked. The retina of the eye is affected particularly, having a high oxygen demand, so that the first symptoms of poisoning make themselves apparent through impaired vision. 50 to 70 g (1.76 to 2.46 oz) of methanol constitute a fatal dose.

2.2. Its economic importance

Methanol is an important key product for the production of intermediates and end-products of a higher value. In particular, methanol is an indispensable solvent and extracting agent. About 50% of the methanol produced is converted into formaldehyde (see IV, 4, page 30). Large quantities are also used for the manufacture of methyl esters, especially dimethyl terephthalate, DMT, which is indispensable as an intermediate in the manufacture of polyesters (see IV, 8, page 23).

Methanol is used as an extracting agent in the crude oil industry for removing mercaptans, which are organic sulphur compounds, and for washing carbon dioxide out of synthesis gases.

Around 9.5 million t methanol were produced in the western world in 1977. Methanol production in the USA in 1979 amounted to 3.36 million t and in the FRG to 0.87 million t.

IV Organic raw materials and large-scale products: Industrial processes

Synthesis gas, which contains the two reactants carbon monoxide and hydrogen, is obtained at present from naphtha, ie crude oil hydrocarbons, and from natural gas, by means of a reaction in which steam is added.

In recent years, the raw materials used for the production of methanol in the West have been employed in the following proportions:

> 50% natural gas
> 25% light naphtha
> 10% fuel oil
> 10% coke-oven gas
> 5% coke.

2.3. Industrial manufacture of methanol by synthesis under high pressure

2.3.1. The chemistry of the process

The principle reaction in the synthesis of methanol is the hydrogenation of carbon monoxide in the presence of the mixed catalyst zinc oxide/chromium oxide, ZnO/Cr_2O_3. This process takes place in accordance with the following equation:

Carbon monoxide		hydrogen			methanol	
				catalyst		
$CO_{(g)}$	$+$	$2 H_{2 (g)}$		$\longrightarrow$	$CH_3OH_{(l)}$; $\Delta H = -105.0$ kJ/mol	
28 g		4 g			32 g	

Assuming that complete conversion takes place, 0.875 t (1929 lb) of CO and 0.125 t (276 lb) of H_2 must be employed to produce 1 t (2205 lb) of methanol; 2.83 million kJ (2.69 million Btu) of heat energy are released in the process. Methanol synthesis is thus strongly exothermic.

The reaction is accompanied by a decrease in the volume of the reactants, since 1 volume of methanol vapour is formed from 3 volumes of synthesis gas (1 volume of carbon monoxide and 2 volumes of hydrogen). The formation of methanol is greatly assisted by the application of pressure.

Dimethyl ether is formed in a **side-reaction** between two molecules of methanol, with the elimination of water. Such reactions are described as **inter-molecular elimination of water**.

Methyl alcohol		methyl alcohol		dimethyl ether		water
$H_3C-OH_{(l)}$	$+$	$HO-CH_{3(l)}$	$\longrightarrow$	$H_3C-O-CH_{3(g)}$	$+$	$H_2O_{(l)}$

$$\Delta H = +7.4 \text{ kJ/mol}$$

Because of this, the reaction mixture always contains a small quantity of water and ether as by-products. Particularly if the reaction temperature rises too high, further side-reactions lead to the formation of higher alcohols, such as:

Ethanol H_3C-CH_2-OH
Propanol $H_3C-CH_2-CH_2-OH$
Butanol $H_3C-CH_2-CH_2-CH_2-OH$ and

small quantities of carboxylic acids $R-COOH$.

2.3.2. How the process is carried out

The synthesis of methanol can be subdivided into five process stages (see Figure 1):
1st stage: producing the synthesis gas;
2nd stage: compressing the synthesis gas;
3rd stage: hydrogenation of carbon monoxide;
4th stage: separating the methanol from the reaction mixture;
5th stage: purifying the crude methanol from by-products by distillation.

1st stage
Synthesis gas as water-gas from coke

If steam is passed over incandescent coke, the coke acts as a reducing agent and removes the oxygen from the water, absorbing energy as it does so. This is, therefore, an endothermic process, which occurs at a temperature from 1200 to 1300 K (2192 to 2372°F).

Coke	water	carbon monoxide	hydrogen
$C_{(s)}$	$+\ H_2O_{(l)} \longrightarrow$	$CO_{(g)}$	$+\ H_{2(g)}$

$$\triangle H = +130 \text{ kJ/mol}$$

The mixture of gases formed is called water-gas, since one of the raw materials is water vapour. In order to increase the proportion of hydrogen in water-gas, as required for the methanol synthesis, this gas mixture is passed, together with further steam, over an iron oxide catalyst (Fe_2O_3). In this reaction, part of the carbon monoxide acts as a reducing agent and removes oxygen from the water in accordance with the following equation:

Carbon monoxide	water	carbon dioxide	hydrogen
$CO_{(g)}$	$+\ H_2O_{(g)} \xrightarrow{\text{catalyst}}$	$CO_{2(g)}$	$+\ H_{2(g)}$

$$\triangle H = -42 \text{ kJ/mol}$$

IV Organic raw materials and large-scale products: Industrial processes

The carbon dioxide formed can be removed from the gas mixture.

Synthesis gas from naphtha

The name naphtha is given to petroleum hydrocarbons with an upper boiling range not exceeding 150°C (302°F).

In the first stage of the reaction, part of the hydrocarbon is burnt with oxygen. Carbon dioxide and water are formed, for example:

Methane	oxygen	carbon dioxide	water

$$CH_{4(g)} \quad + \quad 2\,O_{2(g)} \quad \longrightarrow \quad CO_{2(g)} \quad + \quad 2\,H_2O_{(l)}$$

$$\triangle H = -840.4 \text{ kJ/mol}$$

In the second stage of the reaction, both dioxide and steam react with the unreacted hydrocarbons present in the reaction mixture. If necessary, further steam is added to the reaction mixture.

The course of the reaction is shown by the following equations:

Methane	carbon dioxide	carbon monoxide	hydrogen

$$CH_{4(g)} \quad + \quad CO_{2(g)} \quad \longrightarrow \quad 2\,CO_{(g)} \quad + \quad 2\,H_{2(g)}$$

$$\triangle H = +247.3 \text{ kJ/mol}$$

Methane	water

$$CH_{4(g)} \quad + \quad H_2O_{(g)} \quad \longrightarrow \quad CO_{(g)} \quad + \quad 3\,H_{2(g)}$$

$$\triangle H = +205 \text{ kJ/mol}$$

The composition of the resulting synthesis gas, consisting of carbon monoxide and hydrogen, is adjusted so that the 1:2 volume ratio required for the synthesis of methanol is obtained.

2nd stage
The synthesis gas, also known as make-up gas, is compressed to the pressure required, between 200 and 700 bars (2,900 and 10,150 lb/in^2), and is passed through a pressure-resistant heat exchanger, in which the make-up gas is pre-heated by the hot reaction gas mixture leaving the reactor. The gas mixture from the reactor is cooled through giving up its heat. The reaction temperature required is adjusted by regulating the rate of flow of the make-up gas and that of the reaction gases.

3rd stage
The pre-heated synthesis gas (make-up gas) then flows through the reactor, which contains a mixture of zinc oxide (ZnO), copper oxide (CuO) and chromium oxide (Cr_2O_3) as

catalyst. The hydrogenation of the carbon monoxide is carried out at about 380°C (716°F). The temperature in the reactor rises because of the exothermic reaction. In order not to allow the temperature to rise above a certain maximum only 15% of the carbon monoxide fed in is reacted in a single passage of the synthesis gas. If this limitation were not applied unwanted side-reactions would be promoted. The synthesis of methanol is thus a continuous cyclic process.

4th stage

After the reaction that has taken place in the reactor the gas is passed first into the pre-heater and then into a condenser. Products such as methanol, dimethyl ether and other higher-boiling compounds separate out here as a liquid phase. This mixture of liquids, which is called crude methanol, is let down to normal pressure and subjected to various purification processes and then to fractional distillation in distillation columns.

Unreacted synthesis gas, also called recycle gas, flows back to the compressor at the start of the process (see also ammonia synthesis II,3). A fraction of this gas is, however, withdrawn from the cycle in order to prevent the accumulation of inert gases such as nitrogen.

5th stage

In the fractional distillation, dimethyl ether, which boils at $-23.6°C$ ($-10.5°F$), can be removed easily in the first column. This column is operated under a slight pressure in order to produce dimethyl ether in liquid form. In the second column, pure methanol is taken off as a top product. Higher boiling compounds remain as a residue. Methanol is stored in iron tanks, if necessary under a nitrogen atmosphere.

2.4. Chemical reactions of methanol

2.4.1. Reaction with certain metals

While it is not possible to replace hydrogen on the carbon atom of methanol by metals, the hydrogen of the OH group can be replaced by certain metals, for example by Na, Mg and Al.

Methanol		sodium		Na methylate		hydrogen
CH_3OH	+	Na	$\longrightarrow$	CH_3-ONa	+	$\frac{1}{2} H_2$

This reaction takes place very vigorously, liberating hydrogen and heat.

Sodium methylate is used in the synthesis of sulphonamides, barbituric acid, dyestuffs and synthetic resins.

IV Organic raw materials and large-scale products: Industrial processes

2.4.2. Esterification with inorganic acids

Methanol	hydrogen chloride	methyl chloride	water

$$CH_3OH_{(g)} \quad + \quad HCl_{(g)} \qquad CH_3Cl_{(g)} \qquad + \quad H_2O_{(g)}$$

$$\Delta H = -33 \text{ kJ/mol}$$

The reaction occurs in a temperature range from 600 to 680 K (621 to 765°F) and at a pressure between 3 and 6 bar (44 and 87 lb/in²).

The large quantities of hydrogen chloride produced in the chlorination of methane (4 mols of hydrogen chloride for 1 mol of carbon tetrachloride) and the increasing demand for chlorinated derivatives of methane, especially for the manufacture of propellant gases for spray cans, have provided an incentive to utilize the chlorine combined in the hydrogen chloride for obtaining methyl chloride.

Figure 1:   Flow-sheet for methanol synthesis

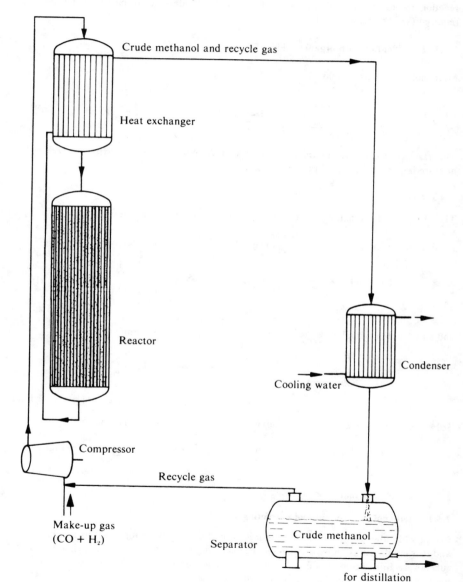

IV Organic raw materials and large-scale products: Industrial processes

Methanol, which is relatively cheap, is esterified with hydrogen chloride in a catalysed reaction and the chloromethane formed (ie methyl chloride) is chlorinated further as required (see IV, 5.2 and 5.4).

2.4.3. Esterification with organic acids

Methanol   benzoic acid   methyl benzoate   water

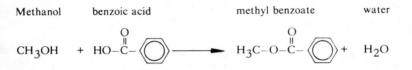

This reaction is catalysed by the addition of acid. Removal of the water from the reaction mixture improves the yield in the reaction.

2.4.4. Oxidation

The oxidation of methanol proceeds through three stages to carbon dioxide

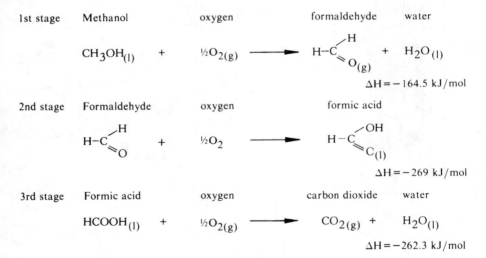

2.5. Uses of methanol

2.5.1. Methanol as a solvent and extracting agent

Mixed with other components, methanol is an important solvent for fats, oils and paints and it is employed as an extracting agent in splitting up mixtures of solid or liquid substances. In addition, it is used as an agent for gas purification. Ten per cent of the methanol produced is consumed in the industries mentioned.

In addition, methanol is used as an additive to motor fuels and rocket fuels and it is used in the USA, mixed with water, as an anti-freeze agent in car radiators.

2.5.2. Methanol as a raw material for formaldehyde

About 50% of the methanol produced is converted into formaldehyde (see IV, 4, page 30). Methanol vapours are passed with air over silver catalysts at 600°C (1112°F).

Methanol	atmospheric oxygen		formaldehyde	water

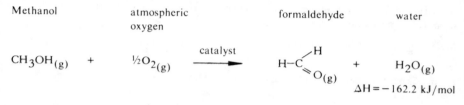

$$CH_3OH_{(g)} \quad + \quad \tfrac{1}{2}O_{2(g)} \xrightarrow{\text{catalyst}} \quad H-C{\overset{H}{\underset{O_{(g)}}{\Big\langle}}} \quad + \quad H_2O_{(g)}$$

$$\Delta H = -162.2 \text{ kJ/mol}$$

In this reaction hydrogen is removed from the methanol and it is therefore termed dehydrogenation. The hydrogen eliminated reacts with the atmospheric oxygen to give water.

2.5.3. Methanol as a methylating agent

Dimethyl sulphate

Methanol reacts with oleum (fuming sulphuric acid) to give dimethyl sulphate.

Methanol	oleum		dimethyl sulphate	water	sulphuric acid

$$2\,CH_3OH \; + \; (H_2SO_4 + SO_3) \longrightarrow \; {\overset{CH_3-O}{\underset{CH_3-O}{\Big\rangle}}}S{\overset{\nearrow O}{\underset{\searrow O}{}}} \; + \; H_2O \; + \; H_2SO_4$$

Dimethyl sulphate is used almost exclusively as a methylating agent, an agent for introducing the methyl group, $-CH_3$, into another molecule.

Carboxylic acid	dimethyl sulphate		methyl ester of carboxylic acid	water	Na methyl sulphate

$$R-C{\overset{\nearrow O}{\underset{OH}{}}} \; + \; (CH_3)_2SO_4 \xrightarrow{\;NaOH\;} \; R-C{\overset{\nearrow O}{\underset{O-CH_3}{}}} \; + \; H_2O \quad NaCH_3SO_4$$

2.5.4. Methylamine from methanol

Methylamines are obtained industrially from methanol vapours and ammonia, with the aid of a dehydrating catalyst.

Methanol		ammonia		methylamine		water
CH_3OH	+	NH_3	$\xrightarrow{\text{catalyst}}$	CH_3-NH_2	+	H_2O

A mixture of monomethylamine, dimethylamine and trimethylamine is formed, and this is separated by distillation under pressure.

Methylamine is used in organic chemistry, inter alia for the synthesis of caffeine and vat dyestuffs.

It can be converted easily into dimethylformamide, DMF,

which is important as a solvent for the manufacture of polyacrylonitrile fibres (see IV, 8, page 25).

Methylamine is also involved in the synthesis of N-methylpyrrolidone.

N-methylpyrrolidone is required as a selective extracting agent for the removal of unsaturated compounds from mixtures of gases.

It is also an additive for tanning substances, photographic developers and rocket propellants.

2.5.5. Dimethyl terephthalate

Dimethyl terephthalate, DMT, the dimethyl ester of terephthalic acid, is an important intermediate in the manufacture of synthetic fibres (see IV, 8, page 23). DMT is obtained by esterifying terephthalic acid with methanol.

Terephthalic acid	methanol	dimethyl terephthalate	water

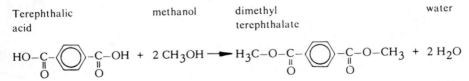

Dimethyl terephthalate is transesterified with glycol, $HO-CH_2-CH_2-OH$, a dihydric alcohol, and this reaction, with polycondensation taking place simultaneously, gives the well-known polyesters.

2.5.6. Methyl methacrylate

The esterification of methylacrylic acid with methanol produces methyl methylacrylate, also known as methyl methacrylate:

$$H_2C = \overset{\overset{\displaystyle CH_3}{|}}{C} - \underset{\underset{\displaystyle O}{\|}}{C} - O - CH_3$$

The polymethacrylates are produced from this by polymerization. They are solid, hard, transparent chemical materials. They can be machined easily. As plastics, the polymethacrylates belong to the category of thermoplastics (see IV, 7, page 10). Safety window glass substitutes are manufactured from them.

2.5.7. Methylcellulose

Methylcellulose is prepared by reacting cellulose with dimethyl sulphate (obtained from methanol).

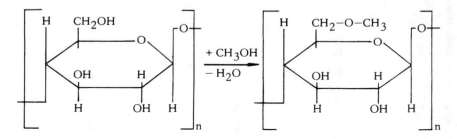

Methylcellulose dissolves in water forming highly viscous colloidal sollutions.

It finds a wide field of applications, inter alia as a wallpaper paste and a binder for dyestuffs, as a binder for plaster casts, as an additive to detergents for holding dirt in suspension (a protective colloid), as a replacement for gelatine in the manufacture of drug capsules and as an additive for improving the quality of liquid soaps.

2.6. Summary of the uses of methanol and its conversion into further products

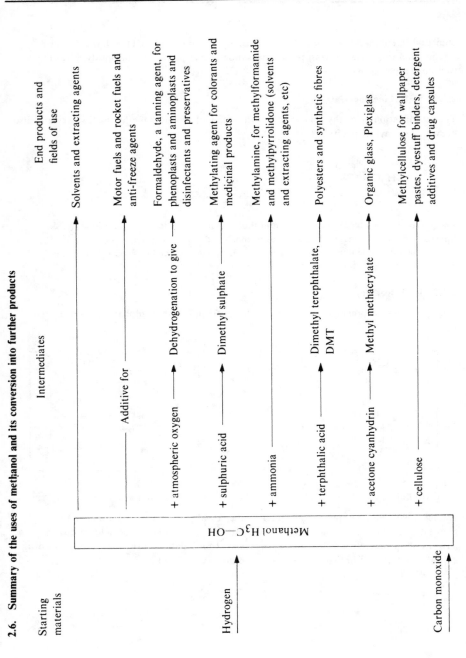

3. The oxo reaction

3.1. Evolution of the oxo reaction

The oxo reaction is a modern process for the manufacture of aliphatic aldehydes and alcohols on a large industrial scale. The starting materials are carbon monoxide, CO, hydrogen, H_2, and hydrocarbons with a terminal double bond, $R-CH=CH_2$.

In 1938, the German chemist Otto Roelen discovered that, in the presence of cobalt-thorium catalysts, ethylene reacts with carbon monoxide and hydrogen under elevated pressures and at an elevated temperature to give propionaldehyde.

| Ethylene | + | carbon monoxide | + | hydrogen | catalyst | propionaldehyde |

$$H_2C=CH_{2\,(g)} + CO_{(g)} + H_{2(g)} \longrightarrow H_3C-CH_2-C{\overset{H}{\underset{O_{(g)}}{\diagup}}}$$

$$\triangle H_{298} = +31.6 \; kJ/mol$$

The reaction product formed is a hydrocarbon derivative containing oxygen, that is to say an oxo compound.

In the early stages of development it was assumed that it was possible in principle to convert alkenes (ie olefins) by this route into aldehydes and also into ketones, ie into oxo compounds.

The name oxo reaction rapidly established itself for this process, although the general validity of this assumption was not confirmed.

Under the conditions of the oxo reaction it is only possible to prepare ketones from ethylene. All the other alkenes produce aldehydes.

In 1940, the German companies Ruhrchemie AG, IG-Farbenindustrie AG and Henkel & Cie. together formed Oxogesellschaft mbH, which set up the first oxo manufacturing plant in the world in Oberhausen-Holten. The plant had a capacity of 10,000 t/year and began manufacture at the beginning of 1945. It produced detergent alcohols with a chain length of C_{12} to C_{14}. The first units operated discontinuously. In 1959 experience gained made it possible to start up a fully continuous large-scale plant, and the capacity of this plant had been extended to 220,000 t/year of oxo products by 1968. Since then, the capacity has been increased again to 320,000 t/year. In 1979, 900,000 t of Oxo-synthesis derivatives were produced in the FRG. As a result of the loss of the patent rights for the oxo reaction after the Second World War, chemical companies elsewhere in the world were also enabled to make use of the know-how for this process. The name oxo products for the products obtained by the oxo reaction has been generally adopted.

IV   Organic raw materials and large-scale products: Industrial processes

3.2. Its economic importance

In 1974, world production of oxo products was 2.7 million t, of which 56% related to Europe and 44% to the rest of the world.

The rapid development of the oxo reaction was assisted by two factors. It has been possible to obtain sufficiently pure alkenes for the oxo reaction from crude oil. It has also been possible to manufacture the synthesis gas, consisting of hydrogen and carbon monoxide, advantageously from crude oil as a raw material.

The opening up of crude oil as a source of raw materials, not only for alkenes but also for synthesis gas, has made it possible to manufacture alcohols at moderate prices and in increasing quantities.

Oxo alcohols have manifold uses. The most important oxo alcohol is 2-ethylhexanol.

$$H_3C-CH_2-CH_2-CH_2-\underset{\underset{C_2H_5}{|}}{CH}-CH_2OH$$

This alcohol is used as the starting material for the manufacture of plasticizers for the big-volume plastic polyvinyl chloride (PVC).

PVC production in the EEC in 1979 amounted to 3.7 million t.

In 1978, 2.7 million t PVC were manufactured in the USA, 1.0 million t in the FRG and 0.4 million t in Great Britain.

The annual rate of increase in PVC manufacture is estimated at 5 to 10 % and the manufacture of oxo products for plasticizer manufacture will also increase in proportion. Plasticizers are produced by reacting the oxo alcohol with phthalic anhydride to form an ester (see IV, 3.5.4, page 29)

3.3.   The chemistry of the process

3.3.1.   Oxo aldehydes

In the oxo reaction, a mixture of straight-chain and branched aldehydes is formed from straight-chain alkenes with a terminal double bond by an addition reaction with carbon monoxide and hydrogen.

The reaction is always carried out under pressures of 200 to 450 bars (2940 to 6525 lb/in²) and at temperatures between 100 and 200°C (212 and 392°F). The chemistry of the synthesis can be explained by the following three partial reactions.

a) An atom of hydrogen adds on to a carbon atom linked by a double bond in the alkene.

Alkene	hydrogen molecule	unstable intermediate product	hydrogen atom

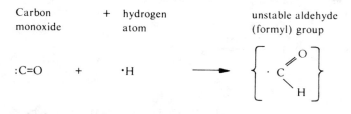

This partial reaction can be regarded as a hydrogenation.

b) The carbon monoxide also reacts with an atom of hydrogen, forming the aldehyde group, also known as the formyl group.

Carbon monoxide	+ hydrogen atom	unstable aldehyde (formyl) group

$$:C=O \quad + \quad \cdot H \quad \longrightarrow \quad \left\{ \cdot C \overset{\displaystyle O}{\underset{\displaystyle H}{\diagup\,\diagdown}} \right\}$$

c) The unstable intermediate product and the unstable aldehyde group react immediately, forming a stable aliphatic aldehyde.

Unstable intermediate product	+ unstable aldehyde group	aliphatic aldehyde

These partial reactions show why the reaction between an alkene, carbon monoxide and hydrogen is known to scientists as hydroformylation.

The partial reactions cannot be separated from one another during the actual course of the reaction, they take place simultaneously and in parallel, so that the oxo reaction or hydroformylation can be described by the equation shown below.

IV Organic raw materials and large-scale products: Industrial processes

alkenes (olefins)   carbon monoxide   hydrogen   straight-chain+ branched alkanals (aldehydes)

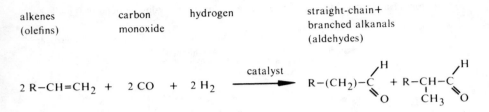

$$2\ R{-}CH{=}CH_2\ +\ 2\ CO\ +\ 2\ H_2\ \xrightarrow{\text{catalyst}}\ R{-}(CH_2){-}\overset{H}{\underset{O}{C}}\ +\ R{-}\overset{}{\underset{CH_3}{CH}}{-}\overset{H}{\underset{O}{C}}$$

The aldehydes are thus the primary products of the oxo reaction; each molecule contains one carbon atom more than the initial olefin. Propene, C_3H_6, is the alkene employed most frequently on an industrial scale for the oxo reaction. The reaction product obtained is a mixture consisting of the isomers butanal (n-butyraldehyde) and 2-methylpropanol (i-butyraldehyde).

propene   carbon monoxide   hydrogen   butanal   2-methylpropanal

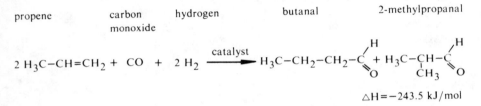

$$2\ H_3C{-}CH{=}CH_2\ +\ CO\ +\ 2\ H_2\ \xrightarrow{\text{catalyst}}\ H_3C{-}CH_2{-}CH_2{-}\overset{H}{\underset{O}{C}}\ +\ H_3C{-}\overset{}{\underset{CH_3}{CH}}{-}\overset{H}{\underset{O}{C}}$$

$$\triangle H = -243.5\ kJ/mol$$

The mixture obtained after the reaction consists of 70% by weight of butanal and 30% by weight of 2-methylpropanal.

The mixtures of isomers are formed because the formyl group

can undergo an addition reaction with either of the two C atoms of the C=C double bond.

The oxo reaction is exothermic. In the example described above, 122 kJ of heat energy are liberated per mol of propene employed.

3.3.2. Oxo alcohols

The oxo alcohols are obtained by hydrogenating the aldehydes formed as the primary product in the oxo reaction. This is effected by hydrogenating the reaction mixture from the oxo reaction to give the corresponding alcohols, by reducing it with catalytically activated hydrogen immediately afterwards, without isolating the products from the oxo reaction mixture.

Mixture of butanal and 2-methylpropanal	hydrogen	mixture of normal butanol and isobutanol

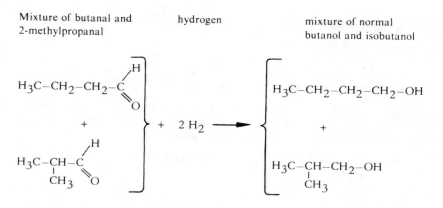

Since these alcohols are obtained from oxo products, they are known as oxo alcohols. From an industrial point of view, they are the most important of the oxo products. They have a chain length ranging from C_3 to C_{13}.

3.4. Manufacturing processes

3.4.1. Manufacture of the raw materials

The starting materials required for the oxo reaction are synthesis gas, consisting of carbon monoxide, CO, and hydrogen, H_2, and alkenes with a terminal double bond. The synthesis gas is obtained from coke, natural gas or naphtha, ie specific petroleum fractions (for details see IV, 3, pages 7 and 8*).

The gaseous olefins, which have a chain length of C_2 to C_4, are at present obtained mainly from light naphtha, a petroleum fraction boiling between 25 and 110°C (77 and 230°F).

The light naphtha is heated at 800°C (1472°F). During this process, the saturated hydrocarbons are cracked, ie split into smaller fragments, hydrogen being evolved at the same time. The required alkenes are thus formed.

In the presence of catalysts, cracking temperatures of 600°C (1112°F) are sufficient.

2-Methylbutane	propene	ethene	hydrogen

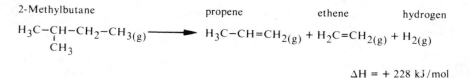

$$\Delta H = + 228 \text{ kJ/mol}$$

The nature and quantity of the cracked products depend on the reaction conditions.

IV Organic raw materials and large-scale products: Industrial processes

When light naphtha is cracked, the mixture formed can have the following composition:

- 20.3% by volume of ethene (ethylene), C_2H_4
- 13.1% by volume of propene (propylene), C_3H_6
- 5.6% by volume of butene (butylene), C_4H_8
- 9.9% by volume of hydrogen, H_2.

The remainder consists of saturated hydrocarbons. The cracked mixture is cooled to a low temperature, in the course of which it liquefies and can thus be separated by distillation. Pure olefins are obtained by this means for the oxo reaction.

Secondary reactions can occur, for example tripropene (tripropylene), C_9H_{18}, is formed by trimerization from propene (propylene), C_3H_6.

propene   tripropene

$$3\ H_3C - CH = CH_2 \longrightarrow H_3C-CH_2-CH_2-CH_2-CH_2-CH_2-CH_2-CH{=}CH_2$$

One molecule of tripropene has been formed from three molecules of propene. Isobutene, C_4H_8, undergoes a similar reaction, giving di-isobutene by dimerization.

isobutene   di-isobutene

$$2\ H_3C - \underset{\underset{\textstyle CH_3}{|}}{CH} = CH_2 \longrightarrow H_3C - \underset{\underset{\textstyle CH_3}{|}}{CH} - CH_2 - CH_2 - \underset{\underset{\textstyle CH_3}{|}}{C} = CH_2$$

These compounds produced in the secondary reactions also contain terminal double bonds and can thus be used in the oxo reaction.

3.4.2. Carrying out the reaction on an industrial scale

The synthesis of oxo aldehydes and oxo alcohols can be subdivided into five stages (see Figure 2):

1st stage: hydroformylation;
2nd stage: regeneration of the catalyst;
3rd stage: isolation of aldehydes by distillation;
4th stage: hydrogenation of the aldehydes to give alcohols;
5th stage: distillation of the mixture of alcohols.

1st stage

The gaseous starting materials for the oxo reaction, such as the alkenes and the synthesis gas, consisting of carbon monoxide and hydrogen, are introduced continuously from below into a vertical high-pressure tube by means of a compressor.

The hydroformylation reaction is carried out at temperatures from 120 to 200°C (248 to 392°F) and pressures of 200 to 300 bars (2940 to 4350 lb/in²) using suitable catalysts.

3 Hydrogenation
(Methanol and oxo syntheses)

The catalyst is metered in in the form of a solution or suspension. Cobalt and cobalt compounds are generally used for this purpose. The catalytic action stems from cobalt hydrocarbonyl, $HCo(CO)_4$, which is formed very rapidly from the cobalt compounds fed in and the synthesis gas under the reaction conditions mentioned. Very small quantities of it are sufficient to catalyse the hydroformylation reaction.

The metals rhodium and ruthenium and their compounds are also catalytically active, but the cobalt catalysts are the cheapest.

Hydroformylation can be applied to nearly all alkenes that have a terminal double bond. The reaction is exothermic and liberates 117.2 to 146.5 kJ per mol (111.2 to 139.0 Btu) depending on the alkene employed.

The heat of reaction is removed by means of cooling coils inserted in the reaction tube and is used to pre-heat the make-up gas mixture entering the reactor.

The product from the hydroformylation reaction is separated into a liquid and recycle gas (unreacted synthesis gas) by reducing the pressure in several stages.

The liquid consists essentially of the so-called oxo aldehydes and the catalyst dissolved in the latter.

2nd stage

When gaseous constituents have been removed from the reaction product by passing it through a gas separator, the catalyst is removed from the liquid phase in a separate stage and is worked up and returned to the oxo reactor as a solution or a suspension.

In the case of aldehydes with fairly long chains the reaction product is treated with hydrogen at 120 to 180°C (248 to 356°F) in order to remove the catalyst. The lower aldehydes, which are sensitive, are freed from catalyst carbonyls using air as an oxidizing agent. The catalysts are obtained in the form of aqueous solutions of their salts.

The removal and regeneration of catalysts is particularly important in the oxo reaction. Many different chemical and technical processes have been developed for this purpose.

3rd stage

The aldehydes formed in the hydroformylation reaction must not only be separated from one another by distillation, they must also be freed from by-products.

Varying small quantities of compounds such as saturated hydrocarbons, alcohols, formic acid, esters, ketones, etc, are produced as by-products.

An example of a side-reaction which is favoured is the hydrogenation of the aldehyde to form a saturated hydrocarbon and also an alcohol.

IV Organic raw materials and large-scale products: Industrial processes

a) Formation of saturated hydrocarbon:

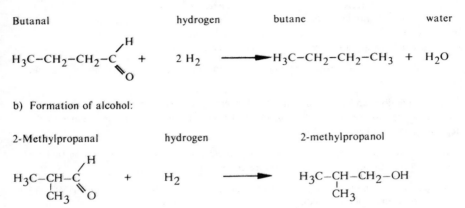

Butanal   hydrogen   butane   water

b) Formation of alcohol:

2-Methylpropanal   hydrogen   2-methylpropanol

Purification and isolation of the oxo aldehydes is only carried out to high standards in the case of the low-boiling products. A little acid is added to the crude aldehydes to break up acetals and trimers and they are then distilled.

4th stage
If alcohols rather than aldehydes are desired as intermediate products for syntheses of more valuable substances, the aldehydes are hydrogenated, ie reduced, with hydrogen in a further stage to give the corresponding alcohols. The starting material for this is a purified mixture of oxo aldehydes, which is fed to a special hydrogenation reactor. Considerable care is required in the hydrogenation of oxo aldehydes, since even traces of unreacted aldehydes can reduce the quality of the alcohol considerably and these traces are very difficult to remove in the subsequent distillation.

After the hydrogenation, the liquid mixture of alcohols is collected in a separator. If propene has been employed in the oxo reaction, the alcohols obtained are:

       n-Butanol $H_3C-CH_2-CH_2-CH_2-OH$ and
       i-Butanol $H_3C-\underset{\underset{CH_3}{|}}{CH}-CH_2-OH$.

The hydrogen that has not been consumed in the hydrogenation reaction is extracted by means of a compressor and recycled to the process.

5th stage
Before the alcohols are separated by distillation, alkali is added to the mixture in order to remove traces of aldehydes. Some 10 to 25% of the quantity employed is left as a residue from the distillation, in the form of a thick oil.

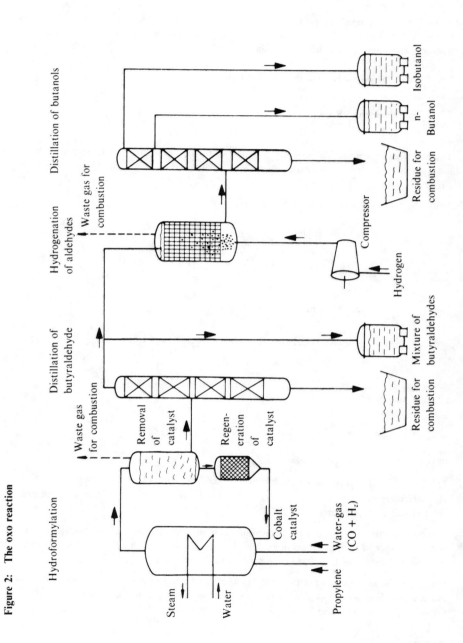

Figure 2:   The oxo reaction

The oxo plant can produce more aldehyde or more alcohol, as required. If less aldehyde is taken off in the first distillation plant, there is an increased yield of alcohol after the hydrogenation and the second distillation.

3.5. Uses of the products from the oxo reaction

3.5.1. Oxo aldehydes,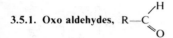

The aldehydes constitute the primary products of the oxo reaction. The structure of the aldehydes or mixtures of aldehydes obtained depends on the alkenes employed in each particular case.

Thus, the following are formed in the oxo reaction:

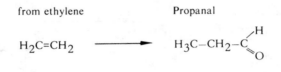

from ethylene   Propanal

$H_2C=CH_2 \longrightarrow H_3C-CH_2-C\overset{H}{\underset{O}{}}$

from propene   Butanal

$H_3C-CH=CH_2 \longrightarrow H_3C-CH_2-CH_2-C\overset{H}{\underset{O}{}}$

and 2-Methylpropanal

$H_3C-\underset{\underset{CH_3}{|}}{CH}-C\overset{H}{\underset{O}{}}$

Physical and chemical properties

The low-molecular aliphatic oxo aldehydes are colourless liquids at room temperature; their viscosity increases with their molar mass. Aldehydes containing 12 C atoms or more are solid at room temperature.

Lower aldehydes have a pungent odour and their vapours irritate the mucous membranes. As the number of carbon atoms increases, the pungent odour falls off; higher aldehydes have an agreeable smell and are even used as perfumes.

Oxo aldehydes are miscible with alcohols, ethers and other organic solvents. Only the low-molecular aldehydes are soluble in water. Compounds containing six C atoms or more are virtually insoluble in water.

By virtue of their functional group, $-C{\Large\lessgtr}{\tiny\begin{matrix}H\\O\end{matrix}}$,

aldehydes are very reactive compounds, which can enter into a large number of chemical reactions, for example reduction, oxidation, addition and polymerization (see III, 6.9). Their reactivity makes them important intermediate products for many syntheses. As a result of the oxo reaction, carried out on a large industrial scale, aldehydes have become cheap industrial chemicals for which there is a continually growing demand. They are very susceptible to oxidation and special precautions and regulations therefore apply to their storage and transport. Containers made of steel, aluminium, glass, ceramic materials or plastics are suitable for storing them. They are transported in rail tank-wagons, road tankers and barrels.

3.5.2. Oxo alcohols, R$-$OH

The catalytic hydrogenation of aldehydes produces alcohols.

Propanal   hydrogen   propanol

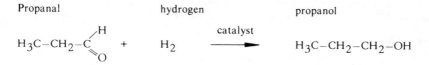

$$H_3C-CH_2-C{\Large\lessgtr}{\begin{matrix}H\\O\end{matrix}} \quad + \quad H_2 \quad \xrightarrow{\text{catalyst}} \quad H_3C-CH_2-CH_2-OH$$

The physical and chemical properties of the alcohols are determined predominantly by the hydroxyl group, $-OH$.

The short-chain aliphatic alcohols are colourless liquids of low viscosity which are soluble in most organic solvents.

Alcohols are much less reactive than aldehydes (see III, 6.3).

The alcohols are by far the most important of the oxo products.

The most important alcohols manufactured by the oxo reaction are:

$CH_3-CH_2-CH_2OH$ n-Propanol

$CH_3-(CH_2)_2-CH_2OH$ n-Butanol

$CH_3-CH-CH_2OH$ i-Butanol
$\quad\quad\ |$
$\quad\ CH_3$

$CH_3-(CH_2)_3-CH_2OH$ n-Pentanol (n-Amyl alcohol)

$CH_3-CH_2-CH-CH_2-OH$ 2-Methylbutanol (i-Amyl alcohol)
$\quad\quad\quad\quad |$
$\quad\quad\quad\ CH_3$

$CH_3-(CH_2)_3-CH-CH_2OH$ $\qquad\qquad\quad \mid$ $\qquad\qquad\quad C_2H_5$	2-Ethylhexanol
$C_7H_{15}-CH_2OH$	i-Octanol
$C_8H_{17}-CH_2OH$	i-Nonanol
$C_9H_{19}-CH_2OH$	i-Decanol
$C_{12}H_{25}-CH_2OH$	i-Tridecanol
$C_{15}H_{31}-CH_2OH$	i-Hexadecanol
$C_{17}H_{35}-CH_2OH$	i-Octadecanol

The oxo alcohols are mainly used, on a large industrial scale, as plasticizer alcohols, solvents and starting materials for surface-active agents and organic intermediates. Oxo alcohols containing four to six C atoms are used preferentially as solvents, while the higher alcohols are mainly used for the manufacture of plasticizers and surface-active substantances. All the oxo alcohols are valuable intermediates for organic chemical syntheses.

3.5.3. Oxidation to carboxylic acids, $R-C\overset{\displaystyle \nearrow O}{\underset{\displaystyle \searrow OH}{}}$

Oxo aldehydes are converted to carboxylic acids even by mild oxidizing agents, in the simplest case air, in the presence or absence of catalysts.

Butanal   oxygen   butanoic acid (butyric acid)

$$H_3C-CH_2-CH_2-C\overset{\displaystyle \nearrow H}{\underset{\displaystyle \searrow O}{}} \quad + \quad 1/2\ O_2 \quad \xrightarrow{\text{catalyst}} \quad H_3C-CH_2-CH_2-COOH$$

The carboxylic acids are weak acids. They are miscible with most organic solvents, but their solubility in water falls off rapidly as their molar mass increases.

By virtue of their carboxyl group, $-C\overset{\displaystyle \nearrow O}{\underset{\displaystyle \searrow OH}{}}$

the carboxylic acids are very reactive (see III, 6.12). They are valuable starting materials for the manufacture of plastics, solvents, driers and lubricants. Some of the acids are converted into carboxylic acid esters, which are used as solvents to a large extent and are also frequently employed as perfumes.

3.5.4. Manufacture of 2-ethylhexanol as an intermediate for plasticizers

The manufacture of 2-ethylhexanol is carried out in a 2-stage process. n-butanal is first converted by means of the aldol condensation into 2-ethylhexenal, which is then hydrogenated in a second stage to give 2-ethylhexanol.

1st stage

n-butanal	2-ethylhexenal	water

$$2\ CH_3-CH_2-CH_2-C\!\!\begin{smallmatrix}H\\ \\O\end{smallmatrix} \longrightarrow CH_3-CH_2-CH_2-CH=\underset{\underset{C_2H_5}{|}}{C}-C\!\!\begin{smallmatrix}H\\ \\O\end{smallmatrix} + H_2O$$

2nd stage

2-ethylhexenal	hydrogen	2-ethylhexanol

$$CH_3-CH_2-CH_2-CH=\underset{\underset{C_2H_5}{|}}{C}-C\!\!\begin{smallmatrix}H\\ \\O\end{smallmatrix} + 2\ H_2 \longrightarrow CH_3-CH_2-CH_2-CH_2-\underset{\underset{C_2H_5}{|}}{CH}-CH_2OH$$

2-Ethylhexanol is the intermediate required for the manufacture of di-(2-ethylhexyl) phthalate (DOP), the standard plasticizer of the plastics industry, particularly for processing PVC. DOP is formed by reacting 2-ethylhexanol with phthalic anhydride.

phthalic anhydride   2-ethylhexanol

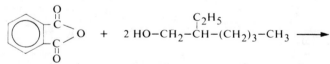

DOP, ester

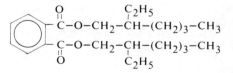

Plasticizers based on 2-ethylhexanol impart good flexibility to plastics over wide ranges of temperature.

IV Organic raw materials and large-scale products: Industrial processes

3.5.5. Formation of acetals

Under the influence of dilute acids, aldehydes react with alcohols to form hemiacetals and acetals:

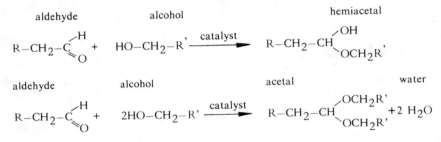

Hemiacetals and acetals contain what is called a "protected aldehyde group". Acetals are prepared as intermediates in reactions in which the very reactive aldehyde group of a compound must not take part in the reaction. Acetals are used, for example, in the manufacture of pharmaceuticals and also that of certain solvents.

3.5.6. The condensation reaction between oxo aldehydes and urea

Macromolecules, the basic units of the aminoplast group of plastics, are formed by a condensation reaction, ie the elimination of water, from successive molecules of urea and aldehyde.

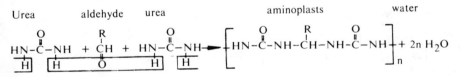

This reaction is also the basis of the production of nitrogenous fertilizers with a long-term action.

3.5.7. The addition reaction between hydrocyanic acid and oxo aldehydes

Aldehydes undergo an addition reaction with hydrocyanic acid (hydrogen cyanide) with the formation of cyanhydrins:

Cyanhydrins are intermediate products for organic syntheses.

3.6. Summary of the uses of oxo products and their conversion into other products

| Starting materials | Intermediate products | | | End products and fields of use |

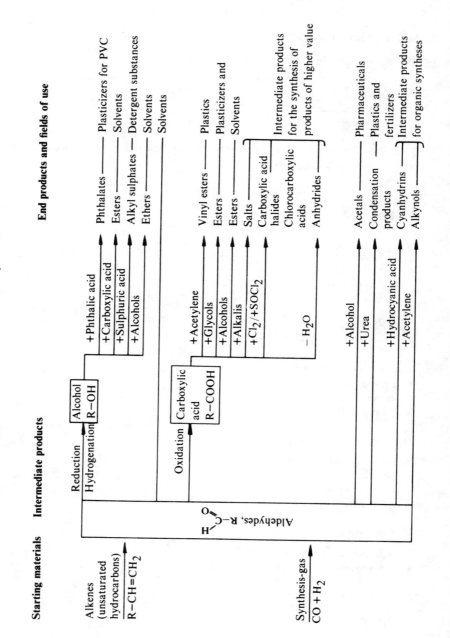

Starting materials

Alkenes
(unsaturated
hydrocarbons)
R−CH=CH₂

Synthesis-gas
CO + H₂

Intermediate products

Aldehydes, R−C⟨O⟩⟨H⟩

Reduction
Hydrogenation → Alcohol R−OH

Oxidation → Carboxylic acid R−COOH

Reactions from Alcohol R−OH:
- +Phthalic acid → Phthalates —— Plasticizers for PVC
- +Carboxylic acid → Esters —— Solvents
- +Sulphuric acid → Alkyl sulphates — Detergent substances
- +Alcohols → Ethers —— Solvents
- Ethers —— Solvents

Reactions from Carboxylic acid R−COOH:
- +Acetylene → Vinyl esters —— Plastics
- +Glycols → Esters —— Plasticizers and
- +Alcohols → Esters —— Solvents
- +Alkalis → Salts
- +Cl₂/+SOCl₂ → Carboxylic acid halides
- Chlorocarboxylic acids
- −H₂O → Anhydrides

Intermediate products for the synthesis of products of higher value

Reactions from Aldehydes:
- +Alcohol → Acetals —— Pharmaceuticals
- +Urea → Condensation products —— Plastics and fertilizers
- +Hydrocyanic acid → Cyanhydrins —— Intermediate products
- +Acetylene → Alkynols —— for organic syntheses

IV   Organic raw materials and large-scale products: Industrial processes

4. Self-evaluation questions

31 Hydrogenation reactions are

A   reactions in which chemical compounds are reacted with hydrogen
B   always reduction reactions
C   endothermic reactions
D   always catalytic reactions

32 Which combinations of hydrogenated substance and resulting product are correct for industrial-scale hydrogenation?

Hydrogenation

Hydrogenated substance	$\longrightarrow$	Resulting product
A nitrogen	$\longrightarrow$	ammonia
B coal	$\longrightarrow$	petrol
C carbon monoxide	$\longrightarrow$	methanol
D vegetable oil	$\longrightarrow$	raw materials for margarine production

33 Methanol is

A   miscible with water in any ratio
B   a primary alcohol
C   hygroscopic
D   inflammable

34 The complete oxidation of methanol leads to

A   carbon dioxide and water
B   formaldehyde
C   formic acide
D   acetic acid

35 Methanol synthesis

A   is greatly facilitated by the application of pressure
B   uses producer gas as a starting product
C   uses ethane as a starting product
D   is a catalytic process

36 Oxo synthesis is used in the direct production of

A   saturated hydrocarbons
B   carboxylic acids
C   aliphatic aldehydes
D   aliphatic alcohols

37 Which substances are used as starting products in oxo synthesis?

A   carbon dioxide
B   hydrogen
C   alkenes with a double bond in the end position
D   carbon monoxide

38 The reaction of acetaldehyde
with hydrogen cyanide, HCN, is a

A   condensation reaction
B   addition reaction
C   substitution reaction
D   oxo synthesis

39 The reduction of acetaldehyde leads to the following product:

A   ethane, C_2H_6
B   ethanol, C_2H_5OH
C   methanol, CH_3OH
D   acetic acid, CH_3COOH

4 Oxidation (Ethylene Oxide, Acetaldehyde, Acetic Acid, Formaldehyde and Phenol)

IV Organic raw materials and large-scale products: Industrial processes

IV Organic raw materials and large-scale products: Industrial processes

Organic oxidation reactions carried out on a large industrial scale

For decades the chemistry of aromatic compounds constituted the major interest of the chemical industry, particularly in the classical disciplines of the synthesis of dyestuffs and medicinal products. With the rise of plastics, the chemistry of aliphatic compounds became increasingly more important. On the raw material side, this rise is characterized primarily by the switch from coal to the use of raw materials obtained from crude oil. Processes for the oxidation of hydrocarbons have acquired an increasing importance in this connection in the last 20 to 25 years. The reasons for this are to be found in the cheap raw materials and the multiplicity of products that can be obtained from oxidation reactions.

Oxygen is introduced into the hydrocarbon molecule in order to increase its reactivity.

The charts on the two following pages are intended to provide a survey of the most important large-scale reactions. They are classified according to the basic principles of the two processes that can be used:
1. Gas phase oxidation (that is to say oxidation of the reactants in the gaseous state), and
2. Liquid phase oxidation.

The multiplicity of processes has only become possible as a result of the development of new technologies and materials and through modern control engineering, which makes it possible to control processes accurately, firstly in order to eliminate danger (oxidation processes are combustion and involve an explosion hazard) and secondly to achieve economic yields.

1. Ethylene oxide (1,2-epoxyethane or oxirane)

Together with the oxygenated ethylene derivatives acetaldehyde and acetic acid, ethylene oxide is also a very important oxidation product of ethylene. Ethylene oxide is the starting material for glycols, polyglycols and all the solvents, plasticizers, etc, derived from these products.

1.1 Manufacturing processes

Nowadays, all the manufacturing processes for ethylene oxide start from ethylene.

There is a whole series of direct oxidation processes for the manufacture of ethylene oxide from ethylene; the Shell process is one of the most modern of these.

The products of gas phase oxidation

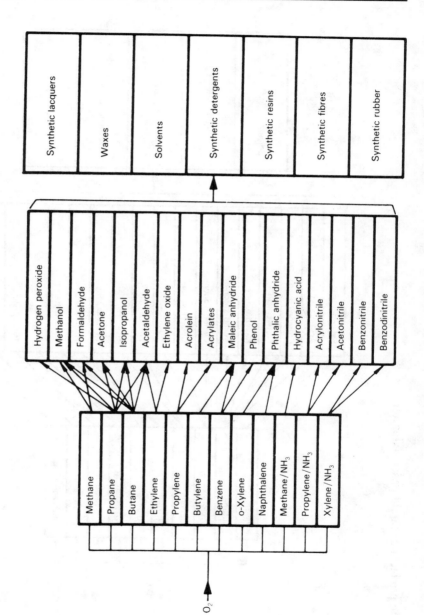

The products of liquid phase oxidation

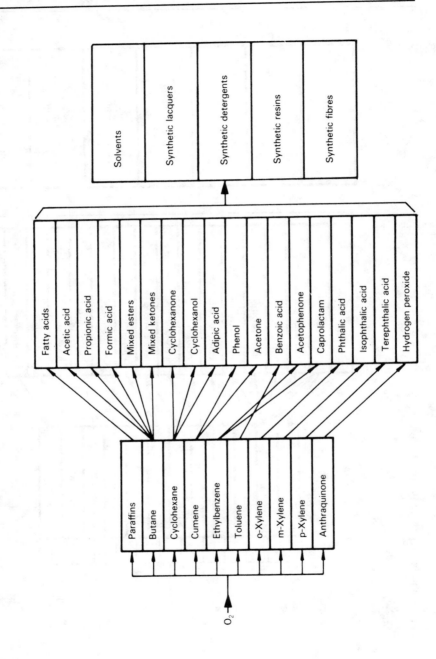

Solvents

Synthetic lacquers

Synthetic detergents

Synthetic resins

Synthetic fibres

Fatty acids

Acetic acid

Propionic acid

Formic acid

Mixed esters

Mixed ketones

Cyclohexanone

Cyclohexanol

Adipic acid

Phenol

Acetone

Benzoic acid

Acetophenone

Caprolactam

Phthalic acid

Isophthalic acid

Terephthalic acid

Hydrogen peroxide

Paraffins

Butane

Cyclohexane

Cumene

Ethylbenzene

Toluene

o-Xylene

m-Xylene

p-Xylene

Anthraquinone

O_2

4 Oxidation (Ethylene oxide, acetaldehyde, acetic acid, formaldehyde and phenol)

1.1.1. Manufacture of ethylene oxide by direct oxidation of ethylene by the Shell process

1.1.1.1. The chemistry of the process

The process is carried out in *one* reaction stage.

Ethylene is oxidized direct to ethylene oxide over a silver catalyst by means of oxygen

$$\underset{\text{Ethylene}}{2\ H_2C{=}CH_2} + \underset{\text{oxygen}}{O_2} \xrightarrow{\text{catalyst}} \underset{\text{ethylene oxide}}{2\ H_2C\underset{O}{\diagdown\diagup}CH_2} \ ; \ \Delta H = -105\ kJ/mol*$$

This is, therefore, a strongly exothermic process, that is to say it liberates heat.

In an unwanted side-reaction, which cannot be suppressed, some of the ethylene burns, forming carbon dioxide.

$$\underset{\text{Ethylene}}{H_2C{=}CH_2} + \underset{\text{oxygen}}{3\ O_2} \longrightarrow \underset{\text{carbon dioxide}}{2\ CO_2} + \underset{\text{water}}{2\ H_2O}$$

1.1.1.2. How the process is carried out

The plant for producing ethylene oxide consists of two sections, the reaction section and the working-up section (see Figure 1, page 9).

In the reaction section, ethylene, oxygen and the unreacted reaction mixture recycled from the working-up section are fed to a reactor.

The reactor contains a catalyst; silver is its active constituent.

The oxidation of the ethylene to ethylene oxide takes place on the catalyst.

The large quantities of heat liberated in the reaction are used for generating steam.

The product leaving the reactor is a mixture consisting of ethylene oxide, carbon dioxide, ethylene, oxygen and inert gases.

In the working-up section, the ethylene oxide is isolated from the reaction mixture by washing with water and subsequent distillation stages.

The residue is recycled to the reactor through a compressor. This fraction is known as recycle gas.

The recycle gas is divided into three part-streams.

The largest fraction is recycled straight back to the reactor. A part-stream is passed through a CO_2 separator, which removes the CO_2 formed in the side-reaction.

---

* The value of ΔH refers to the number of moles represented in the chemical equation.

IV Organic raw materials and large-scale products: Industrial processes

A small fraction of the recycle gas is removed continuously from the plant (waste gas), in order to avoid an accumulation of inert gases (N_2, Ar, CO_2 and H_2O) building up in the plant.

In 1979, the production of ethylene oxide was about 2,394,000 t in the USA, 593,000 t in Japan and 434,000 t in West Germany.

1.1.2. Other manufacturing processes

In the chlorhydrin process, which is carried out in two stages, ethylene first undergoes an addition reaction with hypochlorous acid, ethylene chlorhydrin being formed as an intermediate product.

Ethylene   hypochlorous acid   ethylene chlorhydrin

$$H_2C{=}CH_2 \quad + \quad HOCl \quad \longrightarrow \quad H\text{-}O\text{-}CH_2\text{-}CH_2\text{-}Cl$$

This product is not isolated, but is converted into ethylene oxide by boiling with milk of lime (calcium hydroxide solution).

Ethylene   milk of lime   ethylene oxide   calcium   water
chlorhydrin   chloride

$$2\,H\text{-}O\text{-}CH_2\text{-}CH_2\text{-}Cl \;+\; Ca(OH)_2 \;\longrightarrow\; \underset{O}{H_2C - CH_2} \;+\; CaCl_2 \;+\; 2\,H_2O$$

With the high price of chlorine and the considerable amount of by-products formed, this process has now become uneconomic.

1.2. Physical properties of ethylene oxide

Ethylene oxide is a colourless, poisonous gas which liquefies at 10.7°C (33.8°F) and solidifies at −111.7°C (−179.9°F), both under 1.013 bars (14.5 lb/in²) pressure.

This means that ethylene oxide can only be stored under normal pressure at temperatures below 10°C (58°F).

The storage of ethylene oxide at room temperature requires pressure vessels.

The density of liquid ethylene oxide at 5°C (41°F) is 0.89 g/cm³ (55.5 lb/ft³).

Under normal pressure ethylene oxide vapours mixed with air ignite at 430°C (806°F). Mixtures of air and ethylene oxide containing between 2% by volume and 100% by volume of gaseous ethylene are explosive.

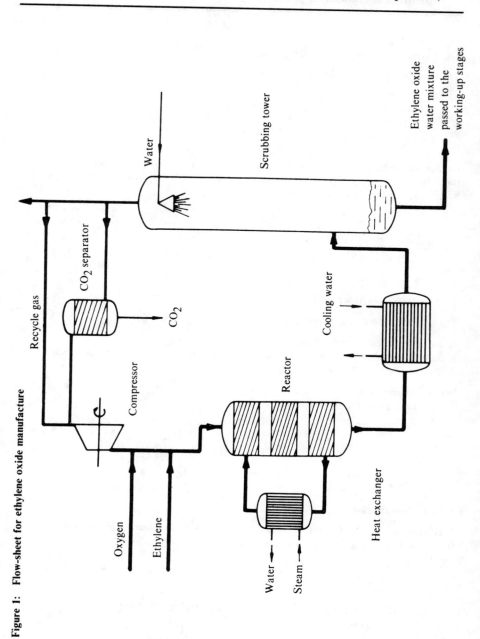

Figure 1: Flow-sheet for ethylene oxide manufacture

1.3. Chemical properties of ethylene oxide

The ethylene oxide molecule has the following structure:

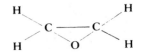

Its high reactivity is due to the oxygen atom, which is linked to the ethylene radical in a three-membered ring of high energy content and low stability. Ethylene oxide always reacts by opening the three-membered ring.

1.3.1. Addition reactions

The addition reactions are divided into:

a. an addition reaction with iself, and
b. addition reactions with other molecules containing reactive H atoms, such as water, alcohols, fatty acids and ammonia.

When ethylene oxide undergoes an addition reaction with itself, its molecules add on to one another by opening the three-membered rings.

1.3.1.1. Dimerization

If ethylene oxide is heated with a little concentrated sulphuric acid or phosphoric acid, it forms 1,4-dioxan, the dimer of ethylene oxide.

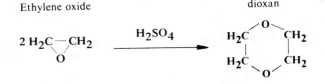

1.3.1.2. Polymerization

If several ethylene oxide molecules link up with one another, molecules with a chain structure are formed, for example:

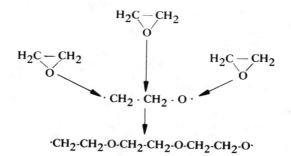

If this self-addition reaction leads to very large molecules (ie macromolecules), the process is called polymerization. In this case polyethylene glycol is formed from ethylene oxide, using tin-IV chloride as catalyst.

1.3.1.3. Addition reaction with water

The addition reaction between water and ethylene oxide gives glycol (monoethylene glycol or MEG). "Glycol" is used both as a trivial name for ethanediol $HO-CH_2-CH_2-OH$ and for dihydric alcohols in general (ie diols).

Ethylene oxide   water   glycol

$$H_2C - CH_2 \;\; + \;\; H_2O \;\; \longrightarrow \;\; HO\text{-}CH_2\text{-}CH_2\text{-}OH$$
$$\diagdown\,O\,\diagup$$

1.3.1.4. Addition reaction with alcohols

The addition reaction with alcohols gives ethylene glycol ethers, which are important as solvents.

Ethylene oxide   alcohol   ethylene glycol ether

$$H_2C - CH_2 \;\; + \;\; R\text{--}OH \;\; \longrightarrow \;\; HO-CH_2-CH_2-O-R$$
$$\diagdown\,O\,\diagup$$

IV Organic raw materials and large-scale products: Industrial processes

1.3.1.5. Addition with ammonia

The addition reaction with ammonia gives ethanolamines:

Ethylene oxide   ammonia   ethanolamine

$$H_2C - CH_2 \overset{\diagdown O \diagup}{} \quad + \quad NH_3 \quad \longrightarrow \quad HO-CH_2-CH_2-NH_2$$

1.4. Ethylene oxide as an intermediate

1.4.1. Glycol

About 60% of all ethylene oxide produced is converted into ethylene glycol (see 1.3.1.3).

A substantial part of this is used as an anti-freeze agent in water-cooled engines.

A further large fraction is used for the manufacture of polyesters. The polyesters of glycol terephthalate are converted into fibres and plastics.

Methyl terephthalate   glycol

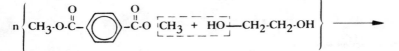

poly-(glycol terephthalate)   methanol

$$\left[-O-\overset{O}{\underset{\|}{C}}-\bigcirc-\overset{O}{\underset{\|}{C}}-O-CH_2-CH_2- \right]_n \quad + \quad (n-1)\,CH_3OH$$

1.4.2. Nonionic detergents

Substances with a detergent action are obtained if an alkylphenol is reacted with an excess of ethylene oxide.

Alkylphenol   ethylene oxide

$$CH_3(CH_2)_m\text{-}CH_2 - \bigcirc OH \quad + \quad n\,CH_2 \overset{\diagup}{\underset{O}{}} CH_2 \longrightarrow$$

alkylphenylpolyethylene glycol ether

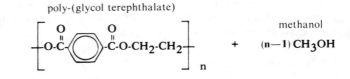

The number n, of ethylene oxide molecules to one molecule of alkylphenol, is 8 to 12. Nonionic surface-active agents manufactured from ethylene oxide are also of importance as emulsifiers and wetting agents in the textile industry.

About 10% of ethylene oxide production is converted into nonionic detergents.

1.4.3. Dioxan

Dioxan is formed by the cyclization of two molecules of ethylene oxide in the presence of acid catalyst (see 1.3.1.1.).

Dioxan is an excellent solvent for resins, waxes, lignin (a principal constituent of wood, together with cellulose), cellulose acetate, fats and colorants.

Chlorination products of dioxan are used as insecticides.

2. Acetaldehyde

2.1. Manufacturing processes

Acetaldehyde is one of the compounds based on ethylene and acetylene. Its high reactivity makes acetaldehyde a key product for the manufacture of chemical products of a higher value. One route of manufacture employs the direct oxidation of ethylene by the Wacker process.

Ethylene	oxygen		acetaldehyde

$$CH_2 = CH_2 \quad + \quad \frac{1}{2} O_2 \quad \xrightarrow{\text{catalyst}} \quad CH_3 - C\underset{H}{\overset{O}{\lessgtr}} \quad ; \Delta H = -244,7 \text{ kJ/mol}$$

2.1.1. Production of acetaldehyde by the Wacker process

2.1.1.1. The chemistry of the process

There are three stages in the Wacker process for the manufacture of acetaldehyde.

1st stage

The actual production of acetaldehyde is effected by oxidizing ethylene with palladium chloride, which is itself reduced to palladium metal in the course of the reaction. Hydrochloric acid is also liberated.

Ethylene	palladium chloride	water	acetaldehyde	palladium	hydrochloric acid
$CH_2{=}CH_2$	$+ \quad PdCl_2$	$+ \quad H_2O$	$\rightarrow \quad CH_3{-}CHO$	$+ \quad Pd$	$+ \quad 2\,HCl$

1.5. Summary of the products made from ethylene oxide

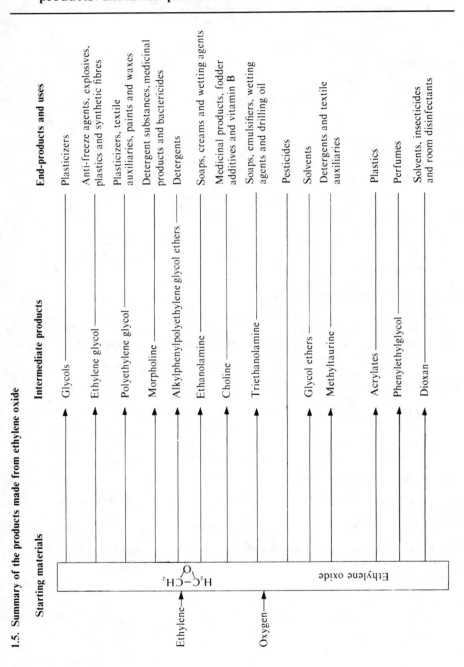

Starting materials

Ethylene →
Oxygen →

Ethylene oxide

H_2C-CH_2
O

Intermediate products

Glycols
Ethylene glycol
Polyethylene glycol
Morpholine
Alkylphenylpolyethylene glycol ethers
Ethanolamine
Choline
Triethanolamine
Glycol ethers
Methyltaurine
Acrylates
Phenylethylglycol
Dioxan

End-products and uses

Plasticizers

Anti-freeze agents, explosives,
plastics and synthetic fibres

Plasticizers, textile
auxiliaries, paints and waxes

Detergent substances, medicinal
products and bactericides

Detergents

Soaps, creams and wetting agents

Medicinal products, fodder
additives and vitamin B

Soaps, emulsifiers, wetting
agents and drilling oil

Pesticides

Solvents

Detergents and textile
auxiliaries

Plastics

Perfumes

Solvents, insecticides
and room disinfectants

4 Oxidation (Ethylene oxide, acetaldehyde, acetic acid, formaldehyde and phenol)

2nd stage

The palladium formed in reaction 1 which is a very expensive metal belonging to the platinum group, is re-oxidized to palladium chloride by means of copper-II chloride.

Palladium		copper-II chloride		palladium chloride		copper-I chloride
Pd	+	2 CuCl$_2$	$\longrightarrow$	Pd Cl$_2$	+	2 CuCl

The palladium chloride thus recovered can be used again in the first stage.

3rd stage

The copper-I chloride formed in the second stage is re-oxidized to copper-II chloride by means of the hydrochloric acid formed in the first stage, and oxygen.

Copper-I chloride		hydrochloric acid		oxygen		copper-II chloride		water
2 CuCl	+	2 HCl	+	$\frac{1}{2}O_2$	$\longrightarrow$	2 CuCl$_2$	+	H$_2$O

The copper-II chloride formed is used again in the second stage and the water formed is used again in the first stage.

When the cycle of the reaction has been completed, palladium chloride, copper chloride and water are once more in their original form. Thus they are not consumed in the reaction, but their presence causes the reaction to take place. Substances of this kind, which guide and accelerate a reaction but emerge from it unchanged, are called **catalysts**.

The catalyst in the Wacker process is an aqueous solution of copper-(I,II) chloride/palladium chloride.

If the three partial reactions are summarized, the following simple equation is obtained:

Ethylene		oxygen		acetaldehyde
CH$_2$=CH$_2$	+	$\frac{1}{2}O_2$	$\xrightarrow{\text{catalyst}}$	CH$_3$-CHO; $\Delta H = -244.7$ kJ/mol

2.1.1.2. How the process is carried out

It has been possible to keep the capital cost of the equipment required for carrying out the process very low, since it proved possible to carry out the three partial reactions "in one pot" (see Figure 2, page 17).

IV Organic raw materials and large-scale products: Industrial processes

Reaction diagram of the Wacker process

Ethylene and oxygen are blown in at the bottom of a reaction tower filled with catalyst solution.

The reaction proceeds with evolution of heat. 5527 kJ (5244 Btu) are liberated per kg of acetaldehyde. The catalyst solution is brought to the boil by the heat of the reaction.

The acetaldehyde formed escapes from the boiling solution in the form of vapour, accompanied mainly by water vapour, together with small quantities of inert gases, by-products and unreacted ethylene and oxygen. A crude aldehyde solution of about 10% strength is then produced in a scrubbing tower by injecting water. This crude aldehyde is purified in a distillation column and is separated into water and pure aldehyde in a second column. The large quantities of effluent produced in this process (10% of pure aldehyde and 90% of effluent are obtained from a 10% aldehyde solution) are purifid in a biological effluent treatment plant. The yield of aldehyde, calculated on ethylene employed, is about 95%.

In 1979, 612,000 t of acetaldehyde were produced in the USA, 583,000 t in Japan and 405,000 t in West Germany.

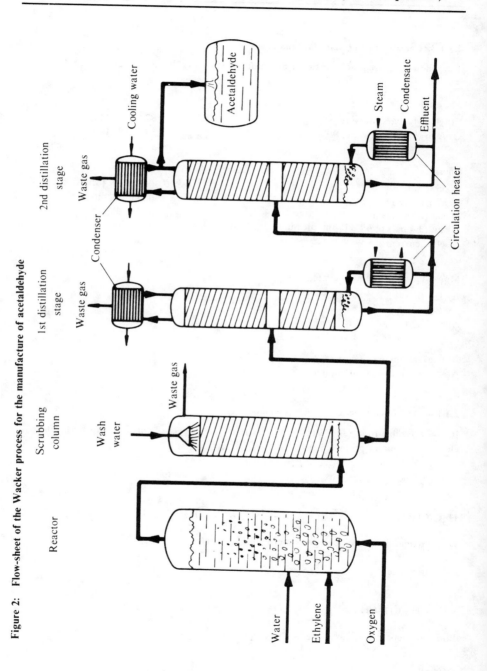

Figure 2: Flow-sheet of the Wacker process for the manufacture of acetaldehyde

IV Organic raw materials and large-scale products: Industrial processes

2.1.1.3. Consumption of materials and energy

The following are required for the manufacture of 100 kg (220.5 lb) of acetaldehyde:

Materials:

Ethylene (7 bars pressure, 99.7% pure)	67 kg (148 lb)
Oxygen (7 bars pressure, 99% pure)	39 kg (86 lb)

Energy:

Cooling water	30,000 kg (66,150 lb)
Demineralized water	300 kg (662 lb)
Steam (12.8 bars or 186 lb/in² pressure)	130 kg (287 lb)
Steam (3.4 bars or 49.3 lb/in² pressure)	30 kg (66 lb)
Electric current	21 kWh

Man-hours (depending on capacity):

At 15,000 t/year	0.19 hour
At 30,000 t/year	0.1 hour
At 60,000 t/year	0.05 hour

Annual maintenance costs:
About 6% of capital investment.

2.1.1.4. The same process -- other products

It is also posible to use the Wacker process to oxidize higher olefins instead of ethylene. Direct oxidation of propylene gives acetone.

Propylene   oxygen   acetone

$$H_3C\text{-}CH=CH_2 \quad + \quad \frac{1}{2}O_2 \quad \xrightarrow{\text{catalyst}} \quad H_3C\text{-}\overset{O}{\overset{\|}{C}}\text{-}CH_3; \Delta H = -268.6 \text{ kJ/mol}$$

Direct oxidation of butylene gives methyl ethyl ketone.

Butylene   oxygen   methyl ethyl ketone

$$CH_3\text{-}CH_2\text{-}CH=CH_2 \quad + \quad \frac{1}{2}O_2 \quad \xrightarrow{\text{catalyst}} \quad CH_3\text{-}CH_2\text{-}\overset{O}{\overset{\|}{C}}\text{-}CH_3$$

Both products are manufactured in large quantities.

4 Oxidation (Ethylene oxide, acetaldehyde, acetic acid, formaldehyde and phenol)

The Wacker process is a good example of how an idea (in this case the oxidation of ethylene with palladium chloride) is capable of revolutionizing a whole series of large-scale production processes in a very short time. It also enables one to realize the great risk that the chemical industry incurs every time a new plant is erected. It can be rendered obsolete overnight by new processes and, instead of earning a profit, only bring a loss.

2.1.2. Other manufacturing processes

The following starting materials are used for the manufacture of acetaldehyde:

Acetylene	$HC \equiv CH$
Ethylene	$H_2C = CH_2$
and Ethyl alcohol	$H_3C - CH_2 - OH$

The oldest processes start from acetylene.

Acetylene   water   acetaldehyde

$$HC \equiv CH_{(g)} \ + \ H_2O_{(g)} \xrightarrow{\text{catalyst}} CH_3 - CHO_{(g)} \ ; \Delta H = -151 \text{ kJ/mol}$$

This process is uneconomic nowadays because of the high cost of acetylene.

In the USA and in other countries with a large production of ethanol, a large part of the acetaldehyde is manufactured by oxidation or dehydrogenation of ethanol.

The oxidation process:

Ethanol   oxygen   acetaldehyde   water

$$2\ CH_3CH_2OH_{(g)} \ + \ O_{2(g)} \xrightarrow{\text{catalyst}} 2\ CH_3\text{-}CHO_{(g)} \ + \ 2\ H_2O_{(g)}$$
$$\Delta H = -346 \text{ kJ/mol}$$

The dehydrogenation process:

Ethanol   acetaldehyde   hydrogen

$$CH_3\text{-}CH_2OH_{(g)} \xrightarrow{\text{catalyst}} CH_3\text{-}CHO_{(g)} \ + \ H_{2(g)} \ ; \Delta H = -69 \text{ kJ/mol}$$

2.2. Physical properties of acetaldehyde

Acetaldehyde is a water-white liquid with a not unpleasant odour. The vapours attack the mucous membranes and make breathing difficult.

It boils at 20.8°C (69.4°F) and freezes at −120.7°C (−185.3°F), both under normal pressure. Its density at 0°C (32°F) is 0.806 g/cm³ (50.3 lb/ft³).

2.3. Chemical properties of acetaldehyde

2.3.1. The aldehyde group

The acetaldehyde molecule has the following structure:

The reactive group in this molecule is the aldehyde group.

$$-\overset{\overset{\displaystyle O}{\|}}{\underset{\underset{\displaystyle H}{|}}{C}}$$

It can undergo a large number of reactions, for example:

2.3.2. Addition reaction

Acetaldehyde   hydrocyanic acid   cyanhydrin

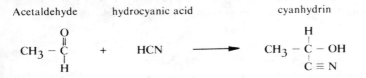

In addition reactions, one end product is formed from two initial products.

2.3.3. Condensation reaction

Acetaldehyde   ethyl alcohol   acetal   water

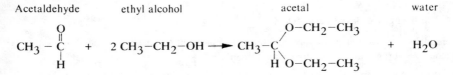

A condensation reaction takes place with elimination of water.

2.3.4. Disproportionation reaction

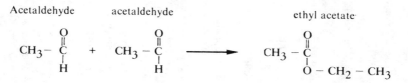

Acetaldehyde   acetaldehyde   ethyl acetate

In a disproportionation reaction, the starting material is converted into products in a higher and a lower stage of oxidation.

A higher stage of oxidation is equivalent to a product richer in oxygen.

A lower stage of oxidation is equivalent to a product richer in hydrogen.

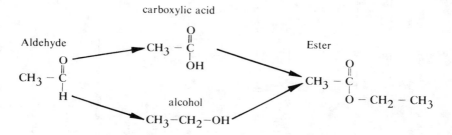

2.3.5. Oxidation

Acetaldehyde   oxygen   acetic acid

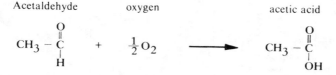

A carboxylic acid is formed when an aldehyde is oxidized.

2.3.6. Reduction

Acetaldehyde   hydrogen   ethanol

$$CH_3 - \overset{\overset{\displaystyle O}{\|}}{\underset{\underset{\displaystyle H}{|}}{C}} \quad + \quad H_2 \quad \longrightarrow \quad CH_3-CH_2-OH$$

An alcohol is formed when an aldehyde is reduced.

2.4. Summary of the products made from acetaldehyde

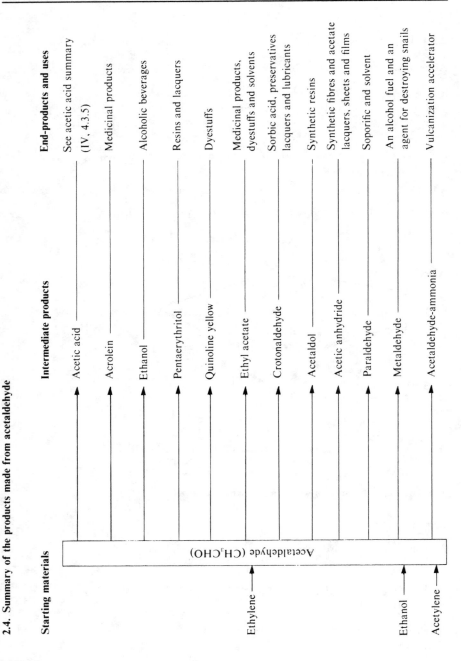

Starting materials	Acetaldehyde (CH₃CHO)	Intermediate products	End-products and uses
		Acetic acid	See acetic acid summary (IV, 4.3.5)
		Acrolein	Medicinal products
Ethylene		Ethanol	Alcoholic beverages
		Pentaerythritol	Resins and lacquers
		Quinoline yellow	Dyestuffs
		Ethyl acetate	Medicinal products, dyestuffs and solvents
		Crotonaldehyde	Sorbic acid, preservatives lacquers and lubricants
		Acetaldol	Synthetic resins
		Acetic anhydride	Synthetic fibres and acetate lacquers, sheets and films
		Paraldehyde	Soporific and solvent
Ethanol		Metaldehyde	An alcohol fuel and an agent for destroying snails
Acetylene		Acetaldehyde-ammonia	Vulcanization accelerator

3.   Acetic acid

3.1.   Manufacturing processes

3.1.1.   Manufacture of acetic acid by oxidation of acetaldehyde

Acetic acid is mainly produced by the oxidation of acetaldehyde with oxygen.

Acetaldehyde   oxygen   acetic acid

$$2\ CH_3 - \overset{\overset{\displaystyle O}{\|}}{\underset{H}{C}} \ + \ O_2 \ \xrightarrow{\text{catalyst}} \ 2\ CH_3 - \overset{\overset{\displaystyle O}{\|}}{\underset{OH}{C}} \quad : \Delta H = -584\ kJ/mol$$

3.1.1.1.   The chemistry of the process

The oxidation takes place at the point where the molecule is reactive, ie at the functional group, the aldehyde group.

Aldehyde group

The oxidation of acetaldehyde is controlled by catalysis. It proceeds by way of intermediate stages.

3.1.1.2.   How the process is carried out

The oxidation reaction takes place in a reactor consisting of a stainless steel tower. Acetaldehyde and catalyst are fed in at the top of the reactor (see Figure 3, page 24).

The oxygen is blown into the reactor at several points which are vertically equidistant.

The entire length of the reactor is covered by a cooling system, which carries away the heat of reaction.

The crude acetic acid is removed at the bottom of the reaction tower and is freed from the catalyst and unreacted acetaldehyde in successive distillation columns.
In 1979, the production of acetic acid was about 1.51 million tonnes in the USA, 0.502 million tonnes in Japan and 0.344 million tonnes in West Germany.

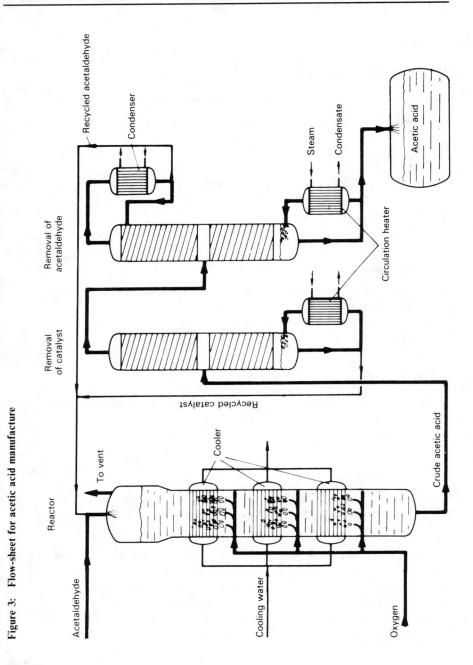

Figure 3: Flow-sheet for acetic acid manufacture

3.1.2. Other manufacturing processes

The oldest method of producing acetic acid is the oxidation of wine caused by bio-catalysts (enzymes).

Acetic acid is formed in this way from ethyl alcohol.

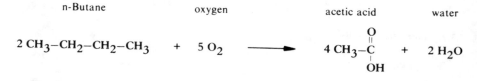

Ethyl alcohol   oxygen   acetic acid   water

$$CH_3-CH_2-OH_{(g)} \quad + \quad O_{2\,(g)} \quad \xrightarrow{\text{enzyme}} \quad CH_3-\overset{\displaystyle O}{\underset{\displaystyle OH}{C}}_{(g)} \quad + \quad H_2O_{(g)}$$

$$\Delta H = -439 \text{ kJ/mol}$$

Even today, this method is still used for the manufacture of high-quality edible acetic acid (wine vinegar).

The oxidation of butane starts from the raw material petroleum. n-Butane (CH_3-CH_2-CH_2-CH_3), a petroleum fraction, is oxidized in the liquid phase with oxygen.

n-Butane   oxygen   acetic acid   water

$$2\,CH_3-CH_2-CH_2-CH_3 \quad + \quad 5\,O_2 \quad \longrightarrow \quad 4\,CH_3-\overset{\displaystyle O}{\underset{\displaystyle OH}{C}} \quad + \quad 2\,H_2O$$

A whole series of by-products, which constitute a disadvantage of the process, are obtained in addition to acetic acid.

3.2. Physical properties of acetic acid

100% pure acetic acid is a water-white liquid with a pungent odour. It solidifies at 17°C and then has the appearance of ice. Anhydrous, pure acetic acid is therefore also called glacial acetic acid.

Its boiling point under normal pressure is 118°C (244°F).

The density of glacial acetic at 15°C (59°F) is 1.055 g/cm³ (65.84 lb/ft³).

Acetic acid is a volatile acid, so that acetic acid vapour as well as water vapour volatilizes from acetic acid solutions.

IV Organic raw materials and large-scale products: Industrial processes

3.3. Chemical properties of acetic acid

3.3.1. The carboxyl group

Acetic acid is a carboxylic acid; the reactive group of a carboxylic acid is the carboxyl group.

 Carboxyl group

In aqueous solution, acetic acid is dissociated into hydrogen ions and acid radical ions.

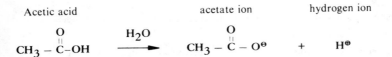

However, since only a small fraction of the acetic acid molecules are dissociated into $H^{\oplus}$ and $CH_3COO^{\ominus}$ ions, acetic acid is classed among the weak acids.

3.3.2. Formation of salts

Like all acids, acetic acid forms salts with alkalis.

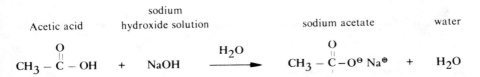

The salts of acetic acid are called acetates. The most important salts are:

Sodium acetate, $CH_3COO^{\ominus}Na^{\oplus}$, added to aqueous solutions in order to keep their pH value as constant as possible (a buffer substance).

Aluminium acetate, $(CH_3COO)_3^{\ominus}Al^{3\oplus}$, used as mordant in dyeing fibres. It is also used in aqueous solution for the treatment of wounds and swellings since it disinfects and contracts the tissue.

4 Oxidation (Ethylene oxide, acetaldehyde, acetic acid, formaldehyde and phenol)

3.3.3. The formation of esters

Carboxylic acids react with alcohols, with the elimination of water, to form esters.

Acetic acid   ethyl alcohol   ethyl acetate   water

$$CH_3 - \overset{O}{\overset{\|}{C}} - OH \; + \; HO - CH_2-CH_3 \longrightarrow CH_3-\overset{O}{\overset{\|}{C}}-O-CH_2-CH_3 \; + \; H_2O$$

3.3.4. Acetylation

The acetyl group can be introduced into organic compounds by way of the acid chloride of acetic acid, acetyl chloride.

$$CH_3 - \overset{O}{\overset{\|}{C}} - Cl$$

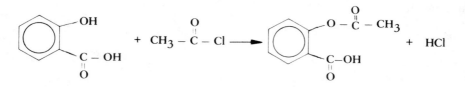

Salicylic acid   acetyl chloride   acetalsalicylic acid (Aspirin)   hydrochloric acid

3.4. Acetic acid as an intermediate

3.4.1. Fibres and filaments made from cellulose acetate

Cellulose acetate, which is soluble, is formed by esterifying cellulose with acetic anhydride. Cellulose acetate is converted into filaments and fibres (see IV, 8, synthetic fibres).

3.4.2. Polyvinyl acetate

Reacting ethylene with acetic acid under oxidizing conditions gives vinyl acetate (oxacetylation)

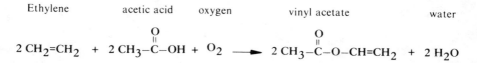

Ethylene   acetic acid   oxygen   vinyl acetate   water

$$2\,CH_2{=}CH_2 \; + \; 2\,CH_3{-}\overset{O}{\overset{\|}{C}}{-}OH \; + \; O_2 \longrightarrow 2\,CH_3{-}\overset{O}{\overset{\|}{C}}{-}O{-}CH{=}CH_2 \; + \; 2\,H_2O$$

Vinyl acetate can be polymerized readily to give polyvinyl acetate. Polyvinyl acetate is a plastic, mainly used as a binder in disperse dyestuffs and as an adhesive, a raw material for paints and a textile finishing agent.

A mixture of polyvinyl chloride and polyvinyl acetate is used for the manufacture of phonograph records (which also contain carbon black as a filler) and also synthetic fibres.

When copolymerized with ethylene, a dry-bright emulsion for floor polishes is obtained. Polyvinyl acetate is also a constituent of hair sprays and is the basic component of chewing gum.

3.4.3. Ethyl acetate

The ethyl ester of acetic acid, ethyl acetate, is obtained by esterifying acetic acid with ethyl alcohol. It is a very valuable solvent. It is also produced by the disproportionation of acetaldehyde (see IV, 2.3.4.).

3.4.4. The use of acetic acid as such

In a dilute form (3 to 3.5% strength), acetic acid is used as edible vingar, for preserving (pickling in vinegar) and as an additive for dyebaths.

3.5. Summary of the products made from acetic acid

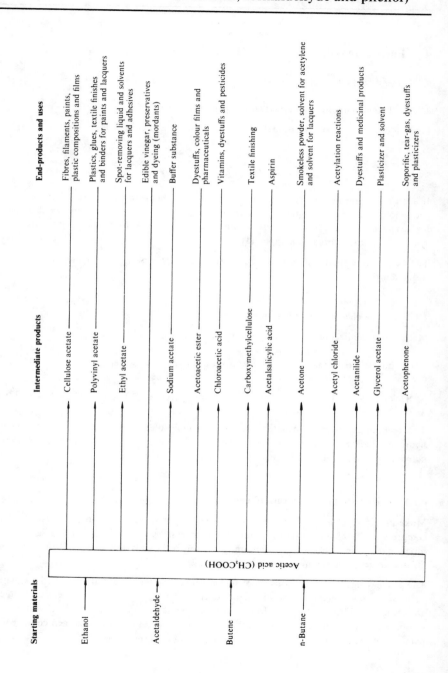

Starting materials

Ethanol

Acetaldehyde

Butene

n-Butane

Intermediate products

Cellulose acetate

Polyvinyl acetate

Ethyl acetate

Sodium acetate

Acetoacetic ester

Chloroacetic acid

Carboxymethylcellulose

Acetalsalicylic acid

Acetone

Acetyl chloride

Acetanilide

Glycerol acetate

Acetophenone

Acetic acid (CH₃COOH)

End-products and uses

Fibres, filaments, paints, plastic compositions and films

Plastics, glues, textile finishes and binders for paints and lacquers

Spot-removing liquid and solvents for lacquers and adhesives

Edible vinegar, preservatives and dyeing (mordants)

Buffer substance

Dyestuffs, colour films and pharmaceuticals

Vitamins, dyestuffs and pesticides

Textile finishing

Aspirin

Smokeless powder, solvent for acetylene and solvent for lacquers

Acetylation reactions

Dyestuffs and medicinal products

Plasticizer and solvent

Soporific, tear-gas, dyestuffs and plasticizers

IV Organic raw materials and large-scale products: Industrial processes

4. Formaldehyde

4.1. The manufacture of formaldehyde

4.1.1. The manufacture of formaldehyde from methanol

Formaldehyde is the simplest aldehyde. Its molecule has the following structure

$$H - \overset{\overset{\displaystyle O}{\|}}{C} - H$$

It is an important intermediate in large-scale organic chemistry. Its outstanding importance is due to its wide use as a starting material for not only low-molecular compounds but also high-molecular plastics, synthetic resins and glues, and as a starting material for the synthesis of further organic compounds.

The oxidation of methanol is a particularly economic process for the manufacture of formaldehyde.

4.1.1.1. The chemistry of the process

Methanol is oxidized catalytically with oxygen to give formaldehyde.

Methanol   oxygen   formaldehyde   water

$$2\,CH_3OH \; + \; O_2 \; \xrightarrow{\text{catalyst}} \; 2\,H - \overset{\overset{\displaystyle O}{\|}}{C} - H \; + \; 2\,H_2O \; ; \; \Delta H = -164.5 \text{ kJ/mol}$$

This reaction is called an oxidation, because the molecules of formaldehyde are poorer in hydrogen than the methanol molecules. The metal oxide catalyst used for the reaction is so active that virtually all the methanol is converted in a single pass.

4.1.1.2. How the process is carried out

Make-up air, together with absorber gas of low oxygen content, is drawn in by a compressor. Methanol is vaporized into the compressed gas in a vaporizer. The gaseous mixture reacts in a tube reactor containing a fixed-bed catalyst. The heat of the reaction is utilized to generate steam (see Figure 4, page 32).

The gas leaving the reactor consists of formaldehyde, water vapour and residual air, with a reduced oxygen content as a result of the reaction. This gas is cooled in a heat exchanger and then flows into an absorber. In the absorber formaldehyde is extracted by washing (absorbed) in counter-current flow with water. The residual air of low oxygen content is partly recycled to the compressor and partly blown off to the atmosphere.

When formaldehyde is dissolved in water considerable quantities of heat are liberated (heat of absorption), which makes it necessary to provide intensive cooling for the absorber.

At the sump of the absorber, an aqueous solution of formaldehyde containing up to 60% of formaldehyde (Formalin) is drawn off and can be sold without further treatment.

4.1.2. Other manufacturing processes

Besides methanol oxidation, processes starting direct from hydrocarbons are used for the manufacture of formaldehyde. Formaldehyde is formed by direct oxidation of methane with atmospheric oxygen over aluminium phosphate contact catalysts. Methanol, acetaldehyde and other oxidation products are formed as by-products. The oxidation of propane or butane produces a complex mixture of oxidation products, containing up to about 25% formaldehyde. The difficulties of the process are caused by the complexities involved in separating the reaction mixture.

In 1979, 1,170,000 t of formaldehyde were produced in the USA, 140,000 t in Great Britain and 489,000 t in West Germany.

4.2. Physical properties of formaldehyde

Formaldehyde is a pungent-smelling gas, which condenses under normal pressure to a liquid at $-21°C$ ($-5.8°F$) and solidifies at $-92°C$ ($-133.6°F$). It is extremely soluble in water. At room temperature 1 litre of water dissolves about 400 litres of formaldehyde gas. The volume of the resulting formaldehyde solution is only slightly greater than the original volume of the water. The formaldehyde of commerce is a 35 to 40% aqueous solution of formaldehyde (Formalin). The density of the 40% solution at 20°C (68°F) is 0.82 g/cm³ (51.2 lb/ft³).

4.3. Chemical properties of formaldehyde

Formaldehyde undergoes the aldehyde reactions already described under acetaldehyde, but is distinguished from the higher aliphatic aldehydes by a few special reactions. For example, formaldehyde and ammonia react to form hexamethylenetetramine (see 4.4.4).

4.4. Formaldehyde as an intermediate

4.4.1. Phenol-formaldehyde resins

Formaldehyde reacts with phenol in a condensation reaction to give plastics and synthetic resins, which are thermoplastic or thermosetting, depending on the degree of condensation (see Phenol, IV, 4.5).

Figure 4: Flow-sheet for formaldehyde manufacture

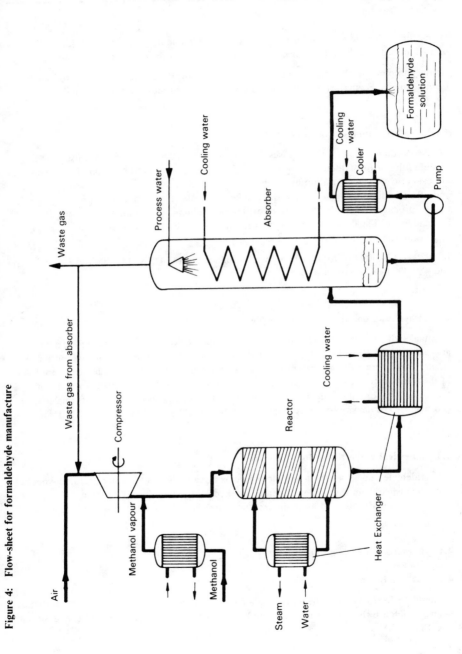

4.4.2.   Urea-formaldehyde resins

Formaldehyde can undergo a condensation reaction with urea to form macromolecules with
a chain structure which are used, for example, as paint resins.

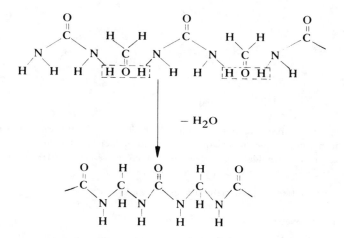

$- H_2O$

The individual chains can be crosslinked with one another by using an excess of
formaldehyde.

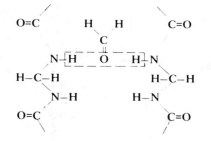

This reaction yields synthetic resins that can be cured (ie can be further crosslinked) and
which are used, after compounding with fillers (for example, chopped up textiles), as
compression moulding materials for the manufacture of articles for domestic use.

4.4.3. Melamine-formaldehyde resins

The melamine molecule has the following structure:

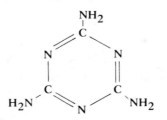

It undergoes a condensation reaction with formaldehyde similarly to urea. In this reaction all the hydrogen atoms of melamine can react with formaldehyde. The preferred molar ratios of melamine to formaldehyde for resin manufacture are between 1:3 and 1:6, depending on the properties desired in the resulting resins.

Melamine resins are the favoured materials for stoving lacquers.

4.4.4.   Hexamethylenetetramine

Formaldehyde reacts with ammonia forming hexamethylenetetramine.

Formaldehyde   ammonia   hexamethylenetetramine   water

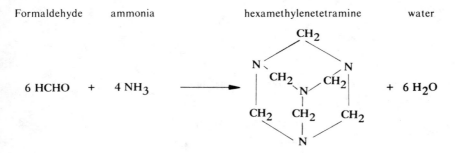

$$6 \ HCHO \quad + \quad 4 \ NH_3 \quad \longrightarrow \quad \quad \quad + \quad 6 \ H_2O$$

Hexamethylenetetramine is used as a preservative for foodstuffs, for curing synthetic resins and, in medicine, for disinfecting the urinary tract and as a remedy for dissolving uric acid in rheumatism and gout.

4.4.5.   Isoprene

An addition reaction between two molecules for formaldehyde and isobutene yields a heterocyclic intermediate product, which can be split by catalytic cleavage into isoprene, formaldehyde and water.

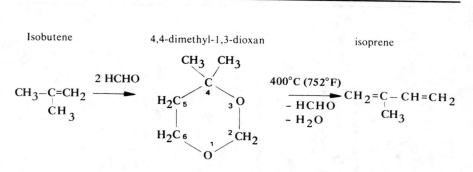

Isobutene   4,4-dimethyl-1,3-dioxan   isoprene

Isoprene is the basic unit of natural rubber. Isoprene can be copolymerized with other monomers; by selecting these and the degree of polymerization, it is possible to vary the properties of the product very widely and to make it suitable for particular applications.

Thus, for example, polymerization of isobutene with isoprene gives butyl rubber, which can be vulcanized and which is used for motor vehicle hoses because of its low vapour permeability.

Isobutene   isoprene   isobutene – isoprene copolymer

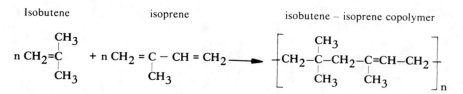

4.5. Summary of the products made from formaldehyde

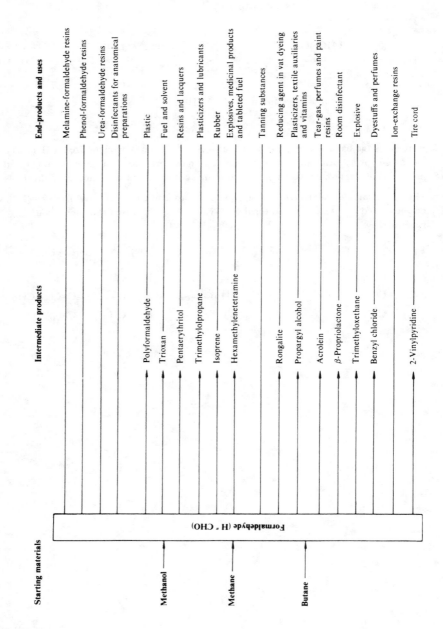

Starting materials	Formaldehyde (H₂CHO)	Intermediate products	End-products and uses
		Polyformaldehyde	Melamine-formaldehyde resins
			Phenol-formaldehyde resins
			Urea-formaldehyde resins
			Disinfectants for anatomical preparations
Methanol		Polyformaldehyde	Plastic
		Trioxan	Fuel and solvent
		Pentaerythritol	Resins and lacquers
		Trimethylolpropane	Plasticizers and lubricants
		Isoprene	Rubber
Methane		Hexamethylenetetramine	Explosives, medicinal products and tableted fuel
			Tanning substances
		Rongalite	Reducing agent in vat dyeing
		Propargyl alcohol	Plasticizers, textile auxiliaries and vitamins
Butane		Acrolein	Tear-gas, perfumes and paint resins
		β-Propriolactone	Room disinfectant
		Trimethyloxethane	Explosive
		Benzyl chloride	Dyestuffs and perfumes
			Ion-exchange resins
		2-Vinylpyridine	Tire cord

5. Phenol

The phenol molecule has the following structure:

It is an intermediate much used in aromatic chemistry. Formerly, phenol was obtained exclusively from coal tar. Now that the demand caused by the increased use of phenol-formaldehyde plastics can no longer be satisfied by coal tar, phenol is made increasingly by synthetic routes.

In 1979, the production of phenol (synthetic) in the USA was 1,352,000 t, in the USSR 487,000 t, in Japan 244,000 t and in Great Britain 179,000 t.

5.1. Manufacturing processes

5.1.1. Manufacture of phenol by the cumene process

5.1.1.1. The chemistry of the process

This synthesis is carried out in three stages.

1st stage

Cumene is obtained from benzene and propylene, using phosphoric acid as catalyst:

Benzene   propylene   cumene

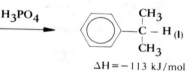

$\Delta H = -113$ kJ/mol

2nd stage

Cumene is oxidized to cumene hydroperoxide by means of air or oxygen.

Cumene   oxygen   cumene hydroperoxide

 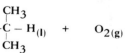 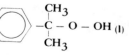

$\Delta H = -117$ kJ/mol

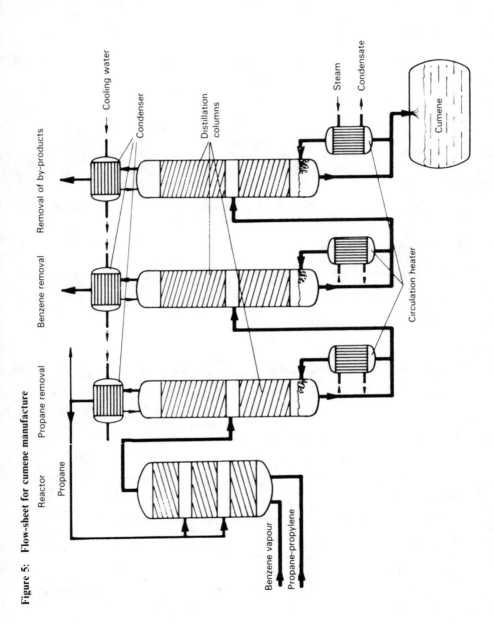

Figure 5: **Flow-sheet for cumene manufacture**

Figure 6: Flow-sheet for phenol manufacture

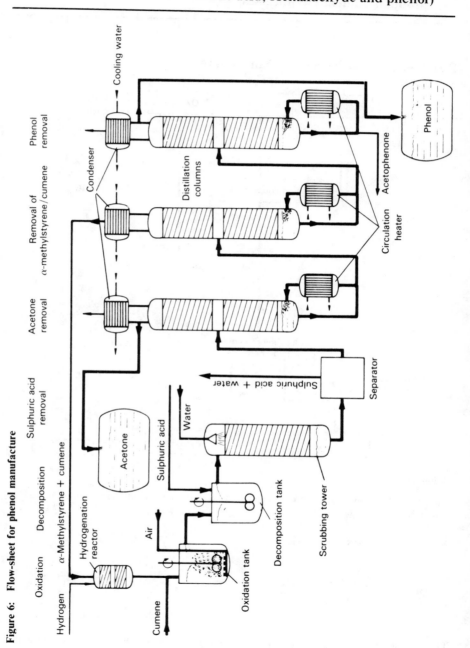

Reaction diagram for phenol manufacture

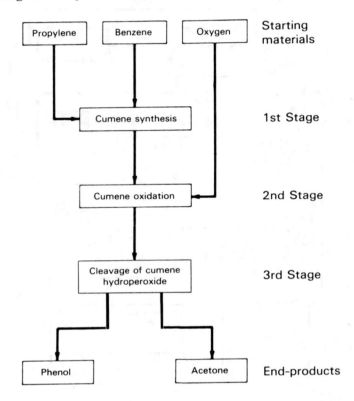

			Starting materials
Propylene	Benzene	Oxygen	

Cumene synthesis — 1st Stage

Cumene oxidation — 2nd Stage

Cleavage of cumene hydroperoxide — 3rd Stage

| Phenol | Acetone | End-products |

3rd stage

Cumene hydroperoxide is split into phenol and acetone by the catalytic action of dilute sulphuric acid.

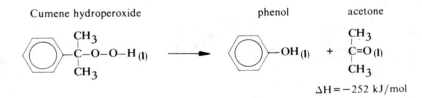

Cumene hydroperoxide　　　　　　　　　phenol　　　　　acetone

$$\Delta H = -252 \ \text{kJ/mol}$$

The acetone obtained as a co-product also finds a ready sale and thus makes a substantial contribution to the profitability of the process.

5.1.1.2. How the process is carried out

1st stage

In the first stage, a propane-propylene mixture (this is cheaper than pure propylene) is mixed with benzene vapour and reacts to form cumene in a fixed-bed reactor. Additional propane is blown into the reactor at various points in order to control the temperature in the interior of the reactor.

In the subsequent purification stages, propane and unreacted benzene are removed from the reaction mixture and 99.9% pure cumene is obtained in a final purification column (see flow-sheet for cumene manufacture, Figure 5, page 38).

2nd stage

The cumene is oxidized to cumene hydroperoxide in an oxidation tank, using air in the presence of an emulsifier (see flow-sheet for phenol manufacture. Figure 6, page 39).

3rd stage

The cumene hydroperoxide is split into phenol and acetone by adding 10% sulphuric acid. (α-Methylstyrene and acetophenone – about 1% of each – are formed as by-products). The acetone is removed by distillation. The sump product from the bottom of the column is freed by vacuum distillation from cumene and α-methylstyrene, which are recycled. The α-methyl-styrene contained in the recycled cumene is oxidized back to cumene over a nickel catalyst at 100°C (212°F). The acetophenone is separated from pure phenol in a further column.

5.1.2. Other manufacturing processes

Sodium phenate is obtained by an alkali fusion of sodium benzenesulphonate at about 300°C (572°F).

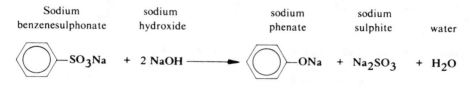

| Sodium benzenesulphonate | sodium hydroxide | | sodium phenate | sodium sulphite | water |

The sodium phenate reacts with hydrochloric acid to form phenol and sodium chloride.

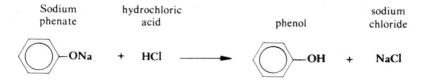

| Sodium phenate | hydrochloric acid | | | phenol | sodium chloride |

IV Organic raw materials and large-scale products: Industrial processes

Another large-scale synthetic route is the hydrolysis of chlorobenzene.

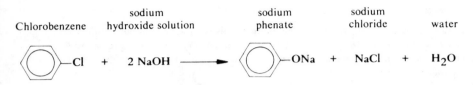

This reaction is carried out at 300°C (572°F) under a pressure of 280 bars (4060 lb/in²), using copper as a catalyst.

The processes mentioned are at present no longer competitive with the cumene process.

5.2. Physical properties of phenol

Phenol has a density of 1.06 g/cm³ (66.2 lb/ft³). It melts at 43°C (109°F) and boils at 181.4°C (358.5°F) under normal pressure.

It has a characteristic odour. It causes acid burns when in contact with the skin. Pure phenol is colourless. On prolonged standing in air, it turns a reddish colour.

5.3. Chemical properties of phenol

5.3.1. The formation of salts

In aqueous solution, the hydrogen atom of the phenolic OH group is split up (dissociated) to a small extent, forming a hydrogen ion:

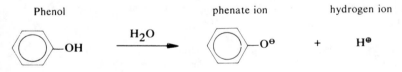

Hence, phenol has a slightly acid reaction and forms salts with alkalis.

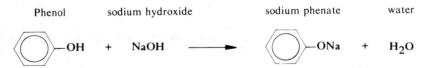

5.3.2.   The formation of esters

Phenol esters are obtained by the action of acid chlorides on phenol.

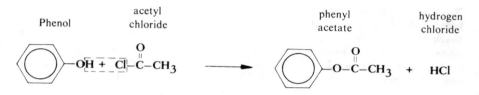

5.3.3.   The formation of ethers

Phenyl ethers are obtained by reacting an aqueous solution of sodium phenate with a
dialkyl sulphate, eg:

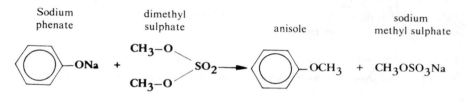

The reaction can also be carried out with sodium phenate and alkyl halides.

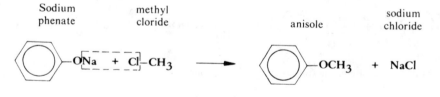

5.3.4.   Substitution reactions in the benzene nucleus

The 'phenolic OH groups makes the benzene nucleus in phenol considerably more reactive
than benzene itself.

Thus, it is possible to nitrate phenol at room temperature with 15% nitric acid, while the
nitration of benzene requires a mixture of concentrated nitric acid and sulphuric acid
(nitrating acid) and heat must be applied.

IV   Organic raw materials and large-scale products: Industrial processes

5.4.   Phenol as an intermediate

5.4.1.   Phenol-formaldehyde condensation products

Nowadays, by far the largest quantity of phenol is converted into polycondensation products formed from phenol and formaldehyde. These are plastics, generally called phenoplasts or Bakelite. If the degree of condensation is low, resins of low melting point, soluble in organic solvents, are obtained. These resins can be converted into a thermoplastic intermediate product by further condensation, which can be effected by heating. This thermoplastic intermediate product can be shaped and cured by the application of heat and pressure. In this third condensation stage, three-dimensional crosslinking takes place. Brittle thermosetting products, stable at high temperatures, are formed.

Thermosetting plastics have a wide field of uses, particularly in the electrical industry.

5.4.2.   Polyamide 6 (Perlon)

Phenol is used as the starting material for the synthesis of ε-caprolactam which is polymerized direct to Perlon (see IV, 8).

5.4.3.   Epoxide resins

4,4,-Dihydroxydiphenyldimethylmethane (bisphenol A) is obtained from phenol and acetone.

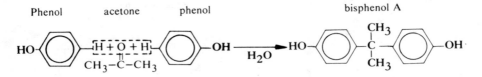

This is subjected to a condensation reaction with epichlorhydrin

$$CH_2 - CH - CH_2$$

(with Cl on the first CH$_2$ and O bridging the CH and CH$_2$)

to form epoxide resins. These are used as paint raw materials, particularly for stoving lacquers, and as casting resins.

5.4.4.   Phenolphthalein

Phenolphthalein is a substance belonging to the triarylmethane series. It is prepared by a condensation reaction (involving the elimination of water) between phenol and phthalic anhydride.

5.4.5. Acetylsalicylic acid (Aspirin)

Salicylic acid is obtained from phenol and carbon dioxide under pressure, at a temperature of 130°C (266°F).

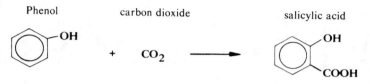

Acetylsalicylic acid, which is known all over the world as aspirin, is formed by acetylating the OH group with acetic anhydride.

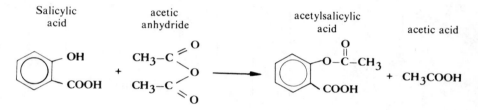

5.5.   Summary of the products made from phenol

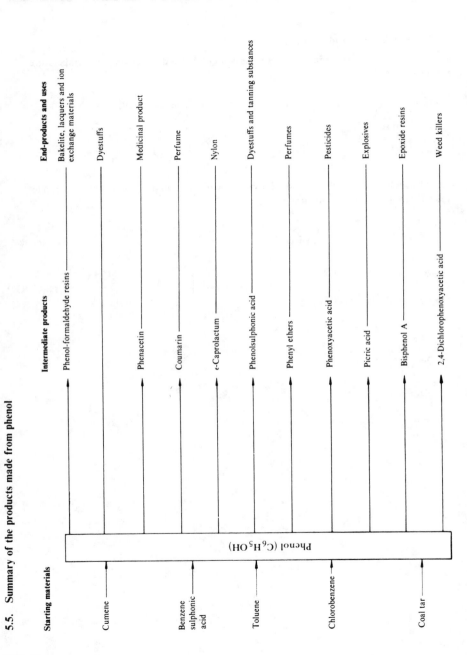

Starting materials	Phenol (C_6H_5OH)	Intermediate products	End-products and uses
Cumene		Phenol-formaldehyde resins	Bakelite, lacquers and ion exchange materials
			Dyestuffs
		Phenacetin	Medicinal product
Benzene sulphonic acid		Coumarin	Perfume
		ε-Caprolactum	Nylon
Toluene		Phenolsulphonic acid	Dyestuffs and tanning substances
		Phenyl ethers	Perfumes
Chlorobenzene		Phenoxyacetic acid	Pesticides
		Picric acid	Explosives
		Bisphenol A	Epoxide resins
Coal tar		2,4-Dichlorophenoxyacetic acid	Weed killers

6.   Self-evaluation questions

40 The starting products used for the industrial-scale production of ethylene oxide are

A   ethane
B   ethene
C   acetylene
D   ethanol

41 The compound H_2O-CH_2 can be termed

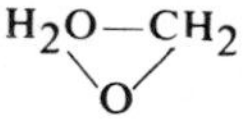

A   oxirane
B   1,2-epoxyethane
C   ethane oxide
D   ethylene oxide

42 Ethylene oxide is obtained

A   endothermically
B   exothermically
C   in the gas phase
D   catalytically

43 Ethylene oxide is

A   liquid at room temperature
B   toxic
C   explosive in admixture with air
D   not highly reactive

44 Addition reactions are typical reactions of ethylene oxide
    In which cases are the added substance and the reaction product correctly stated?

A   water/glycol
B   water/polyethylene glycol
C   ammonia/methylamine
D   ammonia/ethanolamine

45 From which substances can acetaldehyde, CH_3CHO, be manufactured industrially?

A   acetylene
B   ethylene
C   ethanol
D   ethane

IV Organic raw materials and large-scale products: Industrial processes

46 Acetaldehyde is an oxidation product of

   A acetylene
   B ethylene
   C ethanol
   D ethane

47 The oxidation of aldehydes leads to

   A carboxylic acids
   B primary alcohols
   C glycols
   D ketones

48 The reduction of acetaldehyde with hydrogen yields

   A ethylene
   B acetic acid
   C ethanol
   D ethane

49 By which methods can acetone, CH_3COCH_3, be produced?

   A oxidation of propylene
   B dehydrogenation of isopropanol
   C oxidation of butylene
   D dehydrogenation of ethanol

50 Which two identical formulae are applicable to acetic acid?

   A HCOOH
   B $C_2H_4O_2$
   C CH_3COOH
   D CH_3CH_2OH

51 Acetic acid can be manufactured on an industrial scale by

   A hydrogenation of acetaldehyde
   B oxidation of ethanol
   C alcoholic fermentation
   D oxidation of butane

52 Acetic acid is

   A a weak acid
   B known as "glacial acetic acid" in a pure state
   C a highly aggressive compound in undiluted form
   D a volatile acid

4 Oxidation (Ethylene oxide, acetaldehyde, acetic acid, formaldehyde and phenol)

53 Which two products are created in the reaction of a carboxylic acid with an alcohol?

   A   a ketone
   B   a salt
   C   water
   D   an ester

54 Which plastic can be described as a derivative of acetic acid?

   A   polyurethane
   B   polyvinyl acetate
   C   polystyrene
   D   polyvinyl chloride

55 Acetic acid is employed

   A   in diluted form for disinfecting wounds
   B   in concentrated form as culinary vinegar
   C   as a preservative
   D   as an ingredient of dyeliquors for textiles

56 Formaldehyde is

   A   the oxidation product of ethanol
   B   insoluble in water
   C   the simplest aliphatic aldehyde
   D   a compound with the formular HCHO

57 The reaction of formaldehyde with urea produces

   A   a polyester
   B   a condensation production, ammonia
   C   a curable synthetic resin if an excess of formaldehyde exists
   D   a peptide linkage

58 From which starting materials is formaldehyde manufactured on an industrial scale?

   A   methanol
   B   ethanol
   C   methane
   D   ethane

IV Organic raw materials and large-scale products: Industrial processes

59 Which starting materials are used in the production of phenol by the cumene method?

A benzene
B acetone
C propylene
D oxygen

60 Which coupling product occurs in the production of phenol by the cumene method?

A benzene
B acetone
C water
D methanol

61 Phenol is

A a highly aggressive compound
B a compound with a basic reaction
C a starting material for the production of phenoplasts
D less reactive than benzene

62 Phenol serves as a starting product for the manufacture of

A nylon
B epoxy resins
C dyestuffs
D melamine resins

63 Which starting materials can be used in phenol production?

A chlorobenzene
B coal tar
C toluene
D benzosulphonic acid

5   Halogenation

1. Definitions

The name **halogens** is given to the group of elements consisting of fluorine (F), chlorine (Cl), bromine (Br) and iodine (I).

The word halogenation means a chemical reaction in which one or more halogen atoms are incorporated into a molecule. The introduction of fluorine into a compound is called fluorination and the introduction of chlorine is similarly called chlorination. If the halogen atom is introduced into the molecule in place of another atom or group of atoms (for example a functional group), this is known as a **substitution reaction**, ie an exchange reaction. For example:

Methane   chlorine   monochloromethane   hydrogen chloride

$$H - \underset{\underset{H}{|}}{\overset{\overset{H}{|}}{C}} - H_{(g)} \quad + \quad Cl_{2(g)} \quad \longrightarrow \quad H - \underset{\underset{H}{|}}{\overset{\overset{H}{|}}{C}} - Cl_{(g)} \quad + \quad HCl_{(g)}$$

$$\Delta H = -98 \text{ kJ/mol*}$$

One hydrogen atom of the methane has been **replaced** by a chlorine atom. Hydrogen chloride gas is formed as a by-product.

If halogen atoms are incorporated into a molecule containing double or triple bonds between C atoms, this is known as an **addition reaction**. There is **no** by-product split off in such a reaction. For example:

Ethylene   chlorine   1,2-dichloroethane

$$\Delta H = -180 \text{ kJ/mol}$$

The chlorine has been incorporated into the molecule as an **addition**.

2. Aliphatic halogen compounds

Aliphatic compounds have a chain structure. The carbon atoms are attached to one another without the formation of ring systems.

Aliphatic halogen compounds are obtained by replacing the hydrogen atoms by halogen atoms (a substitution reaction).

---

* The value of ΔH refers to the number of moles represented in the chemical equation.

IV Organic raw materials and large-scale products: Industrial processes

2.1. Chloromethanes and chloroethanes

The simplest hydrocarbon is methane,

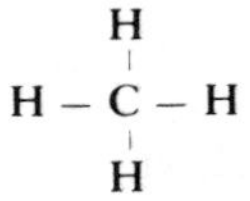

In the chlorination reaction, all the hydrogen atoms of methane can be replaced (ie substituted) successively by chlorine. Products with an increasing Cl content are formed.

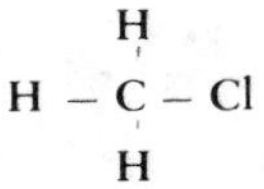

Monochloromethane
= methyl chloride

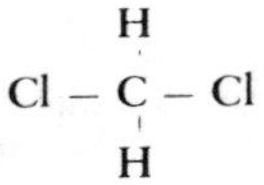

Dichloromethane
= methylene chloride

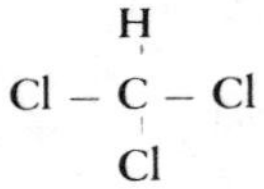

Trichloromethane
= chloroform

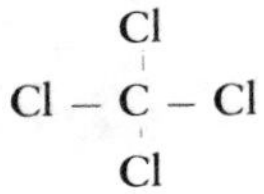

Tetrachloromethane
= carbon tetrachloride

Chloroethanes can be obtained in a similar manner by substitution reactions. For example, 1,2-dichloroethane is formed from 1 mol of ethane and 2 mols of chlorine:

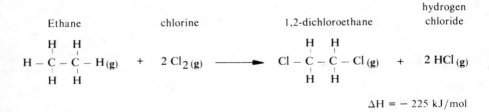

Ethane — chlorine — 1,2-dichloroethane — hydrogen chloride

$\Delta H = -225$ kJ/mol

1,2-Dichloroethane is also formed by an addition reaction between chlorine and ethylene (see page 3).

Vinyl chloride, which is polymerized to give PVC (see IV,7) is manufactured on a large industrial scale from:
1,2-dichloroethane. Like many chloroalkanes, 1,2-dichloroethane is also used as a solvent.

2.2. Methane and ethane containing different halogen substituents

Containing different halogen substituents means that atoms of different halogens are present in one molecule. For example, if carbon tetrachloride reacts with hydrogen fluoride, the product is dichlorodifluoromethane.

Carbon tetrachloride	hydrogen fluoride		dichloro- difluoromethane	hydrogen chloride

$$\underset{\overset{\displaystyle |}{Cl}}{\overset{\overset{\displaystyle Cl}{|}}{Cl - C - Cl}} \quad + \quad 2\,HF \quad \longrightarrow \quad \underset{\overset{\displaystyle |}{F}}{\overset{\overset{\displaystyle F}{|}}{Cl - C - Cl}} \quad + \quad 2\,HCl$$

Dichlorodifluoromethane is especially suited for use:
—as a refrigerant
—for air-conditioning
—as a blowing agent for foamed plastics
—as an industrial solvent
—and for the dry cleaning of textiles.

In a number of European countries, fluorinated hydrocarbons are also used as aerosol propellants.

The methane derivative

$$\underset{\overset{\displaystyle |}{Cl}}{\overset{\overset{\displaystyle H}{|}}{F - C - F}} \qquad \text{Chlorodifluoromethane,}$$

which is obtained when chloroform is fluorinated, yields tetrafluoroethylene if hydrogen chloride is removed by heating.

Chlorodifluoro- methane		tetrafluoro- ethylene	hydrogen chloride

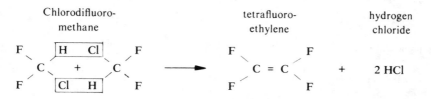

IV Organic raw materials and large-scale products: Industrial processes

Tetrafluoroethylene is used for the manufacture of the plastic polytetrafluoroethylene (PTFE), which is very resistant to heat and chemicals.

Another example of ethane containing different halogen substituents is bromochlorotrifluoroethane, which contains three different halogens in its molecule.

$$
\begin{array}{ccc}
\text{F} & \text{Cl} & \\
| & | & \\
\text{F} - \text{C} - \text{C} - \text{Br} \\
| & | & \\
\text{F} & \text{H} &
\end{array}
\qquad \text{2-Bromo-2-chloro-1,1,1-trifluoroethane}
$$

It is manufactured from hexachloroethane by reaction with hydrogen fluoride and hydrogen bromide in a multi-stage process.

Bromochlorotrifluoroethane is an excellent anaesthetic, which is used a great deal. Anaesthesia is induced rapidly, the patient re-awakens quickly and there is virtually no vomiting.

2.3. Vinyl chloride and PVC (see IV,7)

2.4. Chloroprene

Chloroprene can be produced from 3,4-dichlorobutene by catalytic elimination of HCl, analogously to the production of vinyl chloride from 1,2-dichloroethane.

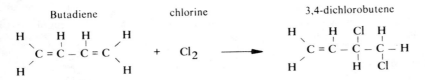

3,4-dichlorobutene is produced by an addition reaction between chlorine and butadiene.

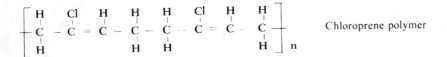

Chloroprene can be polymerized readily to give long thread-like molecules.

$$
\left[
\begin{array}{cccccccc}
\text{H} & \text{Cl} & \text{H} & \text{H} & \text{H} & \text{Cl} & \text{H} & \text{H} \\
| & | & & | & | & | & & | \\
\text{C} & - \text{C} & = \text{C} & - \text{C} & - \text{C} & \cdots \text{C} & = \text{C} & - \text{C} \\
| & & & | & | & & & | \\
\text{H} & & & \text{H} & \text{H} & & & \text{H}
\end{array}
\right]_n
\qquad \text{Chloroprene polymer}
$$

The polymerization product can be crosslinked and has properties similar to those of natural rubber (see IV,7). Its uses include the manufacture of hoses for oil, rubberized fabrics, balloon casings and adhesives.

3. Chlorination of methane on a large industrial scale

3.1. The chemistry of the process

The chlorination of methane is a sequence of substitution reactions.

1st stage

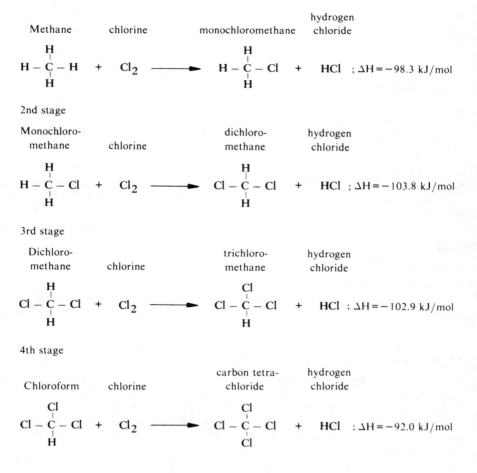

It will be noticed in this sequence of reactions that a further atom of chlorine is consumed to form hydrogen chloride for every atom of chlorine introduced into the molecule. Thus only 50% of the chlorine employed enters the chlorinated product.

In order to recover chlorine, which is valuable, from the by-product HCl, the hydrogen chloride produced is converted by the addition of water into hydrochloric acid and the latter is then split up into chlorine and hydrogen. This is effected by means of electrolysis.

Hydrochloric acid		chlorine		hydrogen
2 HCl	$\xrightarrow{\text{electrical energy}}$	Cl_2	+	H_2

Varying the ratio between the starting products chlorine and methane enables the percentage proportions of the individual chlorination products formed to be adjusted to suit the market.

3.2. Raw materials used

a) Natural gas
Methane is the major constituent of natural gas. In some cases the methane content of natural gas is so high (96 to 97%) that it can be used for the chlorination of methane without further treatment.

b) Refinery operations and high temperature pyrolysis (HTP) Methane is produced in refineries as a by-product when long-chain hydrocarbons are cracked. It is also formed as a by-product in the manufacture of ethylene and acetylene by the high temperature pyrolysis process (see IV, 1, page 23).

Which particular source of methane is used in a given case depends on the local circumstances.

c) Rock salt
The chlorine is obtained by electrolysis of rock salt (NaCl) (see II, 2).

If the hydrochloric acid produced in the chlorination of methane cannot be sold, chlorine is recovered from it by electrolysis.

3.3. How the process is carried out

Recycle gas (80% methane, 10% methyl chloride and 10% inert gases) flows into a steel cylinder lined with nickel (the chlorination furnace), about 10% of chlorine being added in a mixing nozzle. The reaction temperature is 420°C (788°F). This temperature is only slightly below the decomposition temperature of the mixture. A higher temperature would lead to cleavage reactions and the formation of carbon. 99.9% of the chlorine is converted. The gaseous mixture is cooled as it emerges.

Hydrogen chloride is first removed from the gaseous mixture by washing with water in several scrubbing towers arranged in series. Traces of chlorine, hydrogen chloride and carbon dioxide are then removed by washing with sodium hydroxide solution.

After the methane consumed in the reaction has been replaced, the gaseous mixture is compressed to 5 to 6 bars (72.5 to 87 lb/in^2). The chlorination products formed in the chlorination furnace are then liquefied and removed by cooling the gaseous mixture to $-40°C$ ($-40°F$) (see diagram page 10).

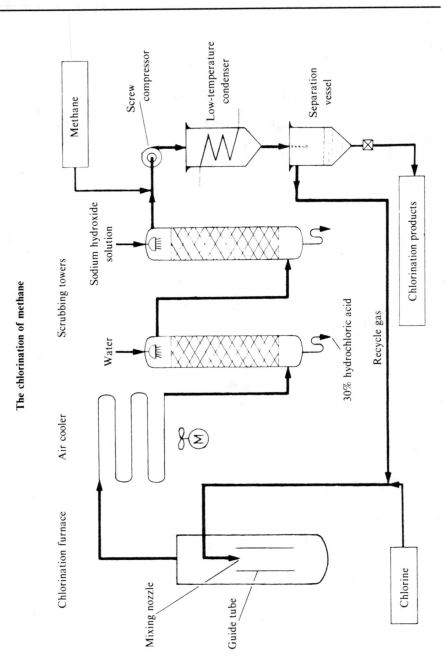

The chlorination of methane

Methane

Screw compressor

Low-temperature condenser

Separation vessel

Sodium hydroxide solution

Scrubbing towers

Water

Chlorination products

30% hydrochloric acid

Recycle gas

Air cooler

Chlorination furnace

Mixing nozzle

Guide tube

Chlorine

3.4. Summary of the products made from the chlorination products of methane

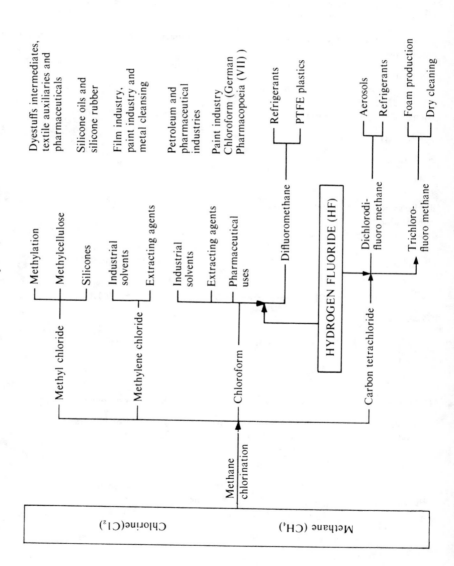

4. Aromatic halogen compounds

4.1. Definitions

If halogen is introduced into an aromatic molecule, for example benzene, an aromatic halogen compound is formed.

If the substitution by halogen takes place on the aromatic nucleus itself, it is called **nuclear substitution**.

For example:

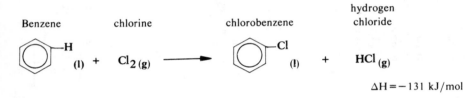

$$\Delta H = -131 \ kJ/mol$$

If the substitution by chlorine takes place in the side-chain, it is called **side-chain substitution**. For example:

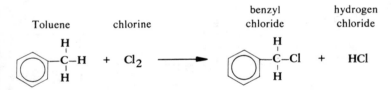

4.2. Chlorobenzene

Chlorobenzene is one of the most important nuclear-substituted halogen compounds.

It is prepared by passing chlorine into anhydrous benzene, using iron chloride as catalyst, at moderate temperatures of 40 to 50°C (104 to 122°F).

The following reaction takes place:

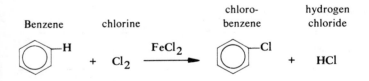

Polysubstituted benzenes are formed as by-products, for example:

| | Benzene | chlorine | o-dichloro-benzene | hydrogen chloride |

Uses:

Chlorobenzene is used as a solvent for oils, fats, resins, rubber, ethylcellulose and Bakelite, as an intermediate in the manufacture of colorants, medicinal products and perfumes and as a heating fluid.

4.3. Dichlorodiphenyltrichloroethane (DDT)

If ethyl alcohol is chlorinated, the product is trichloroacetaldehyde, known as chloral:

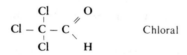

This takes up 1 mol of water on crystallizing in the form of chloral hydrate.

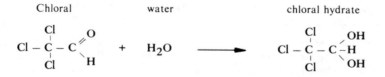

Chloral hydrate is the oldest synthetic soporific. The very effective insecticide DDT is obtained from chloral hydrate and chlorobenzene.

IV Organic raw materials and large-scale products: Industrial processes

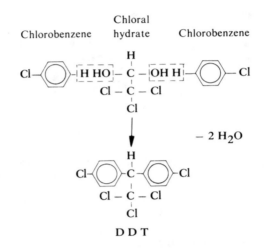

Chlorobenzene   Chloral hydrate   Chlorobenzene

DDT

DDT is a very effective contact poison for insects of all kinds.

In particular, it is employed in large quantities for combating malaria mosquitoes. However, since DDT is not bio-degradable, its unrestricted use would lead to DDT pollution.

For this reason, the use of DDT has already been forbidden or restricted in many countries.

In many countries of the world, DDT saved millions of people from death through malaria. Extensive use of DDT in Ceylon enabled the number of malaria cases to be reduced from 2.8 million in 1946 to 110 in 1961. The following tables gives a general idea of the effectiveness of DDT in controlling the mosquitos that carry malaria.

The fact that in India the number of cases of malaria again rose to 6 million in 1976 is due firstly to neglect in controlling the mosquito by chemical means and secondly because certain species of mosquito have become resistent to DDT. Similar features are also being observed in other tropical developing countries, eg in Honduras, Nicaragua, Guatemala etc.

4.4. Chloramine-T

Compounds in which chlorine is directly linked to nitrogen (N-chloroamines) are called chloramines. Their distinguishing feature is a very high oxidizing power. This is caused by the "active chlorine atom" in the molecule. The content of "active chlorine" is therefore often quoted instead of concentration for chloramines. Owing to the fact that they are not very stable, only a few of the numerous organic chloramines known are actually used.

Country	Year	Cases of disease
Cuba	1962	3 519
	1969	3
Jamaica	1954	4 417
	1969	0
Venezuela	1943	8 171 115
	1958	800
India	1935	> 100 mio.
	1969	185 962
Italy	1945	411 602
	1968	37
Jugoslavia	1937	169 545
	1969	15
Taiwan	1945	> 1 mio.
	1969	9
Ceylon	before 1950	> 2 mio.
	1953*	17
	1968	1 mio.

*1963 control measures discontinued; 1968 recommenced

In particular, they are used as a replacement for chlorine or hypochlorites in chlorination or oxidation reactions. Chloramine-T may be mentioned as an example of chloramines.

It has the following formula:

$$\left[CH_3 - \bigcirc - SO_2 - \overset{\ominus}{N} - Cl \right] \ Na^{\oplus} \cdot \ 3\,H_2O$$

and is the sodium salt of p-toluenesulphonyl chloramide. It is manufactured industrially from the p-toluenesulphonyl chloride formed as a waste product in the manufacture of saccharin (a sweetener), by reacting this compound with ammonia to give p-toluenesulphonamide.

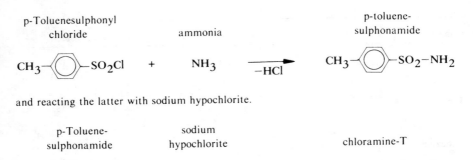

and reacting the latter with sodium hypochlorite.

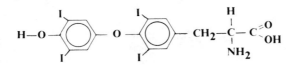

Aqueous solutions of chloramine-T are used as disinfectants, as an antiseptic and as a bleaching agent, for example in the textile industry.

4.5. Thyroxine

The thyroid hormone thyroxine, which has the following formula:

$$H-O-\underset{I}{\overset{I}{\bigcirc}}-O-\underset{I}{\overset{I}{\bigcirc}}-CH_2-\underset{NH_2}{\overset{H}{\underset{|}{C}}}-C\underset{OH}{\overset{O}{\diagup}}$$

is an aromatic amino-acid containing four atoms of the halogen iodine per molecule.

The thyroxine present in blood is responsible for stimulating the metabolism of all the cells in the body (see also II, 9.3.2).

5. Summary of the products made from halogenated compounds

Starting materials	Intermediate products	End-products and uses
	Halogenalkanes ———————————→	Solvents, propellants, refrigerants, anaesthetics and fire extinguishing agents
	Trifluorochloroethylene and tetrafluoroethylene ——→	Plastics
	Vinyl chloride ——————————————→	PVC
	Chloroparaffins ——————————————→	Paint raw materials
	Chlorinated rubber ——————————→	Corrosion-resistant paints
Hydrocarbons	Chloroprene ————————————————→	Oil-resistant varieties of rubber
	Hexachlorocyclohexane ———————→	Insecticides
	Chloral hydrate and chlorobenzene ——→	DDT (Insecticide)
	Chloroacetophenone ————————→	Tear-gas
	Chloroacetic acids ——————————→	Dyestuffs
	Chloromycetin ——————————————→	Antibiotic
	Chloranil ———————————————————→	Dyestuffs
	Chlorobenzaldehyde ——————————→	Dyestuffs

Fluorine ——→
Chlorine ——→
Bromine ——→
Iodine ——→

IV Organic raw materials and large-scale products: Industrial processes

6. Self-evaluation questions

64 In halogenation it is possible for

    A   a halogen molecule to be added to a compound with a double or triple bond
    B   a halogen molecule to be split from a compound
    C   a hydrogen halide molecule to form
    D   an atom or group of atoms in the molecule to be replaced by a halogen atom

65 Which groups consist only of products derived from methane chlorination?

    A   methanol, methyl chloride, trichloromethane
    B   monochloromethane, methylene fluoride, pentachloromethane
    C   methyl chloride, hydrogen chloride, tetrachloromethane
    D   chloroform, dichloroethane, methylene chloride

66 Which structural formula is that of 1,1,1-trifluoro-2-chloro-2-bromoethane?

67 The partial reactions of methane chlorination are

    A   addition reactions
    B   substitution reactions
    C   condensation reactions
    D   addition and substitution reactions

68 Which products are created by nuclear substitution?

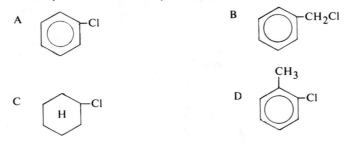

69 In which reactions is a chlorine compound produced by an addition reaction?

A $C_2H_6 + 2 Cl_2 \longrightarrow ClH_2C-CH_2Cl + 2 HCl$

B $CH_4 + 4 Cl_2 \longrightarrow CCl_4 + 4 HCl$

C $H_2C=CH_2 + Cl \longrightarrow ClH_2C-CH_2Cl$

D $HC\equiv CH + HCl \longrightarrow H_2C=CHCl$

70 Which natural raw materials are used in the production of chloromethane?
A natural gas
B coal
C rock salt
D air

71 In substituting chlorination processes, hydrogen chloride is produced as a by-product. Chlorine is then recovered from this HCl. Which statements about this are correct?

A Hydrogen chloride and water together form hydrochloric acid, which yields chlorine and hydrogen during electrolysis.
B The electrolysis of hydrochloric acid is an endothermic reaction.
C The recovery of chlorine increases the price of the chlorination products.
D The conversion of HCl into chlorine is an oxidation process.

72 For which desirable property are chlorinated hydrocarbons notable in comparison with other fat solvents?

A low volatility
B solubility or miscibility in water
C non-flammability or low flammability
D high density

6 Sulphonation and Detergent Substances; Nitration

IV Organic raw materials and large-scale products: Industrial processes

1. Sulphonation

1.1. Definition of the term sulphonation

If **sulphuric acid** is allowed to act on aromatic compounds, for example benzene, a reaction takes place with the formation of benzenesulphonic acid and water.

Benezene   sulphuric acid   benezene-   water
                                      sulphonic acid

This reaction is a **substitution reaction**. A hydrogen atom of the benzene is replaced by a sulphonic acid group. A direct chemical bond between **carbon** and **sulphur** is thus formed.

SO_3H = sulphonic acid group = sulpho group

The chemical process in which the sulpho group is introduced into an organic molecule is called **sulphonation**.

Sulphonic acid, like sulphuric acid is dissociated in aqueous solution, forming hydrogen ions.

$$R - \underset{\underset{\displaystyle H}{|}}{\overset{\overset{\displaystyle H}{|}}{C}} - SO_3H \xrightarrow{\ H_2O\ } R - \underset{\underset{\displaystyle H}{|}}{\overset{\overset{\displaystyle H}{|}}{C}} - SO_3^{\ominus} + H^{\oplus}$$

Molecules containing sulphonic acid groups rank as acids. They have an acid reaction in aqueous solution and form salts with alkalis. These salts are called sulphonates. For example, with sodium hydroxide solution benzenesulphonic acid forms sodium benzenesulphonate.

Benzene-   sodium   sodium   water
sulphonic   hydroxide   benzene-
acid   sulphonate

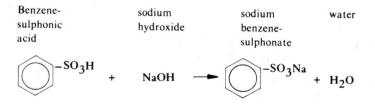

Since the salts of the sulphonic acids in most cases crystallize more readily and, above all, are more soluble in water than the free acids, it is preferable to carry out technical operations with these salts, the sulphonates.

IV Organic raw materials and large-scale products: Industrial processes

The increase in solubility in water caused by introducing sulpho groups into organic compounds is used, inter alia, in the field of dyestuffs and detergent substances.

1.2. Preparation of aliphatic sulpho compounds

When sulphonic acid groups are introduced into organic molecules with a chain structure, aliphatic sulpho compounds are obtained.

The direct sulphonation of paraffins with sulphonic acid is not possible.

Sulphoxidation and sulphochlorination are important industrial processes for the sulphonation or paraffins.

1.2.1. Sulphoxidation

1.2.1.1. The chemistry of the process

In photochemical sulphoxidation, hydrocarbons with a chain length of C_{13} to C_{17} react at room temperature with sulphur dioxide and oxygen. The reactants are brought into a reactive state by UV light.

Paraffin	sulphur dioxide	active oxygen	alkyl-sulphonic acid

$$R-\underset{\underset{H}{|}}{\overset{\overset{H}{|}}{C}}-\underset{\underset{H}{|}}{\overset{\overset{H}{|}}{C}}-H \quad + \quad SO_2 \quad + \quad O \quad \xrightarrow{\text{UV light}} \quad R-\underset{\underset{H}{|}}{\overset{\overset{H}{|}}{C}}-\underset{\underset{H}{|}}{\overset{\overset{H}{|}}{C}}-SO_3H$$

A mixture of different alkylsulphonic acids is formed in the sulphoxidation reaction, since the point of entry of the sulpho group within the paraffin chain is determined by a statistical distribution. For example, in a paraffin chain of 15 C-atoms, the entry of the sulpho group can take place at all the 15 C-atoms, for example:

$$C_{10}H_{21} - \underset{\underset{H}{|}}{\overset{\overset{H}{|}}{C}} - \underset{\underset{H}{|}}{\overset{\overset{H}{|}}{C}} - \underset{\underset{H}{|}}{\overset{\overset{H}{|}}{C}} - \underset{\underset{H}{|}}{\overset{\overset{H}{|}}{C}} - \underset{\underset{H}{|}}{\overset{\overset{H}{|}}{C}} - SO_3H$$

$$C_{10}H_{21} - \underset{\underset{H}{|}}{\overset{\overset{H}{|}}{C}} - \underset{\underset{H}{|}}{\overset{\overset{H}{|}}{C}} - \underset{\underset{H}{|}}{\overset{\overset{H}{|}}{C}} - \underset{\underset{SO_3H}{|}}{\overset{\overset{H}{|}}{C}} - \underset{\underset{H}{|}}{\overset{\overset{H}{|}}{C}} - H$$

$$C_{10}H_{21} - \overset{\overset{\displaystyle H}{|}}{\underset{\underset{\displaystyle H}{|}}{C}} - \overset{\overset{\displaystyle H}{|}}{\underset{\underset{\displaystyle H}{|}}{C}} - \overset{\overset{\displaystyle H}{|}}{\underset{\underset{\displaystyle SO_3H}{|}}{C}} - \overset{\overset{\displaystyle H}{|}}{\underset{\underset{\displaystyle H}{|}}{C}} - \overset{\overset{\displaystyle H}{|}}{\underset{\underset{\displaystyle H}{|}}{C}} - H$$

A mixture of eight different monoalkylsulphonic acids is thus obtained.

1.2.1.2. How the process is carried out

Sulphoxidation by the light/water process can be subdivided into four process stages:

1. Reaction

Hydrocarbons with a chain length of C_{13} to C_{17} react with water, oxygen and sulphur dioxide under high-intensity UV irradiation in a reactor.

The reaction mixture is circulated through a cooler in order to remove the heat of reaction.

2. Phase separation

The resulting reaction mixture passes to a separator, where it forms an aqueous, heavier phase containing the sulphonic acid and sulphuric acid and a lighter phase containing the unreacted hydrocarbons.

The hydrocarbon phase is recycled to the reactor. This is, therefore, a **cyclic process**.

3. Neutralization

The aqueous phase, which contains sulphuric acid as well as the sulphonic acids, is neutralized with sodium hydroxide solution:

Alkylsulphonic acid	sodium hydroxide	sodium alkyl-sulphonate	water

$$R - \overset{\overset{\displaystyle H}{|}}{\underset{\underset{\displaystyle H}{|}}{C}} - \overset{\overset{\displaystyle H}{|}}{\underset{\underset{\displaystyle H}{|}}{C}} - SO_3H + NaOH \longrightarrow R-\overset{\overset{\displaystyle H}{|}}{\underset{\underset{\displaystyle H}{|}}{C}}-\overset{\overset{\displaystyle H}{|}}{\underset{\underset{\displaystyle H}{|}}{C}}-SO_3Na + H_2O$$

sulphuric acid	sodium hyrdoxide	sodium sulphate	water

$$H_2SO_4 + 2\,NaOH \longrightarrow Na_2SO_4 + 2\,H_2O$$

4 Isolating the solid sodium alklsulphonate from the solution

In the last process stage, the solution of sodium alkylsulphonate is evaporated. The end product is a mixture of the various sodium alkylsulphonates, containing sodium sulphate as an impurity.

Flow-sheet: Sulphoxidation

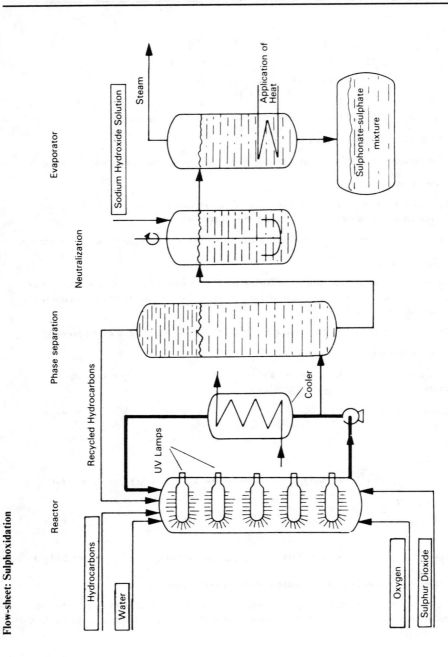

1.2.2. Sulphochlorination

Sulphochlorination is carried out similarly to sulphoxidation, except that the reactive element chlorine is used instead of oxygen.

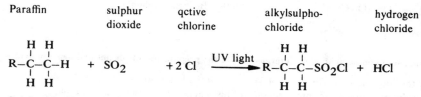

| Paraffin | sulphur dioxide | qctive chlorine | alkylsulpho- chloride | hydrogen chloride |

Sulphochlorination is also a photochemical reaction. The reaction product is a mixture of a wide variety of sulphochlorides, since the entry of the sulphochloride group once again takes place according to a statistical distribution.

Compared with sulphonic acids, sulphochlorides are more reactive. They are used in industry as starting materials for the manufacture of detergents and wetting agents, tanning auxiliaries and plasticizers.

1.3. Preparation of aromatic sulphonic acids

In contrast with aliphatic sulphonic acids, aromatic sulphonic acids can be prepared by direct reaction of aromatic compounds with concentrated sulphuric acid or oleum.

Thus benzenesulphonic acid is obtained from benzene.

| Benzene | sulphuric acid | benzene- sulphonic acid | water |

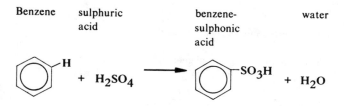

The sulphonic acid group is introduces into aromatic molecules for the following purposes:

a) to render them soluble in water or to increase their solubility in water,
b) to introduce into the molecule, by replacement of the sulphonic acid group, other functional groups that can only be introduced with difficulty or not at all, and
c) in order to prepare, in subsequent reaction stages, derivatives of these sulphonic acids, such as sulphochlorides, sulphonamides and sulphonic acid esters:

$\bigcirc$—SO$_2$Cl   Benzenesulphochloride

$\langle\bigcirc\rangle$—SO_2NH_2 Benzenesulphonamide

$\langle\bigcirc\rangle$—SO_2–O–CH_3 Methyl Benzenesulphonate

1.4. Benzenesulphonic acid as an intermediate in the preparation of phenol and benzoic acid

Besides the Cumene process (see IV, 4.5), some synthetic phenol is prepared via benzenesulphonic acid as an intermediate. In this reaction the sulphonic acid group is replaced by the phenolic group.

1.4.1. Preparation of phenol

The industrial process is carried out by fusing sodium benzenesulphonate with 60% sodium hydroxide solution in a pressure reactor (autoclave) at 270°C (518°F).

| Sodium benzene-bulphonate | sodium hydroxide solution | sodium phenate | sodium sulphite | water |

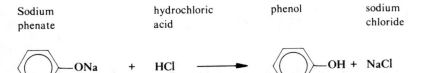

$$\langle\bigcirc\rangle\text{—}SO_3Na \quad + \quad 2\,NaOH \longrightarrow \langle\bigcirc\rangle\text{—}ONa \quad + \quad Na_2SO_3 \quad + \quad H_2O$$

The sodium phenate formed is converted into phenol with acid.

| Sodium phenate | hydrochloric acid | phenol | sodium chloride |

$$\langle\bigcirc\rangle\text{—}ONa \quad + \quad HCl \longrightarrow \langle\bigcirc\rangle\text{—}OH \quad + \quad NaCl$$

The process requires a large consumption of the chemicals sulphuric acid, sodium hydroxide and hydrochloric acid. However, it can be carried out in simple apparatus and is therefore suitable for preparing small quantities.

1.4.2. Preparation and uses of benzoic acid

The **preparation** of benzoic acid from benzenesulphonic acid or its sodium salt is carried out in two stages.

1st stage

The sulpho group is replaced by the nitrile group, -CN.

| Sodium benzene-sulphonate | sodium cyanide | benzo-nitrile | sodium sulphite |

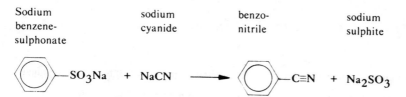

The replacement of the functional group is effected with sodium cyanide, a salt of hydrocyanic acid, HCN.

2nd stage

The nitrile group is saponified, ie split by means of water.

| Benzo-nitrile | water | benzoic acid | ammonia |

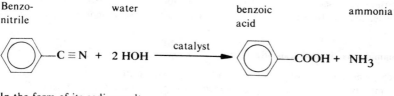

In the form of its sodium salt

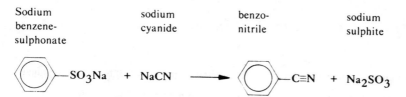

Benzoic acid finds a wide variety of uses as a disinfectant and preservative, for example for fruit juices, syrups, margarine, fish and cosmetics. It is also an important auxiliary for processing rubber.

Sodium benzoate inhibits the corrosion of steel. Wrappings for needles, razor blades and machine parts are therefore impregnated with sodium benzoate. Dissolved in water, it is used as an anti-corrosion agent – for example in refrigerating plants.

The methyl and ethyl esters of benzoic acid are pleasant-smelling liquids. They are used in perfumery as low-priced aromatic essences.

IV Organic raw materials and large-scale products: Industrial processes

Benzoic acid sulphimide

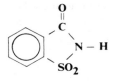

is well known under the name of saccharin. Its sweetening power is 550 times that of sugar.

Since it is eliminated from the body unchanged it has no nutritive value. It is therefore used as a dietetic sweetener for diabetes and cases of obesity and is also used in dietetic jams and beverages.

It is used in the feedstuffs industry as a flavouring component, particularly for feeding pigs.

The cosmetics industry uses saccharin as an additive for toothpastes, mouthwashes and lipsticks.

In the industrial sector, saccharin is used as an additive for electrolytic baths for producing high-gloss nickel coating, for example for umbrellas and bicycle wheel spokes.

1.5. Sulphonamides as chemotherapeutic agents

Sulphonamides, which are derivatives of aromatic sulphonic acids, have acquired considerable importance in chemotherapy. By chemotherapy we mean the treatment of diseases with anti-microbial chemicals, which act against micro-organisms. This action can destroy the organisms or prevent them from growing and multiplying.

In most cases, chemotherapeutic agents have a bacteriostatic action, that is to say they inhibit growth. In this way they facilitate the destruction of the bacteria by the natural defensive forces of the body.

The most effective sulphonamides are derivatives of sulphanilamides in which the hydrogen atoms linked to the nitrogen have been substituted by organic radicals.

Sulphanilamide

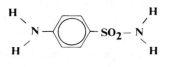

The term "sulpha" is used to designate the radical:

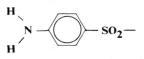

Accordingly, the names of the N-substituted sulphonamides are formed from the prefix "sulpha" and the amine or amide attached to the N-atom.

The following are a few sulphonamides with a therapeutic action:

Sulphaguanidine   Substituent: guanidine

$$H_2N-\langle\bigcirc\rangle-SO_2-NH-\underset{\underset{NH}{\|}}{C}-NH_2 \qquad\qquad H_2N-\underset{\underset{NH}{\|}}{C}-NH_2$$

Sulphaacetamide   Substituent: acetamide

$$H_2N-\langle\bigcirc\rangle-SO_2-\underset{\underset{H}{|}}{N}-\underset{\overset{O}{\|}}{C}-CH_3 \qquad\qquad H_2N-\overset{\overset{O}{\|}}{C}-CH_3$$

Sulphathiourea   Substituent: thiourea

$$H_2N-\langle\bigcirc\rangle-SO_2-NH-\underset{\underset{S}{\|}}{C}-NH_2 \qquad\qquad H_2N-\underset{\underset{S}{\|}}{C}-NH_2$$

The action of the sulphonamides can be explained, in terms of a model, on the basis of their structural similarity to p-aminobenzoic acid.

p-Aminobenzoic acid, abbreviated to PABA, is a basic unit of the growth vitamin folic acid. Certain bacterial causative organisms also require folic acid in order to multiply. If compounds with a similar chemical structure are incorporated in the molecule instead of PABA, substances similar to folic acid are formed, but they no longer have the specific properties of folic acid. This interrupts the growth of the bacteria.

p-Aminobenzoic acid   Sulphanilamide

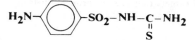

$$H_2N-\langle\bigcirc\rangle-COOH \qquad\qquad H_2N-\langle\bigcirc\rangle-SO_2-NH_2$$

1.6. Sulphonylureas as antidiabetics

1.6.1. Diabetes and remedies for it

Diabetes is caused by insulin deficiency. Insulin is a hormone formed by the pancreas.

Among the industrialized nations, diabetes mellitus has become widespread. 1.5 to 2.5% of the population in the highly civilized industrialized countries suffer from this disease.

IV Organic raw materials and large-scale products: Industrial processes

It is becoming increasingly difficult to maintain adequate supplies of insulin, since it is extracted from pancreas of cattle and pigs. The number of these animals is limited. In this field, genetic engineering appears to be opening undreamed-of possibilities.

The deoxyribonucleic acids are the carriers of genetic information in the cells (see III/7–00). By selective interference in the structures of these DNA (deoxyribonucleic acids) of certain micro-organisms, the latter are rendered capable of producing insulin.

Insulin deficiency can arise from two causes:
a. the production of insulin by the pancreas has been reduced or interrupted;
b. the secretion of insulin, that is to say the release of insulin, by the pancreas is blocked; this is the typical diabetes of old age.

In the first case, extraneous insulin, obtained from animal pancreases, must be supplied to the body continually. This is effected by daily injections of insulin.

As far as the second case is concerned, blockage of insulin secretion, the sulphonylureas have been found to cause the secretion of the insulin produced and stored in the pancreas and to stimulate further production of insulin.

Sulphonylureas can also be administered in the form of tablets that are not broken down in the gastro-intestinal tract.

Sulphonylureas have the general formulae:

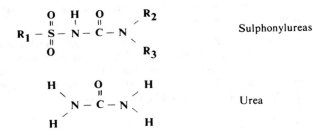

the structure of the radicals R_1, R_2 and R_3 being important for the hypoglycaemic action.

As an example, the structural formula of one sulphonylurea used as an antidiabetic is:

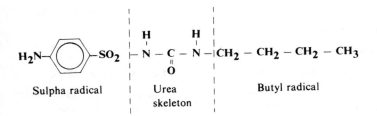

Sulpha radical   Urea skeleton   Butyl radical

This sulphonylurea is called Carbutamid and it is a derivative of sulphanilamide (sulphaurea).

Tolbutamid is another, very similar, antidiabetic.

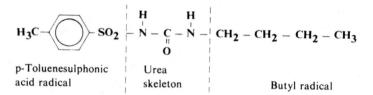

| p-Toluenesulphonic acid radical | Urea skeleton | Butyl radical |

1.6.2. Preparation of sulphonylureas

Sulphonylureas are prepared from sulphonamides in two stages.

1st stage

The sulphonamide reacts with phosgene to form the sulphonyl isocyanate.

| Sulphonamide | phosgene | sulphonyl isocyanate | hydrogen chloride |

$$R_1-SO_2-N \begin{matrix} H \cdots Cl \\ \\ H \cdots Cl \end{matrix} \quad + \quad C=O \quad \longrightarrow \quad R_1-SO_2-N=C=O + 2\ HCl$$

2nd stage

The isocyanate reacts with an amine.

| Sulphonyl isocyanate | amine | sulphonylurea |

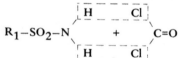 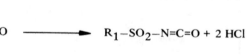

$$R_1-SO_2-N=C=O \quad + \quad H-N\begin{matrix} R_2 \\ R_3 \end{matrix} \quad \longrightarrow \quad R_1-SO_2-\overset{H}{N}-\underset{O}{\overset{\|}{C}}-N\begin{matrix} R_2 \\ R_3 \end{matrix}$$

2. Detergent substances

2.1. Chemical structure of detergent substances

Hydrocarbons, such as petrol and mineral oils, are completely insoluble in water. Acids, such as acetic acid and sulphuric acid, are miscible with water in all proportions.

IV Organic raw materials and large-scale products: Industrial processes

This is connected with the fact that water, as a polar solvent, is best capable of dissolving those substances that are also polar. As a result of mutual attraction, a sheath of water molecules is formed around the dissolved particles, isolating them from one another.

This sheathing with H_2O molecules (hydration) is particularly marked with ions, because they are electrically charged.

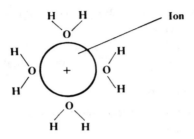

Compounds formed from sulphuric acid and hydrocarbons, for example the aliphatic sulphonic acids, consist of a water-repelling part, the hydrocarbon radical, and a water-attracting part, the sulphonic acid group.

Representation by formula   Representation by symbol

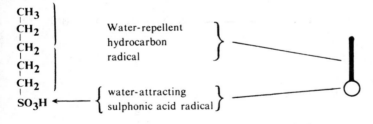

> Water-attracting groups are termed **hydrophilic**
> Water-repellent groups are termed **hydrophobic**

The following are examples of hydrophilic groups:

the carboxylic acid group $-COO^{\ominus} H^{\oplus}$
the sulphonic acid group $-SO_3^{\ominus} H^{\oplus}$
the sulphate group $-O-SO_3^{\ominus}H^{\oplus}$

the quaternary
ammonium group

$$\left[R - \overset{\overset{\displaystyle CH_3}{|}}{\underset{\underset{\displaystyle CH_3}{|}}{\overset{\oplus}{N}}} - CH_3 \right] Cl^{\ominus}$$

polyethylene oxide groups

$$-O-(CH_2-CH_2O)_n-H$$
$$n = 5 \text{ to } 10$$

The hydrophilic character of the polyethylene oxide groups stems from the many oxygen atoms that are incorporated in the chain of the molecule.

Hydrophobic groups are especially the long-chain aliphatic or aliphatic-aromatic radicals, such as:

$$CH_3-(CH_2)_n-$$ alkyl radical
$$n = 10 \text{ to } 20$$

$$CH_3-(CH_2)_m-\langle\bigcirc\rangle-$$ alkylphenyl radical
$$m = 8 \text{ to } 16$$

Aliphatic sulphonic acids with a short hydrocarbon chain are readily soluble in water. As the chain length of the hydrophobic group increases, the solubility in water of the sulphonic acid decreases.

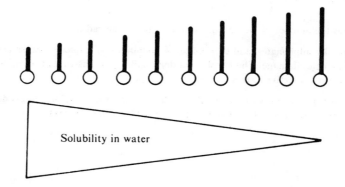

Solubility in water

2.2. Ionic substances

2.2.1. Anionic substances

If the hydrophilic group is an acid group, such as the sulphonic acid group $-SO_3H$, this group forms salts that dissociate in aqeuous solution forming radical ions and metal ions.

IV Organic raw materials and large-scale products: Industrial processes

Since the detergent part, the acid radical ion, is an anion, such substances are called **anionic**.

Symbol   ⬤   (−) indicates the negative charge of the anion

The sodium and potassium salts of anionic compounds have become very popular as detergents, because they are dissociated into ions in aqueous solution to a greater extent than the comparable sulphonic acids and the resulting alkalinity has a favourable effect on the detergent action.

Soaps are the oldest anionic detergent substances. Hard soap consists of the sodium salts of stearic and palmitic acid; soft soap consists of the potassium salts of these acids. These soaps are obtained by saponifying fats.

The most important synthetic anionic substances include the following:

$CH_3-(CH_2)_n-CH_2 - \bigcirc - SO_3^{\ominus} Na^{\oplus}$ Sodium alkylbenzenesulphonate

$CH_3-(CH_2)_n-CH_2-SO_3^{\ominus} Na^{\oplus}$ Sodium alkylsulphonate

$CH_3-(CH_2)_n-CH_2-O-SO_3^{\ominus} Na^{\oplus}$ Sodium alkyl sulphate

"n" indicates the number of CH_2 groups in the hydrocarbon radical.

The sodium alkylsulphonates, and even more the sodium alkyl sulphates, cause particularly little environmental damage. Their biological degradability surpasses that of all other categories of surface-active agents.

The graph on page 17 shows that one half of a linear alkylbenzenesulphonate is degraded biologically after two days, 90% after five days and 100% after ten days.

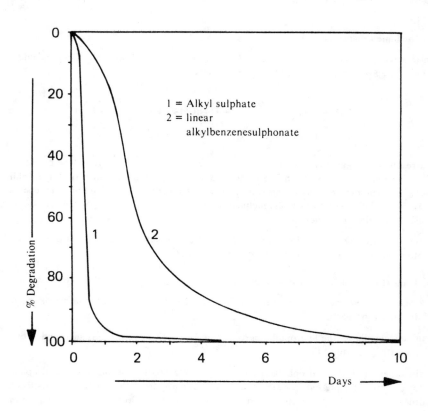

1 = Alkyl sulphate
2 = linear alkylbenzenesulphonate

2.2.2. Cationic substances

If the detergent part of the substance is a cation, that is to say a positively charged ion, the substance is called **cationic**.

Examples of cationic detergent substances are quaternary ammonium compounds, such as an alkyltrimethylammonium chloride:

$$\left[CH_3-(CH_2)_n-\overset{\overset{\displaystyle CH_3}{\displaystyle |}}{\underset{\underset{\displaystyle CH_3}{\displaystyle |}}{\overset{\oplus}{N}}}-CH_3 \right] \quad Cl^{\ominus} \qquad n = 10 \text{ to } 16$$

Cationic substances are poisonous to micro-organisms and thus form the basis of cleansing agents with a disinfecting action.

2.3. Nonionic substances

Substances that do not form ions in aqueous solution are termed nonionic. They include, for example, the alkylphenol polyethylene glycol ethers.

$$CH_3-(CH_2)_n - \langle \bigcirc \rangle - O-(CH_2-CH_2-O)_m-H$$

hydrophobic radical   hydrophilic radical

As a result of its agglomeration of polar oxygen-carbon bonds, the polyethylene glycol radical is capable of forming a hydrate sheath and thus fulfils the function of a hydrophilic group. Nonionic detergent substances are neutral chemically and their action is independent of pH. Since their solutions are less inclined to produce foam, they are particularly suitable for use in washing machines.

Because they are inert physiologically, nonionic detergent substances are used preferentially as emulsifiers in medicinal and cosmetic formulations.

2.4. Properties of surface-active substances

As a result of their chemical structure, surface-active substances have a number of special and distinctive properties.

In an aqueous phase, the hydrophilic part of a surface-active substance is surrounded by a water sheath (see page 14)

On the other hand, the hydrophobic part attempts to leave the aqueous phase. If droplets of oil, for example, are present in water, the oil-attracting hydrophobic parts of the molecule (hydrocarbon radicals) penetrate into the oil droplet, while the hydrophilic part is held in the aqueous phase by its hydrate sheath.

The result of this is that the substance is concentrated, in an ordered manner, in the oil/water interface.

> Substances that are concentrated in an ordered manner in interfaces are known as surface-active substances.

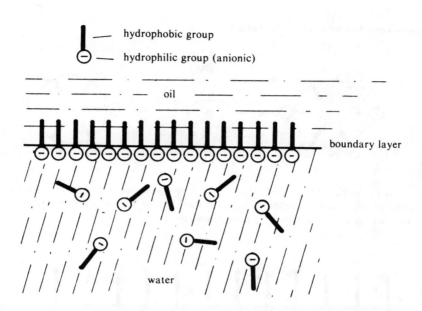

Since ionic surface-active substances dissociate into ions in aqueous solution, the particles that are concentrated in the interfaces are electrically charged:

For example: $R - SO_3Na \longrightarrow RSO_3^{\ominus} + Na^{\oplus}$

Symbols:

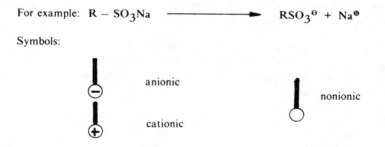

Any droplets of oil or dirt particles present in the water become electrically charged in this way. Because they have a similar charge, the individual particles have a mutual repulsion for one another. This makes agglomeration more difficult, but promotes further dispersion.

Repulsion caused by being similarly charged

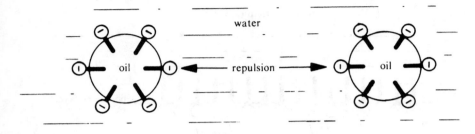

The surface-active particles are also concentrated at the water/air interface. The molecule orients itself in the water/air interface in such a way that the hydrophilic part remains in the water and the water-repellent part projects into the air.

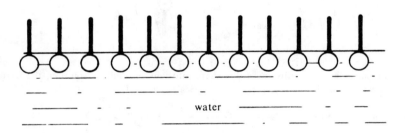

The result of this is to reduce the surface tension of the water. By surface tension we mean the work that must be carried out in order to enlarge the surface of the liquid, that is to say to break up a large drop into several small drops, for example, or to enlarge the surface of the liquid by the formation of foam.

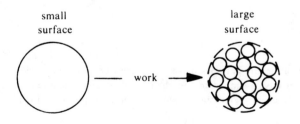

The addition of surface-active substances thus makes it easier for water to penetrate into very small cavities, such as exist in textile fabrics, for example, and makes it easier to wet water-repellent substances such as dirt, fat and oil. These properties make surface-active substances the essential components of washing, rinsing and cleansing agents. Here the sulphonates are superior to conventional soaps, because they form **soluble** salts with the calcium and magnesium salts that are essentially responsible for the hardness of water, and thus do not tend to give textiles a grey colour.

However, since the Ca and Mg sulphonates have hardly any detergent action, polyphosphates are added to detergents in order to bind the calcium and magnesium ions (see II,6).

Besides detergent substances and polyphosphates, the complete detergents on the market contain a number of other active ingredients.

A complete detergent can have the following composition:

Group of active ingredient	Example	% content
1. Surface-active substance	Alkylbenzene sulphate	10 to 15
2. Complexing agent	Pentasodium triphosphate	35 to 45
3. Bleaching agent	Sodium perborate	20 to 30
4. Optical brightener	Stilbene derivatives	0.1 to 0.3
5. Anti-greying inhibitors	Carboxymethyl-cellulose	0.5 to 1.0
6. Softening rinsing agent	Soaps	5
7. Foam regulators	—	3 to 5
8. Stabilizers	—	0.2 to 2
9. Perfume oils	—	0.2

An optimum washing effect is only produced by a combination of the detergent present in the washing liquid, the temperature of the liquid and vigorous agitation of the liquid and the goods being washed (for example in a washing drum).

Besides their use as detergents, surface-active substances are also employed for fire-fighting (foam), for dust suppression (in mining), for the production of emulsions of plastics and bitumen and in the building industry. Special types are used as boring and cutting oils in metal fabrication.

The separation of rock from minerals with the aid of surface-active substances plays a very important part in the mining of ores and coal. The process is called **flotation**.

The material is finely ground and treated with interceptors, which are surface-active substances that coat the surfaces of certain minerals, for example metal sulphides or metal oxides, and thus render them **unwettable**. When brought into water through which air is passed, the ore particles that have been rendered unwettable attach themselves to the air bubbles and reach the surface, while the waste rock (gangue) gets wetted and falls to the bottom. In order to keep the ore particles on the surface of the liquid, a strongly foaming surface-active substance is added to the water. The foam supports the ore particles so that they can be skimmed off.

3. Summary of products made from sulpho compounds

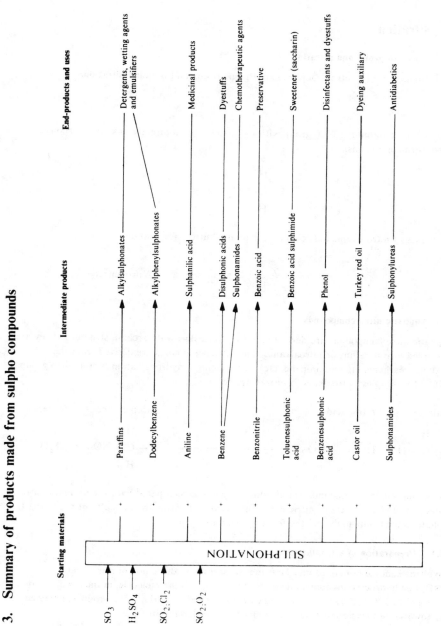

Starting materials		Intermediate products	End-products and uses
SO_3	Paraffins +	Alkylsulphonates	Detergents, wetting agents and emulsifiers
H_2SO_4	Dodecylbenzene +	Alkylphenylsulphonates	
SO_2Cl_2	Aniline +	Sulphanilic acid	Medicinal products
SO_2O_2	Benzene +	Disulphonic acids	Dyestuffs
		Sulphonamides	Chemotherapeutic agents
	Benzonitrile +	Benzoic acid	Preservative
	Toluenesulphonic acid +	Benzoic acid sulphimide	Sweetener (saccharin)
	Benzenesulphonic acid +	Phenol	Disinfectants and dyestuffs
	Castor oil +	Turkey red oil	Dyeing auxiliary
	Sulphonamides +	Sulphonylureas	Antidiabetics

SULPHONATION

4. Nitration

4.1. The nitro group and nitration

The introduction of a nitro group into an organic compound is called **nitration.**

$$-NO_2 = \text{nitro group}$$

Compounds containing a NO_2 group with its nitrogen atom directly linked to carbon, are called nitro compounds.

$$
\begin{array}{c}
H \\
| \\
R-C-NO_2 \\
| \\
H
\end{array}
$$

The names of nitro compounds contain the prefix "nitro", for example

—NO_2 Nitrobenzene

4.2. Aliphatic nitro compounds

Aliphatic nitro compounds are derivatives of hydrocarbons with a chain structure. They are also called nitroparaffins or nitroalkanes. The nitro compounds are derived from the paraffins (methane, ethane, propane, etc) by replacing a hydrogen atom by the nitro group, $-NO_2$. The simplest nitroalkane is nitromethane.

Methane   nitric acid   nitromethane   water

The aliphatic nitro compounds are of minor importance compared with the aromatic nitro compounds and the esters of nitric acid. The reason for this is the difficulty of preparing the aliphatic nitro compounds and the few uses open to them.

4.2.1. Preparation of nitroalkanes

The nitroalkanes are obtained on a large industrial scale by the action of nitric acid (HNO_3) on hydrocarbons containing up to five C-atoms in the molecule, in the gas phase at about 400 to 500°C (752 to 932°F) and with a short residence time. The products obtained are, however, not single substances but a mixture of different nitro compounds.

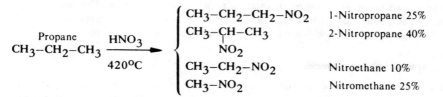

$CH_3-CH_2-CH_2-NO_2$		1-Nitropropane 25%
$CH_3-CH-CH_3$		2-Nitropropane 40%
$\qquad \vert$		
$\qquad NO_2$		
$CH_3-CH_2-NO_2$		Nitroethane 10%
CH_3-NO_2		Nitromethane 25%

Propane
$CH_3-CH_2-CH_3 \xrightarrow[420^{\circ}C]{HNO_3}$

Special separation processes are required to isolate the products of the reaction.

4.2.2. Uses of aliphatic nitro compounds

Nitroalkanes are mainly used as solvents. From a mixture of aromatics and paraffins, nitromethane dissolves virtually only the aromatics, and it is therefore employed as a selective solvent for obtaining aromatics from crude oil fractions. Nitroalkanes are also good solvents for cellulose esters, Buna, vinyl resins and nitrocellulose.

Since nitroalkanes and chloronitroalkanes are capable of dissolving the catalyst aluminium chloride, they are employed as solvents in alkylation reactions (Friedel-Crafts reactions).

Mixed with compounds rich in oxygen, such as H_2O_2 and N_2O_4, nitroalkanes are employed as rocket propellants. Because of their high energy content, nitroparaffins are also an ingredient of special fuels.

Halogenonitroalkanes, such as chloropicrin (trichloronitromethane)

$$Cl - \overset{\overset{\displaystyle Cl}{\vert}}{\underset{\underset{\displaystyle Cl}{\vert}}{C}} - NO_2$$

and Ethide (1,1-dichloro-1-nitroethane)

$$H - \overset{\overset{\displaystyle H}{\vert}}{\underset{\underset{\displaystyle H}{\vert}}{C}} - \overset{\overset{\displaystyle Cl}{\vert}}{\underset{\underset{\displaystyle Cl}{\vert}}{C}} - NO_2$$

are used as insecticides for sterilizing the soil and for fumigating cereals and tobacco.

Aminoalcohols, such as

$$H - \overset{\overset{\displaystyle H}{\vert}}{\underset{\underset{\displaystyle H}{\vert}}{C}} - \overset{\overset{\displaystyle H}{\vert}}{\underset{\underset{\displaystyle H}{\vert}}{C}} - \overset{\overset{\displaystyle H}{\vert}}{\underset{\underset{\displaystyle NH_2}{\vert}}{C}} - \overset{\overset{\displaystyle H}{\vert}}{\underset{\underset{\displaystyle H}{\vert}}{C}} - OH$$

2-Amino-1-butanol

can be prepared from nitroalcohols by catalytic reduction with hydrogen. They are important emulsifiers for the industries producing textile auxiliaries, detergents and polishing agents. Aminoalcohols are also used as stabilizers for melamine resins.

4.3. Aromatic nitro compounds

The preparation of aromatic nitro compounds has been a matter of outstanding importance for a long time. Nitration takes place more readily and more rapidly than in the case of the aliphatic products. Most aromatic nitro compounds are colourless, yellowish or yellow solids. Only a few are, like nitrobenzene, liquid at room temperature.

Nitration is carried out by means of "nitrating acid", a mixture of concentrated nitric acid and concentrated sulphuric acid.

The sulphuric acid activates the nitric acid and thus enables nitration to take place.

4.3.1. Nitrobenzene and its manufacture on a large industrial scale

The chemistry of the process

Nitrobenzene is the best known and most important aromatic nitro compound:

 Nitrobenzene

On reduction it gives aniline, which occupies a position of fundamental importance as a starting material in the chemistry of dyestuffs.

Benzene is converted into nitrobenzene by means of nitrating acid ($HNO_3 + H_2SO_4$).

Benzene   nitric acid   nitrobenzene   water

 + HNO_3 $\xrightarrow{H_2SO_4}$ —NO_2 + H_2O

In general, the yield of pure nitrobenzene amounts to about 85%.

How the process is carried out

Concentrated sulphuric acid and concentrated nitric acid are mixed in a stirred kettle to form nitrating acid.

Benzene and nitrating acid are introduced into the nitrator in the correct stoichiometric ratio. The reaction leading to nitrobenzene is carried out with stirring and by applying heat. The reaction product passes into the separator, in which the crude nitrobenzene collects at the top, as the lighter component, since nitrating acid is not miscible with nitrobenzene or benzene. The residual nitrating acid is drained off. Residual acid is then removed by neutralizing the crude nitrobenzene with sodium carbonate in a scrubber and washing it with water. Since the density of nitrobenzene is higher than that of water, it can be drained off at the bottom. It is then purified by distillation.

Residues of water and benzene are removed in the first distillation stage.

Distillation to give pure nitrobenzene is carried out in the second stage.

The pure nitrobenzene obtained is run into storage containers.

4.3.2. Aniline

Aniline is the trivial name for aminobenzene

It is prepared from nitrobenzene by reduction with hydrogen, using a copper catalyst.

Nitrobenzene   hydrogen   aniline   water

$$\text{C}_6\text{H}_5{-}NO_2 \; + \; 6\,H \longrightarrow \text{C}_6\text{H}_5{-}NH_2 \; + \; 2\,H_2O$$

If the reduction of nitrobenzene is carried out with iron turnings and hydrochloric acid, valuable iron oxide pigments are obtained as well as nitrobenzene.

Flow sheet for nitrobenzene manufacture

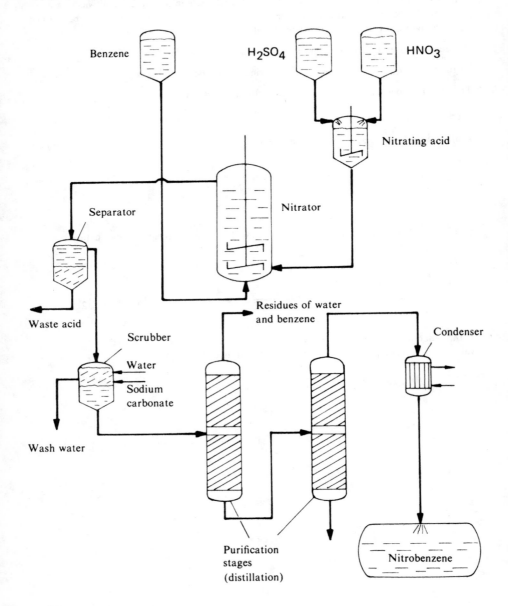

Various crude pigments can be obtained by varying the reaction conditions.

The colours range from yellow through browns and red to black (see II,8).

Aniline has become so important because it is used as the starting material for azo dyes (see IV,9).

Aniline is also a starting material for the synthesis of many pharmaceutical products. One of the simplest aniline derivatives, which, amongst other things is a constituent of many pain-relieving tablets, is phenacetin (p-ethoxyacetanilide).

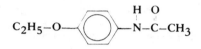

Phenacetin

4.3.3. Trinitrotoluene (TNT)

Trinitotoluene, which is known in industry by the name "Trotyl", is used on a large scale as an explosive. Compared with other explosives it possesses the advantage that it can be handled safely and can only be made to detonate by means of a primer.

In particular, TNT has found widespread use as a mining explosive.

It is manufactured on a large industrial scale by first nitrating toluene under cold conditions to give a mixture of the mononitro compounds.

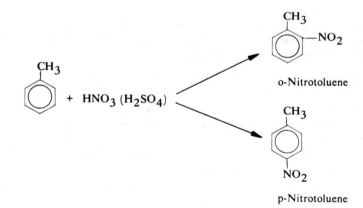

o-Nitrotoluene

p-Nitrotoluene

First the dinitro compound and then the trinitro compound are formed by using stronger
nitrating acid and raising the temperature to 90°C (194°F):

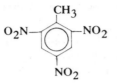

2,4,6-Trinitrotoluene

Trinitrotoluene is also used as a measure of explosive power; this is spcified in "Tonnes of
TNT".

4.3.4. Chloramphenicol

The best known natural nitro compound, and the first to be discovered, is chloramphenicol.

It has the following structural formula:

O_2N—⟨benzene ring⟩— $\overset{\underset{|}{H}}{C}$ – C–H with OH below, CH_2–OH, $\overset{\overset{H}{|}}{N}$ – $\overset{\overset{O}{||}}{C}$ – $\overset{\overset{Cl}{|}}{C}$ – H, Cl

Chloramphenicol was first obtained from Streptomyces venezuela, a species of ray fungus.
It is a **broad spectrum antibiotic** and is used against various typhus diseases and against the
causative organisms of tuberculosis and pneumonia.

This antibiotic is nowadays manufactured solely by synthetic means.

4.3.5. E 605

The highly effective insecticide known by the name E 605 has the following structure:

$$O_2N-\langle\,\rangle - O - P = S \begin{array}{c} O - CH_2 - CH_3 \\ \\ O - CH_2 - CH_3 \end{array}$$

and its chemical name is diethyl nitrophenyl thiophosphate.

Its structure is based on p-nitrophenol from which it is synthesized:

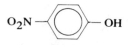

4.4. Esters of niric acid

4.4.1. Definition

Esters of nitric acid are compounds in which the nitro group is linked to an organic radical **through an oxygen atom**. They are also called nitrates and thus do not belong to the nitro compounds.

| Alcohol | nitric acid | | nitric acid ester | water |

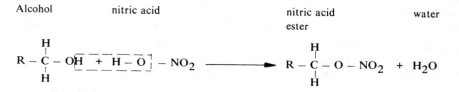

4.4.2. Nitroglycol

The action of nitrating acid (mixed nitric and sulphuric acids) on ethylene glycol gives the glycol ester of nitric acid (nitroglycol).

| Glycol | nitric acid | | nitroglycol | water |

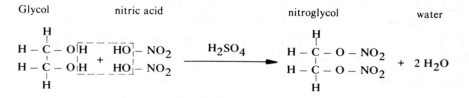

Nitroglycol is the main constituent of the Ammon-Gelit safety explosives, which are permitted for transport by rail, since there is no risk of explosion in the event of their catching fire.

4.4.3. Nitroglycerine and dynamite

Nitroglycerine is the trivial name used (wrongly) for the glycerol ester of nitric acid (glycerol trinitrate). It is prepared analogously to glycol nitrate by means of nitrating acid.

IV Organic raw materials and large-scale products: Industrial processes

Glycerol nitric acid nitroglycerine water

$$
\begin{array}{c}
\text{H} \\
| \\
\text{H} - \text{C} - \text{O} - \\
| \\
\text{H} - \text{C} - \text{O} - \\
| \\
\text{H} - \text{C} - \text{O} - \\
| \\
\text{H}
\end{array}
\begin{array}{c}
\text{H} \quad\quad \text{H} - \text{O} - \text{NO}_2 \\
\text{H} \;+\; \text{H} - \text{O} - \text{NO}_2 \\
\text{H} \quad\quad \text{H} - \text{O} - \text{NO}_2
\end{array}
\;\xrightarrow{\text{H}_2\text{SO}_4}\;
\begin{array}{c}
\text{H} \\
| \\
\text{H} - \text{C} - \text{O} - \text{NO}_2 \\
| \\
\text{H} - \text{C} - \text{O} - \text{NO}_2 \;+\; 3\,\text{H}_2\text{O}\\
| \\
\text{H} - \text{C} - \text{O} - \text{NO}_2 \\
| \\
\text{H}
\end{array}
$$

Nitroglycerine is one of the most important and most used constituents of explosives. Together with nitroglycol, it forms the basis of dynamite and blasting gelatine.

Blasting gelatine is a mixture containing 93% of nitroglycerine and 7% of collodion cotton. Blasting gelatine is one of the most effective mining explosives for blasting. Any nitroglycerine explosive can freeze at temperatures as high as $+12°C$ ($53.6°F$). In a partially thawed state, this material is very dangerous, since it can explode as a result of impact or friction.

In order to reduce the freezing point (to render it safe against frost) about 40% of the nitroglycerine is therefore replaced by nitroglycol, which does not freeze until $-22°C$ ($-7.6°F$); this gives a frost resistance of $-15°C$ to $-20°C$ ($-7°F$ to $-4°F$).

Nowadays, dynamite has become a general term for explosives based on nitroglycerine.

The original dynamite consisted of 75% of nitroglycerine and 25% of kieselguhr. The liquid nitroglycerine was absorbed by the kieselguhr, giving a solid explosive of low sensitivity to impact and friction.

Combined with nitrocellulose and stabilizers, nitroglycerine is an important constituent of propellants, powders and propelling charges for rockets.

In medicine too, nitroglycerine has a certain importance as an agent against asthma and arterial sclerosis and as an agent for enlarging the coronary vessels.

4.4.4. Nitrocellulose

Nitrocellulose is obtained similarly by nitrating cellulose (in most cases obtained from cotton) with nitrating acid.

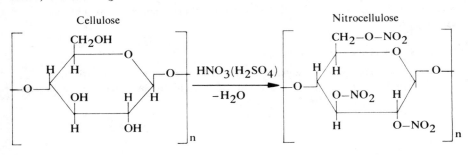

Cellulose → Nitrocellulose

According to this equation, the theoretically achievable proportion of nitrogen in the molecule should be 14.14%.

Nitrocellulose manufactured industrially always contains less nitrogen than this. Thus, not all the OH groups in the cellulose are esterified with nitric acid.

Nitrocellulose reaches the market in varying degrees of nitration. Nitrocellulose containing 12.6 to 13.4% of nitrogen (highly nitrated cellulose) is called "guncotton" and is used as a smokeless powder for small arms. Large quantities of nitrocellulose are also converted into nitro-lacquers.

Cellulose with a lower degree of nitration is called collodion cotton and contains 10 to 11% of nitrogen.

Blasting gelatine is made from collodion cotton, combined with nitroglycerine. Plasticizing collodion cotton with camphor gives the thermoplastic plastic material celluloid, which is used for the manufacture of films of all kinds, protective casings, table tennis balls and various other things.

5.   Summary of products made from nitro compounds

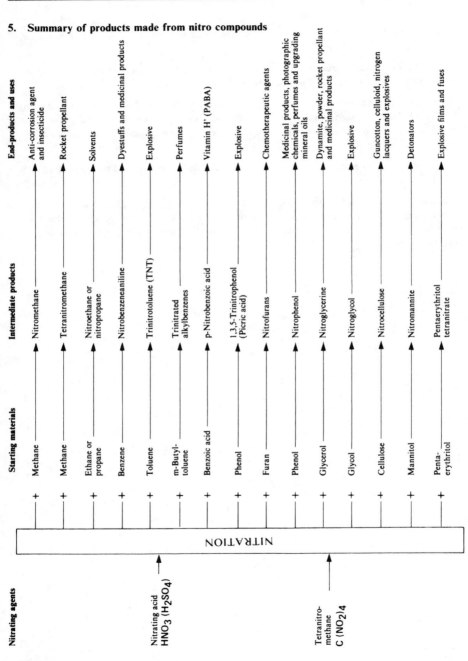

6.   Self-evaluation questions

73 Which statements about sulphonation are correct?

A   Sulphonation takes place with sulphurous acid, H_2SO_3.
B   Sulphonated compounds are acids.
C   The introduction of sulpho groups into organic compounds makes these latter more readily water-soluble.
D   The reaction of alkanes with sulphuric acid produces aliphatic sulpho compounds.

74 Sulphonation is a

A   substitution reaction
B   addition reaction
C   neutralization process
D   esterification process

75 The reaction

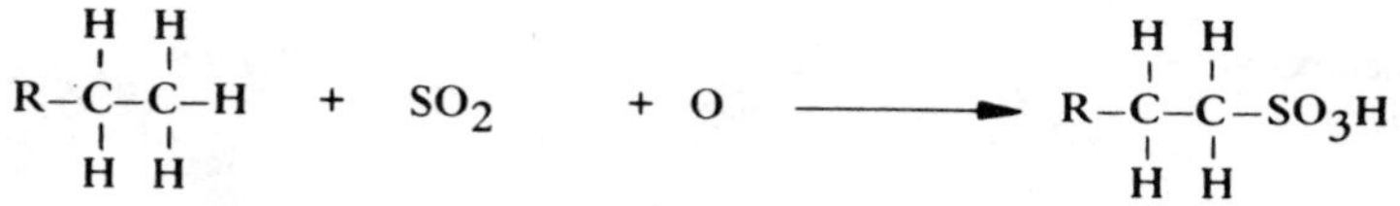

$$\begin{array}{c} H\ \ H \\ |\ \ \ | \\ R-C-C-H \\ |\ \ \ | \\ H\ \ H \end{array} \ +\ SO_2\ \ \ +\ O\ \longrightarrow\ \begin{array}{c} H\ \ H \\ |\ \ \ | \\ R-C-C-SO_3H \\ |\ \ \ | \\ H\ \ H \end{array}$$

is

A   an addition reaction
B   usable for the production of aliphatic sulpho compounds
C   a sulphonation process

76 Sodium benzoate   ⬡— COONa is used

A   as a preservative
B   as a disinfectant
C   as an anticorrosive agent
D   in the rubber industry

IV Organic raw materials and large-scale products: Industrial processes

77 In sulphochlorination

    A   a saturated hydrocarbon is reacted with sulphur dioxide and chlorine

    B   a hydrogen atom of an alkane is replaced by an SO_2Cl group

    C   a hydrogen atom of an alkane is replaced by an SO_2 group and a second hydrogen atom is replaced by a chlorine atom

    D   hydrogen chloride is released

78 By approximately what factor is the sweetness of benzoic acid sulphimide (saccharin) greater than that of sugar?

    A   5

    B   50

    C   500

    D   1000

79 Sulphonamides are important as

    A   detergents

    B   insecticides

    C   chemotherapeutic agents

    D   herbicides

80 Which groups in a molecule or ion are hydrophilic?

    A   all water-repellent groups

    B   carboxyl groups

    C SO_3Na groups

    D   alkyl groups

81 How does the water-solubility of aliphatic sulphonic acids change with an increase in the length of the hydrocarbon chain?

    A   not at all

    B   increases

    C   decreases

    D   varies to an irregular extent

82 Which substances are classed among the anionic detergents?

    A   soaps

    B   sodium alkyl sulphonates

    C   alkylphenyl polyethylene glycol ethers

    D   alkyltrimethyl ammonium chloride

83 What advantages do the sulphonates have over conventional soaps as detergent components?

A They are better bleaching agents.
B They are more effective in hard water.
C They are less conducive to greying of textiles.
D Unlike soaps they are anionic.

84 Nitrogenation of aromatic compounds is carried out with

A pure nitric acid
B pure nitrous acid
C a mixture of nitrous acid and concentrated sulphuric acid
D a mixture of concentrated nitric acid and concentrated sulphuric acid

85 Aniline  NH_2 is obtained by

A the reaction of benzene with ammonia
B the reduction of nitrobenzene
C the hydrogenation of benzene
D the dehydrogenation of nitrobenzene

86 Which compounds are esters of nitric acid?

A $R- CH_2- O - NO_2$

B $HO-$⟨◯⟩$-NO_2$

C nitrocellulose
D nitroglycerin

87 Nitroglycerin is

A a constituent of asthma drugs
B a constituent of dynamite explosives
C an ester
D a constituent of cellulose

IV Organic raw materials and large-scale products: Industrial processes

88 Which industrial processes are photochemical reactions?

    A   methane chlorination
    B   sulphonation
    C   sulphoxidation
    D   sulphochlorination

7 Plastics, Synthetic Resins and Synthetic Rubber

1. From a substitute to a tailor-made chemical material

Little attention was paid in 1909, when the Belgian chemist Baekeland, the inventor of Bakelite, reported his experiments with phenol and formaldehyde, which finally led to "plastics", in the Chemikerzeitung. The change came with the First World War. In Germany, which was short of raw materials, it was necessary to replace at least some imports by plastics. An example of this was the manufacture of methyl rubber at the Farbenfabriken Bayer in Germany.

The efforts to achieve self-sufficiency in the years after 1933 also led to the rapid construction of plants for the manufacture of plastics and synthetic rubbers.

The manufacture of synthetic rubber on an industrial scale was achieved in the middle of the 1930s, at about the same time in Germany, the USA and the Soviet Union. In Germany, the industrial synthesis of rubber was initially carried out by polymerizing **bu**tadiene using sodium (**na**trium) as the initiator. This is how the name Buna arose.

However, the mass production of articles made of plastics did not begin until about the end of the 1950s. Owing to their wide range of properties and ways in which they can be fabricated, plastics are used not only to replace paper, wood, building materials, aluminium, glass or steel, but also for numerous end uses that are not possible with conventional materials.

Amongst the outstanding properties of plastics, one may mention the ease with which they can be shaped, the ease with which solid plastics can be machined, their low density and also their good resistance to chemicals, particularly acids and alkalis, and their electrical insulating properties.

The science of macromolecular chemistry was founded by the Freiburg Professor and Nobel prizewinner Hermann Staudinger in 1922; after the structure of naturally occurring macromolecules (cellulose, protein and rubber) had been revealed, it became possible to modify natural materials by reactions directed towards a specific purpose or to construct plastics having desired properties by fully synthetic routes.

World production of plastics in 1950 was only 1.5 million t; by 1974 it had already risen to 44 million t and estimates for 1980 vary between 80 and 120 million t.

The production of plastics in 1979 amounted in
the USA to	19.600 million tonnes
Japan to	8.209 million tonnes
West Germany to	7.255 million tonnes
the USSR to	2.911 million tonnes
Great Britain to	2.647 million tonnes.

Figure 1: Manufacture of plastics in leading countries, in million tonnes

Legend:
- 1970
- 1974
- 1978

no comparable figure available for 1974

Million tonnes
16
14
12
10
8
6
4
2

USA — Japan — W.-Germany — Italy — USSR — France — United Kingdom

Figure 2:   Forecast of world consumption of materials in million m³

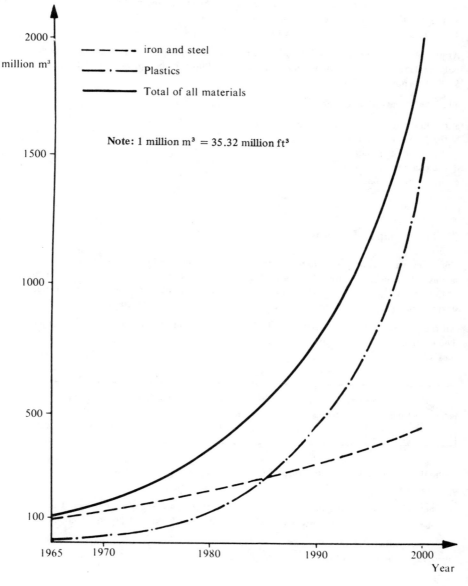

IV Organic raw materials and large-scale products: Industrial processes

The conditions required for the great expansion in plastics were as follows:

1. an adequate knowledge of the chemistry and process technology involved;
2. crude oil and natural gas as cheap or relatively cheap starting materials;
3. the development of efficient techniques for fabrication;
4. the relatively little energy required to shape thermoplastic utensils.

At present (1980), the prices of the large-tonnage plastics are about $1,000 per t high-density polyethylene, $900 per t low-density polyethylene, $800 per t polypropylene and below $700 per t polyvinyl chloride.

The largest individual consumers of plastics are:

a the building industry, including furniture;
b the packaging industry;
c the automobile industry;
d the engineering industry;
e the electrical industry.

2. The groups of plastics

The term plastics comprises materials **produced by wholly synthetic or partly synthetic means**. The synthetic materials consist of giant molecules, which contain thousands or tens of thousands of carbon atoms linked to one another and to other atoms, mainly hydrogen, oxygen, nitrogen and chlorine. Another name for these giant molecules is **macromolecules**.

All the important synthetic routes to plastics start from oil fractions or from natural gas. The elements oxygen and nitrogen come from water and air, while chlorine comes from rock salt.

> For the most part, plastics are produced by wholly synthetic routes. Small, simple basic molecules, called monomers, are linked by means of chemical reactions to form the macromolecule of the plastic, called the polymer.

Although the fundamental structure of the silicones is inorganic they are also included amongst plastics.

Synthetic fibres and also synthetic rubber, on the other hand, are often not included amongst plastics in the narrower sense, although they are polymeric organic compounds.

The term **synthetic resins** comprises raw materials for plastics (in an unshaped state, as powders or liquids) and their precursors. These are used in two ways: firstly, to produce adhesives, lacquers and paints and as binders, and secondly for conversion into moulding materials and laminated plastics by the application of pressure and heat, or into cast resins by the addition of curing agents and accelerators.

As far as **semi-synthetic plastics**, prepared from naturally occurring polymers, are concerned, the cellulose compounds obtained from wood or cotton are still the only products of importance.

The following are the four most important factors so far as the fabrictation and use of plastics are concerned:

1) the shape of the macromolecules and the extent to which they can be deformed;
2) the size of the macromolecules;
3) the arrangement of the macromolecules within the plastic;
4) the chemical structure and the intermolecular forces of attraction between the macromolecules.

Since the shape of the macromolecules and the extent to which they can be deformed are crucial factors in deciding the fabrication and properties of synthetic materials, three main groups of plastics are distinguished:

1) thermoplastic materials;
2) thermosetting materials;
3) elastomers.

Properties such as elasticity, plasticity, stiffness and the capacity to form films or fibres or the capacity for swelling, etc, are essentially determined by the shape of the macromolecules and the extent to which the latter can be deformed.

2.1. Thermoplastic materials (Plastomers)

These consist of threadlike macromolecules, which can be unbranched or branched and have a length 100 times to 1000 times that of their diameter.

Figure 3: Section of the chain of unbranched low-pressure polyethylene (cup model)

IV Organic raw materials and large-scale products: Industrial processes

Because the threadlike molecules are not rigid structures, they are rarely stretched out in a straight line such as is shown in Figure 3. The chemical structural formula of low-pressure polyethylene also illustrates its threadlike structure:

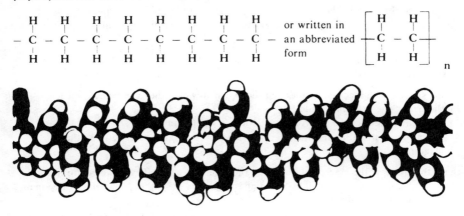

Figure 4: Section of the chain of polystyrene (cup model)

Structural formula:

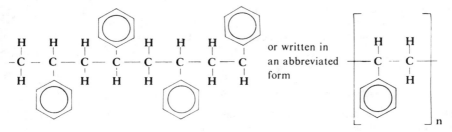

The threadlike molecules of synthetic and natural macromolecules are less than one millionth of a millimetre thick. In a splinter of wood 1 mm thick, there is room for 20 million cellulose molecules side by side. The "tiny macromolecules" form solids because of the action of intermolecular forces of attraction and because loops and coils are formed. In addition, the molecule chains can be completely disordered (amorphous), like the filaments in cotton wool.

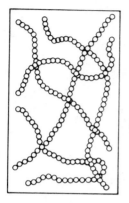

Figure 5: An amorphous thermoplastic.
In reality, the macromolecules fill up the space in an extremely dense molecular network, and are packed about one hundred times more tightly than the filaments in cotton wool. The molecular chains are coiled and/or folded.

Because of strong forces of attraction between polar sections of molecules, parts of the molecular chains can also be in a very ordered state, for example arranged parallel to one another. The term crystalline areas or crystallites is then used. The degree of **crystallinity** of a plastic means the percentage of the chains arranged in an ordered state.

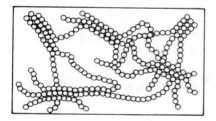

Figure 6: A partly crystalline thermoplastic.
The partly crystalline thermoplastic consists of a dense network of molecular chains, some coiled and some lying parallel to one another

The name **thermoplastic** describes the property of this group of plastics, of being capable of plastic deformation when heated, that is to say of becoming deformable above their softening point and of becoming solid again on cooling, with retention of their new shape.

In terms of quantity and value, the thermoplastics are the most important plastics.

The largest group of thermoplastics consists of macromolecules containing a basic framework of C-atoms linked to one another.

IV Organic raw materials and large-scale products: Industrial processes

Table 1: Important thermoplastics

A. Polyolefins	Basic chemical unit
Polyethylene, PE	$-CH_2-CH_2-$
Polypropylene, PP	$-CH_2-\underset{\underset{CH_3}{\vert}}{CH}-$

B. Polymers containing halogens	
Polyvinyl chloride, PVC	$-CH_2-\underset{\underset{Cl}{\vert}}{CH}-$
Polytetrafluoroethylene, PTFE	$-CF_2-CF_2-$

C. Styrene polymers	
Polystyrene, PS	$-CH-CH_2-$
ABS copolymers, ABS	Copolymers and blends of the components acrylonitrile/butadiene/styrene
MBS copolymers, MBS	Copolymers formed from methyl methacrylate/styrene/butadiene
ASA copolymers, ASA	Copolymers formed from acrylonitrile/styrene/acrylates

D. Other homopolymers containing a $-C-C-$ chain	
Polyvinyl acetate, PVA	$-CH_2-\underset{\underset{O-COCH_3}{\vert}}{CH}-$
Polymethyl methacrylate, PMMA	$-CH_2-\underset{\underset{COOCH_3}{\vert}}{\overset{\overset{CH_3}{\vert}}{C}}-$

E. Polymers containing hetero-atoms (O and N) in their chains, so-called hetero-polymers

Polyacetals,
 for example
 polyoxymethylene, POM

$$-CH_2-O-$$

Polyamides, PA
 for example
 polyamide 6, PA 6

$$-NH-(CH_2)_5-\overset{\displaystyle O}{\overset{\displaystyle \|}{C}}-$$

Polyesters
 for example
 polycarbonates, PC

$$-R-O-\overset{\displaystyle O}{\overset{\displaystyle \|}{C}}-O-$$

and polyethylene glycol
 terephthalate, PETP

Cellulose esters,
 for example cellulose acetate,
 CA

  cellulose propionate,

 CP

  cellulose acetobutyrate,

 CAB

The Big Three are:
1. Polyolefins (polyethylene and polypropylene)
2. Polyvinyl chloride
3. Polystyrene

At present (1975), these represent about two-thirds of all plastics consumed (excluding elastomers) in West Germany.

2.2. Thermosetting materials.

> Thermosetting materials are curable plastics. During fabrication, for example during shaping, chemical reactions take place under the influence of initiators ("curing agents") or by the application of pressure and heat, and these reactions produce huge, spatially crosslinked molecules through interlinking.

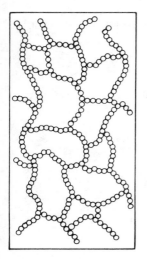

Figure 7:   Section of the amorphous three-dimensional network of a cured thermosetting material

Since the crosslinking takes place in all directions of the space, an enormous number of close-meshed networks are formed, linked to one another and coiled. In the ideal case, an article made from a thermosetting plastic consists of a single, gigantic, spatially crosslinked molecule.

Cured thermosetting materials are glassy, hard and infusible. The cured articles made from them are no longer plastic, but can only be shaped by machining. Thermosetting plastics that are to be shaped without machining must therefore be supplied in the form of liquid or fusible intermediates. These intermediates are available commercially in a dissolved form or in the form of powders, or in a ready-made form as emulsions, dispersions, pastes, casting resins, compression moulding materials (ie precondensates containing fillers) and adhesives.

Bonding materials by means of adhesion has become a process of very great industrial importance thanks to modern thermosetting adhesives.

Synthetic resins are in most cases fairly small macromolecules. Their chemical structure provides an opportunity for crosslinking, for example by $- C = C -$ bonds (carbon-carbon double bonds).

Fillers (resin binders), such as ground minerals and wood flour, short-fibre asbestos, cellulose flocks, chopped up paper and textiles and glass fibres, are incorporated in order to reduce the brittleness of the cured synthetic resins.

The following survey summarizes important groups of synthetic resins from which the corresponding thermosetting plastics are manufactured:

Table 2: Thermosetting plastics

Phenoplasts:	Phenol (Cresol or Resorcinol)-formaldehyde	PF
Aminoplasts:	Urea-formaldehyde resins	UF
	Melamine-formaldehyde resins	MF

Reactive resins:	Unsaturated polyesters in styrene	UP
	Expoxide resins	EP
Polyurethanes cured by crosslinking:		PU
Silicones:	Silicone resins	SI

2.3. Elastomers

Formation of a three-dimensional network by chemical attachment also takes place in elastomers. However, the degree of crosslinking is much less, so that the molecular chains remain able to move between the points of attachment.

> Elastomers are plastics with a wide-meshed three-dimensional crosslinked network. They are rubber-elastic to rigid-elastic.

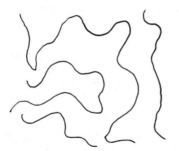

Figure 8: Unvulcanized rubber (caoutchouc)

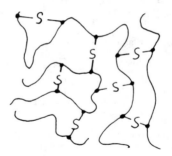

Figure 9: Vulcanized rubber

The vulcanization of (mixtures of) caoutchouc (rubber) with sulphur produces a material with wide-meshed crosslinking: rubber.

In hard rubber (vulcanite) there are more sulphur bridges than in soft rubber.

Under tensile or compressive forces, the coiled portions of the chains can slide over one another and stretch between the points at which they are fixed.

The macromolecules cannot, however, flow past one another, as is possible with thermoplastics above their softening range.

When the deforming force is relaxed, the chain sections once more recover their original coiled arrangement as a result of their thermal agitation. The object has its old shape once again. It can be deformed elastically.

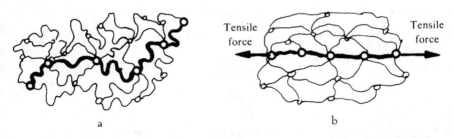

Figure 10: The position of the rubber molecules in the (a) unstretched and (b) stretched state of the rubber.

Rubber-elastic behaviour is also shown by plasticized PVC, copolymerized high-pressure polyethylene and special grades of polyurethane. In the case of the polyurethane grades, which are not crosslinked, this behaviour is due to the very strong intermolecular forces of attraction.

It is a distinguishing feature of non-crosslinked elastomers that they can be processed like thermoplastics. The following survey summarizes important elastomers (R = rubber):

Table 3: Elastomers

Natural rubber	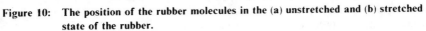	cis-1, 4-Polyisoprene

Synthetic rubber	Styrene-butadiene copolymer,	SBR
	Acrylonitrile-butadiene copolymer,	NBR
	cis-1, 4-Polybutadiene,	BR
	Isobutylene-isoprene copolymer (butyl rubber),	IIR
	Poly-2-chlorobutadiene or chloroprene rubber,	CR
	Ethylene/vinyl acetate copolymer,	EVAC
	Ethylene/propylene copolymer.	EPR
Polyurethane	Polyester/polyurethane	PUR
Polyvinyl chloride	PVC containing partially crosslinked plasticizers	Plasticized PVC

Silicone rubber

$$-O-\underset{\underset{O}{|}}{\overset{\overset{R}{|}}{Si}}-O-\underset{\underset{R}{|}}{\overset{\overset{O}{|}}{Si}}-O-$$ R is an organic radical, such as $-CH_3$.

The silicones are macromolecules with an inorganic basic structure (silicon and oxygen). The outstanding properties of the silicone rubbers are their stability to heat and to weathering.

In conclusion, Table 4 gives a simplified summary of the behaviour of plastics to temperature changes.

In contrast with substances consisting of small molecules, like water, or of ions in an ionic lattice, such as sodium chloride (common salt), or of atom cores and electrons in an atomic lattice, like iron, all of which melt and boil at quite definite temperatures, the behaviour of macromolecular substances is different.

Table 4:   The behaviour of plastics at different temperatures

Thermoplastics, non cross-linked thread-like molecules	brittle state	glass transition temperature	range in which they can be used	softening	decomposition
Elastomers, slightly crosslinked	brittle state	glass transition temperature	range in which they can be used	softening in certain circumstances	decomposition
Thermosetting materials strongly cross-linked	range in which they can be used				decomposition

When heated, thermoplastics soften slowly over a range of temperature. Particularly when pressure is applied, they can undergo plastic deformation in this condition. However, they have no definite melting points. Below a certain temperature range ("glass transition temperature" or "second-order transition temperature") they become hard and brittle.

On the other hand, thermosetting plastics and elastomers that are not thermoplastic decompose if heated too strongly (in most cases above 200°C, or 392°F), without melting beforehand.

The range of temperatures within which it is possible to use plastics can be extended, and the transitions into the brittle state and into the melt state can be made less definite, by copolymerization and by means of suitable additives such as plasticizers and fillers.

3. Synthetic routes to plastics

The most important plastics are manufactured by complete synthesis. The starting materials are simple, low-molecular compounds, called monomers. Many of these small molecules are combined by means of chemical reactions to form the macromolecule, the polymer.

The individual molecules that link up to form a macromolecule must be at least bifunctional, ie each molecule can link itself to at least two neighbouring molecules.

Bifunctional
monomers

A **trimer,** which can undergo further growth with additional monomers to form the **polymer.**

Bifunctional monomers produce macromolecular chain molecules, thermoplastics.

It is also possible for different kinds of monomers to react to form a polymer. If at least one out of two components is a trifunctional or tetrafunctional compound, molecules with a three-dimensional network are finally formed.

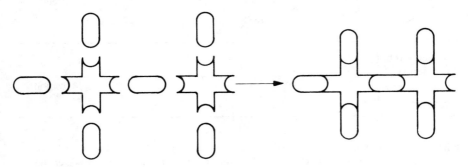

Bifunctional and
tetrafunctional
monomers.

An oligomer (oligo means few), which can undergo further growth with additional monomers to form the **polymer.**

The reactive points or groups that enable compounds to be monomers for polymerization
are as follows:

a carbon-carbon double bonds, $- C = C -$, such as are present in olefins (ie alkenes), or
b) functional groups, such as:
   - OH in alcohols and phenols,
   - COOH in carboxylic acids,
   - NH_2 in amines, and
   - $N = C = O$ in isocyanates.

The following three types of reaction make it possible to build up
macromolecules suitable for plastics:
1. Polycondensation (known since 1910)
2. Polymerization (known since 1930)
3. Polyaddition (known since 1940)

3.1. Polymerization and polymerization plastics

Polymerization is the reaction used most frequently to produce plastics. The monomer
molecules must contain at least one double bond in the molecule, like the olefins and vinyl
compounds.

In **homopolymerization,** molecules of only one kind of monomer come together to form the
polymer. The double bond present in the molecule of the monomer disappears in this self-
addition process. The formation of linear, unbranched polyethylene may be mentioned as an
example:

$$
\begin{array}{cccc}
\underset{\underset{H}{|}}{\overset{\overset{H}{|}}{C}} = \underset{\underset{H}{|}}{\overset{\overset{H}{|}}{C}} &
\underset{\underset{H}{|}}{\overset{\overset{H}{|}}{C}} = \underset{\underset{H}{|}}{\overset{\overset{H}{|}}{C}} &
\underset{\underset{H}{|}}{\overset{\overset{H}{|}}{C}} = \underset{\underset{H}{|}}{\overset{\overset{H}{|}}{C}} &
\underset{\underset{H}{|}}{\overset{\overset{H}{|}}{C}} = \underset{\underset{H}{|}}{\overset{\overset{H}{|}}{C}}
\end{array}
\qquad \text{Ethylene}
$$

+ Initiator

$$
- \overset{\overset{H}{|}}{\underset{\underset{H}{|}}{C}} - \overset{\overset{H}{|}}{\underset{\underset{H}{|}}{C}} - \overset{\overset{H}{|}}{\underset{\underset{H}{|}}{C}} - \overset{\overset{H}{|}}{\underset{\underset{H}{|}}{C}} - \overset{\overset{H}{|}}{\underset{\underset{H}{|}}{C}} - \overset{\overset{H}{|}}{\underset{\underset{H}{|}}{C}} - \overset{\overset{H}{|}}{\underset{\underset{H}{|}}{C}} - \overset{\overset{H}{|}}{\underset{\underset{H}{|}}{C}} -
\qquad \text{Polyethylene}
$$

or, written simply:

$$n \ CH_2 = CH_2 \longrightarrow \left[CH_2 - CH_2 \right]_n ; \Delta H = -106 \ kJ/mol \ C_2H_4$$

Ethylene, a
gaseous monomer

Polyethylene, a
solid polymer

> Polymerization is a consecutive linking up of many simple unsaturated molecules to form threadlike macromolecules. In this process, the double bond in the monomer is opened up.

Substances that initiate polymerization by opening up double bonds are called **polymerization initiators.** Polymerization by ring opening is a special type of polymerization, for example:

Ethylene oxide

Polyethylene oxide

$$n \ H_2C - CH_2 \xrightarrow{\ H_2O \ } \left[CH_2 - CH_2 - O \right]_n$$

> The number of monomer molecules linked to form a macromolecule is called the **degree of polymerization.**

This applies also when the macromolecule has not been built up by polymerization, but by polycondensation or by polyaddition.

Plastics never consist of macromolecules of a uniform size. Instead, macromolecules of varying chain lengths are formed in polymerization. The average degree of polymerization and the average molar mass constitute yardsticks for the average size of the macromolecules formed.

The properties of a polymer also depend on the proportions in which the various chain lengths are present (the molecular mass distribution).

The desired average size of the macromolecules and also their shape (for example branched or unbranched threadlike molecules) can be achieved by selecting:

a) suitable polymerization initiators, at a suitable concentration;
b) polymerization catalysts (for example Ziegler catalysts);
c) the reaction pressure;
d) the reaction temperature;
e) the duration of the reaction.

Because of termination reactions, the macromolecules do not grow to an infinite size.

3.1.1. Polyethylene, PE

The colourless compound ethylene, $H_2C = CH_2$, which is a gas at room temperature, is the simplest and most important monomer of all.

Polyethylene is the most important large-tonnage plastic, taking precedence over polyvinyl chloride and polystyrene.

In 1979, the production of polyethylene amounted in
the USA to	5.628 million t
Japan to	2.165 million t
West Germany to	1.591 million t
the USSR to	0.582 million t
Great Britain to	0.427 million t.

3.1.1.1. Manufacture

Ethylene is obtained from oil products by dehydrogenation, by thermal cracking.

Ethylene, which is a gas, is polymerized by high-pressure, medium-pressure or low-pressure processes to give grades of polyethylene with a wide range of properties. Even today, about two-thirds of all polyethylene is still produced by the high-pressure process:

A little oxygen is added to purified ethylene under a high pressure (300 to 2000 bars, or 4350 to 29,000 lb/in^2). At the reaction temperature (about 200°C, or 392°F), the oxygen reacts with part of the ethylene to form peroxides, which initiate the polymerization reaction in the compressed ethylene gas. The polymerization is an exothermic reaction; it lasts only a few minutes. Operating at high pressures and removing the considerable heat of polymerization (104.7 kJ/mol, or 99.34 Btu) is a major technical problem and requires a considerable outlay on equipment.

The PE formed is liquid at the reaction temperature.

IV   Organic raw materials and large-scale products: Industrial processes

As shown in Figure 11, the unpolymerized ethylene is recycled to the reactor, which consists of long, thin tubes.

Figure 11:   Manufacture of polyethylene by the high-pressure process

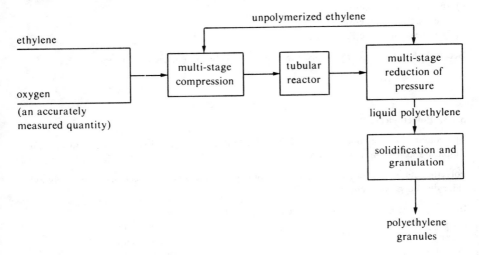

The polyethylene formed in the high-pressure process is a branched threadlike molecule with a molecular mass of 10,000 to 40,000 g/mol. High-pressure polyethylene has a lower density than low-pressure polyethylene.

The medium-pressure process produces PE with a very slight degree of branching.

In the low-pressure process, polymerization is not carried out in the gas phase, but in a solvent. From a technical point of view, this method of polymerization is much simpler; it is carried out in ordinary stirred vessels.

A Ziegler catalyst is added to the solvent (diesel oil or xylene fractions). Very pure ethylene is then passed in. The polyethylene formed is precipitated in the form of flakes. The polymer has a low degree of branching; its molecular mass is between 40,000 and several million g/mol. About one-third of all polyethylene is produced by the low-pressure process.

3.1.1.2. Properties

The following table shows the great extent to which the properties of polyethylene depend on the structure of its macromolecules:

Table 5: Structure and properties of PE

	High-pressure process (ICI)	Medium-pressure process (Phillips)	Low-pressure process (Ziegler)
Number of side-groups in 1000 C-atoms of the chain	about 20	about 2	about 5
Degree of crystallinity	6–70%	85–95%	75–85%
Density, g/cm^3	about 0.92	about 0.96	about 0.95
Softening range	from about 100°C (212°F)	130°C (266°F)	120°C (248°F)
Tensile strength, kg/cm^2 (lb/in^2)	about 140 (1990)	about 400 (5690)	about 250 (3560)
Extensibility, %	500	20	100

Table 5 shows that the crystallinity and the density decrease, as a result of the blocking effect of the molecular chains, as the degree of branching increases.

Although the intermolecular forces of attraction in (non-polar) polyethylene are small, the areas of crystallinity increase the dimensional stability under heat, the tensile strength, the hardness and the stiffness, but also the brittleness, to a marked extent. On the other hand, as the crystallinity increases, the impact strength and the transparency fall off.

Grades of polyethylene with very high molecular weights can also be produced by the low-pressure process. Although these polymers are more difficult to fabricate, because they have a higher melt viscosity, this is compensated by the fact that mouldings with even better mechanical properties and reduced brittleness can be made from them.

Another means of modifying polyethylene, besides the degree of branching, the average molecular weight, and the molecular weight distribution, is to carry out reactions on the finished polyethylene, for example subsequent chlorination.

Copolymerization is an important process for altering the properties of a plastic in a desired direction.

A copolymer is produced by the joint polymerization of two or more different monomers. The different monomers are incorporated into a polymer chain.

Statistical copolymerization: $\quad -E-P-P-P-E-E-P-E-$

The sequence of the various monomers in the threadlike molecule is "statistical" (in this case synonymous with "fortuitous").

Block copolymerization: $\quad -E-E-E-E-P-P-P-P-P-P-$

$$E = \text{Ethylene}, \quad P = \text{Propylene}$$

The statistical copolymers formed from ethylene and a great deal of propylene have elastomeric properties, since the methyl side-groups, which are incorporated at random,

$$-\overset{|}{\underset{\underset{\text{P}}{CH_3}}{CH}}-CH_2-CH_2-CH_2-\overset{|}{\underset{\underset{\text{P}}{CH_3}}{CH}}-CH_2-\overset{|}{\underset{\underset{\text{P}}{CH_3}}{CH}}-CH_2-CH_2-CH_2-$$

$$\quad\quad\quad\quad\quad\quad\quad\quad E \quad\quad\quad\quad\quad\quad\quad\quad\quad\quad\quad\quad\quad\quad E$$

do not allow the formation of extensive regions of crystallinity. The introduction of the small side-chains thus introduces centres of imperfection into the polymer chain. The macromolecules are in a curled and coiled form. The greater the proportion of comonomer, the greater is the amorphous content of the copolymer. Special elastomers (not containing built-in double bonds) are prepared from these products by crosslinking with peroxides.

Reducing the regions of crystallinity also makes itself evident in an increased compatibility with fillers.

3.1.1.3. Uses

The various types of polyethylene have one common feature, which forms the basis for their wide use:

high quality and low manufacturing cost.

The injection moulding process (see page 40) is used to fabricate high-pressure polythene in particular, producing household utensils (bowls and buckets) for example, and also packaging film. Low-pressure polyethylene is employed where the material is subjected to a greater stress, for example for beer-crates.

Tubes, cable sheathings, sheets and films are produced by the extrusion process (see page 40). Ethylene copolymers are mainly used for fabricating film of a higher quality.

Bottles, boxes and (from high-molecular PE) storage tanks are produced by the blow-moulding process (see page 40). Continuous sacks are blown by a similar method "(the air balloon principle)"; these are cut up to give film.

Their chemical structure gives polyethylene and the polyolefins a decisive advantage in waste disposal. Being hydrocarbons, they burn with a waxy smell to form non-toxic carbon dioxide, CO_2, and water vapour.

3.1.2. Polyvinyl chloride, PVC

In terms of quantity manufactured, PVC is the second most important thermoplastic. This is surprising, if consideration is restricted to the properties of unmodified PVC.

Pure PVC is not very stable to heat, has moderate dimensional stability under heat, can only be fabricated with difficulty, becomes brittle at low temperatures and ages rapidly, losing its strength and turning yellow.

However, the properties of PVC can be improved decisively by the addition of stabilizers that prevent or retard the chemical ageing processes (ie degradation processes) caused by the action of light and heat, and by the addition of plasticizers, lubricants, coloured pigments and fillers (for example, carbon black for the manufacture of phonograph records). The manufacture of copolymers of vinyl chloride and of chlorinated PVC and stereospecified polymerization (see page 25) yield further improvements.

3.1.2.1. Manufacture

Vinyl chloride monomer is mainly manufactured by the gas phase chlorination of ethylene (the dichloroethane process):

Ethylene   chlorine   1,2-dichloroethane

$$
\underset{H}{\overset{H}{C}} = \underset{H}{\overset{H}{C}} \ + \ Cl_2 \longrightarrow H - \underset{Cl}{\overset{H}{C}} - \underset{H}{\overset{Cl}{C}} - H \quad ; \Delta H = -175.3 \ kJ/mol*
$$

Dichloroethane is cracked at high temperatures in the gas phase:

          vinyl   hydrogen
1,2-Dichloroethane   chloride   chloride

$$
H - \underset{Cl}{\overset{H}{C}} - \underset{H}{\overset{Cl}{C}} - H \xrightarrow[\text{catalyst}]{500^\circ C} \underset{H}{\overset{H}{C}} = \underset{H}{\overset{Cl}{C}} \ + \ HCl \quad ; \Delta H = +72.8 \ kJ/mol
$$

After the cracking reaction, dichloroethane is immediately injected in order to chill the vinyl chloride and to extract it by scrubbing. The vinyl chloride is then isolated by fractional distillation.

In the production of vinyl chloride from ethylene, hydrogen chloride gas, HCl, is formed as a by-product.

---

* The value of ΔH refers to the number of moles represented in the chemical equation.

IV Organic raw materials and large-scale products: Industrial processes

Further vinyl chloride is, therefore, often obtained from this HCl, by using acetylene in a combined process.

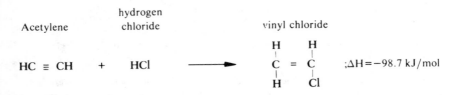

At room temperature vinyl chloride is a colourless, combustible gas with a slightly sweetish odour (boiling point $-14°C$; $6.8°F$). If moisture, oxygen and light are excluded, it is stable in the gaseous and liquefied states.

The reactivity of vinyl chloride, $H_2C=CHCl$, is due to the carbon-carbon double bond. It undergoes rapid polymerization when exposed to high energy radiation (UV or X-rays, see also radio-nuclides) or when peroxides are added.

Vinyl chloride   polyvinyl chloride
 (Monomer)   (polymer)

$$n\ H_2C = \underset{\displaystyle Cl}{CH} \longrightarrow \left[H_2C - \underset{\displaystyle Cl}{CH} \right]_n \qquad ;\Delta H = -96.3\ kJ/mol\ of\ monomer$$

The polymerization is an exothermic reaction: 96.3 kJ/mol are liberated. The chains grow by a repeated *head-to-tail addition reaction* of the monomers. The monomer molecules arrange themselves in the order front-back-front-back, as the position of the chlorine atoms in the threadlike molecule shows:

$$-CH_2-\underset{\displaystyle Cl}{CH}-CH_2-\underset{\displaystyle Cl}{CH}-CH_2-\underset{\displaystyle Cl}{CH}-CH_2-\underset{\displaystyle Cl}{CH}-$$

A head-to-head addition reaction is a rare and exceptional event with PVC. Polyvinyl chloride is a white solid. It is usually fabricated in the form of powder.

The following diagram shows how the chlorine atoms in the threadlike molecule of PVC can exist in three different spatial (ie steric) arrangements in relation to the axis of the chain.

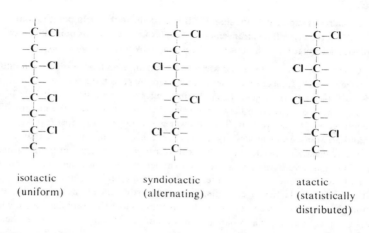

Figure 12: **The chain structure of polyvinyl chloride**

Stereospecific catalysts make it possible to synthesize, for example, predominantly syndiotactic PVC. This is called **stereospecific polymerization.**

3.1.2.2. Properties and uses

The mechanical properties of PVC can be understood by considering the structure of its macromolecules and their arrangement close to one another.

The polarity of the $C-Cl$ bonds produces strong electrostatic forces of attraction between adjacent chains. As a result, regions of crystallinity are formed in PVC. This leads to the hardness and the high second-order transition temperature (glass transition temperature), about $+70°C$ ($158°F$), of unplasticized PVC.

Stereospecific polymerization makes it possible to arrange the otherwise atactic chain structure in a very predominantly syndiotactic pattern. However, the sections with a regular chain structure render possible a particularly close packing of these chain segments. As a result, the intermolecular forces are fully effective. Regions with a crystalline structure (ie a higher degree of order) are formed to an even greater extent, which results in improved mechanical properties, a higher softening range and an even greater stability towards solvents. A similar improvement is achieved by post-chlorinating PVC, that is to say by treating it subsequently with chlorine. In contrast with PVC, post-chlorinated PVC of high chlorine content (approx. 70% by weight of chlorine) is no longer flammable.

Above the glass transition temperature the cohesion of the PVC molecule chains is loosened as a result of the increasing thermal agitation. The linkage between the chains is loosened, particularly in the amorphous regions. The material becomes rubber-elastic.

If even more energy is supplied, if the material is heated above the yield point (about 160°C or 320°F), the crystalline arrangement is also broken up. At this point PVC can undergo plastic deformation without stress, for example by injection moulding.

If rigid PVC is intimately mixed with plasticizers, for example on heated rollers, a modified PVC is obtained, which, for example, does not undergo the glass transition until a temperature of about −30°C (−22°F) is reached. The polar molecules of the plasticizer partially neutralize the forces of attraction between the PVC chains. This gives the macromolecules greater mobility, which manifests itself in a lower glass transition temperature, an improved solubility and greater ease of processing. Plasticizer concentrations of more than 20% give the PVC rubber-elastic properties in a temperature range covering room temperature. This plasticized PVC has a wide variety of uses as a rubber-elastic to leather-like chemical material. PVC pastes, which are used inter alia for coating fabrics (protective coatings), are formed at even higher plasticizer contents. While good resistance to acids, alkalis, alcohols and oils and physiological acceptability are features of rigid PVC, problems arise when plasticizers are incorporated. The smaller molecules of the plasticizer can be leached out slowly by fats, oils, tars and solvents, for example.

Migration and leaching out is, however, rendered impossible if suitable plasticizers can be incorporated into PVC by copolymerization. This process is called **internal plasticization.**

If monomers such as propylene, vinyl acetate or acrylates are incorporated into the polymer chain, it is possible to alter the properties of PVC in a controlled manner: for example to reduce its softening point and to improve its solubility.

Conversely, it is possible to alter the properties of polyacrylonitrile and synthetic rubber for example, in a controlled manner by copolymerization with vinyl chloride.

Blends (ie mixtures) of rigid PVC and soft, rubber-like chlorinated polyethylene are impact-resistant. In the polyethylene, the regions of crystallinity that would otherwise be present are prevented, to a considerable extent, from forming, because the structure is disturbed by the chlorination and the admixture of PVC. The rubber-like, chlorinated polyethylene acts as a (polymeric) plasticizer.

Break-resistant films and impact-resistant pipes, for example, can be made from this material.

Thus, the broad spectrum of uses of PVC extends from gramophone records through floor coverings and impact-resistant pipes to raincoats.

An undesirable property of PVC, which is due to its chemical composition, is the formation of HCl (hydrogen chloride gas) when PVC is burnt. With water (for example condensate or the moisture of the atmosphere), hydrogen chloride forms hydrochloric acid. In refuse incineration plants, attempts are therefore made to neutralize the hydrogen chloride formed, for example by adding burnt lime.

3.1.2.3. Plasticizers

Like pure PVC, other high polymers too are in many cases so brittle that they cannot be fabricated and used until they have been "plasticized".

> Plasticizers are high-boiling, non-volatile liquids. They render the plastic pliable and flexible.

A good plasticizer should have as many as possible of the following important properties and effects:

1. a powerful plasticizing action at as low a concentration as possible;
2. only a slight reduction in the strength and surface hardness of the plastic;
3. as small a tendency as possible to migration and leaching out by solvents such as water, gasoline, fats, etc;
4. resistance to chemicals, fastness to light and heat stability;
5. as resistant as possible to combustion;
6. no colour of its own;
7. good compatibility with fillers and pigments;
8. freedom from odour and taste;
9. should be physiologically acceptable (for example non-toxic).

Suitable plasticizers and suitable plasticizer concentrations have been selected from the large number of known plasticizers by systematic serial tests on a case by case basis.

The range of uses of a plasticizer are not immediately obvious from its chemical composition.

Various esters of phthalic acid and of phosphoric acid are important plasticizers, for example:

Diethyl phthalate	Di-(2-ethylhexyl) phthalate	Trioctyl phosphate
Dibutyl phthalate	Dioctyl phthalate	Tricresyl phosphate

Compared with the high polymers in which they are incorporated, the plasticizer molecules are small, for example:

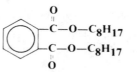 Dioctyl phthalate

Polymeric plasticizers are also important, for example for use in ABS copolymers and ABS blends (see page 29).

IV Organic raw materials and large-scale products: Industrial processes

3.1.3. Polystyrene and ABS plastics

3.1.3.1. Manufacture

Polystyrene is the third important large-tonnage plastic. The monomer is prepared from benzene and ethylene. The reaction is carried out at 95°C (203°F) under normal pressure:

Benzene   ethylene   ethylbenzene

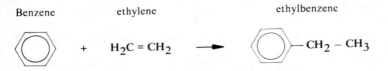

The ethylbenzene is dehydrogenated catalytically in the gas phase at 650°C (1202°F).

            styrene,
Ethylbenzene   ie vinylbenzene   hydrogen

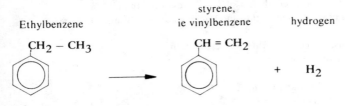

Styrene is a water-white liquid, which is readily polymerized.

  Styrene   polystyrene
  (monomer)   (polymer)

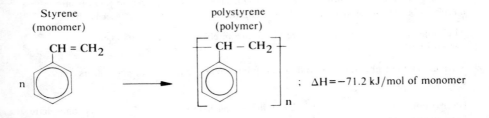

3.1.3.2. Properties and uses

Polystyrene is a transparent plastic. Its chain structure is atactic (see Figure 4, page 8).

As a result of its irregular chain structure, the bulkiness of the benzene rings and its lack of polarity, it has only slight intermolecular forces. Polystyrene is amorphous. The softening point of polystyrene is just below the boiling point of water.

Polystyrene is a pure hydrocarbon. For this reason it is resistant to water, acids, alkalis, oils and alcohols. It is not resistant to petrol, benzene, ethers, esters or ketones.

Pure polystyrene can be fabricated very easily by the injection moulding process. The mouldings do not suffer from ageing, but are somewhat brittle. Polystyrene is an important material for high-frequenty engineering.

In addition to its important uses as an injection moulding material, polystyrene is also employed as a foam plastic. This is prepared by polymerizing styrene, for example in the presence of low-boiling petroleum ether. The petroleum ether occluded evaporates at higher temperatures and thus acts as a blowing agent. Polystyrene can also be foamed by means of steam to produce large block shapes. Foamed polystyrene is of importance as an insulating material in the building industry and also as a packaging material, for example.

Copolymers of styrene with acrylonitrile have a better resistance to solvents and a markedly higher dimensional stability under heat, thus being resistant to boiling, for example, and are not so readily flammable as pure polystyrene. Polystyrene/acrylonitrile copolymers, SAN, are used for the manufacture of tableware and drinking vessels. Copolymers with butadiene are less brittle and can even be impact-resistant.

3.1.3.3. ABS plastics

The impact-resistant copolymers, blends and graft polymers formed from the components acrylonitrile/butadiene/styrene, the ABS plastics, have become particularly important.

In these plastics, elastic polybutadiene (glass transition temperature about $-85°C$, or $-121°F$) is incorporated in a rigid copolymer formed from styrene and acrylonitrile (glass transition temperature about $100°C$, or $212°F$).

Figure 13: ABS plastics consist of two phases:

 soft, elastic polybutadiene (inclusion)

rigid styrene/acrylonitrile copolymer (matrix)

While the pure, rigid polymer splinters in a brittle manner under a blow or impact, the mixture stands up to the impact. The momentum of the impact is transferred to the elastic phase, which acts as a kind of crushed zone and converts the impact energy into heat.

Styrene or styrene/acrylonitrile is grafted onto the polybutadiene rubber in order to ensure, or to improve, compatibility between the soft, inner phase and the harder, external phase. The effect of the graft polymer is to cement both phases in the submicroscopic region.

IV Organic raw materials and large-scale products: Industrial processes

> By a **graft polymer**, we mean a branched copolymer in which the branches are composed of a monomer other than that forming the main chain.

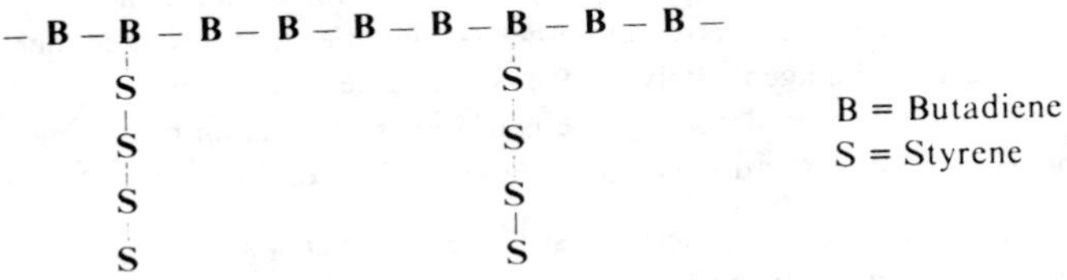

$$- B - B - B - B - B - B - B - B - B -$$

with styrene side chains:

S	S
S	S
S	S
S	S

B = Butadiene
S = Styrene

Figure 14: Polybutadiene chain with styrene graft

The two-phase impact-resistant plastics are used at temperatures between the glass transition temperatures of the soft and hard components.

Impact-resistant grades of thermoplastics are acquiring an increasing importance. Besides the ABS plastics, mention should also be made of impact-resistant PVC and impact-resistant polypropylene. They are also produced by an admixture of soft, rubber-like polymers.

3.1.3.4. Methods of polymerization

There are various techniques for the polymerization of gaseous or liquid monomers:

1. polymerization in the gaseous state under pressure;
2. solution polymerization;
3. bulk polymerization (mass polymerization);
4. emulsion polymerization;
5. suspension polymerization.

Gaseous monomers such as ethylene and propylene can be converted into plastics by **polymerization in the gas phase** under pressure (see high-pressure processes under PE, page 20).

An example of **solution polymerization**, in the course of which the insoluble polymer is precipitated from the solution, is afforded by the low-pressure process for the manufacture of polyethylene (see page 20).

The heat of polymerization constitutes the greatest difficulty in polymerization. This heat energy must be removed as quickly as possible, since otherwise undesirable degradation and crosslinking reactions will take place in the polymer.

Only a few liquid monomers polymerize so slowly that they can be polymerized in bulk, ie undiluted. In the **bulk polymerization** of styrene, for example, initially about one-third of the monomer employed is polymerized in stirred vessels at 80°C (176°F). The viscous mixture is then separated into solid polymer and liquid monomer in a drum drier and the monomer is recycled.

The polystyrene obtained by this process is very pure and has a high molecular mass. Its average molecular mass is about 350,000 g/mol (771 lb) and there is only a low content of low-molecular material in the polymer.

Emulsion polymerization is the preferred process for polymerization. It is used for liquid monomers that are almost completely insoluble in water, such as vinyl chloride and styrene.

In the emulsion polymerization of styrene, for example, the latter is broken down into very fine droplets by vigorous stirring. Suitable emulsifiers (for example soap solution) are added to ensure that these droplets remain finely dispersed. Polymerization is started by adding water-soluble initiators (peroxides). The water-soluble initiator only causes the polymerization of the monomer dissolved in the water; this monomer gradually dissolves in the water from the fine droplets. This gives rise to the formation of very fine particles of solid polymer, which remain distributed (ie dispersed) in the water. As in the case of solution polymerization, the heat of polymerization liberated can be removed easily by means of the liquid.

The polystyrene, which in this process always contains small quantities of the emulsifier, is isolated in the form of a powder by means of drum driers or by precipitation.

The dispersion formed in the copolymerization of styrene and butadiene has properties similar to those of natural rubber latex*.

In **suspension polymerization**, polymerization is also carried out in an aqueous emulsion. However, since the polymerization initiator (for example benzoyl peroxide) is insoluble in water, it is the individual droplets of monomer, for example styrene, that polymerize.

The polymer then settles out in the form of small grains or beads. It is filtered off, washed and dried.

The polymer obtained by this process can be melted and extruded into ribbons, which are reduced to the desired particle size (granules) by means of roller cutters.

3.1.4. Synthetic rubber

By far the most important varieties of synthetic rubber are the copolymers of 1,3-butadiene

$$H_2C = CH - CH = CH_2,$$

In terms of quantity manufactured, the styrene/butadiene copolymer is far and away the most important synthetic rubber. Synthetic rubber is manufactured by emulsion

---

* The word latex (plural: latices) originally meant an emulsion of natural rubber, such as is obtained by cutting the bark of rubber trees.

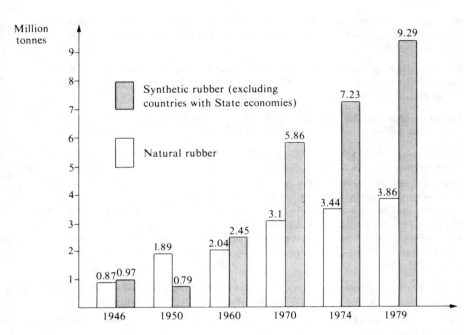

Figure 15:   World production of natural and synthetic rubber

polymerization. It is processed further by vulcanization (ie crosslinking) and by the incorporation of additives.

The discovery by Ziegler and Natta of catalysts for sterospecific polymerization reactions was an important milestone in the development of synthetic rubber. Grades of synthetic rubber with greatly improved properties have been manufactured in this way since 1955.

Economic reasons, in particular, were a decisive factor in the rapid increase in production.

The starting materials for the manufacture of synthetic rubber, butadiene, ethylene, benzene, propylene and recently isoprene, have become considerably cheaper as a result of being produced by petrochemical routes. And there had been no change in general since the rise of crude oil prices. In contrast with natural rubber, which has a chain with a uniform three-dimensional structure, the polybutadienes have a variety of structures inside the individual chains. However, Ziegler catalysts have made it possible to synthesize stereospecifically homogeneous cis-1,4-polybutadiene for example, which is similar to natural rubber (compare page 14):

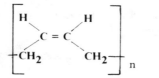 cis-1,4-polybutadiene

For reasons concerned with the quality of the rubber, it is necessary to mix most of the grades of synthetic rubber with natural rubber. In addition, fillers, particularly carbon black, but also silicates and highly disperse silica, are added to the rubber mixtures for processing. These fillers should not be regarded as extenders, but rather as substances that improve the quality.

Examples of important rubber auxiliaries that are added to the rubber mixtures, are plasticizers, vulcanization accelerators, antioxidants and reinforcing agents.

Now that it is possible to produce pure isoprene at a moderate price from crude oil and to convert it into "synthetic natural rubber", industry is less dependent on the raw materials market, in particular on natural rubber from South East Asia.

The purity of the starting material makes it possible to produce cis-1,4-polyisoprene in a quality superior to that of the natural product.

A completely new route for the production of elastomers has been found in polyurethane chemistry, using polyaddition reactions. Toluylene diisocyanate and the polyethers made from propylene oxide are particularly important products in this field. When subjected to polyaddition reactions they yield crosslinked elastomers. Besides these, there are also linear polyurethane elastomers. There would also seem to be good prospects for the ethylene/propylene grades of rubber.

These elastomers all enjoy the decisive advantage that the expensive vulcanization process, which is otherwise necessary for further processing, is eliminated.

Elastomers consisting of chain-like, non-crosslinked macromolecules also have the special advantage that they can be processed in a thermoplastic manner.

3.2. Polycondensation and polycondensation plastics

By condensation we mean the combination of two molecules to form a new molecule, with the elimination of simple molecules such as water, hydrogen chloride or ammonia.

> Polycondensation means the successive combination of polyfunctional molecules, with the formation of a simple by-product (for example water).

IV Organic raw materials and large-scale products: Industrial processes

For polycondensation to result in the formation of macromolecules, it is necessary for each of the reacting molecules to contain at least two functional groups.

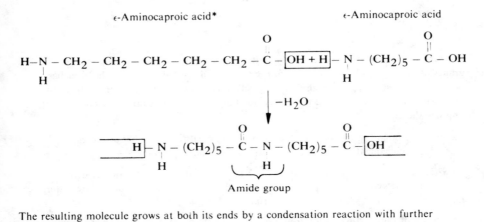

ε-Aminocaproic acid*   ε-Aminocaproic acid

Amide group

The resulting molecule grows at both its ends by a condensation reaction with further molecules of ε-aminocaproic acid to form the macromolecule. Written simply:

ε-Aminocaproic acid   Polyamide 6
(monomer)   ("6" because the monomer
                  contains 6 C-atoms in the molecule)

$$n\ H_2N - (CH_2)_5 - COOH \longrightarrow \left[HN - (CH_2)_5 - CO \right]_n + (n-1)\,H_2O$$

However, the polycondensation process enables not only identical molecules, but also two different kinds of molecules, to be combined with one another:

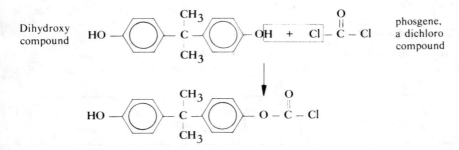

Dihydroxy compound   phosgene, a dichloro compound

---

* ε = Epsilon. The NH_2 group is attached to the fifth C-atom, counting from the carboxylic acid group. ε is the fifth letter in the Greek alphabet.

The resulting molecule can grow further at both its ends by a further condensation reaction to form the macromolecule:

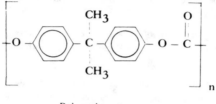

Polycarbonate

The following are plastics of industrial importance, produced by polycondensation:

Polyamides,
polyesters,
phenoplasts,
aminoplasts,
silicones.

3.2.1. Polyamides and polycarbonates

As the linear structure of their molecule chains shows, these compounds can be processed in a thermoplastic manner. These plastics have good resistance to heat, because the threadlike molecules are bulky (in the case of polycarbonates) or have a polar structure (the amide group in the case of polyamides).

High-quality articles can be manufactured from them. The average chain lengths of these thermoplastics are relatively small. In the case of Polyamide 6, the average molecular mass is between 15,000 and 60,000 g/mol (33 and 132 lb) depending on the grade; for the polycarbonate types the figure is between 25,000 and 100,000 g/mol (55 and 220 lb). If the macromolecules have a higher average chain length, this means a higher melt viscosity and a less favourable flow behaviour when injection moulded, for example. On the other hand, the grades with a higher molecular weight are stiffer and have a better dimensional stability under heat, for example.

The polycarbonates in particular display a range of important properties seldom found in other materials.

3.2.2. Phenoplasts and aminoplasts

Phenoplasts (for example phenol-formaldehyde resins) and **aminoplasts** (for example urea-formaldehyde resins) are typical thermosetting plastics.

These resins are prepared by heating the reactants in stirred kettles (usually under an atmosphere of nitrogen as a protective gas), until the condensation reaction commences in

IV Organic raw materials and large-scale products: Industrial processes

the melt. Like polymerization, the polycondensation reaction is exothermic. The H_2O eliminated in the condensation reaction escapes from the melt in the form of water vapour. The reaction is brought to an end before excessive crosslinking has taken place (for details see IV, 4, formaldehyde, page 30 and phenol, page 37).

The distinctive features of completely crosslinked fabricated articles made from phenoplasts and aminoplasts (containing fillers) are essentially good resistance to heat and chemicals and low cost.

An interesting application discovered for plastic emulsions consisting of urea-formaldehyde resins is the improvement of soils from the point of view of growth conditions. Heavy, hard soils can be loosened by means of the flocks of plastic.

3.2.3. Polyester resins

Reactive resins, for example the crosslinkable **polyester resins**, are prepared by subjecting saturated and unsaturated dicarboxylic acids to a condensation reaction with polyhydric alcohols. The C=C bonds in the unsaturated acids make it possible to crosslink the resins, in some grades with added styrene, to give thermosetting polymers. The polyester resins are used as lacquer resins and — particularly with glass fibres and glass fibre mats — as casting resins.

3.2.4. Silicones

Silicones are polymers with an inorganic basic structure.

They are synthesized (the so-called Rochow direct process) from elementary silicon (obtained from quartz sand) and monochloromethane (methyl chloride):

Silicon methyl chloride   dimethyldichlorosilane

$$Si \ + \ 2\,CH_3Cl \ \longrightarrow \ Cl-\underset{\underset{\textstyle CH_3}{|}}{\overset{\overset{\textstyle CH_3}{|}}{Si}}-Cl$$

Dimethyldichlorosilane reacts with water to form a silanediol, from which polymers with a chain structure can be produced by polycondensation.

dimethyldichlorosilane   water   dimethylsilane- hydrogen
                  diol   chloride

$$Cl-\underset{\underset{\textstyle CH_3}{|}}{\overset{\overset{\textstyle CH_3}{|}}{Si}}-Cl \ + \ 2\,H_2O \ \xrightarrow{\text{Catalyst}} \ |\,HO-\underset{\underset{\textstyle CH_3}{|}}{\overset{\overset{\textstyle CH_3}{|}}{Si}}-OH \ + \ 2\,HCl$$

Since silanediol is a bifunctional compound, linear macromolecules can be formed from it

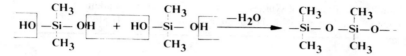

Crosslinked silicones can be prepared from silanetriols.

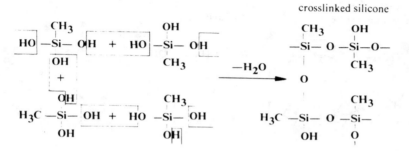

If the macromolecules contain a C_6H_5 group (ie a phenyl group) instead of the CH_3 group (the methyl group), the products are termed phenylsilicones instead of methylsilicones.

Different products are obtained, depending on the size and shape of the macromolecules. It is a prominent feature of silicone chemistry that silicones can be converted into a large number of products, each of which is exactly suitable for a particular end use. The basic products are silicone oils, silicone resins and silicone rubber. From these it is possible to produce, inter alia, pastes, emulsions and dispersions. The special properties of the silicones are as follows:

a) good resistance to temperature;
b) water-repellent action (hydrophobic properties);
c) adhesive-repellent action (this is important for self-adherent sheeting for example);
d) constant viscosity irrespective of temperature changes;
e) absence of colour and odour;
f) harmlessness from a physiological point of view.

3.3. Polyaddition and polyurethanes

The third chemical process used on a large industrial scale for the production of macromolecules as plastics, is essentially limited to the production of polyurethanes and epoxide resins. Polyaddition is similar to polymerization in that no by-products are eliminated in the reaction forming the macromolecules.

However, while the atomic grouping of the monomer is retained in the polymer in the case of polymerization, in the case of polyaddition there is a migration of H-atoms.

> Polyaddition is a means of synthesizing macromolecules by an addition reaction at double bonds with the simultaneous migration of H-atoms.

Mobile H-atoms are supplied by alcohols, for example; the reactive double bonds are supplied by isocyanates.

Diisocyanate,
hexamethylene diisocyanate

Dihydroxy compound,
1,4-butanediol

$$O = C = N - (CH_2)_6 - N = C = O \quad + \quad HO - (CH_2)_4 - OH$$

$$\left[\begin{array}{c} H \\ | \\ C - N - (CH_2)_6 - N - C - O - (CH_2)_4 - O \\ \| \quad \quad\quad\quad\quad\quad\quad \| \\ O \quad\quad\quad\quad\quad\quad\quad O \end{array} \right]_n$$

linear polyurethane

This linear polyurethane is a thermoplastic substance. It is distinguished from the other linear, unbranched polyamides by the advantage of having a low water absorption. Another notable feature is its high impact strength, even at low temperatures.

If the dihydroxy compounds have very long chains, elastic materials that can be deformed in a thermoplastic manner are formed.

The rubber-elastic properties of polyurethane grades are due to strong intermolecular forces of attraction. At the temperature of use, stable linkages are formed. When the material is processed by thermoplastic methods, however, the chains detach themselves from one another; the material can then flow and be shaped. When the moulding is cooled, the intermolecular forces of attraction become fully effective once more.

Special distinguishing features of the polyurethane elastomers are high tensile strength and abrasion resistance and resistance to weathering, solvents and chemicals.

Crosslinked polyurethanes are formed if dihydroxy or polyhydroxy compounds (ie polyols) are used in the reaction. Smaller and more highly branched polyhydroxy compounds produce a more close-meshed and solid molecular network, when all the OH groups have been saturated with diisocyanates by polyaddition.

On the other hand, if polyols consisting of small macromolecules are used, elastomers that are crosslinked at wide intervals can be prepared (via intermediate stages) by means of diisocyanates.

The final crosslinking reaction is often controlled by adding low-molecular diols (dihydric alcohols) or diamines containing mobile (and thus reactive) H-atoms. These products are therefore called crosslinking agents.

In most cases three types of basic unit are involved in the synthesis of crosslinked polyurethanes:

i. polyisocyanates;
ii. polyhydroxy compounds;
iii. crosslinking agents.

The manufacture of a wide variety of polyhydroxy compounds and a large number of diisocyanates and polyisocyanates makes such a multiplicity of basic units available, that it is possible to assemble suitable components to form the desired product like pieces from a construction set.

Thus the spectrum of possible uses for the polyurethanes ranges from adhesives through extremely hard and abrasion-resistant lacquers to gear wheels.

The most important products from a commercial point of view are, however, the *polyurethane foam plastics*. Isocyanates react with water forming carbon dioxide gas; a crosslinking reaction takes place at the same time:

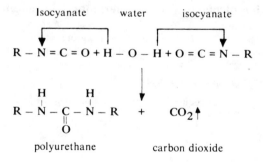

R above denotes the "radical of the macromolecule" and this "radical" can be crosslinked.

As can be seen in the equation for the reaction, the mobile H-atoms that react with the isocyanate come from the H_2O.

If the reaction mixture consists of a diisocyanate, a polyhydroxy compound and water, the diisocyanate is confronted by two reactants simultaneously. It reacts with both of them.

A polyurethane and carbon dioxide are thus formed simultaneously. During the formation of the polyurethane, it is expanded by the CO_2 to produce a foam.

The foaming is assisted and controlled by using propellant gases (mainly halogenated hydrocarbons), in addition.

If, for example, the metered reaction mixture is run into a mould while still in a liquid state, this mould is completely filled on completion of the reaction. The desired moulding is obtained in one operation (so-called foaming in the mould).

By selecting the components suitably, it is possible to prepare foams with a wide range of hardness. All the intermediate stages from a rigid foam to a soft, elastic foam can be obtained.

The so-called structural foams are a recent development. These are grades of polyurethane capable of being foamed into mouldings with a solid external skin that gives way gradually to a porous core.

The uses of polyurethane foams range from soft upholstery material and the insulating material for refrigerators to coated fabrics, and from structural foams for instrument casings in cars to structural members of sandwich construction for highly stressed components and for building construction.

In 1979, the production of polyurethanes in the USA was about 129,000 t.

4. The fabrication of plastics, using thermoplastics as an example

Since the thermoplastics are by far the most important group of chemical materials, only typical fabrication techniques for this group of plastics will be described here. However, in order to provide a survey covering the fabrication of all groups of plastics, the following summary lists all the techniques of importance:

The following processes do not require the application of pressure:
casting,
dipping,
spreading,
coating,
impregnating,
sintering.

The following processes employ pressure to shape the plastic:
injection moulding,
extrusion,
blow moulding,
rolling (ie calendering),

drawing,
bending,
stamping,
all methods of fabrication by machining.

Foaming is a special process.

Thermoplastics are shaped very largely by **injection moulding, extrusion and blow moulding**. In these processes, thermoplastics are fabricated in three stages:
1. melting;
2. shaping;
3. subsequent solidification with cooling.

Since this method of fabrication is carried out at much lower temperatures, and therefore with a lower outlay of energy, than the shaping of steel, aluminium or glass, it is much cheaper.

However, like all plastics, thermoplastics are poor conductors of heat. When heat is supplied, they warm up only slowly and in limited regions. Although this factor is an advantage for hot air welding or high-frequency welding, it is a disadvantage if, as was still the case a few years ago, thermoplastic material is to be melted for injection moulding by means of an external supply of heat.

1. The thermoplastic material is melted
Until a few years ago, macromolecular granules or powder used to be melted by heating.

The slow rate of melting of the granules or powder was a limitation, not only on the efficiency of the injection moulding process, but also on the quality of the shaped parts.

A fundamental change was brought about by the invention of the screw extruder. This machine acts like a mincing machine and conveys the thermoplastic by means of a rotating screw, plasticizing it as it travels. The plastic thus becomes uniformly free-flowing, not predominantly as a result of external heating, but as a result of frictional heat.

In extrusion and blow moulding, the first stage of fabrication also consists in plasticization by means of a screw.

2. Shaping is the second stage
In injection moulding, the plastic composition is injected into the mould through a die. After the moulding has solidified, the mould is opened and the moulding is pushed out. Injection moulding machines operate in strokes.

Extruders (single-screw or twin-screw), on the other hand, operate continuously. The plasticized composition is forced through a die.

Depending on the shape of this die (for example a tubular die, a profile die or a slit die), a continuous tube, a continuous cable covering, a continuous window frame profile or a continuous strip of film or sheet is pressed out.

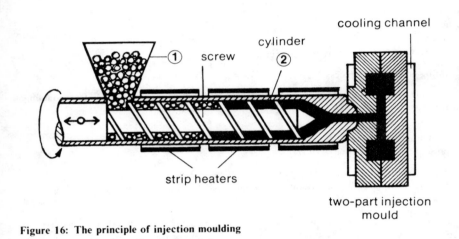

Figure 16: The principle of injection moulding

In blow moulding processes, a hollow profile (tube or hose) is inflated by blowing to form a fuel tank, or, continuously, to form a continuous wide hose, — similarly to a balloon.

3. The shaped plastic part solidifies with cooling
The faster mouldings cool, the better is the quality of the product and the efficiency of the process. Thus in injection moulding the moulds are cooled, and in blow moulding internal cooling is applied, for example with carbon dioxide, in order to increase the rate of output.

A development of modern automatic injection moulding machines is the multi-component machine in which sandwich mouldings, for example, are produced. A plastic melt is first injected into the mould; this solidifies on the wall of the mould under a second plastic, following the first, and thus forms a coating. This process makes it possible, for example, to injection mould gear wheels in which the less stressed interior consists of a solid, but light plastic, while the more highly stressed outer surface is composed of an abrasion-resistant material with good anti-friction properties.

5. New developments in the field of plastics

There are two aspects to this subject. From the point of view of the raw materials manufacturer it means the development of new types of plastics and the modification of known plastics. From the point of view of the machinery manufacturer it means the development of new manufacturing techniques and the improvement of known techniques.

Since fabrication of plastics began, there has always been very close collaboration between raw material manufacturers, manufacturers of machinery and plastics fabricators, and this has led to substantial advances.

In the case of some new plastics, the relationship between the plastic and a specific technique of fabrication has been so close that the raw material manufacturer became the machinery manufacturer too and supplied the fabrication technique together with the new plastic.

From the point of view of the raw material manufacturer, the main emphasis is on a broad development of the known plastics for new fields of application. It can hardly be expected that spectacular, novel plastics will be discovered.

The fundamentally new polymers that have been reported in recent years are not large-tonnage plastics, but interesting plastics for special fields: polysulphones, polyimides, polyphenylene oxides (PPO), polybenzimidazoles, poly-p-xylylene and polymers containing

sulphur bridges such as 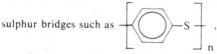 ,

which are prepared from p-dichlorobenzene and sodium sulphide. Like the polyimides, the fluorine containing polymers and the novel metal powder/polyester mixtures, these sulphur-containing polymers are amazingly resistant to heat. A second target is the development of novel synthetic materials that can be subjected to a **high mechanical stress**. Plastics that have a programmed service life and destroy themselves by means of additives are also an interesting new development.

There are almost boundless possibilities for varying the chemical and spatial structure of the macromolecules and thus the properties of the plastics already known. The methods of effecting this are known:

a) copolymerization;
b) polycondensation and polyaddition of various mixtures of monomers;
c) controlling this reaction to form macromolecules with desired average molecular weights;
d) mixing and graft-polymerizing mixtures of plastics.

These variations are now being worked through increasingly on a systematic basis. A considerable scientific effort is thus being applied to develop many new grades from existing ranges of products.

Particularly in the case of the thermoplastics, it has been possible to make advances by producing high-molecular grades. The distinguishing features of these are special rigidity and toughness. The invention of rigid, low-pressure polyethylene and rigid polypropylene has made it possible to manufacture large and robust containers for liquids from these materials.

Other recent developments are the application of an anti-static finish to plastics and the manufacture of grades of low flammability and low combustibility. There is a very great need for these in the building industry and in the electrical industry.

IV Organic raw materials and large-scale products: Industrial processes

The lamination of materials play an important part in extending the properties of plastics and enabling them to conquer new markets. The range of laminated materials is large. It ranges from laminated films for packaging, where one film supplies the mechanical strength and another supplies impermeability to aroma, through fabrics coated with polyurethane, metal coatings deposited electrolytically or chemoelectrolytically on ABS copolymers, and chemical materials reinforced with glass fibres, to sandwich units.

The example of the building industry shows the markets that are still open in this field. Although the consumption of plastics increases continually in this field, the building shell is almost completely unaffected. But it is precisely in the shell of a building where there are opportunities of making a decisive reduction in building costs, if the load-bearing structures consist of prefabricated steel and concrete units lined with plastic sandwich units that can be assembled rapidly and arranged in different ways.

From the point of view of the machinery manufacturer, the problem of plastics fabrication has been solved in principle. Radical, spectacular changes can hardly be expected in the foreseeable future. The emphasis in development is directed towards systematically improving and refining known techniques. This work is carried out in close collaboration with the technical service departments of the plastics manufacturers.

One new process for fabricating thermoplastics is still in the development stage; this is high-pressure plasticization, in which pressures are used that have hitherto only been customary in the field of steel processing. It is still too early to estimate the effect of this new development.

The trend towards computer controlled automatic machines and towards entire fabricating lines, including labelling and printing machines, can, however be clearly seen.

6. Self-evaluation questions

89 Which group contains the elements from which plastics are mainly built?

A carbon — sulphur — oxygen — hydrogen — chlorine
B hydrogen — oxygen — chlorine — nitrogen — sulphur
C hydrogen — oxygen — iodine — nitrogen — carbon
D carbon — hydrogen — oxygen — nitrogen — chlorine

90 Which two of the stated raw material sources are most important for plastics synthesis?

A coal
B crude oil
C air
D earth

91 Which substances are classed as plastics?

A synthetic resins
B polyolefins
C silicones
D proteins

92 Which group of plastics is most important in terms of quantity and value?

A thermoplastics
B thermosetting plastics
C elastomers

93 The chemical building block belongs to

A polyethylene
B PVC
C polystyrene
D polyurethane

94 Which type of plastic has a structure in principle like that in the sketch?

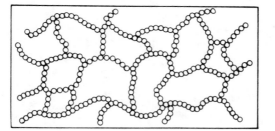

A amorphous thermoplastic
B partially crystalline thermoplastic
C cured thermosetting plastic
D vulcanized rubber

IV Organic raw materials and large-scale products: Industrial processes

95 The term "crystallinity" of a plastic connotes

   A   the number of monomers involved in a macromolecule
   B   the number of its molecules in an ordered arrangement
   C   its brittleness
   D   its elasticity

96 Which polymers belong to the thermoplastics?

   A   polyvinyl chloride
   B   phenol-formaldehyde resins
   C   polyethylene
   D   synthetic rubber

97 Which building block(s) does polypropylene contain?

   A $-CH_2-CH_2-CH_2-$

   B $-CH_2-\underset{\underset{\displaystyle CH_3}{|}}{CH}-$

   C $-CH_2-\underset{\underset{\displaystyle OOCH_3}{|}}{CH}-$

   D $-CH_2-\underset{\underset{\displaystyle Cl}{|}}{CH}-$

98 Which building block(s) does a polyester contain?

   A $-CH_2-O-$

   B $-NH-(CH_2)_5-\underset{\underset{\displaystyle O}{||}}{C}-$

   C $-R-O-\underset{\underset{\displaystyle O}{||}}{C}-O-$

   D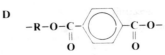

99 Which two plastics belong to the so-called "big three" — the three most important plastics in terms of quantity and value?

   A   polyolefins
   B   polyurethane
   C   polyamide
   D   polyvinyl chloride

100 Which plastics are known as heteropolymers because they do not contain carbon atoms in their basic structure?

   A   polytetrafluoroethylene
   B   polyvinyl chloride
   C   polypropylene
   D   polyamide

101 Which statements about plastics are correct?

 A   In elastomers the cross-linkage between the individual molecule chains is much
     more powerful than in thermosetting plastics.
 B   Thermosetting plastics cannot be melted once they have been cured.
 C   Thermoplastics consist of macromolecules which can slide past one another under
     the influence of heat and pressure.
 D   Silicones contain an inorganic basic building block.

102 Which materials are similar in their temperature characteristics?

 A   thermoplastics and elastomers
 B   thermoplastics and thermosetting plastics
 C   elastomers and thermosetting plastics
 D   metals and thermosetting plastics

103 In which order do the degrees of cross linkage of the individual plastics groups
increase?

 A   thermosetting plastics — elastomers — thermoplastics
 B   thermoplastics — thermosetting plastics — elastomers
 C   elastomers — thermosetting plastics — thermoplastics
 D   thermoplastics — elastomers — thermosetting plastics

104 The degree of polymerization of a plastic represents a statement on

 A   the chain length of the macromolecule
 B   the cross linkage of the individual macromolecules
 C   the number of functional groups in the monomer
 D   the number of monomers linked to form a macromolecule

105 Thermoplastics are brittle at low temperatures. As the temperature rises their
behaviour changes until they decompose. Which is the correct order?

 A   working temperature — softening — glass transition temperature
 B   softening — glass transition temperature — decomposition
 C   softening — glass transition temperature — working temperature
 D   glass transition temperature — working temperature — softening

106 Which groupings do compounds have to contain in order to be monomeric starting
materials?

 A   Two $-C=O$ groups per molecule
 B   One $-C=C-$ linkage per molecule
 C   Two $-NH_2-$ groups per molecule
 D   Two $-COOH-$ groups per molecule

IV   Organic raw materials and large-scale products: Industrial processes

107 Which types of reaction lead to the build-up of macromolecules?

A   polyaddition
B   polysubstitution
C   polycondensation
D   polymerization

108 Which type of reaction was the first to be carried out in the production of plastics?

A   polyaddition
B   polysubstitution
C   polycondensation
D   polymerization

109 When molecules of only one monomer come together to form a polymer, this is known as

A   copolymerization
B   addition polymerization
C   homopolymerization
D   telomerization

110 Which processes for the manufacture of polyethylene yield a product with only relatively few side-chains?

A   low-pressure processes
B   medium-pressure processes
C   high-pressure processes

111 Which process for the manufacture of polyethylene is technically the least expensive?

A   low-pressure process
B   medium-pressure process
C   high-pressure process

112 A guide to the average size of the macromolecules formed to produce a plastic is

A   the density of the plastic
B   its molecular mass
C   its degree of crystallinity
D   its degree of polymerization

113 Which plastic is obtained by polycondensation?

A   polyamide
B   polyester
C   polycarbonate
D   synthetic rubber

114 The term "thermoplastics" connotes plastics that

    A   withstand an operating temperature of up to 300°C (572°F).
    B   can be formed when heated
    C   can be formed when heated and retain this property afterwards
    D   are soft and pliable even at a temperature of −20°C (−4°F).

115 Plasticizers are

    A   involved in the chemical structure of macromolecules
    B   substances that make plastic supple and pliable
    C   low-boiling liquids
    D   high-boiling liquids

116 The preferred process for polymerization is that of

    A   polymerization in the gas phase
    B   solvent polymerization
    C   block polymerization
    D   emulsion polymerization

117 When caoutchouc is vulcanized, this produces

    A   vulcanized rubber
    B   a plasticized plastic
    C   a black-coloured plastic
    D   a cross-linked plastic

118 In the production of plastics by polycondensation

    A   not only are the macromolecules formed but also small molecules are invariably created as a byproduct
    B   the macromolecules have to be condensed to liquids by cooling
    C   at least bifunctional monomers have to be employed
    D   cross-linked plastics always form

119 Pressureless methods of plastics processing are those of

    A   injection moulding
    B   casting
    C   extrusion
    D   drawing

IV Organic raw materials and large-scale products: Industrial processes

120 Which plastics processing operation is illustrated in the drawing below?

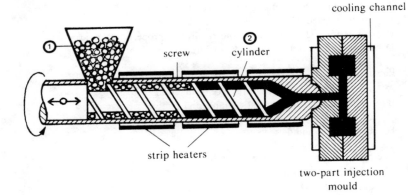

cooling channel

screw   cylinder

strip heaters

two-part injection mould

A injection moulding
B extrusion

C blow moulding
D thermoforming

121 Which statements relating to the sketch in Question 120 are correct?
   A The raw material granules are shaken in at (1).
   B The cylinder (2) rotates and plastifies the raw material.
   C The forward motion of the screw (from left to right in the drawing) fills the closed mould with the plastic melt.
   D After the moulding has cooled the mould opens and the injection moulding is released.

122 Which factors ensure that a molten thermoplast flows well when it is processed by injection moulding?
   A high pressure, with which the melt is forced forward
   B composition from molecules of the largest possible size
   C greatest possible intermolecular forces of attraction
   D smallest possible intermolecular forces of attraction

8 Synthetic Fibres

In contrast with natural fibres, the name "synthetic fibres" covers all fibrous materials made from synthetic and natural starting products.

They owe their name to the fact that they are produced almost exclusively by the chemical industry from simple raw materials or natural, fibre-forming substances — the "textile raw materials" in the narrower sense.

According to the most recent directives (International Standards Organization), the term "artificial silk", which has remained in use until the present time, should no longer be used, although it is still found occasionally in books and periodicals.

1. The commercial importance of synthetic fibres

If the development of the manufacture of synthetic fibres is considered from its beginnings until today, it is evident that a mass product has evolved from a luxury article — the first artificially manufactured filaments, "artificial silk", were at first prohibitively expensive. The preconditions for this change were provided by the chemical industry by developing cheap raw materials and by advances in chemical research and process technology. In addition to this the demand for textiles grew appreciably faster than the production of natural fibrous materials, such as cotton and wool, so that there was from the beginning an open market for synthetic fibres.

At the turn of the century world production of wool was 730,000 t for a world population of 1.6 thousand million people, since then the population of the world has nearly trebled — 4.2 thousand million people and the production of wool has risen to 1.6 million t; so that the per capita production of sheep's wood has remained nearly constant in these 70 years, although the demand for "woolly" textiles has risen.

The production figures for synthetic fibres compared with the production of cotton and wool, which are set out in Figure 1 on page 5 show, firstly, the growth of world production since 1950 (I).

2. Properties of natural and synthetic fibres and the principles on which they are constructed

2.1. The general requirements for fibres

Although textile fibres — whether of natural occurrence or manufactured by chemical means — are mainly used for the production of textiles for clothing and for the home, they are also coverted into a variety of industrial articles.

IV Organic raw materials and large-scale products: Industrial processes

---

> **Fibre** is a general term for textile raw materials of natural or synthetic origin. Most natural fibres have a limited length — also known as staple — and are therefore termed **staple fibres**. Synthetic fibres are produced initially in unlimited lengths and are therefore termed "**filaments**" or **continuous filament fibres**.

Both for the various processing stages of the textile industry, such as spinning, weaving or knitting and especially for the end products — textiles for clothing, the home or industrial purposes — the starting materials must satisfy certain requirements:

Fibres and filaments should have a certain **strength and elasticity**, they should possess considerable **suppleness** and **abrasion-resistance** and their weight should be as low as possible, and they should impart these properties to the finished products made from them. Because of their geometrical dimensions and shape – their length is many times their small diameter of about 1/100 to 1/10 mm (4/10,000 to 4/1,000 m), their cross-section can be round, oval, bean-shaped or profiled and the fibre can be crimped to a greater or lesser extent – the fibres in yarn and fabrics can occlude small cushions of air, depending on how they are processed, and, as a result, they have good **heat-insulating properties**; these are required particularly for textiles used for clothing. To enable the finished products to have a wide variety of designs, it should be **possible to dye** the fibres and filaments.

The properties of the fibres and filaments should enable the finished products to retain their shape and to be crease-resistant and to be looked after as easily as possible — ie washed or cleaned.

2.2. Properties of natural and synthetic fibres

The composition and origin of each type of fibre give it quite definite and specific properties, both physical and chemical, some of which are particularly pronounced and are very valuable for the particular textile application. Because of the different requirements that textile fibres have to meet, various properties are mutually exclusive, for example high strength and great elasticity. The use of **one** type of fibre for a particular application is therefore always a comprosmie between the good, ie desirable, properties of this fibre and its less desirable properties.

Important properties of some natural and synthetic fibres are explained below and are contrasted in Table 1 on page 7.

Properties:	Definitions:
Strength	The tensile strength is determined by the force required to break the fibre. For textile raw materials it is quoted as the breaking length in kilometres, meaning that the weight

Figure 1:   World production of synthetic fibres

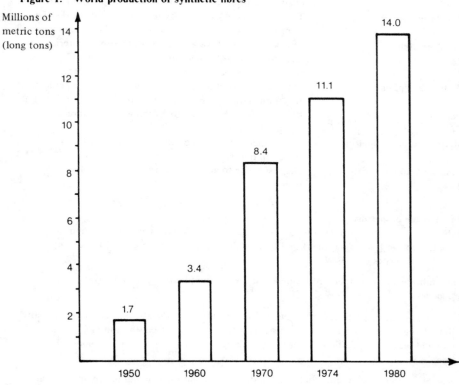

World production of cotton, synthetic fibres and wool in percent

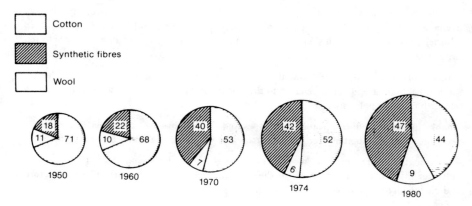

	of this length of fibre would be sufficient to break the fibre.
Elongation	The extension of the fibre by tension until it breaks. The elongation is quoted as a percentage of the initial length.
Moisture content	The percentage of water present in the fibre material. It is measured in air of a normal moisture content (65% relative atmospheric humidity).
Water absorption	The water absorption of the fibre is quoted as a percentage after immersing the fibre in water and then centrifuging it.
Density	This is the weight per unit of volume of the fibre material. It is quoted in g/cm^3.
Melting point (or melting range)	The temperature at which the fibre material melts. It is quoted in degrees centigrade, °C.
Resistance to chemicals	The resistance to acids and bases may be quoted as an example.

Tailor-made properties

Before the development of synthetic fibres, the only textile fibres available to man were wool and cotton (including linen) and to a small extent silk, and man had to be content with the properties of these fibres.

The multiplicity of synthetic fibres makes it possible to obtain a greater variation in properties.

These fibres can be produced in a large number of different variations — white or in many different colours, or with a textured surface, with graduated strengths and elongations, bulky and elastic, to mention only a few possibilities — in other words as tailor-made fibres.

These properties of synthetic fibres are established during their manufacture and are thus adjusted to satisfy the requirements for particular end uses.

This is also the main reason for the versatile uses to which synthetic fibres are put in **technology**. It is precisely for extreme requirements that no natural fibre can satisfy that synthetic fibres have proved themselves particularly suitable. Some examples will help to emphasize this point:

Automobile and aeroplane tires have been able to meet the severe demands placed on them in starting, braking and landing thanks to the textile ply they contain, made of high-strength rayon or polyamide.

Thanks to synthetic fibres, **fishing-nets, ropes** and **ships' cables** can be made from rot-resistant yarns with better strength properties and with a substantially lower weight.

Plastic-coated fabrics made from synthetic fibres — particularly from high-strength polyester filaments — can be used for the manufacture of **tarpaulins, air-inflated structures** and chemically-resistant **containers for liquids**.

Table 1

The properties of natural and synthetic fibres

Properties / Type of fibre	Strength ($\frac{cN}{tex}$)	Elongation (%)	Moisture content (%)	Water absorption (%)	Density (g/cm³)	Melting point (range) (°C)	Resistance to acids/alkalis
Cotton	15 – 35 (50)*	7 – 10	7 - 8	35 – 75	1·53	carbonizes	non-resistant/good
Wool	10 – 16	25 – 48	15	28 – 42	1·30	carbonizes	good/non-resistant
Silk	30 – 45	18 – 24	11	54	1·25	carbonizes	good/non-resistant
Staple viscose (rayon)	20 – 30 (40)*	15 – 20 (30)	13	80 – 125	1·52	carbonizes	non-resistant/good
Acetate	13 – 17	16 – 31	6·5	50	1·31	carbonizes	non-resistant/good
Polyamide	35 – 55 (75)*	15 – 31	4	10 – 14	1·15	Nylon 255 Perlon 215	moderate/very good
Polyester	40 – 60 (68)*	10 – 25	0·5	1 – 2	1·38	255–256	good-very good/good
Polyacrylo-nitrile	20 – 30 (50)*	15 – 18	1	2·5 – 3	1·18	softens and decomposes 200–250	very good/moderate
Structural steel (comparison)	27	8					
*Special types							

IV Organic raw materials and large-scale products: Industrial processes

Special fibres have also been developed for **space travel**; examples of their use are incombustible and heat-resistant space suits for astronauts.

The specific properties of all textiles, natural and synthetic, make them valuable for a particular use. Completely novel textiles can be produced by **blending fibres**, ie by the correct selection and common processing of synthetic and natural fibres or of different synthetic fibres. It is thus possible to combine the properties of different fibres in such a way that the finished product has optimum suitability for the particular requirements it has to meet.

2.3. The shape of fibres and their chemical structure

2.3.1. Chain-like macromolecules

Considering first the external dimensions of natural fibres (staple fibres, and also silk), we find that all fibres have one thing in common: the length of the fibre is many times — a thousand or several thousand times — its thickness (diameter).

Examples:

	Length*	Thickness*
Cotton	10 to 50 mm	10 to 30 μm*
Wool	20 to 120 mm	17 to 60 μm
Silk	100 to 1000 mm	10 to 40 μm

The fine structure of a fibre (cotton) shows that it consists of long bundles of **fibrils**. The relationship between the dimensions of the fibrils is similar to that of the fibres and each fibril is composed of many fine structural units: the molecular bundles of many "giant molecules", the **macromolecules** or **polymers** (see Figure 2).

> Polymers or macromolecules are giant molecules that have been formed by chemical bonds from many (up to several thousand) simple molecules. Polymer means many-membered.

Investigation of these macromolecules by chemical means (a field in which much of the exploration and description was carried out by H. Staudinger, who received the Nobel Prize for his work) has shown that macromolecules have a chain-like structure and are composed of smaller units, monomers, which recur many times within this chain.

---

* These are approximate values; the abbreviations denote:
1 mm is 1 millimetre and = 0.0394 in.
1 μm is 1 micrometre or 1 millionth (1/1000 000) of a metre and = 39.4 millionths of an inch.

Figure 2

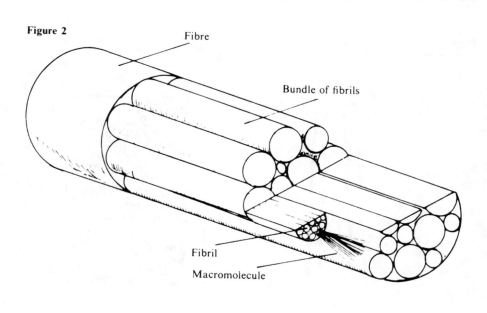

Fibre

Bundle of fibrils

Fibril

Macromolecule

> Monomers are simple, small molecules that can be assembled to form
> polymers by means of chemical reactions.
> Monomer means single-membered.

In the case of cotton fibres (cellulose molecules), chain molecules are mainly formed from carbon atoms alternating with oxygen atoms, while in wool fibres (protein molecules) the chains are formed from carbon atoms and nitrogen atoms.

These molecules have a length of a thousand or several thousand atoms and a "thickness", or rather width, of 2 to 5 atoms; ie there is a relationship between length and "thickness" similar to that in the fibres that are built up from them.

If these natural fibres are contrasted with synthetic fibres, as in Figure 3 on page 10, it is possible to see the great similarity between the external shape and the internal structure.

Figure 3

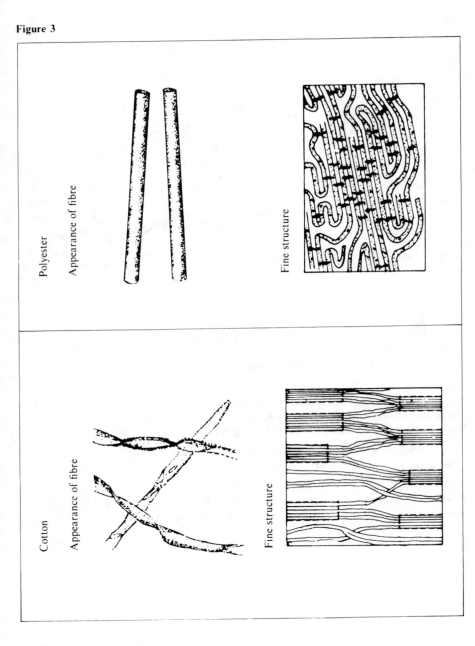

Table 2 Survey of chain structures

Chain structure — representation in symbols	Elements in the chain	Type of fibre
1. ⚫⚫⚫⚫⚫⚫⚫⚫⚫⚫⚫	carbon	eg polyacrylonitrile
2. ⚫—C—O—C—O—C—O—C—O	carbon and oxygen	eg cellulose fibres or polyesters
3. ⚫—N—⚫—N—⚫—N	carbon and nitrogen	eg protein fibres, wool and polyamide
4. —O—⚫—N—⚫—N—⚫	carbon, oxygen and nitrogen	eg polyurethane

2.3.2. Simplified presentation and enlarged detailed sections

The chain structures shown in the table are to be understood as simplified models, from which it is not possible to form an idea of the number of carbon, oxygen or nitrogen atoms and their "environment", ie the hydrogen atoms or other groups that are linked to the atoms forming the chain.

However, inside a chain containing one thousand or more atoms, it is possible, after an easily discernible number of 6 to 12 atoms, to establish that there is a regular recurrence of this section. An enlarged detailed representation of the relevant section thus makes it possible to form an accurate idea of the chemical structure of the chain molecule.

Three types of fibre — cotton, wool and polyester — will be used as examples:

1. Cotton (Cellulose fibre)

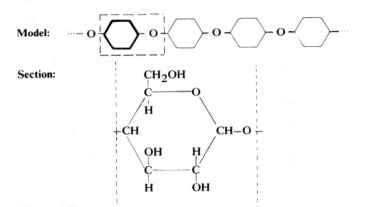

2. Wool (protein fibre)

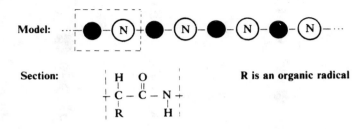

R is an organic radical

3. Polyester

Model:

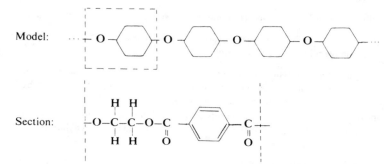

Section:

In each case these sections also represent the basic unit of the chain molecule, the monomer, from which the chain molecule has been formed by multiple linking up of monomers.

2.3.3. The requirements for a synthetic fibre

A synthetic fibrous material, a textile raw material in the narrower sense, ie the composition required for spinning, must satisfy the following conditions:

1. it must consist of chain-like macromolecules with a chain length of at least one thousand atoms;
2. the molecular structure should permit the close assembly of many macromolecules to form molecular bundles; this property can also be achieved by means of what are called polar groups in the macromolecule;
3. the molecular bundles, which are initially present in the solid material in a confused tangle, should orient themselves when spun, ie during the shaping of this material or later, by means of a mechanical stretching process (elongation process), in the longitudinal axis of the fibre.

Before spinning, the fibre material has a largely **amorphous structure** (amorphous means without shape) and the molecular bundles are in a disordered state — tangled or coiled.

When they are spun and, above all, when they are stretched (see page 38 of this chapter), the molecular bundles orient themselves along the axis of the fibre — the tangled and coiled molecular bundles are stretched and for the most part lie parallel to one another. Ordered areas are formed, interrupted by a few disordered areas.

3. The Production of synthetic fibrous materials

3.1. Synthetic fibrous materials obtained from natural polymers, cellulosic synthetic fibres

As already mentioned in 2.3, the polymeric natural fibres have the properties desired. They are chain-like macromolecules that can assemble to form molecular bundles.

Cellulose, which is a vegetable structural material, is an important starting material for textiles, but it is not directly soluble either in water or in other solvents. It must first be modified by means of chemical reactions so that a soluble cellulose compound, capable of being processed, is formed. If this cellulose compound — eg an ester — remains unchanged after the material has been brought into the form of a fibre, a **cellulose ester fibre** is obtained.

If, on the other hand, the cellulose is recovered in a pure form from the cellulose compound during the production of the fibre, **regenerated cellulose fibres** are obtained.

The natural polymeric material **cellulose** is mainly obtained from wood of different species of trees, such as white poplar, pine, spruce and beech, but is also obtained from straw. The cell walls of plants are made of pure cellulose. It consists of 44.4% of carbon, 6.2% of hydrogen and 49.4% of oxygen and is a white substance, insoluble in water. Cellulose is present in wood to the extent of about 50 to 60% (spruce wood contains $\approx 58\%$ cellulose; beech wood $\approx 54\%$ cellulose), together with the other constituents lignin and beta-cellulose. Lignin is the binder which coats the cellulose to form wood.

On the other hand, cotton and cotton linters (a waste product from cotton spinning; very short cotton fibres) are nearly pure cellulose: alpha-cellulose content 98.5 to 99%.

3.1.1. Treatment of cellulose

In cellulose pulp factories cellulose is obtained from wood by crushing the wood and boiling (solubilizing) it in suitable acids (eg sulphurous acid) or liquors (eg sulphite liquor), in order to dissolve the accompanying substances, such as lignin, while the alpha-cellulose is not attacked.

Since the resulting cellulose is not entirely pure, the product obtained from wood is termed **cellulose pulp**.

Outline of process:

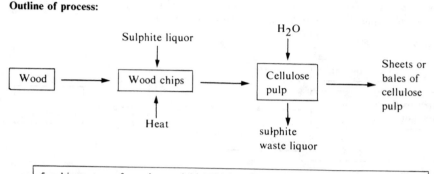

> 5 cubic metres of wood can yield 1000 kg of cellulose pulp, from which 950 kg of cellulose fibres can be obtained.

3.1.2. Regenerated cellulose fibres

The cellulose pulp supplied by pulp factories, and also cotton linters, can be subjected to chemical change, and thereby rendered soluble, by two different processes.

1. Regenerated cellulose fibres manufactured by the **viscose process**

The cellulose pulp is treated with dilute sodium hydroxide solution and thus converted into soda cellulose (also called alkali cellulose). The latter is then treated with carbon disulphide (sulphided), an orange-yellow cellulose xanthogenate (xanthate) being formed, which is again dissolved in dilute sodium hydroxide solution to give a syrupy, viscous composition, the spinning solution or **viscose**. This composition has given its name to the whole process.

> Cellulose + sodium hydroxide solution ⟶ alkali cellulose

> Alkali cellulose + carbon disulphide ⟶ xanthogenate

The **xanthogenate** is the soluble textile raw material for the cellulose synthetic fibres viscose rayon and staple rayon.

2. Regenerated cellulose fibres manufactured by the **cuprammonium process**
Refined pulp or cotton linters is treated with a solution of copper oxide in aqueous ammonia — cuprammonium, which has a deep blue colour because of the copper. A blue composition, which is soluble in the aqueous ammonia, is formed by a chemical reaction from cellulose, copper oxide and ammonia. The viscous blue solution, **the blue composition**, is the actual solution.

IV Organic raw materials and large-scale products: Industrial processes

The name of the process and of the synthetic fibres obtained by the process is derived from the chemicals required for the solution, above all the copper oxide.

Scheme of treatment:

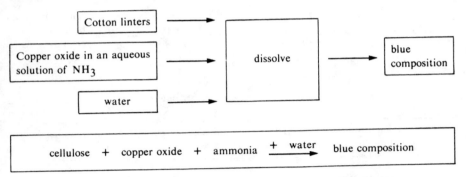

The blue composition **cuprammonium cellulose** is the soluble textile raw material for the cuprammonium cellulose synthetic fibres.

When these dissolved cellulose compounds are spun (see page 33 of this chapter), chemically pure cellulose is re-formed; a **cellulosic** synthetic fibre described as a regenerated **cellulose fibre** is formed.

3.1.3. Cellulose ester fibres

Cellulose is the starting material for these synthetic fibres too.

In this process, cellulose is esterified with acetic acid to form **cellulose acetate**, each member of the cellulose chain molecule having three alcoholic OH groups, which react with acetic acid.

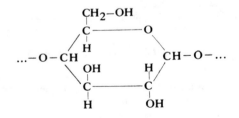

Initially, ester groups are formed from all the OH groups. The triacetate is formed and this is hydrated, ie takes up water, after the esterification and is processed further in this form.

There is 60 to 62% acetic acid present in the triacetate in the form of ester. If some of this bound acetic acid is removed by saponification, the product obtained is 2½-acetate, which still contains 53 to 56% of bound acetic acid.

The two esters — the triacetate and the 2½-acetate — have different solubility characteristics:

2½-acetate dissolves in acetone.

Triacetate dissolves in methylene chloride (and chloroform).

After the esterification and hydration with water, the cellulose acetate is precipitated, washed and dried. The spinning solution is produced by dissolving it in acetone or methylene chloride.

Scheme of treatment:

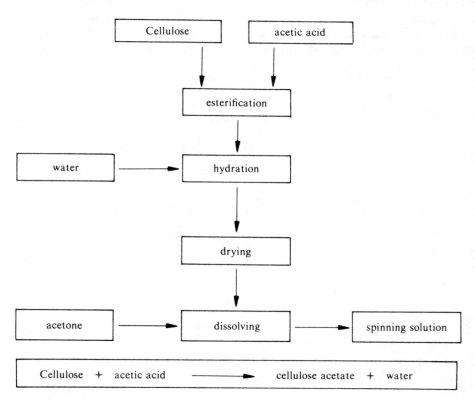

IV Organic raw materials and large-scale products: Industrial processes

The compounds **cellulose triacetate** or **2½-acetate**, which are produced by esterification or by esterification and partial saponification, are the soluble textile raw materials for the cellulose ester fibres known as triacetate and acetate.

When solutions of these cellulose compounds in organic solvents are spun, the solvent evaporates and the cellulose acetate becomes a fibre:

Cellulose acetate fibres are produced.

3.2. Man-made synthetic fibres

In the case of man-made synthetic fibres, the monomers are first prepared from simple chemical raw materials by chemical reactions and the polymers are produced by linking many monomer molecules together.

3.2.1. Polycondensate fibrous materials

This group of synthetic fibres includes every type of product in which the polymer is formed by **polycondensation** from small molecules of the same kind or of different kinds, with the elimination of by-products such as water or alcohol.

> **Polycondensation** is the progressive linking up of many bifunctional molecules. This linking up takes place between molecules
> a) of the same kind, or
> b) of different kinds
> with the elimination of by-products, such as water, alcohol, etc.

a) Bifunctional molecules of the same kind

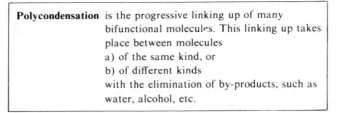

aminocarboxylic acid

Polycondensation:

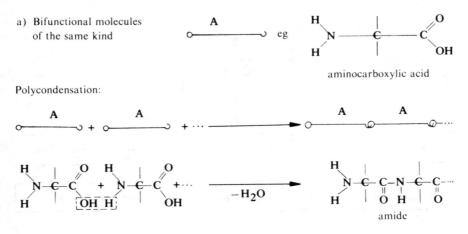

amide

b) Bifunctional molecules
   of different kinds

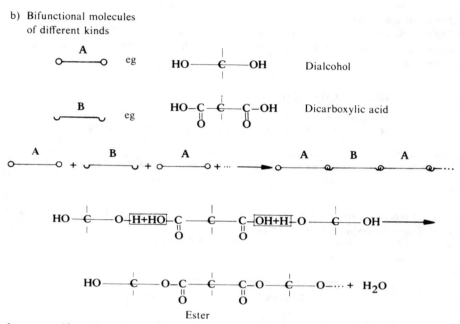

Ester

Let us consider polyamide and polyester as examples of this group of fibrous materials.

1. Polyamide, PA

Amides are a class of organic substances that are formed from an acid (-COOH group) and an amine (NH_2 group) with the elimination of water, eg

Carboxylic acid   amine   amide   water

$$R - \underset{O}{\underset{\|}{C}} - \boxed{OH} \; + \; \boxed{H} - \underset{H}{\underset{|}{N}} - R \longrightarrow R - \boxed{\underset{O \;\; H}{\underset{\| \;\; |}{C} - N}} - R \; + \; H_2O$$

The grouping characteristic of amides is the $- \underset{O \;\; H}{\underset{\| \;\; |}{C} - N} -$

group, which recurs frequently in the **polyamides** (by

selecting suitable bifunctional starting materials).

Depending on the starting materials used for their manufacture, various polyamides are obtained, of which the best known is Nylon.

Polyamide 6,6 (Nylon)
The monomer for the manufacture of nylon is nylon salt, a salt formed from 1 mol of
adipic acid (a dicarboxylic acid) and 1 mol of hexamethylene diamine, two bifunctional
compounds.

Adipic acid
$$HOOC - CH_2 - CH_2 - CH_2 - CH_2 - COOH$$

Hexamethylenediamine
$$H_2N - CH_2 - CH_2 - CH_2 - CH_2 - CH_2 - CH_2 - NH_2$$

A bifunctional compound is characterized by having **two functional groups** in the
molecule, such as two carboxylic acid groups or two amino groups.

The two functional groups in the molecule can also differ from one another, as in the case
of an aminocarboxylic acid (see page 18).
Adipic acid and hexamethylenediamine are prepared from a common precursor —
cyclohexanol — as follows:

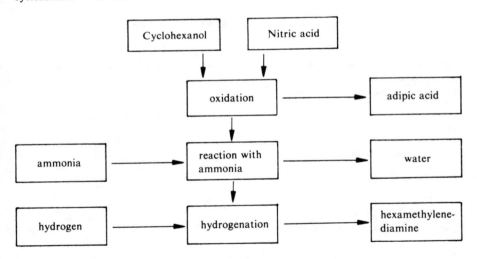

Nylon salt is subjected to polycondensation in an aqueous solution in pressure vessels
(autoclaves) at 280°C to 200°C (536°F to 554°F) under a pressure of about 20 bars, in the
course of which the water distils off.

The molten polyamide (polyhexamethylene adipate) is cast in the form of a ribbon and cut
up into chips.

The melting point of Polyamide 6.6 is 250°C (482°F).

$$n \text{ Nylon salt} \xrightarrow[\text{Pressure}]{\text{Heat}} \left[A - H \right]_n + nH_2O$$

The degree of polymerization n in this case is about 150 to 200. The linking group between the individual molecules is the $- \underset{\underset{O}{\|}}{C} - \underset{\underset{H}{|}}{N} -$ group, known as the "amide" group, which has given its name to this type of fibre.

The designation **Polyamide 6,6** denotes the fact that the chain contains a nitrogen atom after every six successive carbon atoms. Each set of six C-atoms in the chain is derived alternately from the diamine and from the dicarboxylic acid.

Section from the macromolecule:

$$\ldots \underset{\underset{H}{|}}{N}-CH_2-CH_2-CH_2-CH_2-CH_2-CH_2-\underset{\underset{H}{|}}{N}-\underset{\underset{O}{\|}}{C}-CH_2-CH_2-CH_2-CH_2-\underset{\underset{O}{\|}}{C}- \ldots$$

The polyamide chips produced by this process are the textile raw material for the polyamide fibre **Nylon.**

2. Polyester

Polyesters used for fibres are macromolecular compounds formed from a dicarboxylic acid and a dialcohol. The two monomers unite with one another, with the elimination of water, to form an ester.

Formation of an ester:

$$\underset{acid}{R_1-\underset{\underset{O}{\|}}{C}- OH} + \underset{alcohol}{HO-R_2} \longrightarrow \underset{ester}{R_1-\underset{\underset{O}{\|}}{C}-O-R_2} + \underset{water}{H_2O}$$

IV Organic raw materials and large-scale products: Industrial processes

The monomer used for the manufacture of the polyester polyethylene glycol terephthalate is di-(ethylene glycol) terephthalate, formed from the dicarboxylic acid

Terephthalic acid

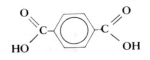

and the dihydric
alcohol (dialcohol) ethylene glycol (abbreviated to glycol).

$$HO - CH_2 - CH_2 - OH$$

It is prepared as follows from the starting materials para-xylene and glycol:

Process flow-sheet:

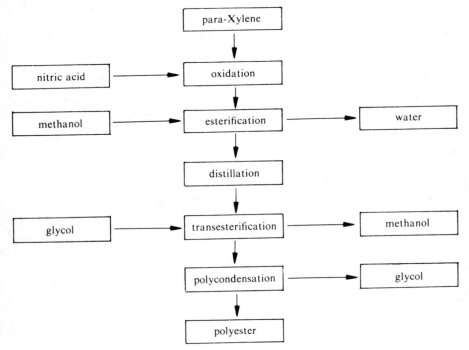

Para-xylene is oxidized with nitric acid to give terephthalic acid, which is first reacted with methanol to form the ester dimethyl terephthalate, DMT, since it is not possible to prepare the glycol diester direct.

After purification, the DMT is "transesterified" by distillation with glycol to form the glycol diester and the latter is then immediately subjected to polycondensation in vacuo.

para-Xylene   nitric acid   terephthalic acid

$$H_3C-\langle\bigcirc\rangle-CH_3 \quad + \quad HNO_3 \quad \xrightarrow{\text{oxidation}} \quad \overset{HO}{\underset{O}{}}C-\langle\bigcirc\rangle-C\overset{OH}{\underset{O}{}}$$

methanol terephthalic acid   methanol

$$CH_3O\boxed{H+HO}-\underset{O}{\overset{\parallel}{C}}-\langle\bigcirc\rangle-\underset{O}{\overset{\parallel}{C}}-\boxed{OH+H}-O-CH_3 \quad \xrightarrow{\text{esterification}}$$

dimethyl terephthalate DMT   water

$$CH_3-O-\underset{O}{\overset{\parallel}{C}}-\langle\bigcirc\rangle-\underset{O}{\overset{\parallel}{C}}-O-CH_3 \ + \ H_2O$$

DMT   + glycol $\xrightarrow[210°C]{\text{transesterification}}$ di-(ethylene glycol) terephthalate   +   methanol

In the polycondensation of the monomer di-(ethylene glycol) terephthalate to form the polymer, one molecule of glycol is eliminated from two adjacent molecules of glycol diester:

di-(ethylene glycol)   polycondensation   polyethylene
terephthalate   glycol   + glycol
                 at 270 to 280°C/vacuum   terephthalate

$$\left[-\underset{O}{\overset{\parallel}{C}}-\langle\bigcirc\rangle-\underset{O}{\overset{\parallel}{C}}-O-CH_2-CH_2-O-\right]_n$$

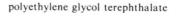

polyethylene glycol terephthalate

the melting point of the polyester is 255°C to 256°C (491°F to 493°F).

IV Organic raw materials and large-scale products: Industrial processes

The macromolecule of the polyester formed from terephthalic acid T and glycol G can be represented diagrammatically in the following way (see page 13).

$$\ldots\; - T - G - T - G - T - G - T - \;\ldots$$

The linking group between T and G is the ester group, which has given its name to the polymer.

The degree of polymerization n is 100 to 150, ie the macromolecule contains 100 to 150 units of T and of G linked to one another.

The polycondensate is produced in the form of a melt and emerges from a slot die; after solidification it is cut up into polyester chips, the textile raw material for polyester fibres.

3.2.2. Polymerization fibrous materials

This group of man-made synthetic fibres includes products in which the polymer is prepared by **polymerization**.

> **Polymerization** means the progressive linking up of many simple, unsaturated molecules (containing a double bond) to form chain molecules by opening the double bond.

The polymerization reaction is initiated by the application of energy (light or heat) or so-called initiators or activators. Threadlike macromolecules are finally formed:

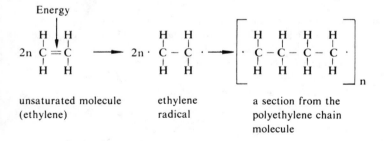

| unsaturated molecule (ethylene) | ethylene radical | a section from the polyethylene chain molecule |

Synthetic fibre materials of this type include polyvinyl chloride, polypropylene, polyacrylonitrile, etc.

Let us consider polyacrylonitrile as an example of this group of fibrous materials.

Polyacrylonitrile

Simple petrochemical starting materials — propylene in this case — are used for the manufacture of the monomer, acrylonitrile (also known as vinyl cyanide):

Propylene is converted into acrylonitrile in a single step by a catalytic reaction with ammonia and oxygen at 450 to 500°C (842 to 932°F), under a pressure of 3 to 4 bars (44 to 60 lb/in²). The gases from the reaction are absorbed in water and the solution containing the product is then distilled.

Propylene   ammonia   oxygen   acrylonitrile   water

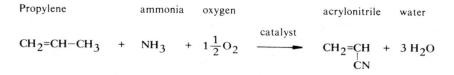

About 60%, is converted into synthetic fibres, 12% is used as a constituent of copolymers for synthetic rubber and a further 12% is used similarly for ABS polymers (compare IV,7). In addition, the copolymer formed from acrylonitrile and styrene is an important synthetic material.

The acrylonitrile monomer, or mixture of monomers, is polymerized into polyacrylonitrile by adding activators. In a modern process this polymerization is carried out in dimethylformamide (DMF), which is the solvent used for polyacrylonitrile.

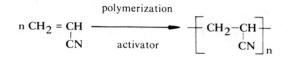

The degree of polymerization n is about 1000 to 2000. Pure polyacrylonitrile has a softening point of 235°C (455°F), but it decomposes (carbonizes) before melting.

3.2.3. Polyaddition fibrous materials

This group of fibres includes polymers prepared by **polyaddition**.

> **Polyaddition** is the progressive linking up of many bifunctional molecules. It takes place between molecules of different kinds as the result of a rearrangement of reactive hydrogen atoms, without the formation of by-products.

IV Organic raw materials and large-scale products: Industrial processes

Bifunctional molecules of different kinds

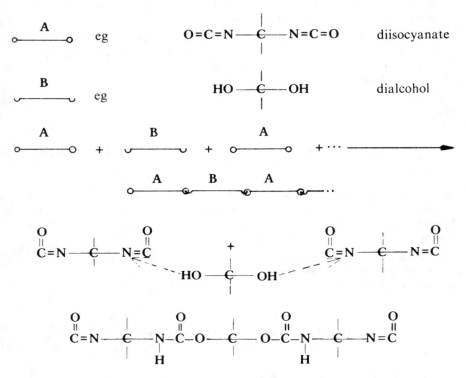

A urethane molecule, which can grow further at both its ends

The only products of importance as fibrous materials in this class are the polyurethanes.

Polyurethanes

Polyurethanes are polyaddition products formed from diisocyanates and dihydric alcohols.

A urethane is formed from an isocyanate and an alcohol as shown below:

Ethyl isocyanate   methanol   urethane

$$CH_3-CH_2-N=C=O+ \quad H-O-CH_3 \longrightarrow CH_3-CH_2-N-C-O-CH_3$$
$$\qquad\qquad\qquad\qquad\qquad\qquad\qquad\qquad\qquad\qquad H \quad O$$

The two monomers required for polyurethane are hexamethylene diisocyanate and butylene glycol (1,4-butanediol)

$$O=C=N-(CH_2)_6-N=C=O \qquad HO-(CH_2)_4-OH$$

As the result of a polyaddition reaction, these two bifunctional molecules form chain molecules:

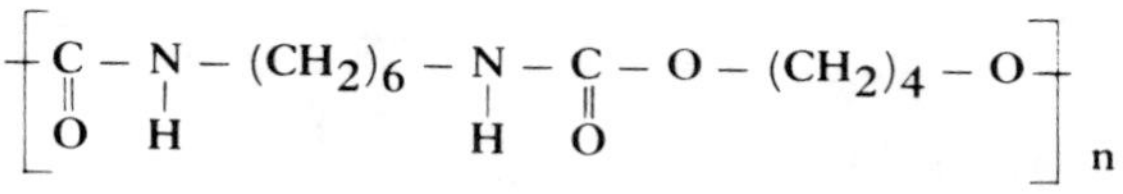

$$\left[\begin{array}{c} C - N - (CH_2)_6 - N - C - O - (CH_2)_4 - O \\ \| \quad | \qquad\qquad | \quad \| \\ O \quad H \qquad\qquad H \quad O \end{array} \right]_n$$

Polyurethane

Hexamethylene diisocyanate is prepared from hexamethylenediamine (see also manufacture of Nylon) and phosgene, a compound formed from carbon monoxide CO and chlorine Cl_2.

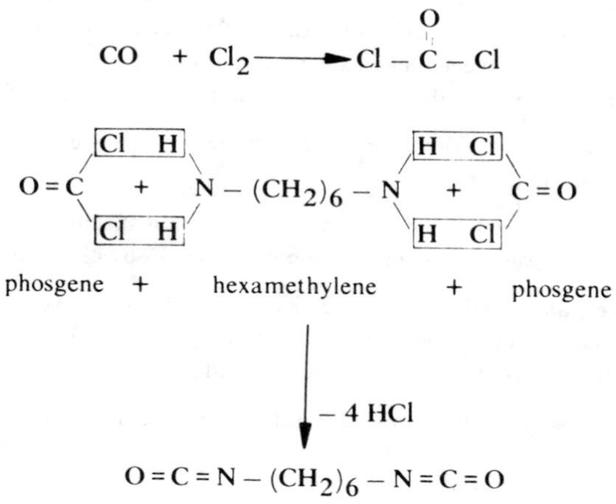

$$CO \ + \ Cl_2 \longrightarrow \overset{\displaystyle O}{\underset{\displaystyle |}{Cl - C - Cl}}$$

$$O=C \overset{\boxed{Cl \quad H}}{\underset{\boxed{Cl \quad H}}{}} \ + \ N-(CH_2)_6-N \ + \ \overset{\boxed{H \quad Cl}}{\underset{\boxed{H \quad Cl}}{}} C=O$$

phosgene + hexamethylene + phosgene

$$\downarrow \ -4 \ HCl$$

$$O=C=N-(CH_2)_6-N=C=O$$

Hexamethylene diisocyanate

This diisocyanate is reacted with the dihydric alcohol (1,4-butanediol) in a polyaddition reaction to give polyurethane.

Polyurethane can be melted without decomposition. Its melting point is about 250°C (482°F). The degree of polymerization n is 40 to 50.

IV   Organic raw materials and large-scale products: Industrial processes

The polyurethane macromolecule formed from diisocyanate D and 1,4-butanediol B can be represented diagrammatically as follows:

$$\ldots \; -D-B-D-B-D-B-D-B- \; \ldots$$

the linking group between D and B being in each case the urethane group

$$-\underset{H}{\overset{}{N}}-\underset{O}{\overset{}{C}}-O-$$

which has given its name to the polymer.

Polyurethane is the textile raw material for **elastomers**, ie rubber-elastic continuous filament yarns, but is also the starting material for **foam plastics**.

Summary

The chain-like structure of the polymers is a feature **common** to all these textile raw materials. Carbon atoms form the main constituent of the chain skeleton. Depending on the manufacturing process, the molecule chains contain not only carbon but also oxygen or nitrogen, as in the case of polycondensation and polyaddition products.

The differences between the individual textile raw materials are caused by the structure of the **linkage unit** within the polymers.

Thus, in polymerization fibres formed from olefins, the chain skeleton contains only carbon. These products are very resistant to degradation of the chain by chemicals.

Polycondensate fibres, on the other hand, such as polyamide and polyester, can be attacked by chemicals under certain conditions, such as high temperatures. The chain skeleton can be ruptured at the linkage point and the chain can be degraded.

This is also the reason why textile fibres have different chemical properties.

Table 3　Linkage units in polymers

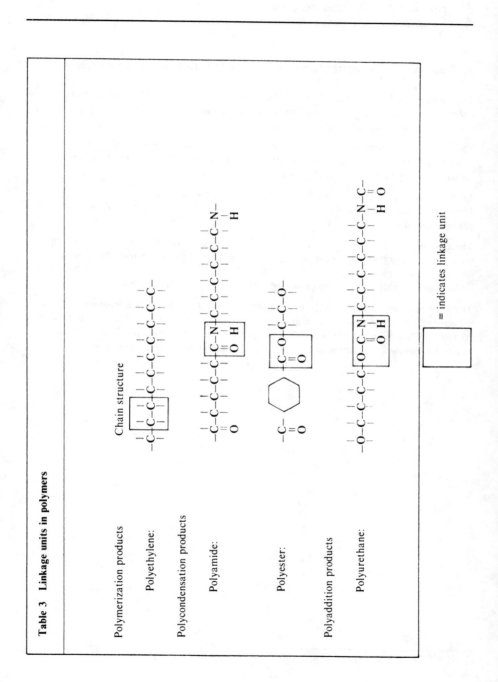

Chain structure

Polymerization products

　Polyethylene:

Polycondensation products

　Polyamide:

　Polyester:

Polyaddition products

　Polyurethane:

□ = indicates linkage unit

4. The conversion of textile raw materials into filaments and fibres

The various textile raw materials effect the spinning compositions. Depending on their chemical properties they are in the form of aqueous solutions or solutions in organic solvents or in the form of melts.

These polymeric materials have to be given the correct geometrical dimensions by the spinning process. They should have a virtually unlimited length and a very small cross-section, a diameter of only a few thousandths of a millimetre.

Each type of spinning composition requires a special process for the production of filaments. However, all the processes have certain common process stages, which are illustrated in the **process flow-sheet**, Figure 4 on page 31.

The spinning composition (a solution or a melt) must be conveyed from a **stock tank** by means of a pump (usually a gear pump) through a filter to the **spinneret**. The spinning composition is then forced in a uniform stream through the fine orifices of the spinneret under pressure; this leads to the formation of filaments (spinning process).

The filaments* are consolidated, are assembled to form a "thread", in a **medium** suited to the special characteristics of the spinning composition and are drawn off at speeds between 50 and 1000 m/minute by **draw-off devices** (guide rollers and wind-up rollers) and are wound up on **spinning bobbins** or are subjected to further processing in the form of **tow**.

---

* The spinning composition emerging from an orifice of the spinneret becomes a filament.

Figure 4   From the spinning composition to the filament

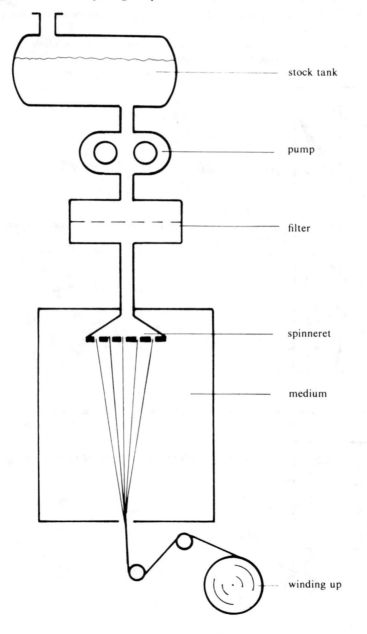

stock tank

pump

filter

spinneret

medium

winding up

Figure 5   The three most important spinning processes

Depending on the spinning composition and the medium, three different spinning processes can be distinguished:

1. The wet spinning process: the spinning composition is in the form of a solution (aqueous or in organic solvents)

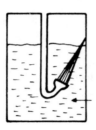

Medium: precipitation bath

precipitation bath

2. The dry spinning process: the spinning composition is in the form of a solution in an organic solvent

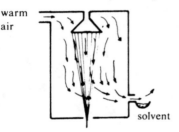

warm air

Medium: hot air in a spinning tunnel

solvent

3. The melt spinning process: the spinning composition is in the form of a melt

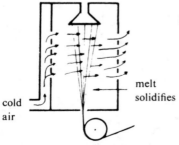

Medium: cold air in a spinning tunnel

melt solidifies

cold air

4.1. Wet spinning

Viscose rayon as an example

The "viscose" spinning composition, a viscous solution of cellulose xanthate in dilute
sodium hydroxide solution (see 3.1.1), is allowed to mature for a suitable time and is then
pumped from the stock tank through a filter to the spinneret and forced through the
spinneret orifices into the **coagulating bath** (spinning bath). The fine jets of the viscose
immediately react with the chemicals in the coagulating bath, mainly sulphuric acid
together with sodium sulphate and zinc sulphate, pure cellulose being **regenerated**: the
cellulose **coagulates** into a filament.

In this reaction, carbon disulphide, hydrogen sulphide and sodium sulphate are formed in
the coagulating bath.

The coagulating liquor is circulated and in the course of this it is made up to strength.

Figure 6
Wet spinning

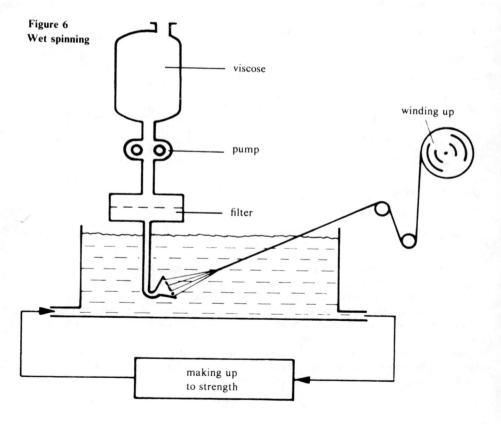

There must be a definite **constant** ratio, the spin draught, between the speed at which the viscose jet emerges from the spinneret and the draw-off speed of the filament, so that fibres with a constant diameter are formed.

> The speed of emergence and the spin draught determine the diameter of the fibres.

The number of filaments in the thread is determined by the number of orifices in the spinneret.

After the actual spinning process, ie the formation of filaments, the residues of chemicals must be washed out of the filament material (continuous filament yarn or tow) in a washing process and the "thread" or the "ribbon" must be dried. The so-called stretching (see 5.1) is also carried out during these further processing stages.

Figure 7 Wet spinning and subsequent stretching

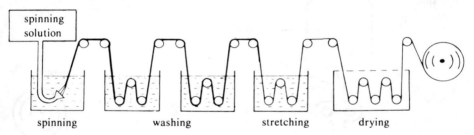

spinning   washing   stretching   drying

4.2. Dry spinning

Polyacrylonitrile as an example

Polyacrylonitrile in the form of powder is dissolved in the organic solvent dimethylformamide (DMF) to form a viscous spinning composition.

This spinning composition is pumped from the stock tank through a filter to the spinneret and is forced through the spinneret orifices into a spinning tunnel, a tall, closed, cylindrical tube. The solvent DMF (boiling point: 150°C, or 302°F) rapidly evaporates at about 200°C (392°F) from the fine jets of the spinning solution in the heated spinning tunnel under the influences of the hot air blown in as a medium (downwards as in Figure 8, page 35, or in the opposite direction). This consolidates the filaments, which, after the evaporation of the solvent, consist solely of the polymeric fibre substance – polyacrylonitrile. The filaments are assembled to form a thread and are drawn off over guide rollers at a draw-off speed of 200 to 800 m/minute (60 to 240 ft/minute) and are wound up.

Figure 8   Dry spinning polyacrylonitrile

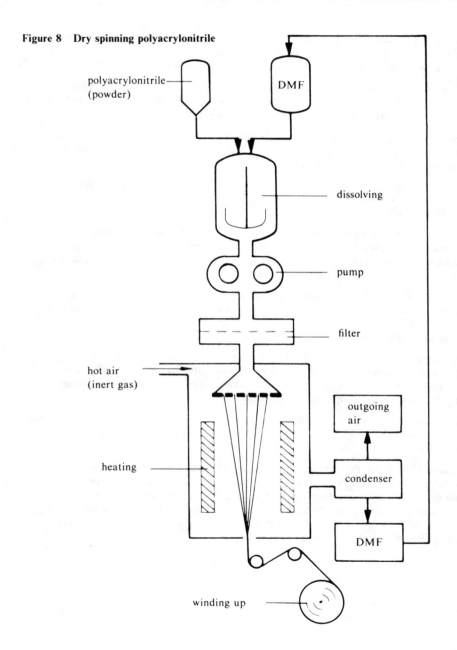

polyacrylonitrile (powder)

DMF

dissolving

pump

filter

hot air (inert gas)

heating

outgoing air

condenser

DMF

winding up

The winding up procedure completes the spinning process. Subsequent operations such as washing and drying the filament are superfluous in the dry spinning process, since the fibre material in the finished filament is free from solvent and is therefore also "dry".

The solvent is recovered from the solvent/air mixture in a condenser and is used again for the dissolving process. Like other synthetic fibres, polyacrylonitrile is subsequently **stretched** (see 5.1) in the further processes applied to the spun filament.

The dry spinning process thus requires a smaller outlay than the wet spinning process, but can only be employed if the polymeric fibre material, the actual textile raw material, can be dissolved in a readily volatile solvent.

Cellulose acetate and polyvinyl chloride can also be spun by the dry spinning process.

4.3. Melt spinning

It is essential for this spinning process that the textile raw material — the eventual fibre material — can be melted without decomposition and above all without chain degradation, and forms a viscous melt, the spinning composition, above its melting point (or, more correctly, above its melting range).

The textile raw materials that satisfy these conditions are primarily polyamides, polyesters and polyurethane. It is also possible to spin polyethylene and polypropylene by this process.

Polyester as an example

After being put through a drying process, the polyester chips must be stored under dry nitrogen in the storage bin, since, at the high melt temperature of 280°C to 290°C (536°F to 554°F), both moisture (water) and oxygen (air) cause cleavage (chain degradation) of the polyester molecule by hydrolysis or oxidation.

While maintaining the above conditions, the chips are fed from the storage bin to a **heating grid**, on which they are melted and then flow in the form of melt to the pump via a small buffer vessel.

The melt is forced into the **spinning tunnel** through a special spinning head, consisting (viewed from below) of a spinneret, stainless steel sieves forming a filter and a sand filter. Since the melt is forced through the orifices of the spinneret plate under a very high **spinning pressure** (about 70 to 75 bar, or 1015 to 1090 lb/in^2) and at a temperature of 200°C to 290°C (536°F to 554°F), the orifice plate must be very robust and the orifices must be somewhat larger than for the wet spinning process.

Figure 9 Melt spinning polyester

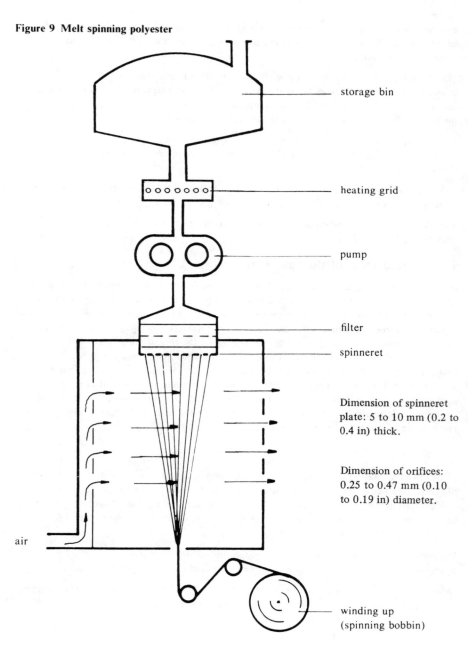

storage bin

heating grid

pump

filter

spinneret

Dimension of spinneret plate: 5 to 10 mm (0.2 to 0.4 in) thick.

Dimension of orifices: 0.25 to 0.47 mm (0.10 to 0.19 in) diameter.

air

winding up (spinning bobbin)

Two pumps are used for generating the pressure and metering the material; this has the advantage that the melt is already under pressure and can be delivered at an accurate rate by the metering pump.

The thin jets of melt forced from the spinneret into the spinning tunnel solidify immediately to form filaments under the influence of the cold air which is blown in and are drawn off at a speed of about 1000 m/minute (3300 ft/minute) and wound up on a spinning bobbin or combined with other filaments to form tow.

The spin draught also produces an orientation of the macromolecules in a longitudinal direction; this is an important requirement for imparting the properties desired later in the fibres.

The strength of the filaments is determined by metering at a constant draw-off speed. The winding up process concludes the spinning process proper.

Subsequent operations are included under after-treatment, once again particularly the stretching process.

5. After-treatment of the spun filaments

Except in the case of regenerated cellulose fibres, the filaments that emerge from the spinning apparatus cannot be subjected to textile processing in this form. They only obtain their essential properties, such as strength and elasticity, as the result of an after-treatment process, the **stretching process**. In addition, for the production of a continuous filament yarn, the filaments running parallel to one another in the thread must be twisted together, in order to make a "closed" thread. If, on the other hand, spun fibres are to be manufactured, the filaments produced by a large number of spinnerets are combined and are cut or crushed and crimped. In addition, the properties of the filament are improved by applying a coating of a dressing to synthetic fibres. The dressing is an aqueous solution or emulsion of derivatives of fats or oils, which make the filaments more pliable and easier to process; the dressing also protects the fibres from picking up a static charge.

5.1. Stretching

Depending on whether they are to be processed into continuous filament yarn or into spun fibres, the filaments obtained from **one** spinneret in the form of **thread**, or the filaments produced from **several** spinnerets in the form of **tow**, are stretched on a stretching machine between two roller systems running at different speeds, ie are drawn out to a multiple (3–5 times, or even 10 times) of their original length.

Depending on the fibre material, man-made synthetic fibres are stretched under hot conditions (polyester or polyacrylonitrile) or under cold conditions (polyamides).

In this process the individual filaments become narrower, the fibre material becomes more dense and the fibre attains the desired **strength** (stretching ratio = 1 : x).

The winding up speed in the stretching process is between 200 and 1000 m/minute, on average 400 m/minute.

Figure 10 Stretching

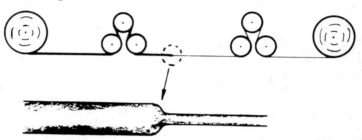

In order to gain a better understanding of these processes, it is necessary to consider the fine structure of the fibre.

Before the spinning process the macromolecules — assembled to form molecular bundles — in the fibre material are in a tangled, disordered and confused state. In the spinning process the spin draught produces a certain pre-orientation, ie alignment of a number of molecular bundles in the axis of the fibre. However, many of the very long macromolecules in the spun, unstretched fibre are still in a **disordered**, tangled state — in a spiral or coiled form: the fibre material is still largely amorphous (ie shapeless or not crystalline).

As a result of the stretching process these disordered molecules and molecular bundles are arranged in an ordered form by means of a relatively low tensile force.

They arrange themselves for the most part parallel to one another in the direction of the axis of the fibre.

They also remain in this condition after the stretching process. The molecules are now very closely packed and can adhere strongly to one another by lateral forces (intermolecular forces of attraction) between the molecular chains. Crystalline areas, called **crystallites**, are formed within the fibre. This is how the filaments acquire their **strength**.

IV Organic raw materials and large-scale products: Industrial processes

Figure 11 The effect of stretching

a)

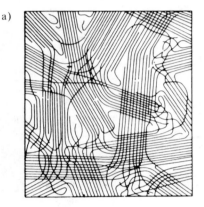

a) disordered molecular bundles
before stretching

b)

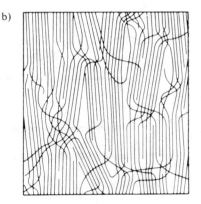

b) ordered (oriented) molecular
bundles **after** stretching

In the stretching process the narrowing of the fibres does not take place gradually over a certain length of fibre, but at a particular point — in the hot stretching process immediately after the heated pin over which the filament runs — a transition shaped like a bottle neck is set up between the thicker, unstretched part of the fibre and the thinner, stretched part of the fibre.

The stretching ratio differs for individual fibre materials. It must always be such that the fibres are given the correct strength **and** elongation values for the further processing they will receive and for the use to which they will be put.

Thus a distinction is drawn, for example, between standard polyester yarns for the clothing trades and **high-strength** polyester yarns for industrial uses, such as safety belts, which must be particularly strong, but must still allow a certain amount of elongation. This is because, if a further tensile force is allowed to act on stretched filaments or fibres, this force must be substantially greater than the force required for stretching if it is to produce a further lengthening of the fibre. If this tensile force is increased to a suitable extent, a value will eventually be reached at which the fibre breaks. Until this force — the **breaking strength** — is reached, the fibre is no longer extended to a multiple of its initial length, but only by amounts between 15% and 35% of its initial length.

The fibre needs this **elongation** not only for further textile processing but also for the ultimate use of the finished textiles; it ensures elasticity and flexibility.

Stretching ratios

Polyamide	1:4	(up to 6)
Polyester	1:3.6	(up to 6)
Polyacrylonitrile	1:4	(up to 7)

5.2. Twisting

If the spun filaments are intended for further processing into a continuous filament yarn, the individual filaments in a thread must be twisted together after stretching, in order to form a closed thread. This twisting — also known as laying — is given to the continuous filament yarns on a twisting machine, which, in the textile processing industry, usually includes the previous stretching process and is described as a draw-twister.

The filament is drawn off from the spinning bobbin obtained after the spinning process, stretched in the desired ratio between the stretching roller systems and is then twisted by the twisting device and at the same time wound up on spools known as "cops".

A higher degree of twist can be given to yarns in a second twisting process: **twisting at the head**. To ensure that the twist is retained during subsequent processing and that the continuous filament yarns do not untwist again when unwound from the bobbin, a heat treatment, usually steam, is applied to the bobbin, ie the twist is **fixed**. At the same time as this treatment is carried out, the continuous filament yarn can **shrink** a little, so that there is no danger that the finished textiles will alter in shape in a subsequent heat treatment at a temperature below that of fixing.

Rewinding the fixed continuous filament yarn on to commercial bobbins, such as **cheeses**, completes the **process of manufacturing continuous filament yarns**.

Figure 12 Stretching and twisting

bobbin shapes

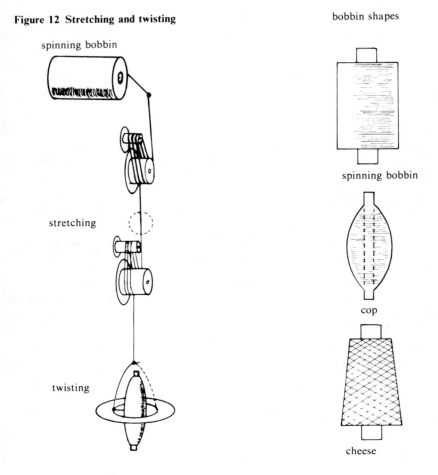

spinning bobbin

stretching

twisting

spinning bobbin

cop

cheese

5.3. Crimping and cutting

In order to produce **spun fibres** for carpet yarns or mixed yarns containing natural fibres, the filaments must be cut to **staple length**, ie cut or crushed into small pieces of definite length.

> **Spun fibres** are staple fibres made from filaments that were initially continuous, which can be processed into **yarn** by a mechanical spinning process, like wool or cotton.

Cutting is carried out on mechanical cutting machines by means of a cutting roller or crossed blades.

For this operation, the filaments obtained from several spinning points are combined to form tow and are stretched and cut.

Spun fibres are, of course, initially very smooth and have no crimp. They are therefore **crimped** by various methods to improve their properties for subsequent processing into yarn and to make their external shape match that of natural fibres.

Crimping is carried out either **before** or **after** cutting, depending on the textile raw material.

Thus in the case of cellulosic spun fibres, for example, the tow is first cut and the spun fibres are crimped by being suspended in water.

In the case of polyamide spun fibres, the tow is stretched and is then crimped in a heated **compression chamber** by mechanical compression and is later cut.

It is also possible, however, to remove the crimped tow and to transfer it to the mechanical spinning process, for **crushing** to staple length shortly before the mechanical spinning process.

This concludes the process for manufacturing spun fibres.

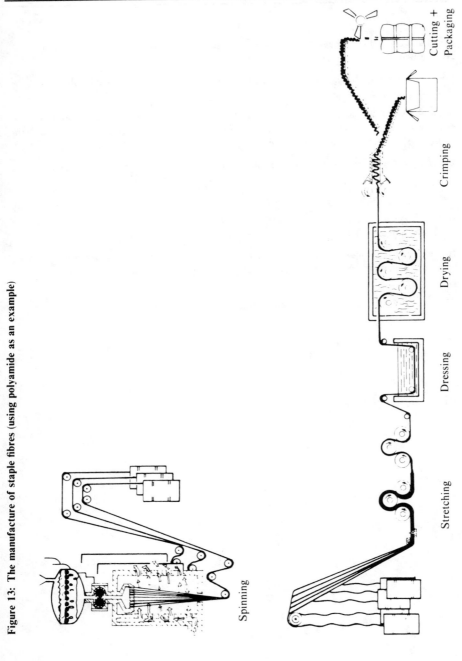

Figure 13: The manufacture of staple fibres (using polyamide as an example)

Spinning Stretching Dressing Drying Crimping Cutting + Packaging

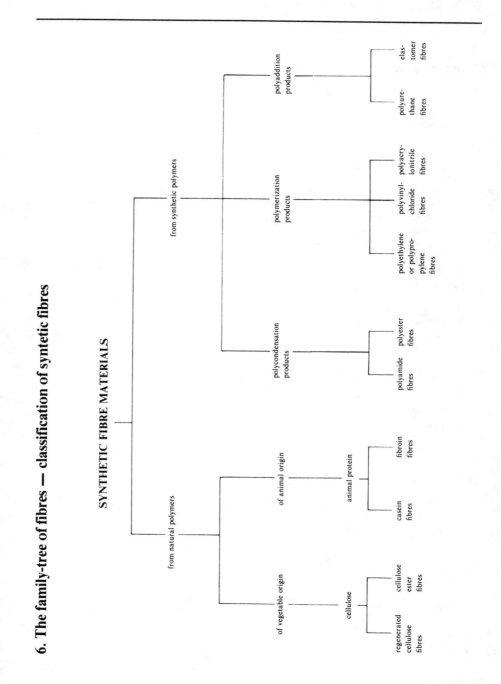

6. The family-tree of fibres — classification of syntetic fibres

SYNTHETIC FIBRE MATERIALS

from natural polymers

of vegetable origin

cellulose

regenerated cellulose fibres

cellulose ester fibres

of animal origin

animal protein

casein fibres

fibroin fibres

from synthetic polymers

polycondensation products

polyamide fibres

polyester fibres

polymerization products

polyethylene or polypropylene fibres

polyvinyl-chloride fibres

polyacry-lonitrile fibres

polyaddition products

polyure-thane fibres

elas-tomer fibres

IV Organic raw materials and large-scale products: Industrial processes

7. Test questions

123 Man-made fibres are produced in the first instance in the form of

A staple fibres
B continuous filament

124 As a measurement of its strength, a fibre is said to have an elongation at break of 15 km. This means that

A the fibre stretches to 15 times its initial length before breaking under load
B the fibre would break under its own weight if it were allowed to hang freely and were 15 km long
C the fibre will break when a weight is attached which corresponds to the weight of a 15 km length of the same fibre

125 Which type of fibre contains not only carbon atoms but also nitrogen atoms in the molecular chain?

A wool
B polyamide
C cellulose
D polyester

126 Which compound serves as a starting material in the production of polyamide 6,6 (nylon)?

A p-xylene
B glycol
C methanol
D cyclohexanol

127 Which fibres are manufactured by polyaddition?

A polyacrylonitrile
B regenerated cellulose
C polyglycol terephthalate
D polyurethane

128 Which drawing describes the dry spinning process?

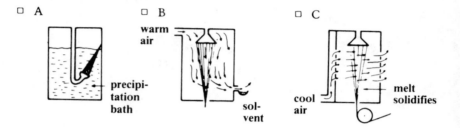

129 In which vessel is the spun fibre stretched?

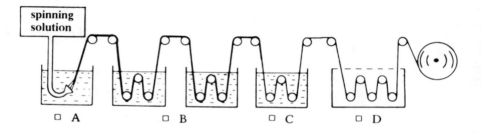

130 When a fibre is stretched

    A   its strength is reduced
    B   crystalline regions are formed within the fibre
    C   the original length is increased by several times
    D   the fibre substance becomes denser

131 Which fibres are of vegetable origin?

    A   cellulosic fibres
    B   asbestos fibres
    C   casein fibres
    D   polyvinyl chloride fibres

132 Which fibres also contain oxygen atoms in the molecular chain?

    A   cellulosic fibres
    B   polyurethane fibres
    C   protein fibres
    D   polyethylene fibres

9   Colorants, Dyestuffs and Pigments

IV Organic raw materials and large-scale products: Industrial processes

1. Definition of the terms colorant, dyestuff and pigment

Colour, dyestuff, colorant and coloured pigment are terms often used by the layman in the same sense and there is, therefore, no sharp distinction between their meanings. These terms have been defined and distinguished unambiguously from one another by the Technical Standards Committees for Colour, Pigments and Fillers of the German Standards Committee. These demarcations are also a benefit to industry.

Thus Colour is defined in DIN 5033 as an impression on the sense conveyed through the eye and the term "Colorant" is defined by DIN 55 945 as a collective name for all substances that impart colour.

Colorants are subdivided into Pigments and Dyestuffs.

> A pigment is an organic or inorganic, coloured or colourless **colorant** that is virtually **insoluble** in solvents and/or vehicles. A **dyestuff**, on the other hand, is an **organic** colorant that is **soluble** in solvents and/or vehicles.

In the case of pigments, DIN 16 515 draws a distinction between inorganic and organic pigments. Inorganic pigments include the natural inorganic pigments (earth pigments), synthetic inorganic pigments (mineral pigments), metallic pigments (bronzes) and carbon pigments (carbon black) (see II, 8. Inorganic Pigments and Fillers).

Organic pigments are subdivided into natural and synthetic pigments. Natural organic pigments are of animal vegetable origin.

Organic dyestuffs are also subdivided into natural, ie animal and vegetable dyestuffs, and synthetic dyestuffs.

In the following article, the terms colorant, dyestuff and pigment will be used in their DIN meanings.

2. Evolution from natural colorants to synthetic colorants

2.1. The first mineral colorants

The first colorants used by mankind are of mineral origin. The oldest coloured pictures that have come down to us are those depicting animals in the caves of Altamira in Northern Spain and in Lascaux in Southern France. They are estimated to be about 10,000 to 20,000 years old. These cave drawings are illuminated in white, yellow, brown, red and black shades. These colorants are naturally occurring, inorganic pigments, prepared from chalk, manganese ores, iron-containing aluminium silicates and iron oxides, etc.

2.2. Organic colorants

It is not possible to establish when vegetable and animal juices were first employed as colorants.

2.2.1. Colorants of animal origin

All we know is that the art of dyeing was important in Japanese, Chinese and Indian cultural groups thousands of years ago.

Purple is mentioned in Phoenician legends in the 6th century BC

Purple (porphyral in Greek) displays colour shades between red and violet and is secreted by the muricidae (purple fish) as a glandular secretion.

It was necessary to process 12 000 muricidae in order to obtain a single gram of this valuable dyestuff. The fish were caught at the sea coast in nets, cut into pieces, salted and boiled for several days. Depending on the colour shades required, after trial dyeings the decoction was concentrated or diluted.

In the Old Testament, investiture with the purple is synonymous with the assumption of royal power.

The chemical structural formula of purple was established as 6,6-dibromoindigo by P. Friedländer in 1909.

The preparation of purple had its last heyday in the 13th century in Palermo in the time of the Hohenstaufen Emperors, as a means of avoiding the expense of Kermes, a scarlet dyestuff. The Kermes scarlet dyestuff is obtained from insects that live on Kermes oak trees, particularly in Southern Europe and in the Orient.

In classical antiquity, the Syrians, Greeks and Romans used the scarlet dyestuff for dyeing wool and silk red. The dyestuff was obtained by scraping the female insects off the oak leaves in April and May, drying them and working them up into particles of scarlet.

In conquest of Mexico by Cortez in 1512, the Spaniards discovered a cramine red dyestuff used by the Aztecs, which proved to be more luminous and more stable than the dyestuff from the Kermes insects. The source of the dyestuff for carmine red was the cochineal insect, which lives as a parasite on cacti. These insects were swept off the cacti, dried in the sun and then ground into a powder. A flourishing trade developed in the dyestuff powder, which came to an end when the chemical formula of carmine red and a means of synthesizing it were discovered in 1894.

2.2.2. Colorants of vegetable origin

Vegetable colorants have also been known from ancient times for dyeing materials and fabrics. Cloths dyed red with madder have been found in Egyptian mummy tombs.

Madder has also been cultivated in Central Europe since the times of Charlemagne (800 A.D.). This plant developed into a remunerative agricultural product. In 1968 the annual harvest of madder roots in France was 50 million kg (110 million lb); at a 1% content of dyestuff, this yielded 500,000 kg (1,102,500 lb) of dyestuff.

The madder roots were crushed, dried, boiled and separated from the extract. The principal colouring component of this madder dyestuff is alizarin, which is the Arabic name for the red root.

The first synthesis of the dyestuff alizarin was achieved by the German, Graebe in 1869. As a result, the madder root was very soon priced out of the market.

The red trousers of the French uniform in the war of 1870/71 were all still dyed with the madder root extract.

However, in 1871 the new synthetic alizarin dyestuff was put on the market, production being at the rate of 15,000 kg (33,000 lb) a year.

Treatment with various metal oxides, such as aluminium oxide or chromium oxide, makes it possible to obtain an insoluble colour lake on the material to be dyed.

In the days before the synthetic dyestuffs industry, a blue dyestuff was obtained from the woad plant. In Germany it was cultivated mainly in the neighbourhood of Erfurt. The woad plant is a cruciferous shrub with a yellow blossom. In the course of two years it grows to a height of 1 metre (3.28 ft). The sap of the leaves contains a substance that rapidly turns blue in the air. This plant has been cultivated since ancient times in order to obtain this useful blue dyestuff, the only one of its kind.

The discovery of the sea route to india by way of South Africa by the Portuguese, Vasco da Gama in 1498 enabled continually increasing quantities of the blue dyestuff indigo to reach Europe at the beginning of the 16th century. Accordingly, the cultivation of the woad plant rapidly declined, since the indigo plant was thirty times more productive and its dye extracts produced a much clearer blue than those of the woad plant.

2.3. Development of synthetic colorants

2.3.1. Mauveine and fuchsine

Mauveine, the first synthetic colorant, was prepared by W. H. Perkin in England in 1856.

It is a basic azine dyestuff and forms brilliant black crystals. It was used for dyeing wool. In the presence of acids, mauveine turns a pruple-red colour.

Mauveine is the oldest dyestuff to be manufactured on an industrial scale. It is not of any importance nowadays.

Three years later, in 1859, fuchsine, the second synthetic colorant, as discovered by Verguin in France. Fuchsine belongs to the triarylmethane dyestuffs and forms green crystals with a metallic lustre. Fuchsine dissolves in water to give a red colour. It was employed for dyeing wool and silk. Nowadays, fuchsine is no longer used as a textile dyestuff for wool or silk, since it is not sufficiently resistant to acids and alkalis.

In the paper industry, on the other hand, fuchsine is a very important dyestuff. The following structural formula shows its chemical composition:

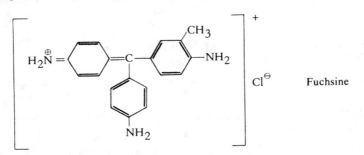

Fuchsine

The synthethesis of the dyestuffs mauveine, fuchsine and alizarin revolutionized the growth of the dyestuffs industry, in which Germany soon acquired the leading position.

2.3.2. Indigo

In 1880 Adolf von Baeyer succeeded in synthesizing indigo. The synthetic, blue indigo dyestuff was employed for dyeing textiles in Germany for the first time in 1897. With this, the importance once enjoyed by the natural colorants had come to an end for ever.

Faster and more brilliant colorants were synthesized and developed for a wide variety of dyeing and printing applications.

2.3.3. Azo dyestuffs

The first diazonium compound was prepared by the German chemist P. Griess in 1858 from aniline ⬡—NH_2 which had been discovered in 1841.

| Aniline hydrochloride (anilinium chloride) | nitrous acid | benzenediazonium chloride |

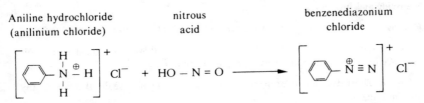

The diazonium compounds opened the way to the large class of azo colorants (see this chapter pages 19 to 21).

2.3.4. Indanthrene®

The development of anthraquinone vat dyestuffs was begun in 1901 with Indanthrene Blue RS.

Indanthrene Blue RS has the following chemical structure:

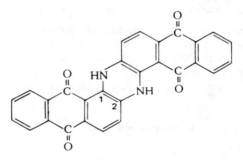

The indanthrene dyestuffs proved to be fast to light and washing. Their name became a protected description of quality for good dyestuff properties in the dyeing of cellulose fibres.The starting materials for this class of dyestuffs, which have a complex structure, had been obtained earlier from coal tar.

2.3.5. Phthalocyanines

The phthalocyanines were discovered by Linstead and de Diesbach in 1828/29. Copper phthalocyanine is the most important representative from an industrial point of view. As a pigment, it is particularly suitable for colouring plastics and it is used considerably as a colorant for lacquers, printing and paints. The size of its molecule and its crystal form are decisively important for its use as a pigment. Copper phthalocyanine colours materials in clear shades ranging from green through turquoise to reddish-tinged blue.

The colour-producing structural unit in its molecule is a cyclic system of conjugated double bonds (see III, 3, page 5). For its structural formula see page 27 of this chapter.

As sulphonic acid ($-SO_3H$) or hydroxyl groups are introduced into the phthalocyanine molecule, it becomes water-soluble and is suitable for dyeing textile fibres.

2.3.6. Reactive dyestuffs

Reactive dyestuffs made a name for themselves after the Second World War because of their fastness. In addition to the actual dyestuff component, they contain special reactive chemical groups, by means of which they form a stable chemical bond with the cellulose or protein fibre to be dyed. The fastness to washing they produce is due to the stability of the fibre-dyestuff bond.

® = Registered trade mark

IV Organic raw materials and large-scale products: Industrial processes

2.3.7. Organic pigments

After a scientific basis had been provided for the coal tar industry by Runge in 1834, the first synthetic products to emerge were textile dyestuffs. The first organic pigments were not produced from these dyestuffs until later on. Water-soluble dyestuffs, either anionic or cationic, were reacted with appropriate precipitation reagents. The know-how of ice colours and later on Napthol AS colours was also turned to good account.

Efforts were made to convert the fastest textile dyestuffs, the insoluble vat dyestuffs, into pigments by suitable finishing processes. Thus the development of pigments ran parallel to that of textile dyestuffs for a long time. Not until the advances made in recent times in our knowledge of pigment physics did the chemistry of pigments become increasingly separate from the chemistry of textile dyestuffs, and it now constitutes a field of research in its own right.

The development of pigments in chronological sequence:

Pigments from textile dyestuffs converted into lakes	up to 1900
Lithol Red	1899
Lithol Ruby	1903
Toluidine Red	1905
Pyrazolone pigments not made into lakes, and Dinitroaniline Orange	1907
Hansa Yellow pigments	1909
Napthol AS pigments	1911
Phthalocyanine and Diarylide Yellow pigments	1935
Vat pigments	since 1950
Quinacridone pigmets	1958
Azo condensation pigments, azo pigments of the benzimidazolone series and of the indolinone series and azomethine metal complex pigments since	1960

3. Production statistics for organic colorants

Until recently, organic colorants were listed under the item coal tar dyestuffs for historical reasons, because before the Second World War coal formed the raw material basis for the organic chemical industry and coal tar provided the intermediates for organic colorants. For the year 1979, world production of dyestuffs and organic pigments is about 965 000 t (excluding carbon black). The distribution of production between various countries in 1979 is shown in table 1.

Half the world production of colorants is manufactured in the USA, the USSR and West Germany. Another important European manufacturing country is Switzerland, whose chemical industry plays a prominent part in the development of colorants.

Table 1

USA	about	160 000 t	ie	16,6%
West Germany	about	147 000 t	ie	15,2%
USSR	about	93 000 t	ie	9,2%
		(estimated)		
PR China	about	82 000 t	ie	8,5%
Japan	about	81 900 t	ie	8,5%
Great Britain	about	61 600 t	ie	6,4%
Switzerland	about	50 000 t	ie	5,2%
France	about	32 000 t	ie	3,3%
India	about	27 000 t	ie	2,3%
Other countries	about	230 000 t	ie	23,8%

In 1979, world production of colorants was absorbed by three large sectors in consumption.

1. About 570 000 t, ie 59,1% of production, were consumed by the textile industry. For synthetic fibres alone, 170 000 t of colorants were consumed

2. About 270 000 t, ie 28,0% were required in the form of organic pigments for printing inks, lacquers and paints and for colouring plastics.

3. The remainder, about 125 000 t, ie 12,9%, is consumed in a large number of special fields such as the paper, leather and fur industries, etc

It is estimated that the chemical industry throughout the world now offers about 10 000 different synthetic colorants.

Every unitary colorant at the present time is registered in the Colour Index.*

Since the Second World War, the expenditure required to develop new colorants has grown at a continually increasing rate. For this reason alone, it is only the large chemical firms who can make a decisive contribution to the development throughout the world of synthetic colorants.

Today it is a major object of research to develop suitable colorants, particularly for synthetic fibres and plastics. Changes in dyeing technology produce further tasks and problems for research in the development of colorants. The properties of printing inks must be adapted to suit new grades of paper and printing machines running at high speed. In the latter field, inorganic pigments are being replaced by very fast organic pigments.

---

* This Index classifies and gives an indvidual description of all synthetic colorants from yellow to black. It consists at present of five volumes, each containing about 1000 pages. The first volume contains the index, the structures of the colorants are set out in the second volume, and the third to fifth volumes describe the individual representatives of the various classes of colorants. The last edition appeared in 1971. A quarterly supplement of the Colour Index is produced

Edited by: The Society of Dyers and Colourists, Bradford, England, and the American Association of Textile Chemists and Colorists.

Publisher: Chorley & Pickersgill Ltd, Leeds and London.

IV Organic raw materials and large-scale products: Industrial processes

The engineering industry has provided a large number of machines which make it possible to carry out dyeing and finishing processes on a semi-continuous or fully continuous basis.

The so-called pad-steam process has made it possible, since the Second World war, to bring together the individual stages of the discontinuous vat dyeing of cotton into a fully continuous procedure. This means that dyestuffs must now also be suitable for dyeing machines running at high speeds.

4. Light and colour

4.1. The spectrum colours of sunlight

All the objects and bodies of our surroundings are visible to us in a variety of colours. The cause of their colour is to be found in the properties of light and the chemical composition of the materials of which the objects and bodies are composed.

If sunlight is allowed to pass through a narrow slit and to fall on a prism, a continuous coloured image is obtained on a screen placed behind the prism (see Figure 1).

Figure 1:

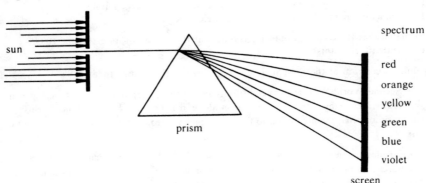

This continuous coloured image is also called a spectrum (Latin: appearance), and the individual colours are termed the spectrum colours of sunlight. The formation of a rainbow depends on the same effect. The drops of water in the air act as double prisms. White light is therefore composed of spectrum colours. The spectum colours differ from one another in their wavelength (see Table 2, page 13).

Light can be desdcribed in physical terms as an electromagnetic wave.

A summary of electro magnetic waves, classified by wavelength or frequency (ie the number of vibrations per second), is shown in Figure 2, page 14.

Of these electromagnetic waves, the human eye can only perceive the region with a wavelength of 400 to 750 nm*. We call this the visible region. Electromagnetic waves of different wavelengths produce a different colour impression in the human eye. Thus, for example, the wavelength range from 500 to 560 nm is registered by the eye as a "green" colour. In this connection see Table 2.

Table 2. The colours of sunlight and complementary colours

Wavelength range nm*	Absorbed (suppressed) spectrum colour	Reflected complementary colour when the corresponding spectrum colour is absorbed	
below 400	ultraviolet	(invisible)	
400 — 435	violet	yellow-green	
435 — 480	blue (indigo)	yellow	
480 — 490	blue or turquoise	orange	Deepening of colour
490 — 500	blue-green	red	
500 — 560	green	purple	
560 — 580	yellow-green	violet	
580 — 595	yellow	blue	
595 — 605	orange	blue or turquoise	
605 — 750	red	blue-green	
above 750	infra-red	(invisible)	

The visible region is bounded below 400 nm by ultra-violet radiation and above 750 nm infra-red radiation begins.

Within the spectrum there are always two colours that complement one another to form white light (see Table 1); they are therefore known as complementary colours.

If the totality of all light rays incident on a body is reflected unchanged, this body appears white.

If all the light rays are absorbed, ie the body does not reflect any rays into the eye of the viewer, the body appears to us black.

If, on the other hand, a body absorbs a colour of the spectrum selectively, it appears to us in the colour complementary to the colour that has been absorbed. For example, if yellow has been absorbed, the body has a blue colour (compare II, 8).

Absorbtion of light consists in part of the light being converted by the molecules of the substance involved into other forms of energy, usually into heat, but also into chemical energy, etc.

---

* Nanometre (abbreviated 1 nm) $= 10^{-9}$ m $= 39.372 \times 10^{-9}$ in.

Figure 2: The electromagnetic spectrum

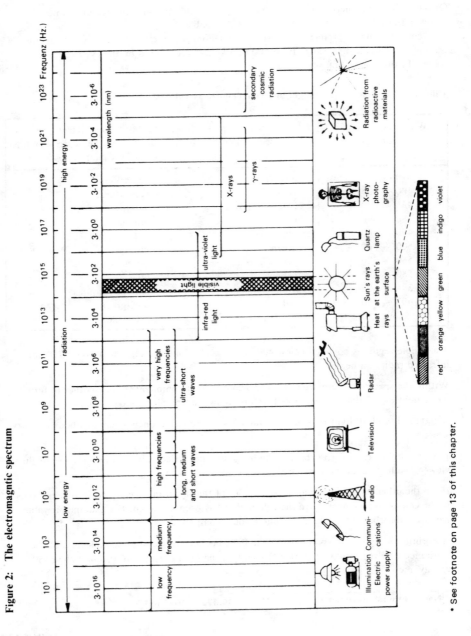

* See footnote on page 13 of this chapter.

4.2. Absorbtion of the colours of the spectrum by chemical compounds

> A chemical compound appears coloured when it absorbs part of the
> colours of the spectrum from the visible range of the sun's spectrum.

Saturated hydrocarbons, such as gaseous methane or liquid octane, are colourless to our
eyes; they absorb only in the ultra-violet range at about 200 nm.

In hydrocarbons with a double bond, however, we find a shift of the absorbtion in the
direction of the visible part of the spectrum.

The result of an accumulation of double bonds, particularly conjugated double bonds (ie
double and single bonds in regular alternation), is to shift the absorption further and
further into the region of the visible spectrum. A typical example of this is afforded by the
polyenes, which are characterized by several conjugated double bonds.

When polyenes with five double bonds are reached, a yellow shade is noticeable, for
example in

$$\underset{H}{\overset{H}{>}}C=C-C=C-C=C-C=C-C=C\underset{H}{\overset{H}{<}} \qquad \text{Decapentaene (1,3,5,7,9)}$$

The group of polyene colorants includes the carotenes, which impart the yellow-red colour
to carrots. The chemical structure of the carotenes contains a large number of double bonds
arranged in a conjugated pattern.

The partners in the interplay caused by light absorption are, firstly, the electron pairs of the
(colorant) molecules and, secondly, the electromagnetic waves of light, ie the light energy.

In the case of organic compounds containing conjugated double bonds or aromatic
compounds, the pre-conditions for this interplay are particularly favourable. These
compounds contain readily mobile electron pairs.

The starting materials for organic colorants must contain, inter alia, double bonds or
aromatic compounds.

Examples of this are the aromatic compounds benzene, phenol, naphtalene and diphenyl,
although these compounds themselves only absorb in the UV range.

Benzene

Phenol

Naphthalene

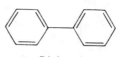

Diphenyl

IV Organic raw materials and large-scale products: Industrial processes

If substituents that also contain double bonds are introduced into organic compounds of this type containing double bonds, the absorption is shifted from the ultra-violet into the visible range of the spectrum: the effect is that visible electromagnetic radiation of the lower energy content is absorbed.

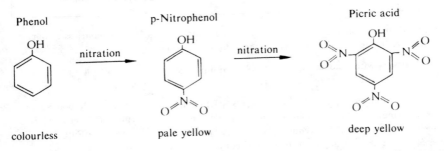

Phenol		p-Nitrophenol		Picric acid
colourless	nitration	pale yellow	nitration	deep yellow

Groups of atoms that cause an effective shift of the selective absorption in the direction of the visible range are called chromophores (colour-carriers). The following groups are important chromophores.

— N = N —	the azo group	>C = C<	the ethylene group
— N = O	the nitroso group	>C = O	the carbonyl group
— N$<^O_O$	the nitro group	>C = NH	the carbimino group

Compounds containing chromophores are known as chromogens (colour-producers). They generally contain several chromophoric groups.

We therefore have the following definition:
Aromatic system + chromophoric group = chromogen

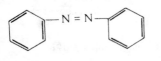   Azobenzene

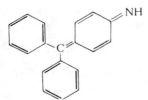   Iminotriphenylmethane

The parent compound of the azo colorants, azobenzene, and the parent compound of the triarylmethane colorants, iminotriphenylmethane, contain one or more chromophoric groups.

Anthraquinone, the parent substance of the group of colorants bearing the same name, also contains two constituents, an aromatic system and a chromophoric group.

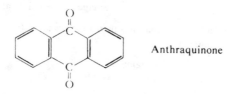

Anthraquinone

The effect of introducing auxochromic groups, also called auxochromes (auxesis, Greek for increase), into the aromatic portion of the coloured chemical compound is to cause, additionally, a shift in the colour, for example from yellow through orange towards red.

The following functional groups have an auxochromic effect:

$-OH$	the hydroxyl or phenolic group
$-NH_2$	the amino group (primary amine)
$-N{<}^{R_1}_{R_2}$	a substituted amino group (tertiaryamine)

The basic components of a colorant molecule are shown below in summary form using para-aminoazobenzene (Aniline Yellow) as an example.

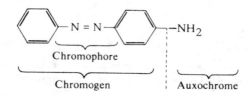

Chromogen + auxochrom = colorant molecule

However, not every coloured compound is a colorant. A colorant is required to adhere in a stable manner, whether by physical or chemical means, to the fibres or woven or knitted fabrics it is used to dye or, in the case of pigments, to be insoluble in and compatible with the appropriate vehicle systems.

IV Organic raw materials and large-scale products: Industrial processes

5. Qualities required in colorants and their quality characteristics

5.1. Qualities required

The quality of a colorant must take into account the requirements of two groups of people: the processor (eg dyer or finisher) and the consumer. The processor expects a colorant to be easy to apply to the material to be dyed and to cause no damage to the material itself.

Depending on the material and the end use, various specified tinctorial and fastness properties are required, such as the following:

for textile dyestuffs:	for pigments
fastness to light	fastness to weathering
fastness to water	fastness to light
fastness to washing	fastness to bleeding
fastness to perspiration	(exudation)
fastness to rubbing	fastness to overspraying
fastness to ironing	etc
fastness to heat	
etc	

Colorants must not be poisonous. Food dyestuffs and cosmetic dyestuffs are subject to very specific statutory requirements.

The multiplicity of special properties that colorants are expected to have arises from the large range of materials to be dyed and the application properties required, some of which are mentioned above. The following materials may be mentioned as examples: wool, silk, cotton and linen, regenerated and synthetic fibres, leather, furs, rubber, paper, plastics, paints, printing inks, wood, fats, waxes and so on.

5.2. Quality characteristics

The quality characteristics of a colorant are determined by the field in which it is used. The choice of a colorant is determined by the material to be dyed or printed.

A dyestuff for lining materials is used for outer garments does not have to be fast to light and need only be relatively fast to washing. Fastness to light and good fastness to washing are required for tablecloths. On the other hand, colorants for dyeing and printing curtains must certainly have good fastness to light but do not need to be particularly fast to washing.

While dyestuffs are dominant in the textile field, pigments are the more important product in the printing industry, the printing ink industry, the paint industry and in the colouring of plastics.

With all colorants, whether dyestuffs or pigments, a large number of different fastness properties are distinguished. These are usually measured in 5-point scales: 1 repressents poor properties and 5 represents very good properties.

1 means poor
2 means moderate
3 means fairly good
4 means good
5 means very good

Fastness to light is an important feature for colorants. Here, the fastness scale is divided into 8 stages, 1 again expressing a poor rating and 8 expressing a very good rating.

The following applies to light-fastness.

1 means very poor
2 means poor
3 means moderate
4 means fairly good
5 means good
6 means very good
7 means excellent
8 means outstanding

Depth of colour is determined by comparing a sample with a standardized pattern. The tests are carried out with a colorimeter. Visual tests are frequently required in addition. This requires considerable practical experience and trained eyes.

The tinctorial strength indicates how much colorant is required for colouring a particular quantity of material to a particular depth of colour.

6. Preparation of a colorant

The production process for the preparation of Hansa Yellow(R) G will be used to illustrate the principles involved. Hansa Yellow G is an organic pigment and belongs to the chemical category of azo compounds.

6.1. The chemistry of the process

The starting for the preparation of azo colorants include aromatic amines; these can be regarded as chemical derivatives of

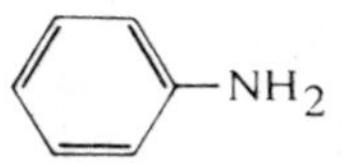

Aniline

IV Organic raw materials and large-scale products: Industrial processes

The starting material for the preparation of Hansa Yellow G is 4-amino-3-nitro-1-methylbenzene.

This compound has the following structural formula

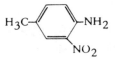

6.1.1. Diazotization

The chromophoric group, in this case the azo group -N=N-, is incorporated by reacting the aromatic amine with nitrous acid in a strongly acid medium.

The reaction of an aromatic amine with nitrous acid to form a diazonium compound is called diazotization.

4-amino-3-nitro-1-methylbenzene	nitrous acid		4-amino-3-nitro-1-methylbenzene-diazonium chloride

$$H_3C-\underset{NO_2}{\underset{|}{\bigcirc}}-NH_2 \quad + \quad HO-N=O \quad \xrightarrow[-2H_2O]{+ HCl} \quad \left[H_3C-\underset{NO_2}{\underset{|}{\bigcirc}}-\overset{\oplus}{N}\equiv N\right]^{+} Cl^{-}$$

In an acid aqueous solution, the resulting diazonium chloride has a slight yellow colour. Diazotization is an exothermic reaction. The diazonium compound decomposes rapidly even at room temperature, so the diazotization reaction is carried out with cooling by ice.

6.1.2. Coupling

The diazonium salt is reacted with a further component, the coupling or azo component. The molecule is enlarged in this reaction, because a second molecule is attached by coupling.

Diazonium salt + coupling component   ⟶   colorant + hydrochloric acid

Coupling 4-amino-3-nitro-1-methylbenzenediazonium chloride and acetoacetic acid anilide in aqueous solution produces the pigment Hansa Yellow G.

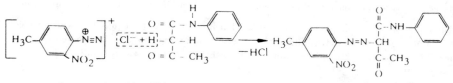

Acetoacetic acid anilide   Hansa Yellow G

6.2. How the process is carried out industrially

6.2.1. Diazotization

The aromatic amine is suspended in aqueous hydrochloric acid in a stirred, acid-resistant
vessel and is stirred for up to several hours. This produces a solution or suspension of the
aromatic amine chloride. The temperature of this is adjusted to about 0°C (32°F) before
the actual diazotization, by adding ice or by means of jacket cooling. Under no circumstances
must the temperature exceed 10°C (50°F). An aqueous solution of sodium nitrite is then
run in until the molar ratio is 1:1 (see figure 3, page 23).

6.2.2. Dissolving the coupling component (azo component)

Parallel with the diazotization, the coupling component is dissolved or suspended in water
in a stirred vessel, the so-called "coupling vat", the pH is adjusted to a definite value and
cooling is applied to reach the temperature required for coupling.

6.2.3. Purifying the diazo component, and the coupling reaction

In most cases the solution of the diazo component must also be filtered, in order to remove
unreacted substances. The process is generally carried out by running the clarified solution
directly into the coupling vat. In order to cause a rapid reaction between the diazo
component and the coupling component (azo component), the mixture is vigorously stirred
during the addition by means of a distribution sprinkler. A suspension (precipitate) of
Hansa Yellow G is formed. The hydrochloric acid formed in the coupling reaction is
neutralized; this keeps the pH value at the desired level. The pH, the temperature and the
rate of adding the diazo component, together with other variables in the coupling reaction,
have a considerable effect on the quality of the resulting pigment.

This concludes the chemical part of the process for preparing Hansa Yellow G.

6.2.4. Working up the product

After the coupling reaction, the colorant must be separated from the mother-liquor, ie the
solution in which the pigment is formed. Filtration is followed by washing, drying and
grinding. Filtration and washing are frequently carried out in the same equipment. The
ginding of colorants involves the comminution of moist or dried manufactured materials
(for example press cakes) into powder.

IV Organic raw materials and large-scale products: Industrial processes

In general, a distinction should be drawn between discontinuous and continuous processing.

Discontinuous processing has the advantage that it employs simple equipment with a wide range of uses, such as suction filters, filter-presses and drying cabinets. It is therefore possible to process nearly all colorants. On the other hand, this process entails very heavy wage costs.

Continuous processing plants are erected when a whole group of colorants or even a special dyestuff or special pigment has reached a certain volume of production, justifying the technological requirements. These plants are less labour-intensive and thus more profitable.

6.2.5. Finishing

Since the properties of a colorant are determined not only by its chemical composition but also, quite materially, by its physical properties, it is frequently necessary to convert the colorant into a form with the physical properties desired for the technology of application. This applies above all to pigments and disperse dyestuffs.

Finishing is thus understood to mean the last operations in a special physical after-treatment for certain colorants. Not all colorants require finishing. Finishing involves special processes, such as grinding, mixing and heat treatment. The purpose of grinding is to give the colorant a certain particle size with an optimum particle size distribution. Particle sizes below 1 μm (1 micrometer or 39.372×10^{-9} in) are frequently demanded, since particle size has a considerable effect on tinctorial strength, hiding power and other important properties.

In general, the very fine division is achieved by grinding a moist material or suspension in special mills and in some cases with the addition of surface-active substances as grinding aids. Stirred ball mills and roll mills are often employed for grinding, and kneaders are used for comminution. The degree of fineness of a colorant is influenced, inter alia, by the surface-active substances added (see IV, 6), the time of grinding, the temperature of grinding and the pH of the material being ground.

The type of crystal structure possessed by a colorant, its particle size and its particle size distribution affect a variety of properties, such as tinctorial strength, fastness to light, hiding power, brilliance and many other properties.

The properties imparted to the colorant in finishing represent the results of working with the product, which have been obtained empirically by many carefully carried out experiments. They form the know-how of each colorant manufacturer.

6.2.6. Formulating a colorant

Although not every colorant is finished, they must nevertheless all be formulated to a commercial grade, ie a specific tinctorial strength, shade, etc. Commercial grades are very often not chemical individuals, but mixtures of colorants. Dyestuff mills are used for formulating.

Figure 3: Flow-sheet for the manufacture of an azo pigment

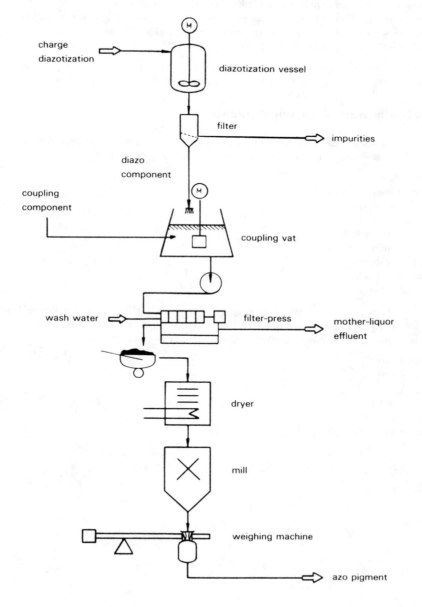

The commercial grade is a sample of colorant which is generally established as a model for quality in the manufacture of the particular colorant. All successive batches are produced in accordance with this sample and, if necessary, are corrected until they conform with the grade sample. Only then is the colorant released for sale. The "Technical Service Department" is responsible for this decision.

Depending on the field in which they will be used and the method or processing, colorants are put on the market as powders, press cakes, slurries and pastes and also in liquid form.

7. Classification of organic colorants

In general, colorants can be classified in accordance with two different principles. Firstly, they can be classified in terms of their chemical structure, for example:

Azo colorants
Anthraquinone colorants
Indigoid colorants
Triarylmethane colorants
Sulphur colorants
Phthalocyanine colorants
Acridine colorants
Quinacridone colorants
Perylene colorants

Secondly, they can be classified in terms of tinctorial considerations, ie in terms of fields of application and methods of dyeing (see 7.2 in this chapter). A few important classes of dyestuffs, ie soluble colorants, will be mentioned here as an example:

Direct dyestuffs
Vat dyestuffs
Reactive dyestuffs
Acid dyestuffs
Cationic dyestuffs
Disperse dyestuffs

7.1. Classification of colorants by their chemical structure (a few important classes of colorants)

	Chemical designation	Examples of use	Colorant example/ name	Formula
7.1.1.	**Azo colorants** Monazo colorants Disazo colorants Polyazo colorants	A Principal representative of organic pigments and dyestuffs; acid, direct, reactive and disperse dyestuffs for dyeing natural and synthetic fibres; pigments for the printing, paint and plastics industries	Fast Red AV (dyestuff) Hansa Red B (Pigment)	
7.1.2.	**Anthraquinone colorants** Hydroxyanthra quinone colorants Anthraquinone disperse colorants Anthraquinone vat colorants Anthraquinone reactive colorants	Used as vat and reactive dyestuffs for dyeing and printing cellulose fibres to meet very high specifications for fastness; as acid dyestuffs for dyeing wool; as disperse dyestuffs for dyeing synthetics	Indanthrene Red 5 GK (dyestuff)	

IV Organic raw materials and large-scale products: Industrial processes

	Chemical designation	Examples of use	Colorant example/ name	Formula
7.1.3.	**Indigoid colorants** Vat dyestuffs Indigosols[R] Anthrasols[R] Pigments	Used for dyeing and printing cellulose fibres to meet very high specifications for fastness for the manufacture of high-grade industrial paints	Indigo (dyestuff) Permanent Red-Violet MR (pigment)	
7.1.4.	**Triarylmethane colorants** Aminotriarylm ethane and hydroxytriaryl methane colorants	Used as cationic dyestuffs for dyeing and printing polyacrylonitrile fibres; as acid dyestuffs for wool; as colour lakes and internal salts they are important pigments	Crystal Violet (dyestuff)	
7.1.5.	**Sulphur colorants**	Used for dyeing cotton (working clothes) similarly to the vat dyestuffs	Immedial Pure Blue B (dyestuff)	

	Chemical designation	Examples of use	Colorant example/ name	Formula
7.1.6.	**Phthalocyanine colorants**	Used as pigments for paints, printing inks and plastics and for printing cellulose fibres	Hostapermblue B2G[R] Pigment	
7.1.7.	**Acridine and Quinacridone colorants**	Cationic dyestuffs; some are used for dyeing leather and for coir and similar fibres;	Acridine Orange NO (dyestuff)	
		used as pigments for high-grade industrial and automobile paints	Red E3B[R] (pigment violet 19) (pigment)	

7.2. Classification of some soluble colorants in terms of their use in the practice of dyeing

7.2.1. Direct dyestuffs

Direct dyestuffs comprise a group of dyestuffs that are capable of being absorbed on to vegetable fibres direct from an aqueous solution without a special pre-treatment. The direct dyestuffs are bound to the cellulose fibre by intermolecular forces. The adhesion of the dyestuff molecules is due to their structure. This group includes a large number of azo dyestuffs.

IV Organic raw materials and large-scale products: Industrial processes

7.2.2. Vat dyestuffs and solubilized vat dyestuffs

Vat dyestuffs are insoluble in water. They can be reduced easily and converted into a soluble form in alkaline liquours (ie dyebaths) by adding sodium dithionite $Na_2S_2O_4$; zinc dust, iron-II salts or other reducing agents. The reduced form of vat dyestuffs is often colourless or less intensely coloured than the actual vat dyestuff, and the reduced vat dyestuffs are therefore also described as leuco-compounds*. The process of reduction is called vatting and the dyestuff solution formed is called a vat. The fibre material to be dyed is put into the vat and is oxidized after dyeing. The oxygen effects a re-oxidation to give the original, water-insoluble dyestuff.

Vat dyestuffs are very fast to washing and light and have high level of fastness ("Indanthrene" dyestuffs).

Solubilized vat dyestuffs
The leuco-compounds, ie the reduced form of vat dyestuffs, can be esterified with chlorosulphonic acid, $Cl-SO_3H$, in organic solvents in the presence of metal powders. Pyridine is frequently used as the organic solvent. Leuco vat ester dyestuffs are formed (solubilized vat dyestuffs). In industrial processes, the reduction and esterification are carried out in a single operation. The solubilized vat dyestuffs of the indigo and anthraquinone series are obtained in this way. They are used for dyeing and printing vegetable fibres.

Since the leuco-esters are stabilized leuco-compounds, after they have been applied to the material to be dyed, they must be developed by re-oxidation to form the vat dyestuff.

7.2.3. Reactive dyestuffs

The reactive dyestuffs have only been developed since 1945. In addition to the actual dyestuff component, they contain special reactive groups or components, through which they can easily form a chemical bond with fibrous materials.

In principle, all organic dyestuffs are suitable as the dyestuff component. During dyeing, the reactive component forms a true chemical bond, for example with the free hydroxyl groups of cellulose or with the amino (-NH_2), carboxyl (-COOH) and thiol (-SH) groups of wool, silk and polyamide fibres.

The following reactive components are of importance:

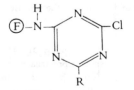

The monochlorotriazine component

---

*leukos (Greek) = colourless or white

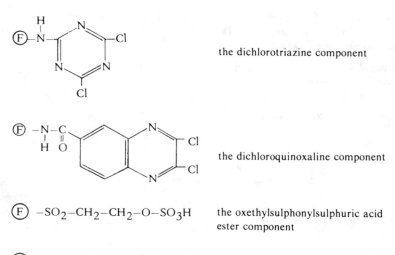

the dichlorotriazine component

the dichloroquinoxaline component

(F) $-SO_2-CH_2-CH_2-O-SO_3H$ the oxethylsulphonylsulphuric acid
ester component

(F)   represents the dyestuff component

The monochlorotriazine and dichlorotriazine groups react with the cellulose fibre to form cellulose esters, the detachable groups, eg chlorine, being eliminated in the form of HCl.

In alkaline solution the oxethylsulphonylsulphuric acid ester group first forms a vinyl group and this then reacts with the cellulose fibre to form a cellulose ester.

(F) $-SO_2-CH_2-CH_2-O-SO_3H$ $\xrightarrow[-Na_2SO_4]{Alkali}$ (F) $-SO_2-CH=CH_2$

Reactive dyestuff   Vinylsulphonyl group

(F) $-SO_2-CH=CH_2$ + HO —☐   (F) $-SO_2-CH_2-CH_2-O-$☐

cellulose fibre   dyed cellulose fibre

7.2.4. Acid dyestuffs

Acid dyestuffs can be applied to animal fibres in the presence of acids.

They are soluble in water and are absorbed direct on to the fibres. Since wool, silk and hair have a similar chemical structure – a feature common to all of them is a large number of amide groups $-N-C-$ – they can all be dyed and printed with acid dyestuffs.
         H O

IV Organic raw materials and large-scale products: Industrial processes

The solubility in water of these dyestuffs is due to the presence of groups that impart solubility in water, the hydrophilic groups (see IV, 6, page 14). Examples of hydrophilic groups in an acid dyestuff molecule are:

the carboxyl group R-COOH
the sulphonic acid group R-SO$_3$H and
the hydroxyl group R-OH

In aqueous solution, the dyestuff is in a dissociated form

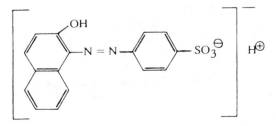

β–Naphthol Orange, an acid dyestuff

The dyestuff is in this case the anion, so that the group is also called anionic dyestuffs. Wool, silk and hair consist of protein, which has suitable free amino groups. In an acid dyebath, the molecules of these animal fibres form reactive ammonium cations. In a simplified form, this process can be represented by the following equation:

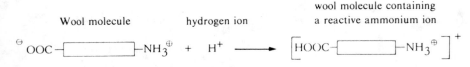

The dyestuff anion and the cationic molecule attract one another because of their opposite charges and form a stable salt bond, so that the dyestuff adheres to the fibre.

The group of acid dyestuffs includes azo dyestuffs, anthraquinone dyestuffs, azine dyestuffs, triarylmethane dyestuffs, nitro dyestuffs, etc.

7.2.5. Cationic dyestuffs

Cationic dyestuffs contain NH$_2$– of N(R)$_2$– groups as auxochromes. They include a few azo dyestuffs. Cationic dyestuffs were amongst the first synthetic organic dyestuffs to be produced commercially. The have become unimportant because of their poor fastness to light on wool and cotton. Only since the advent of polyacrylonitrile fibres have they once more become important. Cationic dyestuffs have a high affinity for polyacrylonitrile fibre and their distinctive features are very good fastness to light and good brilliance.

7.2.6. Disperse dyestuffs

Disperse dyestuffs are only sparingly soluble in water and are therefore applied to the fibre from aqueous dispersion. They are used for dyeing acetyl cellulose fibres. Disperse dyestuffs include water-insoluble monoazo, disazo, anthraquinone and quinophthalone dyestuffs. The dispersed dyestuffs are applied to polyester fibre 120 to 130°C (248 to 266°F) or, in the presence of a carrier, at the boil. Carriers are dyeing accelerators, which are added to the dyebath to promote the diffusion of the dyestuffs into the surface of the fibre. The colouring substance of the disperse dyestuff diffuses into the "softened" fibre surface and becomes dissolved in the fibre. Another method of dyeing with disperse dyestuffs is the Thermosol process. The fibre is impregnated with an aqueous dispersion of dyestuff and is then dried and treated with hot air at 200 to 220°C (392 to 428°F) for about 30 seconds. At this temperature the surface of the fibre is softened and the dyestuff diffuses into the interior of the fibre.

8. The dyeing of textile fibres

8.1. Classification of textiles fibres

In order to survey colorants available for dyeing textile fibres, it is appropriate to classify the multiplicity of types of fibres into groups on the basis of common properties. This approach enables fibres to be classified in five groups (see IV, 8).

1. Cellulose fibres
   Cotton, linen, hemp and jute
2. Regenerated fibres (see IV, 8 page 15)
   These are regenerated cellulose fibres; the most important representatives are viscose and acetate fibres.
3. Animal fibres
   Wool, silk and hair (in the textile industry rabbit hair in particular is especially important for the manufacture of fur felt hats).
4. Synthetic fibres (see IV, 8)
   Polyamide fibres
   Polyester fibres
   Polyacrylonitrile fibres
5. Glass fibres
   The glass fibre has an inorganic composition. Its main constituent is silicon dioxide, SiO_2, together with varying amounts of aluminium trioxide, Al_2O_3, and boron trioxide, B_2O_3. Glass fibres have become increasingly important for non-combustible textiles.

Dyestuffs have a predominant position in the dyeing of textile fibres. Pigments are not very important for this field, but they hold a considerable share of the market for printing textiles.

8.1.1. Dyestuffs for cellulose fibres

The following are suitable for dyeing and printing cellulose fibres:

direct dyestuffs, also called "substantive" dyestuffs,
vat dyestuffs
reactive dyestuffs,
azo developed dyestuffs,
sulphur dyestuffs and
solubilized vat dyestuffs.

Direct, vat and reactive dyestuffs have been discussed briefly under 7.2. of this chapter.

Naphthol AS$^{(R)}$ combinations (azo developed dyestuffs) A Naphthol AS combination is the product of coupling a diazotized dyestuff base on the fibre with naphthol. Many water-soluble textile dyestuffs are so readily soluble in water that they are not only soluble in the form of dyestuff powder, but are also still soluble after being applied to the fibre by dyeing or printing, and therefore often exhibit rather poor fastness to wet processing. This problem can be eliminated if a water-insoluble dyestuff is produced on the fibre. This can be done by first applying a Naphthol AS type to the fabric and then applying a diazotized aromatic amine; the two together produce an insoluble dyestuff.

Sulphur dyestuffs constitute an extensive group of sulphur-contianing synthetic organic dyestuffs. They are obtained by fusing a wide variey of organic compounds with sulphur of polysulphides.

They are insoluble in water and are converted into the water-soluble leuco form in a procedure similar to vatting by reduction, eg with sodium sulphide, Na$_2$S. This leuco form is then readily absorbed on to the cellulose fibre. The original dyestuff, which has good fastness to light and washing, is formed on the fibre by re-oxidation with oxygen. The distinctive features of sulphur dyestuffs are the cheapness and simplicity with which they can be manufactured. The blue and black types are extensively used for working clothes.

8.1.2. Dyestuffs for regenerated fibres

The basic structure and the chemical composition of viscose fibres are fundamentally the same as those of cellulose fibres (cotton). Thus the dyestuffs used for cellulose fibres are also suitable for dyeing and printing regenerated cellulose fibres. These are direct dyestuffs, vat dyestuffs, Naphthol AS combinations, sulphur dyestuffs and reactive dyestuffs.

8.1.3. Dyestuffs for fibres of animal origin

Acid dyestuffs are particularly suitable for dyeing and printing animal fibres (see 7.2.4. in this chapter).

8.1.4. Dyestuffs for synthetic fibres

In the dyeing and printing of synthetic fibres, a distinction must be drawn between polyamide, polyester and polyacrylonitrile fibres.

Polyamide fibres
As in the case of animal fibres, the chemical structure is characterized by the amide group
$-\underset{\underset{H}{|}}{N}-\underset{\underset{O}{\|}}{C}-$ However, only the free amino groups are suitable for the salt-like bond between

fibre and dyestuff. Acid dyestuffs, such as are used for wool, silk and hair, are suitable for polyamide fibres. Disperse dyestuffs are particularly suitable for dyeing polyester fibres, while cationic dyestuffs are suitable for polyacrylonitrile.

9. Organic pigments – the fields in which they are used

9.1. Importance of organic pigments and their properties

In recent years there has been a continuous increase in the importance of organic pigments, caused, above all, by the steeply rising consumption of coloured, graphic printing inks, coloured plastics and coloured paints. A rising trend can also be detected in the other fields in which organic pigments are used: the spin-dyeing of synthetic and regenerated fibres, the printing of textiles by means of pigments, the colouring of cosmetic articles and office requisites and the colouring of paper and leather, etc.

Organic pigments of very diverse classes of compounds can be used: monoazo dyestuffs derived from β-naphthol and β-naphthol arylides, monoazo dyestuffs of the acetyl – acetanilide series and those from pyrazolone, monoazo dyestuffs converted into lakes and disazo dyestuffs based on 3,3'-dichlorobenzidine, naphthalene tetracarboxylic acid, thioindigo, quinacridones, carbazole, tetrachloroisoindolinone, sulphonated triphenylarylmethanes and others.

In addition to the chemical structure of pigments, their physical condition (crystalline form, particle size and particle size distribution) is of prime importance for application technology (see II,8). Attention is given to this factor during the manufacture of the pigments. The manufacturing process is controlled in such a way that pigment crystallites are formed with optimum dispersibility, tinctorial strength, fastness properties and effect on the rheological* properties of the pigmented systems. Under the conditions used for drying pigments isolated from an aqueous medium, such crystallites assemble to form loose agglomerates and these can readily be split up again into their primary particles during the process of dispersion in triple roll mills, ball mills or dissolvers.

The size of the primary particles is between 1 μm and 0.01 μm; the specific surface of the pigments ranges from 8 to over 100 m² per gram (2,400 to 30,000 ft²/oz).

---

*Flow properties

IV Organic raw materials and large-scale products: Industrial processes

The process of dispersion is particularly simple for the pigment processing industries if the pigment manufacturer supplies pigment preparations that have either been obtained by flushing processes, thus eliminating drying, or in which the agglomerating effect of the drying process has been counteracted by pre-dispersion in a suitable medium. In recent times, a number or easily dispersible pigments have also been put on the market, in which the tendency to agglomeration has been reduced by coating the surface of the pigment with suitable substances during manufacture. In all these cases the requirements of pigmentation are met merely by dispersing the preparation by stirring it into the medium to be coloured.

9.2. Organic pigments for printing inks

The manufacture of inks requires mainly pigments with a high tinctorial strength and a low hiding power. These requirements are satisfied more fully by many organic pigments than by inorganic pigments.

The principal pigments suitable for multicolour printing are those based on 3,3'-dichlorobenzidine (yellow), β-hydroxynaphthoic acid pigments converted into a calcium lake (red) and β-modified phthalocyanine (blue). In addition, particularly for packaging printing inks, many other pigments, usually of deep colour, with moderate to average fastness to light are used.

In printing ink factories pigments are processed in the form or powder or as a preparation by dispersion, depending on the viscosity of the printing ink varnish and the dispersibility of the pigment, by means of triple roll mills or ball mills, stirred ball mills in the case of pumpable dispersion batches, or by means of dissolvers. In printing processes the dispersion is effected as required in letterpress and offset printing varnishes, in gravure and flexographic printing varnishes or in rotary printing varnishes.

To simplify matters for the pigment processor, surface-treated, readily dispersible pigments and pigment preparations are marketed; these are used mainly for gravure printing.

Preparations for special gravure printing contain about 20 to 50% of pigment and a vehicle consisting of nitrocellulose, maleate resins or vinyl chloride/vinyl acetate copolymers.

9.3. Organic pigments for paint systems

Inorganic pigments are mainly used for paints, principally because of their superior hiding power and good fastness to light and weathering. Organic pigments are always suitable whenever the tinctorial strength and purity of colour shade of inorganic pigments are inadequate, whenever transparent lacquerings, for example lacquerings with a metal effect, are desired or whenever it is desired to combine the advantages of inorganic pigments with those of organic pigments.

Dispersion in the various vehicles is effected by means of triple roll mills, stirred ball mills, sand mills or dissolvers.

The following types of pigment preparations for paint systems are marketed:

a. aqueous pigment pastes for aqueous emulsion paints;
b. dispersions in medium oil alkyd resin solutions as tinting pastes;
c. universal tinting pastes for pigmenting aqueous and non-aqueous paint systems;
d. pigments in chip form dispersed in nitrocellulose for nitro lacquers and nitro combination lacquers.

9.4. Organic pigments for colouring plastics

Both inorganic and organic pigments are used for colouring plastics. The principal materials coloured are PVC, rubber, polyolefins, polystyrene, unsaturated polyesters and polyurethanes.

For PVC, organic pigments based on 3,3'-dichlorobenzidine, pigments based on β-hydroxynaphthoic acid and converted into a lake, monoazo pigments made from benzimidazolone, condensed azo pigments, derivatives of perylenetetracarboxylic acid, phthalocyanine pigments, carbazole pigments, thioindigo pigments, pigments based on tetrachloroisoindolinone and quinacridone pigments are used.

The pigments used for colouring rubber are selected from the large number of organic pigments suitable for PVC, on the basis of particularly high tinctorial strength and resistance to the stresses caused by vulcanization.

In the case of polyolefins, the number of suitable pigments is even more limited, because of the high processing temperatures. For low-pressure polyolefins (processing temperature up to 300°C or 572°F), only phthalocyanine, quinacridone and perylenetetracarboxylic acid pigments are suitable.

Depending on the type of plastic, pigments are either dispersed in a thermoplastic material at temperatures above its softening point or are added in the form of a finished dispersion to the reaction mixture during the preparation of a thermosetting plastic. In order to facilitate the process of dispersion, which is carried out using extruders, twin roll mills, kneaders or dissolvers, a number of pigment concentrates (master-batches or preparations), in fairly large pieces, in form or powder or granules or in paste form, are offered to the pigment-processing plastics industries by pigment manufacturers.

The preparations contain the appropriate pigments finely dispersed in various dispersing media suited to the particular field application, eg plasticizers, vinyl copolymers, polyolefins and polyolefin waxes.

9.5. Organic pigments for spin-colouring

In spin-colouring the pigments imparting the colour are also spun; finely dispersed in the spinning composition, and thus become uniformly embedded in the fibre material. The pigments are added to the spinning composition in the form of aqueous dispersions (viscose

colouring), dispersions in solutions of the material to be spun (the spin-colouring of
cellulose acetate, polyvinyl chloride or polyacrylonitrile) or in the form of a pigment
concentrate in a plasticized material (the melt-spin colouring of polypropylene).
Accordingly, aqueous pastes and pigments pre-dispersed in the appropriate media are
marketed as preparations. The pigments on which they are based belong to different classes
of organic pigments and are selected on the basis of the dispersing medium, colour shade
and fastness.

9.6. Organic pigments for miscellaneous applications

As well as the three main fields of use, printing, paints and plastics (spin-colouring is
included in plastics colouring), organic pigments are also employed for many other purposes
in identical or similar processes of application technology: for colouring paper and paper
coatings, pigmenting leather dressings, colouring soaps, lipsticks and nail varnish, colouring
the leads of coloured pencils and indian inks, preparing artist's colours and poster colours
and for shoe polishes, floor waxes and wood stains, etc.

10.   Dyestuffs for non-textile uses

The colorants used for non-textile purposes also include dyestuffs, which in contrast with
pigments, disolve in the media to be coloured and owe their use precisely to this solubility.
This group includes the hydrocarbon-soluble dyestuffs for fats, the alcohol-soluble zapon
dyestuffs and the fluorescent dyestuffs that are readily soluble in transparent plastics such
as polystyrene, polymethacrylate and unsaturated polyesters.

The dyestuffs for fats, which are mainly monoazo dyestuffs made from β-napthol and
pyrazolone, are used for dyeing waxes, colouring fuels, mineral oils and lubricating greases
and for colouring phenolic resin compression moulding compositions and, in some instances,
also as dyestuffs for transparent polystyrene and polyester.

The zapon dyestuffs, which are prepared from a wide range of water-soluble dyestuffs by
reacting the sulphonic groups with long-chain amines, are used for coloured, transparent
foil-coating lacquers based on alcohol-soluble nitrocellulose (zapon lacquers) and for metal
effect lacquers. However, they are losing importance in this field more and more to certain
water-soluble metal complex dyestuffs, which are soluble in esters and ketones as well as in
alcohols.

The fluorescent dyestuffs are representatives of various chemical classes, naphthoic acid,
rhodamine, xanthenic acid and thioxanthenic acid derivatives; they only fluoresce when they
form a true solution in the media concerned.

The consequence of their good solubility is, poor fastness to migration. Attempts have been
made to compensate for this disadvantage by embedding the fluorescent dyestuffs in highly

crosslinked resins based on phenol/melamine and employing the resulting material, ground into the form of powder, as fluorescent pigments (daylight fluorescent colours).

Fluorescent dyestuffs and fluorescent pigments are being used to an increasing extent for the production of brilliant, fluorescent marking and warning paints and for coloured fluorescent prints on advertising posters and packaging articles.

11. Optical brighteners

11.1. Their optical properties

Optical brighteners constitute a special type of substance. They differ from colorants in one decisive property. While colorants of various colours do not reflect all the colours of the incident light spectrum, but absorb a part and are thus coloured, optical brighteners do not absorb at all within the range of the visible spectrum. In this respect optical brighteners are therefore not colorants.

They are, however, used in all situations where it is desired to make objects appear as white ("clean") as possible. Optical brighteners do this by converting incident invisible UV radiation into visible light on reflection. An optical brightener thus transforms incident electromagnetic radiation into electromagnetic radiation of a longer wavelength (compare Table 1, page 11). This phenomenon is called fluorescence. Objects whose surface has been treated with optical brighteners can thus reflect more visible light than they have received.

> Optical brighteners are organic substances that convert invisible UV radiation from daylight into visible, blue light.

The blue shade produced by optical brighteners is desirable, since the physiological impression "dirty, grey white" is produced by a yellowish-tinged white. Blue and yellow are, however, complementary colours (see Table 1, page 11): the yellow shade of the "dirty white" and the blue shade produced by the optical brightener are perceived physiologically as a pure, clean white. This effect forms the basis of so-called optical brightening.

The fact that optically brightened textiles, paper, etc, in reality have a (desired) blue tinge is shown by a paper handkerchief lying in the snow.

11.2.   Their chemical structure and commercial importance

At present there are about 4000 trademarks for optical brighteners on the market. From a chemical point of view; they can be traced back to about 50 different compounds and these, in turn, to a few chemical parent substances. One of these parent substances is stilbene:

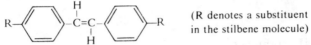

(R denotes a substituent in the stilbene molecule)

Optical brighteners were used as an additive in detergents, in the paper industry and in the textile and plastics industries, including spinning and compositions.

In 1979, the production of optical brighteners in the USA was about 15 274 t.

12. Self-evaluation questions

133   A dye is a

  A   dyestuff which is insoluble in solvents.
  B   dyestuff which is soluble in solvents
  C   dyestuff which is insoluble in solvents and binders
  D   coloured, black or white dyestuff

134   The most important consumer of dyes in terms of quantity is the

  A   automotive industry
  B   building industry
  C   textile industry
  D   leather industry

135   The reason why materials appear to be coloured is

  A   the composition of the white light from various spectral colours
  B   the selective absorption of spectral colours
  C   the interaction between the light and the electrons of the dyestuff molecules
  D   an optical illusion on the part of the viewer

136   When white light is split up into its componenet hues, in which order do these appear?

  A   red – green – yellow – orange – blue – violet
  B   yellow – orange – red – green – blue – violet
  C   violet – blue – green – yellow – orange – red
  D   blue – violet – green – orange – yellow – red

137   Which combination of absorbed spectral colour and reflected complementary colour is correct?

  A   blue – yellow
  B   green – orange
  C   red – blue-green
  D   violet – yellow-green

138   Which statements relating to "visible light" are correct?

  A   Violet light contains more energy than red light of the same brightness
  B   A body appears black to the human eye if all light rays are reflected
  C   The spectral colours can be characterized by their frequency
  D   Coloured materials convert some of the incident light into other forms of energy.

IV Organic raw materials and large-scale products: Industrial processes

139 Which compound absorbs light in the visible spectral range?

A

B OH
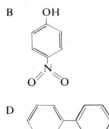

C

D

140 A dyestuff molecule always consists of

    A an aromatic system and chromophoric group
    B a chromophoric group and auxochromic group
    C a chromophoric group and chromogenic group
    D an auxochromic group and chromogenic group

141 The chromophoric groups include

    A $-N=N-$

    B

    C $-NH_2$

    D $C=O$

142 The auxochromic groups include

    A $C=C$

    B $-OH$

    C

    D

9 Colorants, dyestuffs and pigments

143 Which properties must textile dyes possess?

A non-toxicity
B weather fastness
C light fastness
D wash fastness.

144 Which dyes become linked to the fibre by chemical reaction?

A direct dyes
B disperse dyes
C acid dyes
D reactive rays.

145 The water-solubility of dyes can be increased by introducing

A hydrophobic groups
B SO_3Na groups
C COOH groups
D OH groups.

10 Answers to the Self-evaluation Questions on Sections 1–9

10. Answers to the test questions on sections 1–9

1.	A,B	11.	C,D
2.	B	12.	B
3.	A,D	13.	A,D
4.	B,C,D	14.	A,B,C
5.	A,D	15.	B,D
6.	B,D	16.	B
7.	A,B,D	17.	B
8.	A,B,C,D	18.	B,C,D
9.	B,C,D	19.	A
10.	D	20.	A,D

21.	A,D	26.	B,C
22.	A,B,C,D	27.	A
23.	A,D	28.	A,D
24.	A,C,D	29.	A,B
25.	A,B	30.	C

31.	A,B,D	36.	C,D
32.	A,B,C,D	37.	B,C,D
33.	A,B,C,D	38.	B
34.	A	39.	B
35.	A,D		

40.	B	52.	A,B,C,D
41.	A,B,D	53.	C,D
42.	B,C,D	54.	B
43.	B,C	55.	C,D
44.	A,B,D	56.	C,D
45.	A,B,C	57.	C
46.	B,C	58.	A,C
47.	A	59.	A,C,D
48.	C	60.	B
49.	A,B	61.	A,C
50.	B,C	62.	A,B,C
51.	B,D	63.	A,B,D

64.	A,C,D	69.	C,D
65.	C	70.	A,C
66.	A,D	71.	A,B,D
67.	B	72.	C
68.	A,D		

IV Organic raw materials and large-scale products: Industrial processes

73.	B,C	81.	C
74.	A	82.	A,B
75.	A,B,D	83.	B,C
76.	A,B,C	84.	D
77.	A,B,D	85.	B
78.	C	86.	A,C,D
79.	C	87.	A,B,C
80.	B,C	88.	C,D

89.	D	95.	B
90.	B,C	96.	A,C
91.	A,B,C	97.	B
92.	A	98.	C,D
93.	C	99.	A,D
94.	C		

100.	D	112.	B,D
101.	B,C,D	113.	A,B,C
102.	A	114.	B
103.	D	115.	B,D
104.	A,D	116.	D
105.	D	117.	A,D
106.	B,C,D	118.	A,C
107.	A,C,D	119.	B
108.	C	120.	A
109.	B,C	121.	A,C,D
110.	A,B	122.	A,D
111.	A		

123.	B	128.	B
124.	B,C	129.	C
125.	A,B	130.	'B,C,D
126.	D	131.	A
127.	D	132.	A,B

133.	B,D	140.	D
134.	C	141.	A,D
135.	A,B,C	142.	B,C
136.	C	143.	A,C,D
137.	A,C,D	144.	C,D
138.	A,C,D	145.	B,C,D
139.	B		

References

1. "Römpps Chemielexikon", Band 1–6, 7. Auflage, Franckh'sche Verlags-buchhandlung, W. Keller & Co., Stuttgart, 1975.
2. Winnacker, K. und Küchler, L.: "Chemische Technologie", Band 1–7, Springer-Verlag, Berlin, 1979.
3. Franck, H. G. und Knop, A.: "Kohleveredlung/Chemie und Technologie", Springer-Verlag, Berlin, 1979.
4. Weissermel, K. und Arpe, H.-J.: "Industrial Organic Chemistric", 1. Auflage, Verlag Chemie Weinheim/New York, 1978.
5. "Umwelt und Chemie von A–Z" (Herausgeber: Verband der Chemischen Industrie, Frankfurt), 3. verbesserte Auflage, Verlag Herder, Freiburg, 1978.
6. "Deutsche BP Aktiengesellschaft" (Herausgeber): "Das Buch vom Erdöl", 4. Auflage, Reuter und Klöckner Verlagsbuchhandlung, Hamburg, 1978.
7. Kaiser, R. und Hennig, I.: "Physikalische Chemie für die Sekundarstufe II", Verlag Dr. Max Gehlen, Bad Homburg vor der Höhe, Berlin-Zürich, 1977.
8. Fritz Merten: "Der Chemielaborant"
   Teil 1: "Allgem. Chemie und Analyse", 8. Auflage, 1981,
   2: "Anorganische Chemie", 9. Auflage, 1981,
   3: "Organische Chemie", 8. Auflage, 1981,
   Gebr. Jänecke Verlag, Hannover.
9. "Umweltfreundliche Technik — Verfahrensbeispiele Chemie", 2. Auflage, Zusammenstellung und Bearbeitung: Dr. J. Wiesner; aus dem Lehrprogramm der Dechema "Chemie und Umwelt", herausgegeben von Prof. Dr. D. Behrens (Dechema) und Prof. Dr. G. H. Kohlmeier (Universität Frankfurt), 1978.
10. Hopp, V.: "Der Kohlenstoff, seine Verbindungen als Rohstoffquelle und Energieumwandler", aus dem Lehrprogramm der Dechema "Chemie und Umwelt" Dechema, Deutsche Gesellschaft für chemisches Apparatewesen e. V., Frankfurt, 1980.

Index

Index

Index

Index

Index

Index

Index

Index

Index

Index

Index

Index

Index

Index

Index

Index

Index

Index

Index

Index

DATE DUE

NOV 2 6 89			
GAYLORD			PRINTED IN U.S.A.